RADICAL IONIC SYSTEMS

TOPICS IN
MOLECULAR ORGANIZATION AND ENGINEERING

Volume 6

Radical Ionic Systems

Properties in Condensed Phases

Edited by

A. LUND

University of Linköping, Sweden

and

M. SHIOTANI

Hiroshima University, Japan

KLUWER ACADEMIC PUBLISHERS

DORDRECHT / BOSTON / LONDON

0410065 7

Library of Congress Cataloging-in-Publication Data

```
Radical Ionic Systems / edited by A. Lund and M. Shiotani.
     p.   cm. -- (Topics in molecular organization and engineering)
   Includes index.
   ISBN 0-7923-0988-X (alk. paper)
   1. Ions.  2. Molecular structure.   I. Lund, A. (Anders)
II. Shiotani, M. (Masaru), 1940-   . III. Series.
QD561.M6874  1991
541.3'72--dc20                                          90-48009
```

ISBN 0-7923-0988-X

Published by Kluwer Academic Publishers,
P.O. Box 17, 3300 AA Dordrecht, The Netherlands.

Kluwer Academic Publishers incorporates
the publishing programmes of
D. Reidel, Martinus Nijhoff, Dr W. Junk and MTP Press.

Sold and distributed in the U.S.A. and Canada
by Kluwer Academic Publishers,
101 Philip Drive, Norwell, MA 02061, U.S.A.

In all other countries, sold and distributed
by Kluwer Academic Publishers Group,
P.O. Box 322, 3300 AH Dordrecht, The Netherlands.

"The logo on the front cover represents the generative hyperstructure of alkanes", printed with permission from J.E. Dubois, Institut de Topologie et de Dynamique des Systèmes, Paris, France.

Printed on acid-free paper

TABLE OF CONTENTS

II Anion Radicals and Trapped Electrons

III COMPLEX SYSTEMS

Preface

It is now more than 20 years since the book "Radical Ions" edited by Kaiser and Kevan appeared. It contained aspects regarding generation, identification, spin density determination and reactivity of charged molecules with an odd number of electrons. New classes of reactive ion radicals have been detected and characterised since then, most notably cation radicals of saturated organic compounds. Trapping of electrons has been found to occur not only in frozen glasses but also in organic crystals. The structure and reactions of anion radicals of saturated compounds have been clarified during the last 20 years.

We have asked leading experts in the field to write separate chapters about cation radicals, anion radicals and trapped electrons as well as more complex systems of biological or technological interest. More attention is paid to recent studies of the ions of saturated compounds than to the older and previously reviewed work on aromatic ions. In the case of trapped electrons full coverage is out of the question, and focus is on recent efforts to characterise the solvation structure in ordered and disordered systems.

The form of the chapters is that of a review. It is natural that a proportion of the review deals with the author's own work. The proportion depends on the subject and we have trusted the author's judgement on this issue. In most chapters the emphasis is on studies with modern spectroscopic tools which serve to reveal: a) electronic structure; b) dynamic behaviour, *i.e.*, intramolecular motion; c) kinetics of reaction, and d) reaction mechanisms. This information should be of value for experimental and theoretical workers in physical, organic, and biochemistry, chemical physics, and radiation physics and chemistry.

Some omissions are deliberate: we have excluded all aspects of molecular ions in the gas phase as this would require a separate treatise. It was also decided not to include a theoretical chapter on the structural aspects since the experience from calculations on excess electron species, anions and trapped electrons is rather limited. The situation with cation radicals is much better and calculations are included in several of the chapters. Other omissions may also exist due to overlooking by the editors.

We are indepted to the contributors of the book for the preparation of the camera-ready manuscripts, for their patience in making adjustments to ensure a reasonably homogeneous technical quality, and in many cases for reading and commenting the contributions in related chapters to help improve the presentation. We acknowledge the help of Mari Löfkvist in typing and language correction and also the nice cooperation of the editorial staff of Kluwer Academic Publishers, especially, Drs Janjaap Blom.

Anders Lund   Masaru Shiotani

August 1990

Introduction to the Series

The Series 'Topics in Molecular Organization and Engineering' was initiated by the Symposium 'Molecules in Physics, Chemistry, and Biology', which was held in Paris in 1986. Appropriately dedicated to Professor Raymond Daudel, the symposium was both broad in its scope and penetrating in its detail. The sections of the symposium were: 1. The Concept of a Molecule; 2. Statics and Dynamics of Isolated Molecules; 3. Molecular Interactions, Aggregates and Materials; 4. Molecules in the Biological Sciences, and 5. Molecules in Neurobiology and Sociobiology. There were invited lectures, poster sessions and, at the end, a wide-ranging general discussion, appropriate to Professor Daudel's long and distinguished career in science and his interests in philosophy and the arts.

These proceedings have been arranged into eighteen chapters which make up the first four volumes of this series: Volume I, 'General Introduction to Molecular Sciences'; Volume II, 'Physical Aspects of Molecular Systems'; Volume III, 'Electronic Structure and Chemical Reactivity'; and Volume IV, 'Molecular Phenomena in Biological Sciences'. The molecular concept includes the logical basis for geometrical and electronic structures, thermodynamic and kinetic properties, states of aggregation, physical and chemical transformations, specificity of biologically important interactions, and experimental and theoretical methods for studies of these properties. The scientific subjects range therefore through the fundamentals of physics, solid-state properties, all branches of chemistry, biochemistry, and molecular biology. In some of the essays, the authors consider relationships to more philosophic or artistic matters.

In Science, every concept, question, conclusion, experimental result, method, theory or relationship is always open to reexamination. Molecules do exist! Nevertheless, there are serious questions about precise definition. Some of these questions lie at the foundations of modern physics, and some involve states of aggregation or extreme conditions such as intense radiation fields or the region of the continuum. There are some molecular properties that are definable only within limits, for example, the geometrical structure of non-rigid molecules, properties consistent with the uncertainty principle, or those limited by the neglect of quantum-field, relativistic or other effects. And there are properties which depend specifically on a state of aggregation, such as superconductivity, ferroelectric (and anti), ferromagnetic (and anti), superfluidity, excitons, polarons, etc. Thus, any molecular definition may need to be extended in a more complex situation.

Chemistry, more than any other science, creates most of its new materials. At least so far, synthesis of new molecules is not represented in this series, although the principles of chemical reactivity and the statistical mechanical aspects are included. Similarly, it is the more physico-chemical aspects of biochemistry, molecular biology and biology itself that are addressed by the examination of questions related to molecular recognition, immunological specificity, molecular pathology, photochemical effects, and molecular communication within the living organism.

Many of these questions, and others, are to be considered in the Series 'Topics in Molecular Organization and Engineering'. In the first four volumes a central core is presented, partly with some emphasis on Theoretical and Physical Chemistry. In later volumes, sets of related papers as well as single monographs are to be expected; these may arise from proceedings of symposia, invitations for papers on specific topics, initiatives from authors, or translations. Given the very rapid development of the scope of molecular sciences, both within disciplines and across disciplinary lines, it will be interesting to see how the topics of later volumes of this series expand our knowledge and ideas.

WILLIAM N. LIPSCOMB

I   CATION RADICALS

ELECTRONIC STRUCTURE, SPECTROSCOPY, AND PHOTOCHEMISTRY OF ORGANIC RADICAL CATIONS

THOMAS BALLY

Institute of Physical Chemistry, University of Fribourg, Pérolles, CH-1700 Fribourg, Switzerland

1. Introduction

Although the concept of one-electron oxidation or reduction in condensed phase did not gain general acceptance until the 1930's when the advent of electrochemical techniques made it possible to lend direct experimental proof to this notion, radical ions (henceforth abbreviated as $M^{+\cdot}/M^{-\cdot}$) had aroused the interest of chemists much earlier [1]. In fact, we find that it was the property forming the focus of this article, i.e. the special electronic structure of organic radical ions, as manifested by their sometimes very vivid colors, which often stood at the origin of the fascination with these species, as can be seen from the choice of words used in the following quotes:

> "[Man erhält] aus der blaugrünen Lösung nach längerem Stehen das Additionsprodukt in *prächtig* dunkelstahlblauen Prismen, die einen *lebhaften* grünen Oberflächenglanz zeigen" [2]

> "[Man erhält] in Lösung *prächtig* blaue, in krystallinem Zustand *herrlich* fuchsin-glänzende Addukte" [3]

Actually, the often suprising color changes that accompany one-electron oxidation or reduction continue to fascinate those of us who study radical ions under conditions which allow visual inspection of the samples. Obviously, one would like to be able to rationalize this interesting feature in terms of the simplest possible model of general applicability. Once this is achieved, one can for example try to understand the connection between the excited state electronic structure and the *photochemical* reactions of radical ions [4]. One is beginning to see that these are unexpectedly rich, and the fact that they can usually be induced by visible light secures the interest of the solar energy storage community. However, it is not only for photochemistry that excited state properties must be understood because - unlike in the case of neutrals - most simple *thermal* reactions of radical ions such as valence isomerizations are state symmetry forbidden within the point group containing the symmetry elements shared by reactants and products. Hence, the energy position of the product excited states which correlate adiabatically with the isomeric ground states is needed if one wants to draw meaningful state correlation diagrams in the sense of the Woodward-Hoffman rules which were so succesfully applied in the case of neutrals.

In spite of the fact that the emphasis of this volume is on condensed phase work, a discussion under the title of this chapter cannot avoid to touch on some gas-phase techniques whose results are intimately related to those of the condensed phase experiments described herein and which have contributed greatly to our understanding of $M^{+\cdot}$ electronic structure. Most prominent among them is *Photoelectron* (PE) spectroscopy which has long dominated (and, as we will se, limited) our view

3

A. Lund and M. Shiotani (eds.), Radical Ionic Systems, 3–54.
© 1991 *Kluwer Academic Publishers. Printed in the Netherlands.*

on M^{+} electronic structure. This is mainly by virtue of Koopmans' theorem which provides a highly transparent link between molecular orbital theory and the properties of those M^{+} electronic states which can be attained by simple ejection of an electron from neutral M. The exclusive focus on these states tended to obscure the fact that other states, involving electron promotion into virtual orbitals of M (or M^{+}) can also show up prominently in M^{+} spectra and may in some cases even determine the electronic structure of M^{+} as it manifest itelf tothe human eye.

Most of the work on the excited state electronic structure of organic M^{+} in condensed phase employed *Electronic Absorption* (EA) spectroscopy as the main experimental tool. Besides this, it was mainly in connection with gas phase M^{+} investigations that some *emission* studies in Neon matrices were carried out. These elegant experiments led to detailed insight into the vibronic structure of the ground states and sometimes the first excited states of the M^{+} under investigation. However, they were obviously limited to emissive M^{+}, most of which fall into the two classes of (a) fluorinated benzene derivatives and (b) (oligo)acetylenic compounds. Therefore, the corresponding results, which have been excellently summarized in recent times [5], will only find mention in cases where they complement the more general EA studies which stand at the focus of the present review.

An extensive compilation of M^{+}/M^{-} EA spectra obtained in *frozen glasses* recently appeared in the form of a monograph [6]. We will obviously not attempt to duplicate the efforts which led to this impressive volume, and therefore the present summary of M^{+} spectra obtained in Freon glasses (Fig. 9, Section 4) will only cover results which do not appear in the above compilation. On the other hand, a list of molecules will be provided whose radical cations were investigated in *noble gas matrices* or in *boric acid glasses*.

Earlier reviews of the electronic structure of organic radical ions [7] have concentrated mainly on planar conjugated (mostly benzenoid) hydrocarbons and presented the theoretical tools needed for a proper understanding of this particular family of compounds. One example of such a species will be presented in some detail in order to illustrate the general principles and notions used in discussing the electronic structure of organic M^{+}. After this, emphasis will be laid mainly on insights that were gained through investigations of three other classes of compounds, i.e. complex cations, allylic systems, and polyene^{+}. Finally, an attempt will be made to summarize the knowledge which has accumulated over the past ten years on the *photochemistry* of organic M^{+}, a research field still in its early stages of development which is full of surprises.

2. MO-description of M^{+} electronic structure

Although certain types of electronic transitions in M^{+} can be interpreted readily in terms of a valence bond picture (for example the "charge resonance" transitions in dichromophoric M^{+} or in complex cations, see Section 5.1), a molecular orbital (MO) model is usually taken as a convenient starting point because it allows for a straightforward comparison between PE- and EA-spectroscopic results, often an essential element in discussions of M^{+} electronic structure. Of course one has to keep in mind that the MO model in its simplest form (electronic states described by single configurations) may have to be supplemented by more advanced treatments in order to yield a correct picture.

Starting from a general MO scheme for a neutral, closed shell compound M, four different types of M^{+} electronic configurations can be derived as depicted below (we will use the nomenclature introduced by Hoijtink [8] and used rather consistently by most workers in the field since then [7]). Of these, only the ground configuration (G) and the A-type configurations describe states which can be observed in PE spectroscopy because a single photon cannot simultaneously eject an electron and

5

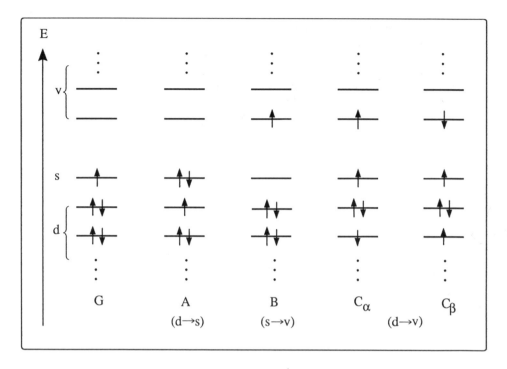

Fig.1: General types of $^2M^{\ddagger}$ configurations

excite another one, a pair of processes which would be required to reach B- and C-type configurations starting from ground-state $M^{\ddagger}$. This limitation is absent in EA spectroscopy of $M^{\ddagger}$ where it
makes no difference in principle whether an electron is promoted from doubly to singly occupied
MO's (d→s) or from there to the virtual ones (s→v or d→v).

2.1. KOOPMANS' THEOREM

If one takes the MO's of neutral M as obtained by a Hartree-Fock (HF) procedure and evaluates the
energy of G- or A-type $M^{\ddagger}$ configurations relative to that of ground state M one finds that the
corresponding expression is identical to that of the HF orbital energy of the MO from which an
electron was ejected (the same is true also in the Hückel approximation). To the extent that a single
configuration MO model is valid one can therefore directly relate HF orbital energies to the energy
differences between ground state M and those $M^{\ddagger}$ states which are described by G- and A-type configurations, i.e. the position of the different bands in the PE spectrum[1].

$$I_\mu = -\varepsilon_\mu, \quad \text{where } \mu = \begin{cases} \text{index of MO - counted downwards from the HOMO} \\ \text{index of PE band - starting at lowest ionization energy} \end{cases}$$

---

[1] Conseqently, G- and A-type configurations are often called Koopmans (K) configurations while those of
B- and C- type are Non-Koopmans (NK) configurations. To the extent that configurations offer good descriptions of the corresponding states, these can also be classified as K or NK.

6

This relation, known as Koopmans' theorem [9], has been (and continues to be) applied with great success in the assignment and interpretation of PE spectra. This success is actually quite astonishing in view of the extreme simplicity of the model. The reason for it lies in the fortuitous cancellation of two neglected energy contributions. Due to the fact that the MO's of M are used to describe $M^{\ddagger}$ states (the so called "frozen orbital" approximation), the energies of the latter are *overestimated* by the amount that would be gained if the electrons were allowed to reorganize themselves after ionization (E_{reor}). On the other hand, the $M^{\ddagger}/M$ energy difference is *underestimated* due to the neglect of the correlation energy contribution (E_{corr}) which is smaller in $M^{\ddagger}$ due to the loss of an electron. In the case of valence ionizations it so happens that $E_{reor}(M^{\ddagger}) \approx -\Delta E_{corr}(M/M^{\ddagger})$, such that $I_{\mu} \approx -\varepsilon_{\mu}$ (see below Scheme).

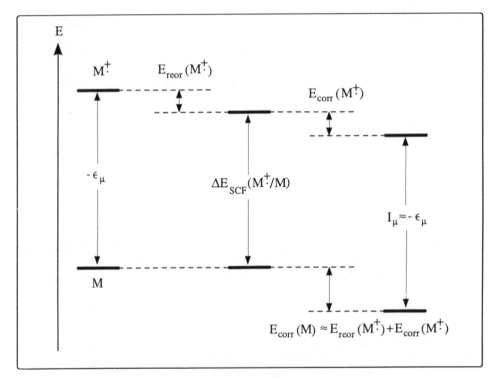

Fig. 2: Cancellation of errors in Koopmans' theorem

The great numerical success of Koopmans' theorem entailed in its wake a certain loss of the distinction between model and reality which led respected scientists to claim that PE spectroscopy "has demonstrated experimentally to chemists, physicists and other sceptics that molecular orbitals really exist" [10]. Since Koopmans' theorem ignores B- and C-type configurations (often called "Non-Koopmans" configurations) and is based on a single configuration MO model it was also widely assumed that this approximation generally offers an adequate description of M electronic structure. As will be shown in the section on polyenes, this is not necessarily true even in cases where $I_{\mu} \approx -\varepsilon_{\mu}$ holds quite well for the purposes of discussing PE spectra.

On the other hand, if one studies $M^{\ddagger}$ electronic structure by EA spectroscopy or related gas-phase techniques such as Photodissociation (PD) spectroscopy which gives essentially the same information, one quickly realizes that Non-Koopmans (NK) configurations cannot be neglected and may in fact play a prominent role in determining the $M^{\ddagger}$ electronic structure even in the energy range containing the first few PE bands. In fact, several cases were found where already the first excited state of $M^{\ddagger}$ arises by electron excitation into virtual MO's. Also, EA spectra of $M^{\ddagger}$ can often not be interpreted satisfactorily on the basis of a simple MO model, i.e. interaction between different configurations (CI) must be taken into account in order to arrive at an adequate picture of the $M^{\ddagger}$ electronic structure.

We note in passing that if K- and NK-configurations mix strongly, then this should have perceptible consequences on the appearance of PE spectra because the probability for detecting a photoelectron corresponding to a given electronic state of $M^{\ddagger}$ must in some way depend on its degree of Koopmans' character, i.e. the weight of K-configurations in the total electronic wavefunction. However, PE band intensities depend on various other instrumental and intrinsic factors, and hence not much attention is usually given to the effect of CI on valence PE spectra. Nevertheless, different investigations [11] have shown that this effect is quite real, and a careful reinspection of PE spectra may often give useful hints for the occurrence of "mixed" states. In any event, a discussion of PE spectra within the limited framework of Koopmans' theorem should never be taken to be more than a starting point if the aim is a comprehensive understanding of the electronic structure of a radical cation under investigation.

2.2. ENERGIES OF $M^{\ddagger}$ EXCITED STATES

We have noted in the introduction that $M^{\ddagger}$ often set themselves apart from their neutral counterparts M by being *colored*. The usual explanation for this phenomenon is, that all electronic transitions in closed-shell M are associated with electron promotion between occupied and virtual orbitals (i.e. HOMO→LUMO promotion) while in $M^{\ddagger}$ additional low-energy transitions arise through excitations within the manifold of the more closely spaced occupied MO's (i.e. HOMO-1→HOMO promotion).

Although this explanation is often to the point, it should be recognized that the HOMO→ LUMO excitation is also shifted to lower energies as a consequence of ionization. A possible rationale for this is, that the energy of the HOMO depends on its occupation number, which expresses itself for example in the observation that radicals usually have lower ionization potentials than related closed shell compounds[2]. There are no easy rules for predicting the resulting change in the HOMO/LUMO gap except in the case of alternant conjugated hydrocarbons, where a guideline is given by the "SDT rule". This rule states that the energy of the HOMO→LUMO excited $^2M^{\ddagger}$ configuration should lie between the arithmetic and the geometric mean of the corresponding $^1M^*$ and $^3M^*$ energies [13]. Because this rule neglects electron reorganization upon ionization and CI between $M^{\ddagger}$ configurations, its numerical success is not overwhelming, but it serves well to show that color in $M^{\ddagger}$ can (and does in fact in certain cases) arise by virtue of a red shift in the HOMO→LUMO transition rather than through intervention of A-type excited configurations unavailable to neutral M.

---

[2] See for example the work of Beauchamp on hydrocarbon radicals R˙ whose first IP's are usually a few eV lower than those of the parent hydrocarbons RH [12]. Note that this comparison should not be taken for more than a trend because the nature of the HOMO differs less between $M^{\ddagger}$ and M than between R˙ and RH.

Even when the energy of the HOMO→LUMO excited configuration is too high for it to appear *per se* in the EA spectrum of a radical cation, it (or other B-type configurations) can influence the position of lower lying A-type configurations through CI. An example of this are the polyene radical ions, where the lowest A- and B-type excited configurations lie very close in energy and where only a strong mixing between them can explain the occurrence of the characteristic long wavelength absorptions in these species (see section 3). This shows that electron correlation can exert an important influence on the electronic structure of M^{+} even in cases where this does not become apparent from the PE spectra.

2.3. QUANTITATIVE PROCEDURES

Quantitative MO models which can be used to calculate spectroscopic properties of M^{+} fall into the two classes of (a) π electron and (b) semiempirical all valence procedures. With a few exceptions which will be discussed below, the commonly available *ab-initio* procedures are not well suited for studying problems of M^{+} electronic structure[3]. Since the above semiempirical programs do not require mainframe computing power, it is fair to say that most of today's chemists have access to facilities which allow them to apply the different models presently available for calculations of M^{+} electronic structure.

It is, however, important to recognize the scope and limitations of the various procedures and to choose the most economical one to treat a particular problem. Thus it makes little sense to apply an all valence model to the radical cation of a planar conjugated hydrocarbon because electronic transitions involving electron promotion between σ- and π-MO's are usually too weak to be detected. On the other hand one must be cautious in applying a single configuration procedure in cases where CI cannot be safely ruled out.

2.3.1. *The HMO-model*

In the case of planar conjugated hydrocarbons, Hückel Molecular Orbital (HMO) theory often proves to be a fully adequate starting point for a discussion of M^{+} electronic structure. From a simple HMO scheme one can deduce a diagram of M^{+} configurations whose energies relative to the ground configuration are given by the energy difference in β units between the starting and the target MO involved in the different electronic excitations (for an example see section 2.4). Such diagrams may contain configurations of all types, the lowest few of which are usually in correct order even at this simple level of theory. For a semiquantitative comparison with experiment one can set $-\beta \approx 3 eV$[4] and evaluate the transition moments for the different excitations which can easily be calculated from Hückel MO's.

---

[3] Very recently, an *ab-initio* program system was announced which was designed specifically for the purpose of calculating spectroscopic energies and properties of excited states of molecules at the Hartree-Fock and post-Hartree-Fock level [14]. We plan to explore the applicability of this program to studies of M^{+} electronic structure.

[4] From linear regressions between PE ionization potentials and Hückel orbital energies, various sets of values clustering around $-\alpha = 6.0$ eV and $-\beta = 2.7$ eV have been deduced for different classes of hydrocarbons [15]. However, these do not take into account the position of NK-states which cannot be observed in PE spectroscopy. In our experience, $-\beta = 3.0$ eV provides a better fit for the purposes of EA spectroscopy of M^{+} (no value of α is needed for obtaining configuration energies relative to ground state M^{+}).

Heterosubstituted π-systems can be treated similarly by using the standard heteroparameters but the HMO procedure is of course only applicable for the purpose of M $\overset{+}{\cdot}$ EA spectroscopy if the ground state of the system under investigation is of π symmetry. Whether this is actually the case can be seen from the ESR and EA spectra as it was demonstrated for the case of N-heteroaromatic compounds [16].

Starting from a Hückel configuration diagram, the next step consists in assessing possible effects of CI. If among the first few configurations none of the same symmetry are close in energy then their ordering will usually not change although their relative energies may be affected by CI with higher lying configurations. Conversely, large CI effects are expected for close lying configurations of similar symmetry. A quantitative evaluation of these effects is possible by using formalisms of the type devised by Pariser and Parr [17]. Although an open-shell variant of HMO-CI procedure was successfully applied by Hoijtink et. al. in their pioneering spectroscopic studies of aromatic hydrocarbon radical anions [18], its use was all but discontinued with the advent of the π SCF/CI procedures described below. Nevertheless, Hückel theory continues to serve very well for a qualitative understanding of the M $\overset{+}{\cdot}$ electronic structure of planar conjugated systems, and there is no need to go to higher levels or theory unless a quantitative evaluation of CI effects is desired or if agreement between theory and experiment is too poor to permit an unambiguous assignment of the spectra under investigation.

2.3.2. *The PPP-CI-model*

In the cases mentioned above, the next method of choice is the open shell version of the Pariser-Parr-Pople (PPP) π SCF/CI procedure. The corresponding closed shell model has been used very sucessfully for the calculation of excitation energies and band intensities in neutral planar π systems [19] and, nonwithstanding the recent advent of much more sophisticated procedures, PPP-CI must still be considered as the best available method within its range of applications, yielding reasonably reliable predictions at almost negligible cost. The same is true of the open shell version which was originally formulated by Ishitani and Nagakura [20] and subsequently employed in a program for the systematic study of open shell π systems in the group of Zahradník and Čársky [7,21]. These authors carefully tested the procedure for possible limitations of the underlying model (for example by using the rigorous Roothaan RHF instead of the simple "half electron" formalism in the SCF part [21a,b] or by admitting doubly excited configurations in the CI part [21c]), but no systematic improvements were found and so the procedure in its original form seems to represent the optimum that one can expect from a simple π-electron model [5].

The main limitation of the procedure in its original form is, that the results are very insensitive to small geometry changes. Taking all bondlenghts equal 1.4Å and all angles equal 120° usually gives adequate results. However, excitation energies for different conformational isomers of, say, polyene radical ions are nearly indistinguishable (in contrast to what is observed experimentally) unless nonzero β terms for nonbonded interactions are introduced. Unfortunately there are no tested guidelines for the choice of such terms and in our experience systematic improvements of the results cannot be achieved in this fashion.

---

[5] The program was later extended for calculations of open shell species with degenerate ground states [21d,e] as well as oxygen- [21f,g], nitrogen- [21h,i], and sulfur-substituted π systems [21k]. Interestingly, the same parameters which were earlier found to give the best results for closed-shell species [21l] were found to be most suitable also for open shell systems.

2.3.3. *CNDO-based models*

When the σ electronic framework of a molecule is involved in electronic excitations, then the above mentioned π electron models are obviously no longer adequate and procedures which take into account all the valence electrons are called for. The first of these is the CNDO/S-CI model developed originally by Del Bene and Jaffé [22a] and applied extensively to the spectroscopy of open shell species by the groups of Zahradnik [21m,n] and Jaffé [22b-e]. For a discussion of the scope and limitations of CNDO-based procedures, a remark on the problems which were encountered in their development is needed.

The main problem with Pople's original CNDO/2 model when applied to planar π-systems containing heteroatoms is that σ and π–MO's are often incorrectly ordered. Del Bene and Jaffé's remedy to this was to introduce different resonance integrals β for the interaction of σ and π-AO's which made it possible to calculate n→π* excitations for planar molecules in reasonable agreement with experiment[6]. However, their prediction of π ground states for the radical cations of pyridine and all diazabenzenes [22d] which was later refuted by experiment [16] shows that the situation is still unsatisfactory if nitrogen lone pairs are involved.

Later, Scholz et. al. carried Del Bene and Jaffé's idea one step further by introducing different exponents ζ for 2s and 2p AO's and using a different expression for the resonance integrals β. The parameters in the resulting procedure (which they called CNDO/ζ [23a]) were optimized such as to give optimal agreement with ab-initio MO energies, which resulted in a correct ordering of σ- and π-excited states in the azabenzenes, but excitation energies which were uniformly too high. The same group later extended the procedure for the treatment of open shell species and introduced CI in a manner analogous to that used in the open shell π-models. Notably, they were the first to use the results of semiempirical calculations to predict relative PE band intensities from the weight of Koopmans' configurations in the CI expansion [23b].

Another drawback of the original CNDO/S-CI method was, that the relative energies of K- and NK-configurations in radical ions were wrong with the latter usually predicted too high. This problem was addressed by Bigelow who proposed a scaling scheme to be applied to CNDO/S configuration energies prior to CI. With this procedure, Bigelow obtained excellent agreement of final state energies and relative PE band intensities for a variety of heterosubstituted aromatic radical cations [11d,24]. Bigelow has shown that EA spectra of M^{+} can also be predicted quite accurately by this method [24d,e].

Finally, it should be mentioned that the neglect of σ-π exchange integrals in CNDO based models makes it impossible to correctly describe the electronic states of *linear* open shell species (all states described by the same configuration come out degenerate [21e][7]) and to account for the fact that spin density occurs in σ-MO's of π-radicals by virtue of spin polarization. This problem can be solved by reintroducing one-center electron repulsion integrals of the type (ij,ij), whereby one arrives at the so called INDO level of approximation on which the models described below are based.

---

[6] Due to the negelect of certain integrals in the CNDO approximation, the results are no longer rotationally invariant (i.e. molecular properties depend on the choice of coordinate axes) if a distinction is made between 2s and 2p AO's in the evaluation of electron repulsion and resonance integrals. In order to overcome this problem, this distinction was abandoned in the original CNDO/2 method. By reintroducing different β's for the interaction between 2s and 2p AO's, rotational invariance is again lost and hence the CNDO/S method is only applicable to planar molecules, where the choice of one axis is self-evident.

[7] For the same reason, singlet and triplet excited n→π* states of planar molecules are degenerate in CNDO.

2.3.4. *INDO and NDDO based methods*

The main application of the original INDO procedure in $M^{+}_{\cdot}$ studies is in the calculation of spin densities where it continues to be applied with great success. Apart from this, the main thrust in the development of INDO-based procedures was aimed at models which are parametrized to give correct *ground state properties* (heats of formation, geometries). From these efforts, which were primarily undertaken by the group of Dewar, resulted the well known family of MINDO procedures [25a-c] which were later replaced by the even more successful MNDO [25d,e] and AM1 [25f] models based on the NDDO approximation[8]. These were incorporated into the widely used MOPAC and AMPAC program packages which allow also calculations on open shell systems both by the UHF and the "half electron" RHF method.

INDO and NDDO-based procedures have also been proposed for electronic structure calculations. In the case of $M^{+}_{\cdot}$, the simplest approach is to perform a calculation of neutral M and to take the energy differences between the HOMO and the lower lying MO's as excitation energies of the $M^{+}_{\cdot}$. Of course, this procedure is only valid for Koopmans' configurations, but Lindholm et. al. showed already in 1972 that a modified INDO procedure which was reparametrized for optimal predictions of ionization potentials (SPINDO [26a]) gives very accurate predictions of $\sigma-$ and π-IP's of planar conjugated hydrocarbons. Unfortunately their procedure was not rotationally invariant and could therefore only be applied to the latter set of compounds. It was later replaced by a procedure designed for the same purpose but based on an entirely different approach (HAM/3 [26b,c]) which can in principle also be used for predictions of $M^{+}_{\cdot}$ EA spectra, both on a Koopmans' as well as on a CI level.

Recently, Clark et. al. have shown that simple Koopmans' calculations using the MNDO model can give amazingly accurate predictions of $M^{+}_{\cdot}$ EA spectra in certain cases, provided that the calculation is carried out at the equilibrium geometry of the radical cation rather than the neutral[9]. They used the MO's of the neutrals also for evaluating transition moments which were of considerable help in the spectroscopic assignments. This shows again that very economical procedures can often yield useful predictions, provided that their trustworthyness in a certain field of application has been properly assessed.

A quantitative assessment of the effects of CI on $M^{+}_{\cdot}$ excited states is possible in a computationally efficient fashion by means of the so called PERTCI method devised by Schweig et. al. [29a][10]. This group is conducting a program for assessing the influence of Non-Koopmans' configurations on PE spectra as well as EA spectra of various organic $M^{+}_{\cdot}$ by application of the PERTCI formalism to wavefunctions obtained by CNDO/S, MNDO as well as their own NDDO based procedure (the

---

[8] In the NDDO approximation, the two-center electon repulsion integrals (ERI's) which are necessary to retain rotational invariance while maintaining the distinction between 2s and 2p AO's in the evaluation of ERI's are retained. Dewar and Thiel use an elegant multipole expansion scheme in the calculation of the two-center ERI's [25e] which avoids an undue proliferation of empirical parameters.

[9] This procedure was termed "ncg" (neutral at cation geometry) [27a]. Predictions were made for $\sigma \rightarrow \sigma$ and $\sigma \rightarrow \pi$ transitions in the radical cations of hydrazines [27a] and tetraalkyl olefins [27b]. Note, however, that the ncg procedure seems to be less universally applicable than one had hoped. It proved to be of little help in a recent study on the effect of pyramidalization in hydrazine radical cations on their EA spectra [28].

[10] In this method, a zero-order many-electron wavefunction is constructed by conventional CI between a reference configuration and those excited configurations which couple most strongly with the former. The contributions of the remaining configurations is subsequently estimated via a second order perturbation method, hence the name PERTCI.

spectroscopically parametrized LNDO/S model [29b]). The results obtained in this fashion are generally in good agreement with experiment [29c-e], but since most of the compounds taken into consideration by Schweig et. al. are planar conjugated hydrocarbons, it is not surprising that his all-valence calculations do not yield any insight which cannot be obtained more economically by the π-electron procedures described above.

2.3.5. *Ab-initio methods*

The procedures which are used in semiempirical methods to account for the effects of correlation on excited state properties can in principle be applied also on the *ab-initio* level of theory. In fact, calculations of excited state potential surfaces by such methods have nowadays become common practice for diatomic and some triatomic molecules [30] and form indeed the basis for most serious spectroscopic assignments in these cases (which include many charged and/or open shell species of astrochemical interest). However, application of such rigorous methods to larger molecules are rather scarce due to the high cost of the involved computations.

Alternatively, methods based on many-body perturbation theory [31] allow an evaluation of excited state energies and transition moments on the basis of single configuration wavefunctions and it is in particular one such method which has gained prominence in the field of $M^{\ddot{+}}$ spectroscopy. In this approach, use is made of the so called *Green's function* (GF) which describes how a general density disturbance such as an electronic excitation (into the continuum in the case of an ionization) propagates through the system under investigtion. After transformation of the GF from the time into the frequency domain, excitation (ionization) energies and transition moments are obtained as solutions [32]. In principle, the method allows the evaluation of these properties in a system of fully correlated electrons but in practice approximations have to be introduced in order to make the problem computationally tractable.

Application of the GF method in conjunction with ab-initio theory has been particularly fruitful for the calculation of band positions and -intensities in PE spectra of molecules up to about the size of benzene [11c]. The main advantage of this approach is that it is not limited to those elements of the periodic table for which well tested parameters are available in the above mentioned semiempirical procedures. However, for the few cases where a comparison is possible, quantitative agreement with experiment turns out to be no better than in the case of semiempirical procedures. Thus, application of ab-initio based models to problems of $M^{\ddot{+}}$ spectroscopy should be reserved to cases which cannot be treated satisfactorily on a lower level of theory.

2.4. AN ILLUSTRATIVE EXAMPLE: ANTHRACENE

The radical cation of anthracene ($AN^{\ddot{+}}$) is sufficiently persistent to be observed under stable conditions in solution. Therefore its EA spectrum has been known for a long time and various parts of it have been observed in $CF_3COOH/BF_3 \cdot H_2O$ [33a], CCl_4 [33b], butyl chloride/isopentane [33c], a Freon mixture [33d], in streched PVC films [33e,f], and recently also in Ar [33g]. Since the full matrix isolation spectrum was never published, we will use this occasion to present $AN^{\ddot{+}}$ as an example to illustrate the notions introduced in the preceding paragraphs without the intervention of complicating factors (strong geometry changes upon ionization, overlapping bands etc.).

The EA spectrum of $AN^{\ddot{+}}$ in Argon is displayed in Fig. 3 where it is plotted under the PE spectrum of AN on the same scale (origin of the energy scale for the EA spectrum coinciding with the 0-0 vibronic componenent of the first PE band) in order to show the presence or absence of coincidences between the two spectra. The example of $AN^{\ddot{+}}$ shows that a considerable amount of information can often be gathered from PE/EA plots without resorting to any computations. Thus, no EA band of AN^{+} is observed in the energy region of the second PE band of AN which indicates that the first excited state of AN^{+} is of different parity than the ground state ($g{\rightarrow}g$ and $u{\rightarrow}u$ electronic transitions are electric dipole forbidden). Conversely, we find an intense EA band coinciding with the third PE band and hence the second excited state must be of opposite parity to the ground state of AN^{+}. Furthermore we can distinguish a number of small peaks in the tail of the first EA band (see expanded inset) which cannot be assigned to vibronic progressions and must hence indicate the presence of an additional excited state of AN^{+}. Since the photoionization cross section is essentially zero in this energy region we conclude that this state involves excitation into virtual MO's. Finally we note that the third EA band of AN (at ≈ 3 eV) coincides with a high energy shoulder rather than with the maximum of the fourth PE band of AN which suggests that there are two states in this energy region, of which the lower lying one has the same parity as the ground state of AN^{+}, while the higher lying one has opposite parity.

In a next step we turn to the lower part of Fig. 3 where the results of theoretical predictions at different levels of theory are graphically displayed. The height of the bars is proportional to the calculated transition moments for electronic excitation in AN^{+} (forbidden transitions are reperesented by dots or circles). Solid bars and dots are for Koopmans' (K) configurations, open bars and circles for Non-Koopmans (NK) configurations. The ground configuration/state is placed at the energy origin and the excited configurations/states appear in ascending energy relative to the former.

In the case of planar conjugated hydrocarbons, HMO theory usually offers a valid starting point for the discussion. At the bottom of Fig. 3 is a configuration diagram obtained on the basis of a Hückel π-MO scheme of AN: the ground configuration (G) of AN^{+} is attained by ejection of an electron from the highest occupied MO π_7 of b_{2g} symmetry.

Ionization from the degenerate MO's $\pi_6(b_{2g})$ and $\pi_5(a_u)$ leads to a corresponding pair of K-configurations at 0.586 β units above the $^2B_{2g}$ ground configuration. Another degenerate pair of K-configurations at 1.0 β is obtained by removal of an electron from $\pi_4(b_{2g})$ and $\pi_3(b_{1u})$. However, below these we find the first B-type configuration of B_{1u} symmetry which arises by HOMO(π_7)$\rightarrow$LUMO(π_8) excitation, thus confirming the above expectation of an NK state in this energy region. At 1.414 β we find four more NK-configurations, two of which are of B-type (excitation from π_7 into π_9 and π_{10}, respectively) and the other two of C-type (excitation from π_5 and π_6, respectively, into π_8, leaving π_7 singly occupied). They are degenerate by virtue of the pairing properties inherent to the Hückel MO-schemes of alternant hydrocarbons.

The rule-of-thumb relationship $-\beta \approx 3$ eV gives a first quantitative estimate of the state energies corresponding to the above configurations, and a comparison with the spectra above shows that the agreement is already surprisingly good. Since the degeneracies in the Hückel scheme are due to the negelect of electron repulsion rather than due to symmetry, they are expected to disappear at higher levels of theory. This is confirmed by PPP/SCF π-electron calculations which yield the configurations depicted in the second column. Since the π-SCF-wavefunctions do not differ significantly from the Hückel MO's, the transition moments remain almost the same. Note, however, that the correct sequence of excited states could already be determined from the spectra in Fig. 3 and the HMO scheme alone, using the parity relationships discussed above.

Finally, all configurations of the same symmetry will interact which may lead to shifts in the energy positions of the individual levels. In general, the effects of this mixing must be evaluated computationally, i.e. by setting up and diagonalizing a CI matrix whose off-diagonal elements describe the extent of interaction between the individual configurations.

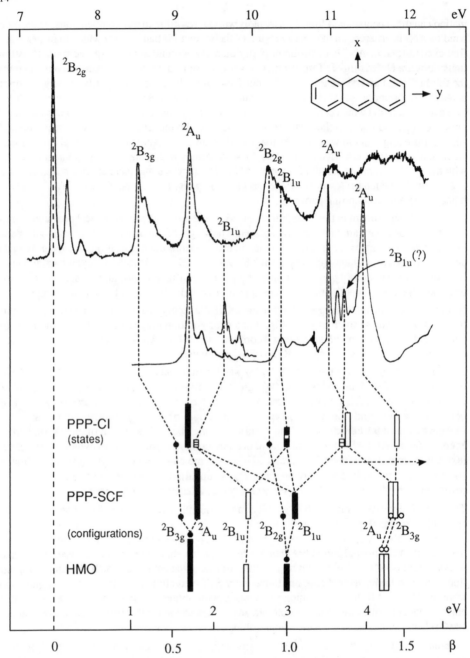

Fig. 3: The electronic structure of the radical cation of anthracene, explanations see text.

However, a qualitative prediction can often be obtained by focussing on *pairs of configurations* which are close in energy and which will therefore interact most strongly. Thus, the two $^2B_{1u}$ configurations at ≈ 2.5 and ≈ 3 eV will interact significantly, a prediction which is borne out nicely by the PPP-CI calculations. The splitting is not symmetric because the resulting upper state is "depressed" through interactions with higher-lying C-type $^2B_{1u}$ configurations not shown on the diagram.

From our experience with PPP-CI we suspect that this procedure generally overestimates extent of interaction between configurations of the same symmetry. Thus, it would be surprising if the calculated 18% admixture of Koopmans' character to the NK $^2B_{1u}$ state would not result in detectable intensity in the PE spectrum (where none is actually found). On the other hand the latter state is predicted too low in energy which may be taken as another indication of the above mentioned overestimation of CI. Therefore, the PPP-CI predictions in the region above 3 eV, where many configurations mix, should perhaps be taken with a grain of salt. Taken together with the calculated transition moments, the agreement between theory and experiments is, however, quite satisfactory[11].

Using the simple methods described above, the spectra of most planar conjugated hydrocarbons, in particular those of benzenoid aromatic systems can be interpreted without further ado. Very often, a juxtaposition of EA spectra of M^{+} and PE spectra of M, together with a simple Hückel calculation, suffices for a qualitative discussion of the M^{+} electronic structure including a correct assignment of all observed bands in the two spectra. Although things may get somewhat more involved if ionization from σ orbitals must be taken into account, the general approach outlined above has proven to be valid also for these cases as will be shown below. We therefore advocate the use of simple and transparent models wherever higher-level procedures are not expected to yield more insight.

3. Experimental Techniques

3.1. GAS PHASE, SOLUTION, POLYMERS

As pointed out in the introduction, there are several gas phase techniques available for the study of M^{+} electronic structure, whose results may supplement those obtained in condensed phase. Next to PE spectroscopy (see above) it is mainly *Ion Photodissociation* (PD) spectroscopy [34] which should not be overlooked in the present context. In this method, gas phase M^{+} are prepared in a beam or in an ICR cell and the appearance of fragment cations is monitored as a function of the wavelength of some source of exciting light. Since the photodissociation cross section for a given M^{+} is usually proportional to the absorptivity over a considerable spectral range, PD spectra can often be compared directly to EA spectra obtained in condensed phase.

Apart from purely spectroscopic applications [35a], PD spectroscopy has proven of invaluable help in understanding M^{+} rearrangements and distinguishing isomeric M^{+} in the gas phase [35]. The method continues to be developed towards higher resolution and recently the first PD experiments on rotationally cooled M^{+} in supersonic jets were reported [36]. Next to PD spectroscopy, other novel techniques which are at least potentially applicable for the study of M^{+} electronic struc-

---

[11] PPP-CI gives exactly three states with u parity between 3 and 5 eV. While the transitions to the two ^2A_u states are predicted to be very intense, that to the intervening $^2B_{1u}$ state should be barely detectable. Thus, it seems reasonable to assign the two intense peaks at 3.6 and 4 eV to the two ^2A_u states with the $^2B_{1u}$ state being perhaps responsible for one of the weak features between them.

ture in the gas phase have begun to enter the field, most prominent among them the velocity modulation technique introduced by Saykally [37a] which was recently used to study the first electronic transitions of small M^+ [37b,c].

Before coming to the solid state methods described below, it should be recalled that many organic M^+ are sufficiently persistent to be observed in *solution*. Thus, the radical cations of many donor-substituted benzenes and polycyclic aromatic hydrocarbons were prepared by chemical or anodic oxidation and subjected to optical and ESR studies [38] long before the advent of matrix isolation. Solution techniques continue to be employed with considerable success in the study of organic M^+ which do not need to be immobilized to prevent bimolecular reactions during the time needed for observation [27,28].

On the other hand, very reactive transients can be studied in solution by *pulsed techniques*. Most prominent among these is *flash photolyis* but since few molecules can be directly photoionized by UV/VIS light, the presence of an electron acceptor is required if M^+ are to be observed. Actually, an entire branch of science has evolved around the disentanglement of the comlexities associated with photoinduced electron transfer processes in solution and the posssible escape of radical ions from the initially formed exciplexes [39]. However, the concomitant appearance of the acceptor radical anion and possibly also metastable triplet states in these experiments makes them less suited as a primary means for the study of M^+ electronic structure.

Alternatively, *pulse radiolysis* may serve for the production of transient M^+ either by direct ionization or by secondary chemical oxidation. It was realized in the early 1970's that this method is particularly appropriate for the study of dimer cations $(M)_2^+$ whose formation requires diffusion [40]. Until today this has remained the only technique whereby the electronic structure of $(M)_2^+$ can be investigated routinely ([41], see also Section 5.1).

A conceptually elegant way to immobilize molecules is to embed them in a *polymer*, either by functionalizing or doping it or by attaching them covalently to a suitable monomer before polymerization. Actually, all three approaches have been used for implanting substrates to be oxidized into polymers, but only one of them was put to systematic use for the study of radical ion electronic structure: Hiratsuka et. al. are conducting a program wherein they combine the streched film technique to obtain partially oriented samples with low-temperature radiolytic methods for the generation of M^+/M^-. This allows them to determine polarization directions of M^+/M^- electronic transitions and thus obtain valuable additional information for the interpretation of the corresponding EA spectra [42].

3.2. FROZEN GLASS TECHNIQUES

Some liquids prefer to form glasses instead of crystals upon cooling and therefore remain transparent over a certain spectral range which is a prerequisite for high quality optical studies. The so called transition temperature where the viscosity of the liquid increases (or that of the glass decreases) very rapidly may be in the hundreds of degrees (°C), as it is the case for most inorganic materials, or far below 0°C as in most organic solvents which form glasses. Note that this parameter is of no significance as far as the ability to immobilize reactive species is concerned, and as long as a substrate does not react with the host material, a glass need not be cooled much below its transition temperature. Thus, it is not astonishing that among the first studies of M^+ in glasses were some conducted leisurly at room temperature, where certain inorganic oxidants form transparent hosts.

In 1955, Evans [43] found that *boric acid glasses* [44] containing aromatic hydrocarbons turned colored upon UV irradiation. He ascribed these changes to the formation of free radicals but later

Hoijtink et. al. showed that spectra of these radicals were identical to those of the corresponding $M^{\ddot{\tau}}$ obtained in solution [45] and extended the boric acid glass studies to a broader range of aromatic hydrocarbons. Ionization was shown to occur via long-lived triplet states [46], but the fate of the ejected electron in the boric acid glass was never consistently clarified. The main advantage of this technique which continues to enjoy a certain measure of popularity (see Section 4.2) is, that it allows to investigate substrates such as large polycyclic aromatic hydrocarbons which are difficult to dissolve in organic solvents. Its main limitation is that the substrate to be studied must survive the 400°C needed to produce a boric acid melt.

Except in the case of very easily oxidizable species such as aromatic amines [47], UV-photons will not suffice to effect ionization in frozen organic solvents. This limitation was overcome by Hamill who pioneered the use of *radiolytic* techniques for the study of radical ions [48]. He also led the search for solvents which are able to act as scavengers for the ejected electron (if $M^{\ddot{\tau}}$ are to be observed) or the positive "holes" (if $M^{\overline{\cdot}}$ are the target of investigation) such as to prevent charge recombination and thus improve the radical ion yield. Most of the early $M^{\ddot{\tau}}$ studies were done in CCl4 which is capable of scavenging electrons by dissociative attachement [49a]. However, this solvent does not form transparent glasses and shows strong absorptions of its own after radiolysis. It was therefore replaced by n- or sec- butyl chloride [49b], mixed sometimes with isopentane to improve the transparency of the glass [49c]. The scheme below summarizes the mechanism of $M^{\ddot{\tau}}$ formation in frozen solvents S containing organic halides RX as electron scavengers (note that S may be identical with RX):

S	$\longrightarrow$	$S^{\ddot{\tau}} + e^-$	(ionization)
$RX + e^-$	$\longrightarrow$	$RX^{\overline{\cdot}}$	(electron attachment)
$RX^{\overline{\cdot}}$	$\longrightarrow$	$R^{\cdot} + X^-$	(dissociation)
$S^{\ddot{\tau}} + S$	$\longrightarrow$	$S + S^{\ddot{\tau}}$	(resonant charge transfer)
$S^{\ddot{\tau}} + M$	$\longrightarrow$	$M^{\ddot{\tau}} + S$	(hole attachement)

The positive charges formed initially on the solvent molecules travel through the matrix by a mechanism which is believed to involve very fast resonant charge transfer between adjacent S molecules [50]. When a substrate molecule M of lower oxidation potential than S is encountered by the migrating "hole" it becomes irreversibly attached. Thus, the solvent's oxidation potential imposes a limit to the range of substrates which can be radiolytically ionized and for this reason, most of today's frozen glass $M^{\ddot{\tau}}$ studies are done using Freons rather than simple alkyl halides. The most popular solvent system for optical studies consists of a mixture of CCl_3F (F11) and CF_3BrCF_3Br (F114B2), first proposed by Sandorfy [51a] but often referred to as the Grimison-Simpson Freon mixture (FM) in reference to the authors who first described its application for $M^{\ddot{\tau}}$ studies [51b]. Its main disadvantage compared to butyl chloride/isopentane is the high glass transition temperature ($\approx$80K) which makes it difficult to conduct controlled annealing experiments without loss of the $M^{\ddot{\tau}}$ due to charge recombination. Occasionally, one finds also reports of optical studies in pure Freons such as those used for ESR spectroscopy of $M^{\ddot{\tau}}$ but unfortunately these generally do not form transparent glasses. All solvents must be carefully purified and dried prior to use.

Usually, a substrate concentration of 10^{-2} M suffices to scavenge most of the positive charges generated during ionization of the solvent. What remains to be clarified is the exact nature of the dissociative electron capture step: since $R^{\cdot}$ and X^- are held in close proximity by the rigid matrix, they will to some degree retain the electronic structure of $RX^{\overline{\cdot}}$ which should have low lying electronic states. An indication of this is that electrons can sometimes be remobilized by IR irradiation with concomitant bleaching of the $M^{\ddot{\tau}}$ absorptions due to charge recombination.

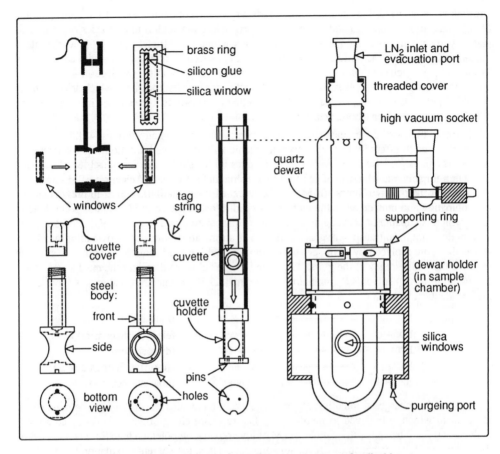

Fig. 4: Appraratus for low-temperature spectroscopy of radical ions

Various devices have been proposed for the M^{+} low temperature spectroscopy experiments. Above we depict the apparatus which evolved in the course of the M^{+} frozen glass studies conducted in Fribourg (see Fig. 4). Special cuvettes were designed with the purpose of avoiding breakage of the silica windows due to expanding solvents and providing easy access of all innner surfaces for cleaning. Thus, each silica window is fixed in a supporting threaded brass ring with a silicon glue which remains flexible at low temperatures. By fitting pairs of these rings of different depth onto the cylindrical steel body, optical pathlengths of 0.5-3 mm can be obtained. The assembled cuvettes are not always completely tight, but this poses no problems because the samples are immediately frozen. On the other hand, this systems allows for very easy maintenance and cleaning. For M^{+} production, a dozen of these steel cuvettes are placed inside a 1000 ml dewar vessel filled with liquid nitrogen which is then exposed to some 0.5 Mrad of γ-radiation. If one wants to take difference spectra (see below) it is important to allow for a precisely reproducible positioning of the cuvettes in the spectroscopic dewar vessel and of the latter in the spectrometer. This is assured by pins in the cuvette bottoms and rings on the dewar vessel's support.

3.3. MATRIX ISOLATION

Although frozen solutions permit an efficient immobilization of embedded substrates, this technique is fraught with certain problems when it comes to studying very reactive species such as atoms, radicals or M^+:

a) The host materials are often not sufficiently inert and may undergo chemical reactions with the substrates.

b) Upon cooling of solutions, even if carried out rapidly, substrates tend to aggregate due to the decreasing importance of the entropy term. In the case of subsequent ionization of the substrate this may lead to the formation of dimer cations $(M)_2^+$ whose spectroscopic properties differ siginificantly from those of the monomeric M^+.

c) The complexity and variety of possible cavity structures in frozen polyatomic solvents together with the often strong interactions between solvent and substrate frequently lead to a blurring of spectral detail and thus to a loss of information.

d) Large parts of the infrared spectral range are obscured by most polyatomic solvents which makes vibrational studies impracticable.

All of the above obstacles can be surmounted if a mixture of the substrate and a noble gas is trapped on a cold surface, a method which forms the basis of the technique of *matrix isolation*. Firstly, noble gases are both chemically inert and transparent from the far UV to the far IR. Secondly the rapid solidification of the gaseous mixture upon its impact on a suitably cold surface prevents agglomeration of the substrate whose molecules are individually isolated at distant sites of the noble gas lattice if sufficiently low concentrations are employed. Thirdly and finally, the variety of cavities and the (physical) interaction with substrates is minimized in solid noble gases.

Since the advent of closed-cycle cryostats in the early 70's, matrix isolation grew into a standard laboratory technique as documented by several monographs and review articles [52] which highlight different aspects of the research on unstable species by electronic, vibrational and ESR spectroscopy in noble gas matrices. Application of this technique to radical ion spectroscopy was pioneered by the group of Andrews (see other chapter in this volume) who also explored several methods for the formation of M^+ in noble gas matrices. Most of their recent studies were done with the technique of "Argon resonance ionization" [53] where the output of an open microwave discharge through Ar interacts with the Ar/substrate gas stream during the process of matrix isolation.

In order to cleanly separate the spectroscopic manifestations of M^+ from those of neutral M it is often necessary to evaluate *difference spectra* before and after ionization (see the example of anthracene above). This is only possible on the same sample if ionization is effected *after* matrix isolation, for example by VUV photoionization through LiF [54] or by the radiolytic techniques which had proven useful in the frozen glass experiments described above. We opted for the latter approach but for practical reasons we had to replace γ- by X-rays[12], which seem to work equally well provided the sample can be brought very close to the tube (see Fig. 5). In order to improve the

---

[12] In principle, standard equipment used for X-ray diffraction can be used. Irradiation times can be shortened by replacing the usual copper target by a tungsten target tube which yields more total X-radiation at similar operating conditions. Irradiation can be effected through 1mm silica windows without much loss in intensity. In our experience, the M^+ yield does not improve significantly after 90 min irradiation at 40 kV/40 mA with the sample held as closely as possible to the X-ray tube.

M^+ yield, it is advisable to add some electron scavenger[13] in equimolar amounts to the substrate before mixing it with a 1000fold excess of noble gas. The mechanism of ion formation is then quite similar to that depicted above with the noble gas playing the role of the solvent.

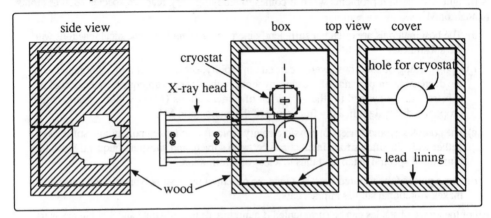

Fig. 5: Setup for X-irradiation of matrix-isolated samples

The main drawback of the radiolytic matrix isolation technique is, that the ultimate charge transfer from the noble gas to the substrate is quite exothermic[14], and that the incipient M^+ cannot easily relieve itself of its excess energy because the noble gas lattice does not provide suitable acceptor vibrations. Instead, this energy may be used to drive a chemical reaction, and therefore parent M^+ of species which can rearrange to lower energy isomers can sometimes not be observed in Ar. In such cases one must resort to the above mentioned Freon glasses which have proven to be very well suited for the stabilization of metastable primary M^+. Alternatively, the exothermicity of the charge transfer may be mitigated by using Xe instead of Ar as a matrix material. Unfortunately, Xe matrices are often not of sufficiently good optical quality to permit EA measurements in the UV region and thus one of the main advantages of the matrix isolation technique is lost.

In sum, matrix isolation has made important contributions to our understanding of M^+ electronic structure in condensed phase. Since noble gas matrices are also transparent in the infrared spectral region, the techniques described above can also be used to study the *vibrational* structure of M^+, a field of research which has not been exploited in a systematic fashion.

---

[13] We use mostly CH_2Cl_2 whose electronic absorption after X-radiolysis at 360 nm can easily be bleached. The groups who employ the photoionization technique for the formation of M^+ in Neon matrices [5,54] claim that in their experiments the ion yield is not improved by added electron scavenger, in contrast to our own experience. However, most of the Neon matrix EA work was done using the so called *waveguide technique* where the probing light is coupled sideways through a slit into the matrix which it leaves through another slit at the opposite end after multiple internal reflections [5b]. In this fashion the sensitivity can be increased greatly compared to normal incidence EA spectrosocpy and therefore much smaller concentrations of substrate are needed to get good spectra.

[14] A good estimate for the charge transfer energy is given by the difference in gas phase ionization potentials of the noble gas and the substrate (solvation energies are generally small in noble gases). Since this difference is very large for organic molecules in Neon, it is preferable to effect ionization in such experiments by VUV light which is only absorbed by the substrate.

4. Summary of M^+ spectra

4.1. SPECTRA IN NOBLE GAS MATRICES

In Fig. 6 and 7 we give a summary of references for all organic compounds whose radical cations were investigated by EA spectroscopy in noble gas matrices. As mentioned previously, most of the work on organic M^+ in Neon matrices was done by emission spectroscopy and some of the fluorescing M^+ investigated in these studies were also subjected to EA spectroscopy (Fig. 6). Ionization in Neon is usually effected by VUV radiation and the samples are probed by the wave-guide technique13. On the other hand, the compounds investigated in Argon matrices (Fig. 7) were generated by Argon resonance or X-ray irradiation in the presence of an electron scavenger and probed by normal incidence EA spectroscopy. In some cases M^+ formation was effected by two-photon ionization with UV/VIS light [55p].

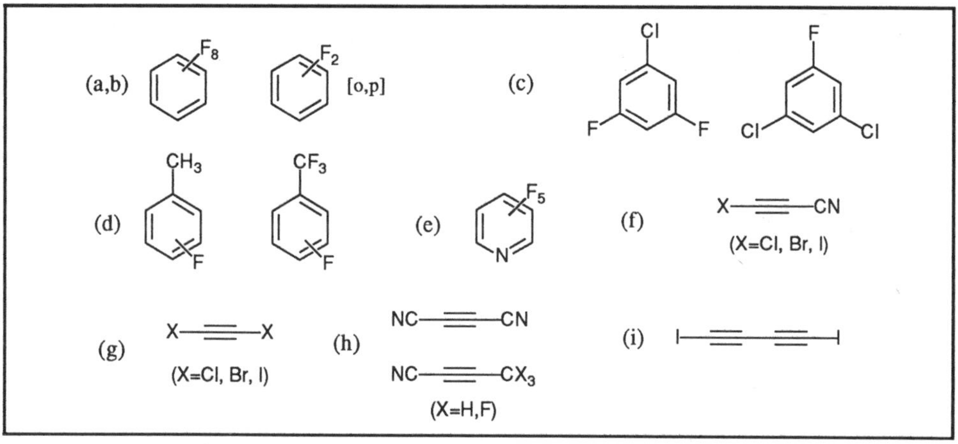

Fig. 6: Molecules, whose radical cations were investigated by EA spectroscopy in solid Neon. Letters correspond to those in reference [54].

The above Neon matrix spectra are unique, i.e. most of these compounds were not studied in other media. On the other hand, many of the compounds in Fig. 7 had already been subjected to EA spectroscopy in frozen glasses or gas-phase PD spectrsocopy. In general, the Ar-matrix spectra did not yield much new insight with regard to the electronic structure of the M^+. Apart from some special cases mentioned below, the energies of the observed transitions correspond reasonably well to what is expected on the basis of the PE spectra of the neutrals. However, the Ar matrix spectra generally reveal more vibronic structure and hence the discussion of Ar matrix M^+ spectra often focussed on this aspect [55].

With respect to the radical cations of *naphthalene* (NA^+) and its derivatives it is interesting to note that the transition into the *first excited state* ($^2A_u \rightarrow ^2B_{1u}$ if y is the long axis of NA) which is expected at ≈ 1700 nm from the PE spectrum of NA but is dipole forbidden in NA^+ is *not* detectable even if the D_{2h} symmetry of NA is lowered by substitution. The corresponding spectral region is not shown in the published papers [55g-i], but our efforts to find an EA band in the NIR region in different NA derivatives were not successful until now. Apparently, a strong perturbation is needed to make the transition moment for the first electronic excitation large enough to result in a detectable EA band.

22

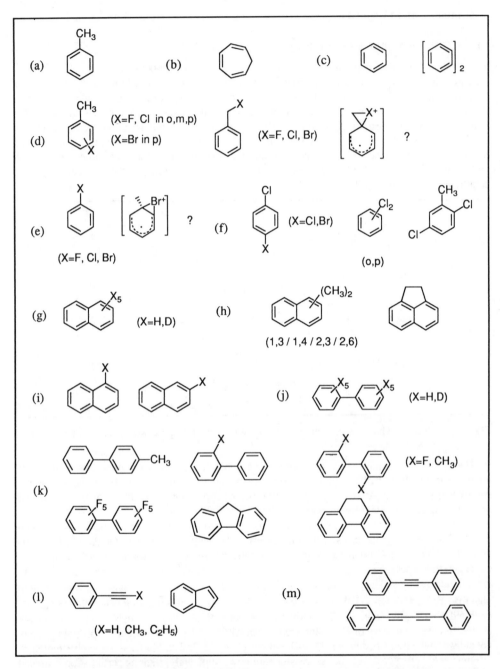

Fig. 7: Molecules whose radical cations were investigated by EA spectroscopy in solid Argon (continued on next page). Letters correspond to those in reference [55] except where otherwise noted.

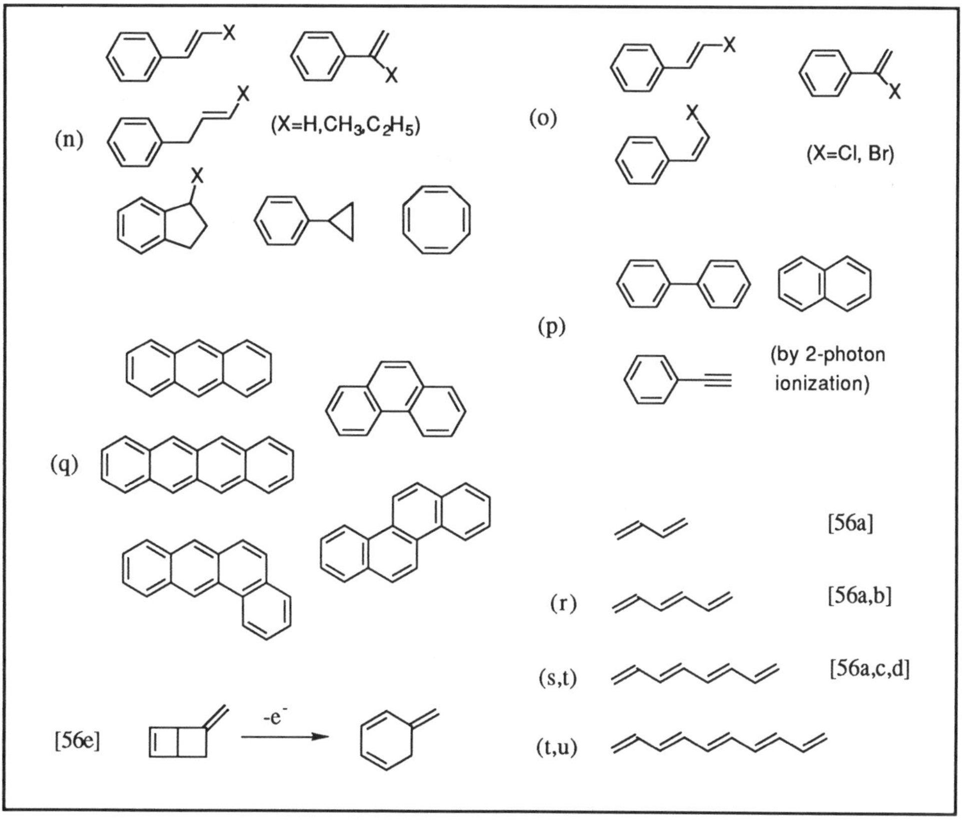

Fig. 7 (continued, caption see previous page)

If two or more π systems are linked by essential single bonds, such as in *styrene, stilbene* or *biphenyl* and their derivatives, then the HOMO is invariably antibonding between the π moieties. Consequently, the π bond order along the connecting bonds increase upon ionization, i.e. they become partial double bonds which brings about a tendency for planarization of the corresponding radical cations. In the case of biphenyls this expresses itself in a mismatch between I_4-I_1 from the PE spectrum of the neutral [57] and λ_{max} for the first electronic transition in the radical cation [55k]. Hence, M^+ spectra are indicative of the molecular structure in such systems.

A frequently recurring feature of Ar matrix experiments is, that ionization is accompanied by rearrangements (see 3.3). In some cases these make it impossible to observe the primary M^+ at all (i.e. cyclopropylbenzene [55n] or 5-methylenebicyclo-[2.2.0]hexene [56e]) while in other cases one must sort out the spectra of different isomers prior to spectroscopic assignments (i.e. polyenes [55r-u,56a-d]). However, such rearrangements, which can often be driven to completion by subsequent photolysis, sometimes yield interesting radical ions whose neutral counterparts are difficult to synthesize and/or handle (i.e. 1-methylene-2,4-cyclohexadiene derivatives [56e]) or do not exist as stable species at all (i.e. simple carbonyl ylids [56f]). The results of such experiments will be summarized in the Section on photochemistry of M^+ and the tables shown there may be regarded as complements to Fig. 7.

24

4.2. SPECTRA IN FROZEN GLASSES

As mentioned in the introduction, a large compilation of M^+/M^- spectra in frozen glasses has recently become available as a monograph [6]. Therefore, Fig.s 8 and 9 list only molecules which do not appear in this book. They stem from the boric acid glass experiments by Khan et. al. (Fig. 8) and from our own investigations in Freon glasses (Fig. 9). The spectra of the polycyclic aromatic hydrocarbon cations in Fig. 8 can usually be well understood on the basis of the standard methods illustrated above on the example of the radical cation of anthracene (AN^+).

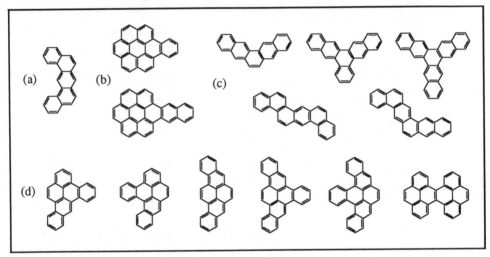

Fig. 8: Aromatic hydrocarbons whose radical cations were investigated in boric acid glasses and which are not listed in [6]. Letters correspond to those in reference [58].

On the other hand, most of the compounds in Fig. 9 (facing page) were investigated because they illustrate special features of M^+ electronic structure which are proper to them. Some of these features will be discussed in more detail in the following sections on complex cations, allylic species and polyenes and we will therefore only mention a few peculiarities of some of the others.

ad (a) This molecule is a prime example for the importance of NK configurations in describing M^+ electronic structure because already its first excited state is dominated by HOMO→LUMO electron promotion. Similarly to AN^+, the correponding electronic transition occurs in an energy region where the PE spectrum is "empty".

ad (b,c,g) These compounds show intense broad bands between 600 and 700 nm which arise through electron promotion from the symmetric to the antisymmetric combination of the two ethylenic π-MO's. If one disregards through-bond coupling then these compounds may be regarded as intramolecular π-complex cations (see Fig 15. and Section 5.1.2 below).

ad (f) The EA spectrum of cyclobutene$^+$ could not be determined unambiguously due to possible interference of the solvent. It seems to show a (pseudo-π)→π absorption at ≈450 nm.

ad (h) The oligoacetylenes have an electronic structure which can be described in similar terms as that of polyene radical cations (see. Section 5.2), except that due to the degeneracy of the π-MO's, the second electronic transition is now associated with a zero transition moment and appearas only weakly in the spectra.

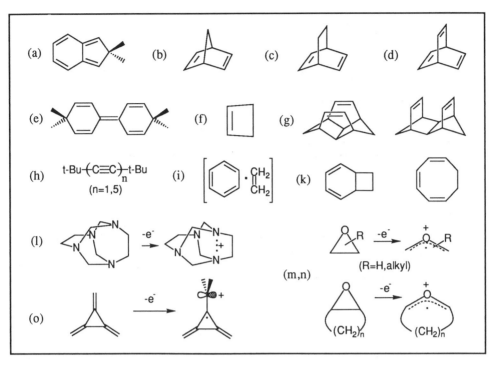

Fig. 9: Molecules whose radical cations were investigated in frozen organic solvent glasses (Freon or butyl chloride/isopentane) and are not listed in [6]. Letters correspond to those in reference [59].

ad (k): The position of the first band in the EA spectrum of cyclooctatriene $\overset{+}{\cdot}$ indicates planarization of the π system in the course of ionization. Again, a PE/EA mismatch could be used to draw conclusions about a geometric change in a radical cation. The spectrum of isomeric [4.2.0]bicycloocta-2,4-diene shows that the cyclohexadiene π-system is essentially unperturbed by the cyclobutane moiety.

ad (l): In this highly symmetric tetramine, a serious mismatch between the PE spectrum of the neutral and the EA spectrum of the radical cation was explained in terms of the formation of a three-electron two-center bond (see section 5.1.1) between two of the four nitrogens. The charge localization was explained as being due to a very strong JT distortion of the first excited ionic state which eventually leads to a crossing with the D_{4h} ground state. Assistance by solvation may be needed to render the charge localized C_{2v} structure more stable than the D_{4h} ground state.

ad (o): A similar effect may be operative in [3]radialene: upon slight warming, the parent radical cation undergoes a conversion which is believed to involve a twisting of one of the double bonds to yield a charge localized carbonium ion. Although this bond rotation is assisted by hyperconjugation of the evolving C^+ moiety with a Walsh MO of the three-membered ring, an interaction with the chlorinated solvent is presumably needed to drive it to completion, whereby a chloronium ion may be formed. The twisted [3]radialene belongs to a class of compounds which were termed "non-vertical" by Roth [60] because upon ionization, it relaxes spontaneously to a structure which does not correspond to a minimum on the neutral hypersurface. Since spin and charge are formally separated it may also be regarded as a "distonic" radical cation [61].

26

4.4. SOLUTION SPECTRA

In spite of the advances in cryogenic technology described above, there are still many novel species whose radical cations are sufficiently stable to be observed more or less leisurly in solution. Most prominent among these are the *alkylated hydrazines* in Fig.11 (facing page) which were studied by Nelsen's group [62]. If we assume that hydrazine$^{+.}$ are (nearly) planar, then their electronic structure distinguishes itself by an electronic transition from the doubly occupied π MO (symmetric combination of the N lone pairs) to the singly occupied π* MO (corresponding antisymmetric combination, see left hand side of Fig.10 below). The position of this π→π* transition is influenced by the geometry (syn- or anti-pyramidalization as well as twisting) of the R_2N-NR_2 moiety and also by the extent and nature of mixing with the σ-MO's of the alkyl substituents [62c]. The strongest red shifts are observed for t-butyl substituted hydrazine$^{+.}$ [62a] which is presumably due to severe twisting as a consequence of steric interactions.

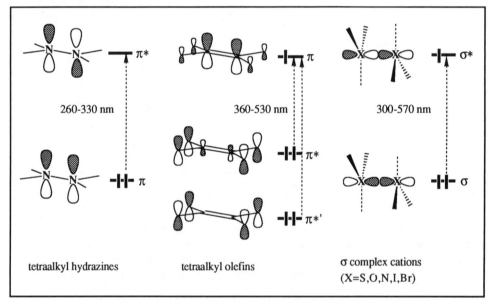

260-330 nm   360-530 nm   300-570 nm

tetraalkyl hydrazines   tetraalkyl olefins   σ complex cations (X=S,O,N,I,Br)

Fig.10: Electronic transitions for tetraalkyl hydrazines (cf. Fig 11), tetraalkyl olefins (cf. Fig. 12) and σ complex cations (cf. Figs 13 and 14). Explanations see text.

Another group of compounds investigated by Nelsen et. al. are the *tetraalkylated olefins and dienes* in Fig. 12. Their radical cations usually show several absorptions in the visible and near-UV region which are thought to involve electron promotion from alkyl σ MO's having a nodal plane containing the ethylenic moiety to the singly occupied π-MO of the olefin (so called "hyperconjugation transitions" [27a]). Depending on the arrangement of alkyl groups, the most prominent EA bands arise through electron promotion from levels labelled π* in Fig. 10 (olefins on the left hand side of Fig. 11) or from those labelled π*' (olefins on the right hand side of Fig. 11). Simple "ncg" calculations (see 2.3.4) proved to be quite reliable in predicting the spectra of such species provided an appropriate scaling procedure was employed. In the case of the two dienes on the bottom of Fig. 11 the strong red shift in the homoadamantene derivative indicates substantial twisting of the diene π-system.

27

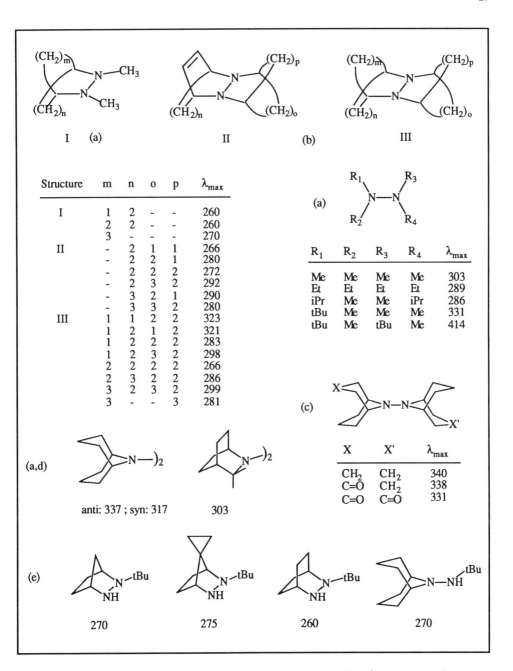

Fig.11: Alkylated hydrazines whose radical cations were studied in solution. λ_{max} values are in nm, letters correspond to those in reference [62].

28

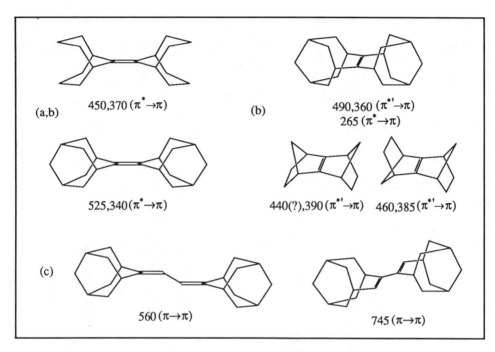

Fig.12: Tetraalkylated olefins and dienes whose radical ions were observed in solution. Numbers indicate λ_{max} in nm, letters correspond to those in reference [63]. For designation of π^* and $\pi^{*'}$, see Fig. 10 above.

A last group of stable radical cations emanated from the work of Alder at al. who synthesized and studied a series of fascinating *multiply bridged diamines* (Fig.13) whose lone pairs are pointing towards each other. As a conseqence, oxidation results in intramolecular complex cations of surprising stability which contain a partial σ bond between the two nitrogen atoms (for a discussion of the electronic structure of complex cations, see 5.1). σ→σ* excitation in these species (c.f. Fig. 10 right side) gives rise to broad absorptions peaking at 470-500 nm.

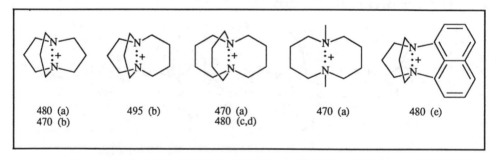

Fig.13: Multiply bridged diamine radical cations. Numbers indicate λ_{max} in nm, letters correspond to those in reference [64].

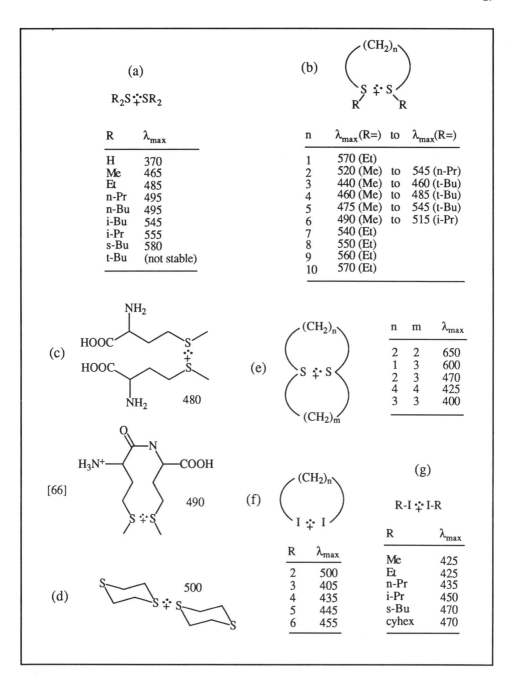

Fig.14 (first part): Homonuclear 2c-3e⁻ bonded S∴S and I∴I complex cations. Numbers indicate λ_{max} in nm, letters correspond to those in reference [65] except where otherwise indicated (continued on next page)

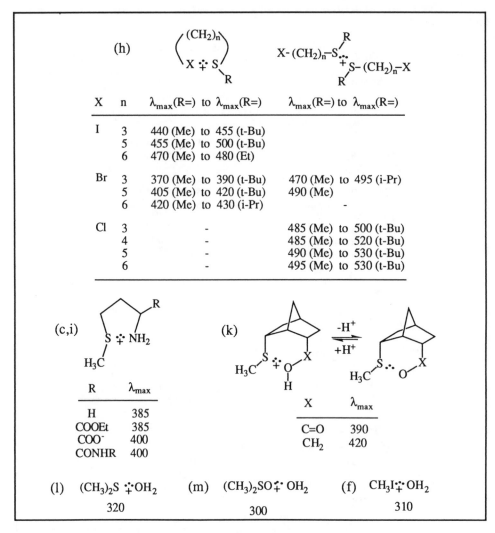

Fig.14 (second part): Heteronuclear 2c-3e⁻ bonded S∴X complex cations (X=I, Br, N, O). Numbers indicate λ_{max} in nm, letters correspond to those in reference [65].

In principle, all compounds containing atoms with lone pairs can form two center three-electron (2c-3e) bonds of the type depicted above. Two classes of molecules which are particularly prone to show this behaviour are *sulfides* and *iodides* shown in Fig.14. However, most of the corresponding complex radical cations are not stable enough to be observed under static conditions in solution and hence the technique of *pulse radiolysis* was applied for their investigation [65]. The electronic structure of these inter- and intramolecular complex cations will be discussed in Section 5.1 and we therefore limit ourselves at this point to a compilation of the λ_{max} values of the various homonuclear (Fig. 14, Part 1) and heteronuclear σ-complex cations (Fig. 14, second part) observed by Asmus et. al. [65].

5. The electronic structure of complex radical cations, allylic radical cations and polyene^{+}

5.1. COMPLEX CATIONS

Whenever a species with an open electronic shell encounters a closed shell neutral, a covalent stabilizing interaction will result from the interaction between their frontier MO's (FMO's) as depicted in Fig. 15. Consequently, the FMO's are displaced in energy which entails shifts in the electronic transitions of the constitutent species such as the characteristic low-energy emissions of excited state complexes (excimers or exciplexes) between planar π systems. In addition, such complexes show new types of electronic absorptions which correspond to electron promotion from the highest doubly occupied to the singly occupied MO[15].

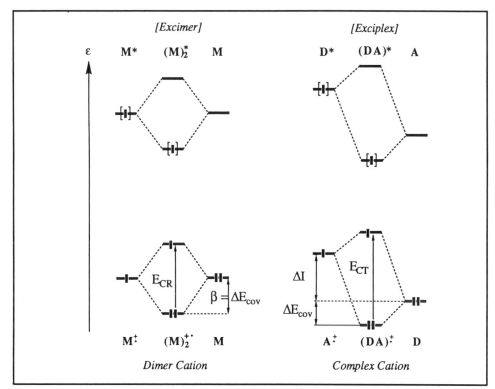

Fig. 15: Interaction of frontier MO's in complexes between open-shell and closed shell species (electrons in brackets are for neutral complexes only)

---

[15] It is interesting to note that such absorptions were never detected in neutral exciplexes or excimers although there is no a priori reason why they should not occur. In fact, there are actually two such transisions to be expected, one analogous to that of complex cations and the other arising through electron promotion from the higher lying of the two singly occupied MO's to the LUMO of the excited state complex.

In the case of radical cation complexes, the importance of these transitions lies in their relation to thermochemical properties as shown in Fig. 15: If we first take the case of dimer cations $(M)_2{}^{+\cdot}$, then the amount of covalent bonding ΔE_{cov} should be proportional to the energy E_{CR} of the so called Charge Resonance transition[16]. In a simple Hückel picture, $\Delta E_{cov} \equiv {}^1/_2 E_{CR} \equiv \beta_{HMO}$ (the Hückel resonance integral) and hence the relationship between the two quantities is linear.[17] In the case of heterocomplex cations $(M \cdot N)^{+\cdot}$, the energy difference ΔI between the two interacting FMO's enters as an additional parameter in the determination of ΔE_{cov} and the energy of what is now a Charge Transfer absorption, E_{CT}. By simple algebra it can be shown that:

$$E_{CT} = 2 \cdot \sqrt{\left(\frac{\Delta I}{2}\right)^2 + \beta^2} \quad \text{and} \quad \Delta E_{cov} = \frac{1}{2}(E_{CT} - \Delta I)$$

The above expression shows that, even in the Hückel approximation, ΔE_{cov} drops rapidly as ΔI increases, an effect which will be enhanced if overlap is included. Since electrostatic terms also contribute to the stabilization of the complex cations, their dissociation energies (DE) cannot be related directly to ΔE_{cov} or E_{CT}. However, in dimer cations $(M)_2{}^{+\cdot}$, DE's should increase in proportion to E_{CR} while in the case of $(M \cdot N)^{+\cdot}$, a dependence of DE on ΔI is anticipated. Both expectations are confirmed by the available experimental and theoretical data.[18]

So far we have not specified the nature of the species involved in complex cation formation, and in prinicple various combinations of molecules can be envisaged to form such species. However, significant covalent bonding is only expected if the HOMO's of the two partners are free to overlap through space $(\beta \neq 0)$[19] and are not too dissimilar in energy $(\Delta I << \beta)$. The first condition can only be met if the HOMO's are of π- or lone-pair nature and it is therefore not surprising that all known complex cations involve compounds of the above two kinds which will be treated separately below.

5.1.1. σ-bonded complex cations

If the HOMO's of the molecules which constitute a complex cation are of lone pair type, then their interaction leads to a bonding (σ) and an antibonding (σ^*) MO which are occupied by a total of three

---

[16] This term originates from a valence-bond description of complex cations [67]. Within this theoretical framework, interaction between a ground state molecule and its radical cation leads to a splitting into two states $R_{\pm}$ via charge exchange:

$$^2R_{\pm} = (^2M^{+\cdot}/N) \pm (M/^2N^{+\cdot})$$

In the case of dimer cations (M=N) the $^2R_+ \rightarrow {}^2R_-$ electronic transition entails no net charge transfer and hence it was termed "charge resonance" absorption

[17] As Baird has pointed out some time ago, inclusion of overlap leads to $\Delta E_{cov} < E_{CR}$ because the upper level is raised more than the lower one is depressed. In fact, ΔE_{cov} drops to zero if the overlap integral is smaller than $^1/_3$ [68].

[18] Clark has recently calculated the dissociation energies of all possible complex cations between the hydrides of N, O, F, P, S and Cl as well as Ne and Ar and their corresponding radical cations and was able to fit the results to the empirical relation

$$D_{AB} = \frac{1}{2}(D_{AA} + D_{BB}) \cdot e^{-\lambda_A \lambda_B \Delta I} \quad [69a]$$

which shows a similar dependence on ΔI as ΔE_{cov}. Clark has also pointed out that it is mainly due to this ΔI dependence that complexes between neutral radicals (which have comparatively high-lying HOMO's) and closed shell compounds are only marginally stable, at least in the gas phase where charge transfer is not favored by solvation [69a,c].

[19] Of course, if the $M^{+\cdot}$ and the M (or N) moiety are part of the same molecule, they can also interact via intervening σ bonds, a factor which must always be taken into account. However, the term *complex cations* should be reserved for cases where interaction is dominated by through space interaction.

electrons (cf. right hand side of Fig. 10). The resulting bonds are termed two-center three-electron (2c-3e) or $\sigma\sigma^*$ bonds. According to the most recent calculations [69a], their strength can vary anywhere between 1-2 (for complexes of second and third row hydrides with noble gases) and ≈ 50 kcal/mol (for $HF\overset{+}{\cdot\cdot}FH$)[20].

Most of the data on the electronic structure of σ-bonded complex cations come from the studies summarized in Fig. 13 and 14. In the case of the intramolecular $N\overset{+}{\cdot\cdot}N$ complexes in Fig. 13, the extent of interaction seems to be rather independent of the number of bridging methylene units while in the $S\overset{+}{\cdot\cdot}S$ and $I\overset{+}{\cdot\cdot}I$ complexes in Fig. 14, a more pronounced variation of λ_{max} with the substitution pattern and/or the structure is noted.

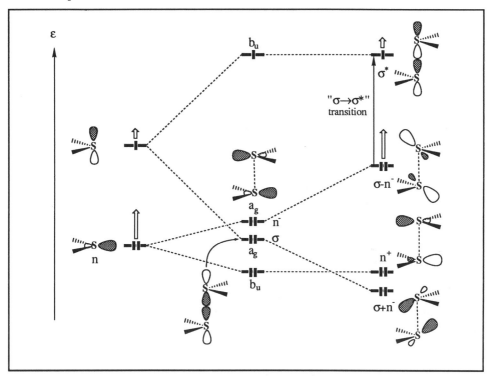

Fig. 16: Electronic structure of $(R_2S)_2^{\overset{+}{\cdot}}$ after Clark [69b]. Open arrows indicate shifts upon alkyl substitution which lead to an increase in λ_{max} of the $\sigma\rightarrow\sigma^*$ transition independent of steric interaction (length of the S···S bond)

It is tempting to explain these variations simply in terms of steric repulsion between substitutents which limit the amount of interaction that can be attained between the lone pairs.[21] However, Clark has shown [69b] that the electronic structure of σ-bonded $(M)_2^{\overset{+}{\cdot}}$ cannot be described correctly if

---

[20] For $(R_2S)_2^{\overset{+}{\cdot}}$, the dissociation energy in aqueous solution was estimated to be ≈ 20 (R=Me) and ≈ 10 kcal/mol (R=i-Pr), respectively [65n].

[21] In fact, we note from Figure 14 (series a) that in the case of R=t-butyl, the steric repulsion is sufficiently large to inhibit complex formation altogether.

one considers only the interaction between the two π-type lone pair HOMO's: Fig. 15 illustrates the role of the subjacent σ lone pairs whose negative combination (n$^-$ in Clarks terminology) is of proper symmetry to mix with the bonding combination of the p-type lone pairs. Since σ lone pairs are more effectively destabilized by alkyl substitution than those of p-type (open arrows in Fig. 15), the energy separation between the levels involved in the $\sigma \rightarrow \sigma^*$ transition of $(M)_2$ $^+$ will decrease as larger alkyl group are attached. This results in a red shift of the $\sigma \rightarrow \sigma^*$ transition, independent of steric interactions which affect primarily the splitting between the p-lone pairs. In this light, the strong λ_{max} shifts in Fig. 14 (series a and g) can be well understood.

In the series of cyclic complex cations (Fig. 14 b,e,f), the spectroscopic data indicate the strongest interaction (shortest λ_{max} for the $\sigma \rightarrow \sigma^*$ transition) for compounds where the heteroatoms are separated by three methylene units. Obviously it becomes more difficult for the lone pairs to overlap when the resulting rings get smaller but the same seems to be true also for larger rings. The thionine derivatives [66] seem to deviate from this rule, but their electronic structure may be determined by factors which are not operative in the simple alkyl sulfides.

Turning to the heterocomplex cations $(M \cdot N)^+$ in Fig. 14 h-m we note that their absorptions are generally at shorter wavelength than those of comparable $(M)_2^+$ as expected from the above Hückel expression for E_{CT}. Since lone pair ionization energies are very similar for sulfides and iodides [70], λ_{max} of $S^+ \cdot S$ and $S^+ \cdot I$ complexes are very close. For the S/Br pairs, ΔI is already more than 1.5 eV and a significant shift to shorter wavelengths is noticeable. Finally, $S^+ \cdot Cl$ complex cations are too unstable to be observed, i.e. ΔI is too large (>2.5 eV [70]) to provide for a sufficient ΔE_{cov}. To conclude this discussion, we wish to point out that the optical absorptions observed for radiolytically generated radical cations of *monomeric* sulfides, iodides and sulfoxides (Fig. 13 l-k) are most probably due to complexes with water molecules. Similarly, the characteristic pair of bands observed for simple sulfides in Freon or butyl chloride matrices [6] may be due to interaction with the halogen atoms of the solvent.

5.1.2. π-bonded complex cations

Another class of complex cations are those between planar π-systems which were actually known much earlier. Their electronic structure can be discussed in the same terms as that of the above σ-bonded complex cations except that now the CR (or CT) transition is between a positive and a negative combination of the respective π–HOMO's (in MO parlance). Such species were first postulated in 1962 by Lewis and Singer who found that upon oxidation of naphthalene with SbCl$_5$ in CH$_2$Cl$_2$ an ESR spectrum was obtained which could only be interpreted in terms of spin and charge being symmetrically distributed over *two* naphthalene units [71]. Later, Badger and Brocklehurst obtained the same species upon controlled annealing of frozen γ-irradiated BuCl solutions of naphthalene and observed for the first time the characteristic CR absorption at 1050 nm [67a].

Subsequently, the same technique was used to investigate a broad range of π-type $(M)_2$ $^+$ comprising olefins and dienes [72a], polycyclic aromatic hydrocarbons [72b] and benzene derivatives [72c] which all show similarly shaped CR absorptions between 800 and 1400 nm. Generally, λ_{max} for the CR transition increases with the size of the π system because the M-M $^+$ interaction becomes weaker as the charge is spread over a larger area. Finally, Badger and Brocklehurst provided a theoretical framework for the rationalization of these absorptions in terms of possible sandwich structures of the different $(M)_2^+$ [67b].

More recently, Kira et. al. carried over the annealing technique to the F11/F114B Freon glasses (see 3.2) [73] and were thus able to observe some new polycyclic aromatic hydrocarbon dimer cations [73b]. Also, they attempted to generate heterocomplex π complex cations (M·N)‡ betweeen pyrene and naphthalene, fluorene, and biphenyl [73c]. Unfortunately they were unable to pinpoint the exact location of the corresponding CR (or, rather, CT) bands because of interference from concomitantly formed (M)$_2^{\ddagger}$ and complexes with the solvent.

On the other hand, there has been (and still is) considerable activity in the field of the *thermochemistry* of π complex cations, both in solution [74] as well as in the gas phase [75], which cannot be reviewed in detail here. Suffice it to say that dissociation energies of dimer cations are typically 4-5 times smaller in solution than in the gas phase, presumably due to the difference in solvation energy between the monomer and dimer radical cations, which makes it difficult to draw correlations between thermochemical and optical data as in the case of the σ-bonded complex cations discussed above.

Recently, we have been able to characterize for the first time a heterodimer cation of the π complex type by optical spectroscopy [59k], i.e. [benzene·ethylene]‡ which is formed cleanly by fragmentation of [4.2.0]- or [2.2.2]bicyclooctadiene (Fig. 9 c,h). To our surprise we found that its CT absorption maximum (λ_{max}=680 nm) can be predicted quite accurately with the simple formula

$$E_{CT} = 2 \cdot \sqrt{\left(\frac{\Delta I}{2}\right)^2 + \beta^2}$$

using $\Delta I = 1.25$ eV from the PE spectra [70] and $-\beta = 0.67$ eV equivalent to $1/2$ E_{CR} of (benzene)$_2^{\ddagger}$ [72c]. Unfortunately, all attempts to induce sytematic shifts in E_{CT} of this complex by varying substituents on the benzene or the ethylene moiety (i.e. changing ΔI) failed because steric repulsion led to a strong reduction in β or prohibited the formation of complex cations altogether [76a]. On the other hand, ESR measurements gave some indications with regard to the structure of this species which seems to be edge-on, as expected from MO considerations [76b]. We plan to study further heterocomplex π cations by means of dissociating suitable formal addition products.

Finally, we wish to point out that, in principle, mixed complex cations between molecules containing π-systems and others with lone pair-HOMO's should exist and that their investigation may reveal interesting new insights. To the best of our knowledge, this large field of complex cation chemistry has barely been touched to date, perhaps with the exception of some recent studies of the very weakly bound complexes between benzene‡ and noble gases [77]. However, it has been known for some time that gas phase M‡ fragmentations may proceed via complex cations of one sort or another [78]. Very often these are hydrogen bonded complexes but in certain cases it seems likely that π-type complexes may be involved.

For example, the distonic cation $CH_2^{\ddagger}$-CH-OH$_2$ was calculated to isomerize via a symmetric ethylene-water complex of substantial stability (17.2 kcal/mol) relative to the separated fragments.[22] This species was described as an ion-dipole complex, which in view of the large ΔI between ethylene and water may be quite accurate. However, if one or both hydrogen atoms in water are replaced by alkyl groups or if the oxygen atom is replaced by sulfur, ΔI should be close to zero and covalent bonding should prevail. We believe that future research in this domain may lead to much new insight into the electronic and molecular structure of complex cations.

---

[22] Note that the Walsh orbitals of small rings may also take the role of π–HOMO's of olefins in complex cation formation: Recent studies have indicated that loss of water from n-propanol may proceed via a [cyclopropane-H$_2$O] complex cation which is bound by $\approx$10 kcal/mol relative to the separated fragments [80].

5.2. ALLYLIC CATIONS

Due to the resonance energy to be gained[23], many radical cations tend to form allylic systems upon ionization, a process which usually involves a bond dissociation. Some examples for this are listed in Fig. 17 most of which are from recent CIDNP [83b-d,f] or ESR work [83e,g-j]. With regard to the electronic structure of allylic systems, we first note that no electronic absorptions are to be expected for allyl cation chromophores above 300 nm [84]. On the other hand, the allyl radical shows a weak absorption around 400 nm (plus a very intense one at 213 nm) [85] and the question arises whether this visible band will appear also in some derivatives and under what form. A brief analysis of the electronic structure of the allyl radical (see Fig. 18) will provide the basis for an answer to this question and an explanation of the spectroscopic features observed in some allylic radical cations.

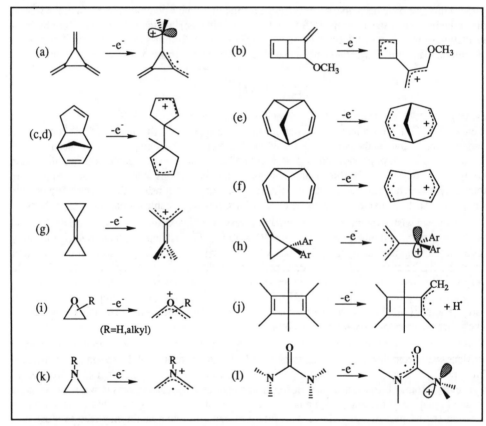

Fig. 17: Molecules which spontaneously form allylic species upon ionization. Letters correspond to those in Ref [83].

---

[23] The resonance energies of the allyl radical and the allyl cation (defined as the radical or cation stabilization energies relative to the n-propyl systems [81]) are evaluated as 13.0 ± 1.5 and 13.4 ± 1.4 kcal/mol, respectively, on the basis of the newest available experimental data [82].

From a simple MO picture one would expect a single, rather intense EA band due to $\pi_1 \to \pi_2$ (a) and $\pi_2 \to \pi_3$ (v) electron promotion, because both are associated with nonzero transition moments which happen to be of the same magnitude and direction. Obviously this prediction is in stark contrast to the observed spectrum, but experiment and theory can easily be reconciled by considering first-order interaction between the degenerate 1^2B_1 and 2^2B_1 pair of excited configurations which leads to a symmetric splitting into two states corresponding to their positive ($^2B_1^+$) and negative combination ($^2B_1^-$). Due to the above mentioned equality of transition moments corresponding to a and v electron promotion, they will cancel exactly for the $^2A_1 \to {}^2B_1^-$ excitation which leads to a zero electronic transition moment[24], while they double for the $^2A_1 \to {}^2B_1^+$ excitation, thus explaining the occurrence of two widely spaced bands of very disparate intensity.

The degeneracy of the excited $^2B_1$ configurations in the allyl radicals can be broken by substitution. For example, replacement of the central carbon atom by more electronegative O^+ or NR^+ will primarily *stabilize* π_1 and π_3 and thus lead to a lowering of the 2^2B_1 (HOMO→LUMO excited) and a raising of the 1^2B_1 (HOMO-1→HOMO excited) configuration (open arrows in Fig. 18). Consequently, their mixing will not be 1:1 and the transition moments will no longer cancel to zero for the $^2A_1 \to {}^2B_1^-$ excitation, which will express itself in an increased intensity (and possibly some shift) of the first EA band. On the other hand, alkyl substitution at the terminal carbon atoms will *destabilize* π_2 more than π_1 and π_3 (filled arrows) which leads to an enhancement of the effect of O^+/NR^+ subsitution.

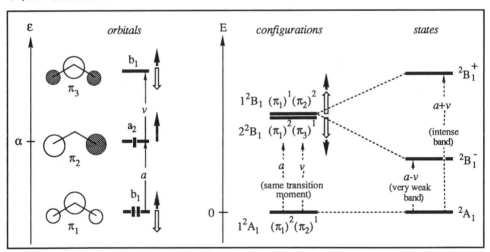

Fig. 18: Electronic structure of allyl radical. Open arrows indicate shifts upon substitution of the central carbon atom by O^+ or NR^+, closed arrows indicate shifts upon alkyl substitution at the terminal atoms (explanation see text).

The above expectation is fully confirmed by experiments on matrix-isolated oxirans which undergo spontaneous ring-opening upon ionization to give oxallyl cations (ionized carbonyl ylids)

---

[24] The $^2A_1 \to {}^2B_1^-$ transition gains some intensity through vibrational borrowing [85].

with λ_{max} between 465 (R=H) and 550 nm (R=CH$_3$) [87][25]. These species are interesting from a theoretical point of view [89a] because due to a pronounced second-order Jahn-Teller effect, deformation of the C-O-C framework from C_{2v} symmetry (I→II or I→III) is predicted to require very little energy[26] which is tantamount to asserting that oxallyl radicals have virtually no resonance energy [89a,b].

$$C_{2v} \qquad\qquad C_s \qquad\qquad C_s$$

$$I \qquad\qquad\qquad II \qquad\qquad\qquad III$$

In fact there is evidence from ESR experiments that in CF$_2$Cl-CFCl$_2$ matrices, ethylene oxide as well as its 1,2-dimethyl and tetramethyl derivatives form structures of type III (the latter only after annealing) [90]. Although a medium effect may be responsible for this particular phenomenon, it indicates that the oxallyl cation framework is indeed very susceptible to deformation. Interestingly, a slight I→II distortion is not expected to have a great effect on the EA-spectrum because in a first approximation it does not affect the energies of the allylic π-MO's (cf. Fig. 18), whereas species III has an entirely different electronic structure which should distinguish itself by a low-energy CT band. Perhaps further EA experiments will help to shed light on the subtle factors which control the structures of ionized oxirans.

Many of the allylic species in Fig. 17 contain both a cation and a radical moiety. The electronic structure of such species can be described in similar terms as those used in section 5.1.2., i.e. they can be viewed as intramolecular complex cations, except that now the two interacting frontier MO's (c.f. Fig. 15) contain altogether only one instead of three electrons. Nevertheless, they are expected to show low-energy CR- (species e and f) or CT-type absorptions (species b and d). So far only two of these species have been investigated by EA spectroscopy, namely the bisallylic cations IV and V obtained after ionization of exo- and endo-dicyclopentadiene, respectively (Fig. 19) [83d].

These cations show intense absorptions at 850 (1.46 eV, IV) and 743 nm (1.67 eV, V), and it is tempting to speculate about the contribution of through-space (TS) and through-bond (TB) terms to the splitting between the π_2 MO's of the two allylic units in IV and V (cf. Fig. 19): The energy differences are too large to be explained solely in terms of TB interaction via the C-C σ-bond linking the two allylic moieties [91], i.e. TS interaction must play an important role. Inspection of molecular models reveals that a conformation where the terminal atoms of the allylic moieties overlap optimally is sterically unfavorable in IV [92] whereas this is not the case in V, in accord with the higher energy CR transition in the latter cation. Similar problems pose themselves also for the other systems in Fig. 17 and we hope that future optical studies will provide more interesting insight into the electronic structure of these interesting bisallylic systems.

---

[25] There has been some controversy among ESR spectroscopists on whether such a ring-opening does indeed take place [86a]. However, the optical spectra of ionized oxirans which are not prohibited from undergoing this process are clearly incompatible with a ring-closed structure and may therefore be taken as an independent proof for spontaneous ring opening, in accord with recent theoretical calculations [86b].

[26] Actually, the first ab-initio calculations on the parent oxallyl cation predicted the most stable geometry to correspond to structure III [89c] or later II (with III lying only 1.9 kcal/mole higher) [89a,b]. Only at the 6-31G*/UMP3 or CAS SCF level did the symmetric structure I emerge as a (shallow) energy minimum [89d]

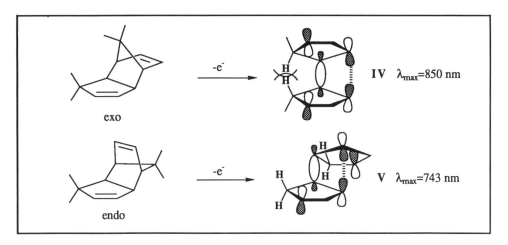

Fig. 19: Electronic structure of the diastereomeric bisallylic cations obtained from exo- and endo-dicyclo-pentadiene, respectively [83c,d]. Note that steric repulsion prevents IV from attaining a conformation of optimal overlap between the termini of the allylic HOMO's.

5.3. POLYENES

The radical cations of linear conjugated polyenes have very characteristic EA spectra showing a weak low-energy and an intense high-energy transition (see Fig. 20 on next page). Along the series of the parent all-trans polyenes one finds furthermore that the disparity between the two band intensities increases while their spacing decreases. Interestingly, this finding does not correspond to what one expects from a simple MO model which predicts a pair of close-lying excited configurations of the same symmetry corresponding to HOMO→LUMO (v) and to HOMO-1→HOMO (a) electron promotion , which are associated with transition moments of similar magnitude and direction.

Similarly to the above case of allyl (see Fig. 18), experiment and theory can be reconciled by allowing for interaction between the a and v excited configurations [56a], which leads to a pair of more widely separated states. Again, the transition moments for a and v electron promotion will add up for the positive combination (upper excited state) while they subtract in the negative combination (lower excited state) which explains the observation of a weak first and an intense second EA band. The increasing dissimilarity between the intensities of the two bands for longer polyenes can be elegantly rationalized on the basis of the same model: Extending the length of the polyene has the twofold effect of (1) narrowing the gap between the a- and v- excited configurations and (2) reducing the angle between the a- and v- transition moments. While (1) results in a more symmetric mixing between the two excited configurations, (2) makes that the directions of the transition moments are better aligned which leads to a more efficient cancellation upon subtraction. In sum, the intensity ratio of the first and second EA band in polyene$^+$ can be taken as a measure of the extent of configuration interaction prevailing in these species.

Since mixing between Koopmans' (K) and Non-Koopmans' (NK) configurations plays an important role in polyene$^+$, this should result in reduced photoionizaton cross sections for the first two excited states. Inspection of the second and third PE bands of polyenes confirms this expectation [11d,56a], although the effect is somewhat less pronounced than expected on the basis of the above considerations.

40

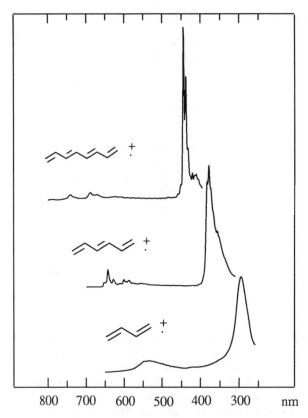

Fig. 20: EA spectra of the first three members of the series of polyene radical cations in Argon. Note that the band intensities in the three spectra cannnot be directly compared due to possible differences in the efficiency of ion formation. Nevertheless it is evident that the relative intensities of the first and second band decreases as the polyene chain gets longer (see text).

Interest in polyene$^{\ddot{+}}$ is due to their possible involvement in biological redox reactions[27] as well as in photoinitiated electron transport phenomena which are crucial to the development of artificial photosynthetic systems for solar energy conversion[28]. Numerous investigations were therefore devoted to the study of retionid [95] as well as carotinoid radical cations [96,97] by ESR and optical spectroscopy. Fig. 21 shows a plot of the first and second excited state energies of different polyene$^{\ddot{+}}$ as a function of the number of double bonds.

From this, one can draw a number of interesting conclusions:

- The intense bands observed in the pulse radiolysis studies of retinoid and carotenoid polyenes are clearly due to the *second* excited state of the corresponding M$^{\ddot{+}}$ contrary to the original assignment by Lafferty et. al. [97b] on the basis of PPP/SCF calculations.

---

[27] For example, it was recently shown that carotenoid radical cations are formed at the photosynthetic reaction center PSII upon photoirradiation of chloroplasts [93].

[28] See for example the elegant work of Moore et. al. on "Triad molecules" consisting of a quinone (Q) and a carotene (C) linked to opposite sides of a porphyrin molecule (P). Excitation of Q-P-C leads to the excited singlet state of the porphyrin (Q-^{1}P-C) which deactivates by electron transfer to the quinone (Q$^{\ddot{-}}$-P$^{\ddot{+}}$-C). The role of the polyene C is to remove the positive charge from the porphyrin to yield a species Q$^{\ddot{-}}$-P-C$^{\ddot{+}}$ which has a lifetime of several microseconds in solution [94].

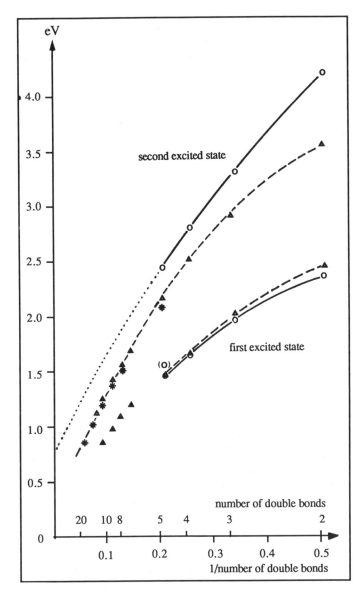

Fig. 21: First and second excited state energies of polyene radical cations. Circles are unsubstituted compounds [56a,55u], asterisks correspond to retinol (5 double bonds [96a]) and a series of carotenes with 8, 9, 11, 15 and 19 double bonds [97c]. The triangles are data from an ongoing study involving polyenes capped by t-butyl groups [99]. The circle in brackets is the second peak of the first EA band of decapentaene$\stackrel{+}{\cdot}$ [55u] (see text).

- The second excited state of polyene$\stackrel{+}{\cdot}$ seems to be much more sensitive to solvent and substituent effects than the lowest one, as revealed by the shifts upon t-butyl substitution at the terminal carbon atoms which are minute for the first but considerable for the second EA band. Note also the important displacements of λ_{max} for the carotenoid radical cations in solvents of different polarity [97a] which indirectly support the above contention.

- The radical cation of decapentaene shows a pair of very weak sharp bands around 800 nm. According to Andrews et. al. the first of these is due to a partially Z-configured rotamer of this species while it is the second which represents the first EA band of the all-trans cation. This assignment is probably wrong because (a) octatetraene⁺̇ also contains a *pair* of peaks in its first EA band [56c] and (b) it is the energy of the *first* peak which falls on the line containing the 0-0 components of the other polyene's first EA bands.

The line for the second excited state of the parent polyene⁺̇ extrapolates to about 0.75 eV for an infinite number of double bonds. As it happens, this is the energy where a strong new absorption arises upon doping of all-trans polyacetylene films with various oxidants [98a]. If one regards oxidatively doped polyacetylenes as assemblies of long polyene⁺̇ then this result is not surprising. However, the electronic structure of conducting polymers is usually described in terms of localized defects (polarons and/or solitons [98b]) leading to a band-gap model which explains λ_{max} in doped polyacetylenes equally well. If this view is correct (as it undoubtedly is, judgeing from its sucess) then a localization of spin and charge must occur somewhere between the carotenoid polyene⁺̇ and doped polyacetylene, i.e. it becomes possible for the charge carriers to induce lattice distortions which trap them at some point on the polyene chain. It is interesting to speculate at what point this phenomenon sets in and why. Perhaps future studies on long polyene⁺̇ will yield more information on this interesting borderline between molecular and solid state electronic structure.

6. Radical cation photochemistry

Contrary to the gas phase, where the principal deactivation pathway after electronic excitation of radical cations is usually dissociation, such behaviour is rarely observed in the condensed media used for studying M⁺̇ (see 3.2/3.3). The main reason for this is that M⁺̇ fragments (A⁺ + B˙ or A⁺̇ + B) are often so reactive that they recombine spontaneously when held in close proximity, as it is the case in solid matrices[29]. As a consequence, it becomes possible to observe *photorearrangements* of radical ions which are normally of minor importance in the gas phase. In fact, most of the M⁺̇ photoreactions described in the literature are of this type and our discussion will focus on them.

The following pages contain a compilation of M⁺̇ photoreactions observed during the last ten years, mostly in studies employing EA spectroscopy in Argon or Freon matrices (for some reason, ESR studies of M⁺̇ photochemistry are rather scarce although much information on the nature of photoproducts could be gained in this fashion). Many of them can be represented as H-shifts, electrocyclic reactions or cis-trans isomerizations, i.e. rearrangements which occur analogously in neutral M. Others involve bond-breaking processes which seem to be unique to M⁺̇. In view of the substantial difference between the electronic structure of M and M⁺̇ this is perhaps not so surprising. However, the question of the relation between electronic structure and photochemistry of M⁺̇ has not been addressed systematically until today.

Fig.22: (following three pages) M⁺̇ photoreactions observed in Argon matrices and/or Freon glasses. All molecules are radical cations, unless otherwise indicated. Numbers in reaction schemes denote wavelength of irradiation in nm. Letters in parentheses correspond to those in reference [101].

---

[29] Exceptions to this rule are *hydrogen atoms* which diffuse even through noble gas matrices. In Argon, photolysis may also lead to loss of *protons* which were found to form complexes $(HAr)_n^+$ [100]. Hence, closed shell cations or neutral radicals are frequently observed as sideproducts upon formation or photolysis of M⁺̇ (see for example the work of Andrews [55a]). Occasionally, two hydrogen atoms may also recombine prior to their escape from the matrix cavity which leads to formal loss of H_2.

(a)

(b)

(c)

(d)

(d)

44

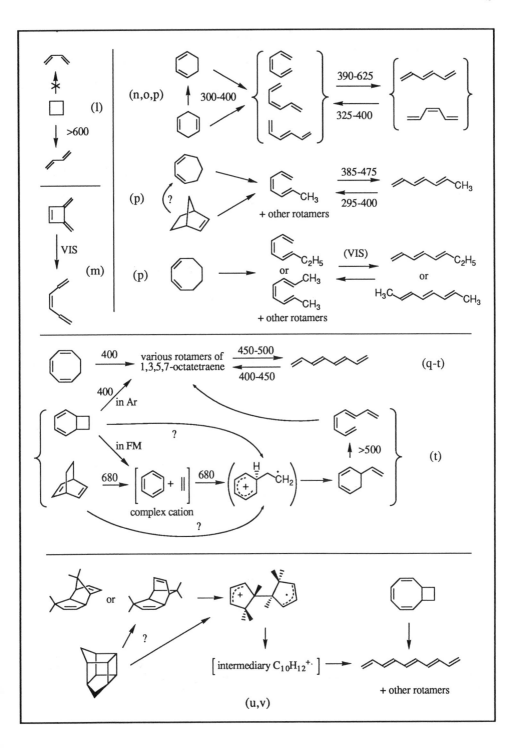

Although photolysis was employed occasionally in the early studies of Shida and Hamill [49a,b] , the first *systematic* study of a photorearrangement dates to 1977 when Shida, Nosaka and Kato reported on the ring opening of 1,3-cyclohexadiene⁺ to hexatriene⁺ which exists in the form of several "rotamers" which can be interconverted by selective photolyses [101n]. This pioneering study laid the ground for a continuing series of investigations on the photochemistry of linear conjugated polyene⁺ and their (poly)cyclic valence isomers [101o-v] which are summarized on the last page of Figure 22. Among the intriguing observations made during these studies we would like to comment briefly on the following:

- A recent study of the "parent" ring-opening reaction cyclobutene⁺→butadiene⁺ has shown that these reactions do not seem to proceed in the usual electrocyclic fashion, because the first observable photoproduct of cyclobutene⁺ is not *cis*-butadiene⁺ (which would be stable under the conditions of the experiment) but the *trans* isomer, which cannot arise by way of a concerted electrocyclic process [101*l*][31]. Interestingly, photochemical *ring-closures* were never observed in polyene radical cations, in contrast to the neutrals.

- The cycloheptadiene⁺→heptatriene⁺ and cyclooctaadiene⁺→octatriene⁺ photorearrangements are unparalleled in the neutrals. These processes probably lead initially to distonic pentadienylium cations which rearrange to the triene cations by H-shifts as shown below :

- A similar distonic species may also be involved in the formation of decapentaene⁺ from the bisallylic radical cation obtained after ionization of dicyclopentadiene (entry (u,v) in Fig. 22; see also section 5.2)

- 1-methylene-2,4-cyclohexadiene⁺ and its derivatives are stable photorearrangement products of alkylbenzene, cycloheptatriene and norbornadiene radical cations (entries b-d, f, h in Fig. 22). It is possible that all these rearrangements proceed through a common distonic intermediate, although its presence has never been proven.

---

[31] Upon excitation of CT complexes between *cis*-3,4-diarylcyclobutenes and TCNE or TCA, stereospecific formation of *cis-trans*-1,4-diarylbutadienes was observed [102]. From this it was inferred that the ring opening of cyclobutene⁺ takes place in a conrotatory electrocyclic fashion. However, the same products could have arisen by way of a very different reaction path [103], as it is suggested by the above cited optical and ESR experiments on the parent compound.

- The readiness with which cyclic conjugation in alkylbenzenes is broken in many $M^{+\bullet}$ photochemical reactions may seem surprising at first sight. However, it must be realized that the π systems of benzene radical cations contain only *five* electrons which do not endow the systems with any special stabilization by virtue of their cyclic delocalization, in contrast to neutral aromatic systems with their cyclic array of *six* π-electrons.[32] Actually, thermochemical estimates have shown that 1-methylene-2,4-cyclohexadiene$^{+\bullet}$ is only some 8.2 kcal/mol less stable than toluene$^{+\bullet}$ in contrast to the neutrals where the difference is 29.2 kcal/mol [56e].

- A number of studies (not represented in detail in Fig 22.) were devoted to the photochemical interconversion of *polyene$^{+\bullet}$ rotamers*. While in the case of hexatriene $^{+\bullet}$, five of the six possible rotamers could be identified [101n-p], only six out of the twenty possible rotamers of octatetraene$^{+\bullet}$ were observed, independent of the source of the polyene$^{+\bullet}$ [56c,d]. In the case of decapentaene, several rotamers were also observed but not characterized individually [55 t,u]. Interestingly, all attempts to obtain the elusive *cis*-butadiene$^{+\bullet}$ have failed so far, although the barrier to rotation in this compound should be sufficiently high to prevent its thermal back reaction to the *trans* isomer at low temperature [101*l*].

Apart from the above, there are two remarks of general nature pertaining to the discussion of $M^{+\bullet}$ photoreactions. The first concerns the role of $M^{+\bullet}$ excited states: do they actually impose a certain course on these reactions or do they simply serve as relay states for accepting energy which subsequently reappears as vibrational ground state energy and drives what is essentially a thermal process?[33] Of course the same question poses itself also in the neutrals but there many cases are known where the products of photochemical reactions differ from those of the corresponding thermal processes which proves that excited states play a decisive role in determining the reaction course. In $M^{+\bullet}$ such examples are still missing and it will be interesting to see whether photorearrangements lead to different products in experiments involving for example electronic vs. multiphoton IR absorption.

If excited $M^{+\bullet}$ actually serve as reactants (in the chemical sense) in photoinitiated reactions, then the question arises whether reactions such as the ring-opening of (poly)cyclic polyene$^{+\bullet}$ valence isomers are subjected to Woodward-Hoffmann type symmetry selection rules. Actually, predictions with regard to the stereochemistry of various pericyclic $M^{+\bullet}$ reactions were made according to such rules [105], but their application turns out to be less straightforward than in the neutrals because most $M^{+\bullet}$ pericyclic reactions are *state-symmetry forbidden* with respect to the symmetry elements shared by reactants and products [106]. As a consequence, the molecules must necessarily *lose all symmetry* at some point on the reaction path in order to cross over adiabatically to the product potential surface. Hence, application of symmetry rules becomes somewhat doubtful and in fact there is evidence from theoretical calculations [103,107] that $M^{+\bullet}$ isomerizations (thermal or photochemical) tend to follow strongly non-concerted pathways.

---

[32] A qualitative way to look at this phenomenon is to say that a cyclic systems with 5 π-electrons lies midway between one with 6 (aromatic) and one with 4 π-electrons (antiaromatic) and is hence neither stabilized nor destabilized by cyclic conjugation. A more quantitative measure of this loss of aromaticity upon removal of an electron from a 6π system may be obtained by comparing the heats of hydrogenation of neutrals (benzene→1,3-cyclohexadiene: +5.6 kcl/mol; 1,3,5-hexatriene→1,3–hexadiene: -26.2 kcal/mol) to the corresponding figures for the radical cations (-17.4 and -20.8 kcal/mol, respectively) [104]. The notion of aromatic *stabilization* therefore has no place in radical ion thermochemistry (see, however, Footnote 4 in [53f] for a defence of other uses of the term).

[33] Note that in noble gas matrices, excess vibrational ground state energy is dissipated only reluctantly by the host medium and is hence available to drive chemical reactions for rather long periods of time.

7. Conclusions

I hope to have succeeded in demonstrating that the electronic spectroscopy and photochemistry of organic radical ions is a multifaceted and fascinating field of research, both on a theoretical as well as on an experimental level. The large body of data which has been accumulated over the past 10-15 years calls for a phase where we should attempt to give explanations for many of the intriguing phenomena described above. However, there is still ample space for "collecting shells on the beach", especially in the field of complex and distonic radical cations, where one has only begun to scratch the surface of what promises to become a major area of research in the title domain. Since electronic spectra often do not contain very much information in these cases, emphasis on the experimental side may shift to the use of ESR (and ENDOR) as well as vibrational spectroscopy for probing the structures of these unusual species which have no neutral analogues. Of course, an understanding of electronic spectra will remain a crucial prerequisite for the successful application of photochemical techniques to the study of $M^{\ddot{\cdot}}$.

ACKNOWLEDGEMENTS

I would like to thank my collaborators S. Nitsche and K. Roth as well as all undergraduate students who have contributed to one or the other aspect of our research discussed in this chapter. I am indebted to E. Brosi, P. Purro and P.-H. Chassot for their technical expertise as well as to the Swiss National Science Foundation for financial support. However, none of this work would have been possible without the continuing support and encouragement by Prof. E. Haselbach which is herewith gratefully acknowledged. Finally, I would like to thank my mentors and colleagues who have helped me during the initial phases of our research and who have contributed in many discussions to some of the thoughts expressed in this contribution. Special thanks go to Prof. T. Shida who has been a most inspiring and obliging competitor, and whose work is held in great esteem in our laboratories.

References

1. For an excellent historical outline on organic radical ions, see: G. A. Russell and R. K. Norris in: Organic Reactive Intermediates (Organic Chemistry Monographs, Vol 26, S. P. McManus ed.), Academic Press, New York (1973). The early days of radical ion chemistry are described entertainingly by: H. D. Roth, Tetrahedron 42, 6097 (1986).
2. H. Wieland and E. Wecker: *Chem. Ber.* 4, 699 (1910), describing the oxidation of di-p-anisylamine with $SbCl_5$ in chloroform. ("[One obtains] form the blue-green solution after prolonged standing the addition product in *magnificent* dark steel-blue prisms which show a *vivid* green surface glitter").
3. W. Schlenk, J. Appenrodt, A. Michael and A. Thal: *Chem. Ber.* 47, 473 (1914), describing the reduction of a tetraphenyl-paraquinodimethane derivative with sodium. ("[One obtains] in solution *magnificent* blue addition products which in the crystalline state show a *marvellous* fuchsia gleam").
4. E. Haselbach and T. Bally, *Pure. Appl. Chem.* 56, 1203 (1984).
5. (a) T. A. Miller and V. Bondybey: *Ann. Rev. Phys. Chem.* 33, 157 (1982); (b) T. A. Miller and V. Bondybey in: *Molecular Ions: Spectroscopy, Structure and Chemistry*, T. A. Miller and V. Bondybey eds., North Holland Publ. Co, Amsterdam (1983); (c) D. Klapstein, J. P. Maier and L. Misev *ibid.*; (d) J. P. Maier, O. Marthaler, L. Misev nad F. Thommen in: *Molecular Ions: Geometric and Electronic Structure*, J. Berkowitz and K.-O. Groeneveld eds., NATO ASI Series B, Vol 90, Plenum Press, New York (1983).
6. T. Shida: *Electronic Absorption Spectra of Radical Ions*, Elsevier, Amsterdam (1988).
7. (a) P. Čársky and R. Zahradník: *Top. Curr. Chem.* 43, 1 (1973); (b) R. Zahradník and P. Čársky: *Prog. Phys. Org. Chem.* , 327 (1973); (c) P. Čársky and R. Zahradník, *Acc. Chem. Res.* 9, 407 (1976).

8. G. J. Hoijtink and W. P. Weijland, *Rec. Trav. Chim.* **76**, 836 (1957).

9. T. Koopmans: *Physica* **7**, 104 (1937); s.a. W. G. Richards: *Int. J. Mass Spectrom. Ion Phys.* **2**, 419 (1969).

10. D. S. Urch: *Europ. Spectrsoc. News* **28**, 55 (1980); quoted from [11a]

11. (a) J. Spanget-Larsen: *Croat. Chim. Acta* **57**, 991 (1984). (b) R. Schulz, A. Schweig and W. Zittlau; *J. Am. Chem. Soc.* **105**, 2980 (1983); (c) L. S. Cederbaum, W. Domcke, J. Schirmer and W. v. Niessen: *Adv. Chem. Phys.* **65**, 115 (1986); (d) R. Bigelow, *Chem. Phys.* **80**, 45 (1983).

12. D. V. Dearden and J. L. Beauchamp: *J. Phys. Chem.* **89**, 5359 (1985) and references 20-26 cited therein.

13. P. Forster, R. Gschwind, E. Haselbach, U. Klemm and J. Wirz: *Nouv. J. Chim.* **4**, 365 (1980).

14. E. R. Davidson: *QCPE Bull.* **9**, 97 (1989). (MELDF: QCPE Program #580).

15. E. Heilbronner and J. P. Maier in: *Electron Spectroscopy, Theory, Techniques and Applications*, C. R. Brundle and A. D. Baker eds., Vol 1, Wiley, New Ork (1977). See page 226f.

16. T. Kato and T.Shida: *J. Am. Chem. Soc.* **101**, 6869 (1979).

17. R. Pariser and R. G. Parr: *J. Chem. Phys.* **21**, 466, 767 (1953).

18. (a) G. J. Hoijtink and P. J. Zandstra: *Mol. Phys.* **3**, 371 (1960); (b) G. J. Hoijtink, N. H. Velhorst and P. J. Zandstra: *ibid.* **3**, 533 (1960); (c) K. H. D. Buschow, J. Dieleman and G. J. Hoijtink, *ibid.* **7**, 1 (1964).

19. R. Zahradník, *Fortschr. Chem. Forsch.* **10**, 1 (1968).

20. A. Ishitani and S. Nagakura, *Theor. Chim. Acta* **4**, 236 (1966).

21. (a) R.Zahradník and P. Čársky: *J. Phys. Chem.* **74**, 1240 (1970); (b) P. Čársky and R. Zahradník: *Coll. Czech. Chem. Commun.* **36**, 961 (1971); (c) P. Čársky and R. Zahradník: *Theor. Chim. Acta* **17**, 316 (1970); (d) J. Kuhn, P. Čársky and R. Zahradník: ibid. **33**, 263 (1974); (e) P. Čársky, J. Kuhn and R. Zahradník: *J. Mol. Spectrosc.* **55**, 120 (1975); (f) P. Čársky, P. Hobza and R. Zahradník: *Coll. Czech. Chem. Commun.* **36**, 1291 (1971); (g) P. Hobza, P. Čársky and R. Zahradník: *ibid.* **38**, 641 (1973); (h) R. Zahradník, S. Hünig, D. Scheuzow and P. Čársky: *J. Phys. Chem.* **75**, 335 (1971); (i) P. Čársky, A. Fojtík and R. Zahradník: *Coll. Czech. Chem. Commun.* **37**, 2515 (1972); (k) R. Zahradník, P. Čársky, S. Hünig, G. Kiesslich and D. Scheuzow: Int. J. Sulfur Chem, Part C **6**, 109 (1971); (l) J. Pancíř, I. Matousek and R. Zahradník: *Coll. Czech. Chem. Commun.* **38**, 3039 (1973); (m) P. Čársky and R. Zahradník: *Theor. Chim. Acta* **20**, 343 (1971); (n) R. Zahradník and P. Čársky: *ibid.* **27**, 121 (1972).

22. (a) J. Del Bene and H. H. Jaffé: *J. Chem. Phys.* **48**, 1807, 4050; **49**, 1221 (1968); (b) H. M. Chang and H. H. Jaffé: *Chem. Phys. Lett.* **23**, 146 (1973); (c) H. H. Jaffé, H. M. Chang and C. A. Masmanidis: *J. Comput. Phys.* **14**, 180 (1974); (d) R. L. Ellis, H. H. Jaffé and C. A. Masmanidis: *J. Am. Chem. Soc.* **96**, 2623 (1974); (e) H. M. Chang, H. H. Jaffé and C. A. Masmanidis: *J. Phys. Chem.* **79**, 1118 (1975).

23. (a) M. Scholz and L. Götze: *Monatsh. Chem.* **111**, 15 (1980); (b) G. Kluge and M. Scholz: Int. J. Quant. Chem. **20**, 669 (1981).

24. All papers by R. Bigelow: (a) *Chem. Phys. Lett.* **100**, 445 (1983); (b) *J. Electron Spectrosc. Rel. Phen.* **35**, 239 (1985); (c) J. Mol. Struct. (Theochem) **122**, 133 (1985); (d) *Chem. Phys. Lett.* **117**, 22 (1985); (e) *Int. J. Quant. Chem.* **29**, 35 (1986).

25. (a) MINDO/1: N. C. Baird and M. J. S. Dewar: *J. Chem. Phys.* **50**, 1262 (1969); (b) MINDO/2: M. J. S. Dewar and E. Haselbach: *J. Am. Chem. Soc.* **92**, 590 (1170); (c) MINDO/3: R. C. Bingham, M. J. S. Dewar and D. H. Lo: *ibid.* **97**, 1285, 1294, 1302 (1975); (d) MNDO: M. J. S. Dewar and W. Thiel: *ibid.* **99**, 4899, 4907 (1977); (e) M. J. S. Dewar and W. Thiel: *Theor. Chim. Acta* **46**, 89 (1977); (f) AM1: M. J. S. Dewar, E. G. Zoebisch, E. T. Healy and J. P. Stewart: *J. Am. chem. Soc.* **105**, 3902 (1985).

26. (a) C. Fridh, L. Åsbrink and E. Lindholm: *Chem. Phys. Lett.* **15**, 282 (1972); (b) L. Åsbrink, C. Fridh and E. Lindholm: *ibid.* **52**, 63, 69, 72 (1977); (c) E. Lindholm and L. Åsbrink: *Molecular Orbitals and Their Energies Studied by the Semiempirical HAM Method*, Lecture Notes in Chemistry Vol 38, Springer, Berlin (1985).

27. (a) T. Clark, M. F. Teasley, S. F. Nelsen and H. Wynberg: *J. Am. Chem. Soc.* **109**, 5719 (1987); (b) S. F. Nelsen, S. C. Blackstock, N. P. Yumibe, T. B. Frigo, J. E. Carpenter and F. J. Weinhold: *ibid.* **107**, 143 (1985).

28. S. F. Nelsen, T. B. Frigo and Y. Kim: *J. Am. Chem. Soc.* **111**, 5387 (1989)

29. (a) H. L. Hase, G. Lauer, K.-W. Schulte and A. Schweig: *Theor. Chim. Acta* **48**, 47 (1978); (b) G. Lauer, K.-W. Schulte and A. Schweig: *J. Am. Chem. Soc.* **100**, 4925 (1978); (c) R. Schulz, A.

50

Schweig and W. Zittlau: *ibid.* **105**, 2980 (1983); (d) R. Schulz, A. Schweig and W. Zittlau: *Chem. Phys. Lett.* **106**, 467 (1984); (e) A. Heidenreich, N. Münzel and A., Schweig: *Z. Naturforsch.* **41A**, 1415 (1986).

30. See, for example P. J. Bruna and S. Peyerimhoff: *Adv. Chem. Phys.* **67**, 1 (1987).

31. A. L. Fetter and J. D. Walecka: *Quantum Theroy of Many-Particle Systems*, McGraw Hill, New York (1973); see also M. F. Herman, K. F. Freed and D. L. Yeager: *Adv. Chem. Phys.* **48**, 1 (1981)

32. L. S. Cederbaum and W. Domcke, *Adv. Chem. Phys.* **36**, 295 (1977).

33. (a) W. I. Aalbersberg, G. J. Hoijtink, E. L. Mackor and W. P. Weijland: *J. Chem. Soc.* 3049 (1959); (b) T. Shida and W. Hamill: *J. Chem. Phys.* **44**, 4372 (1968); (c) T. Shida and S. Iwata: *J. Am. Chem. Soc.* **95**, 3473 (1973); (d) S. Nitsche: *Dissertation Nr 886*, University of Fribourg (1985); (e) H. Hiratsuka and Y. Tanizaki: *J. Phys. Chem.* **83**, 2501 (1979); (f) H. Hiratsuka, Y. Nakamura, Y. Tanizaki and K. Nakajima: *Bull. Chem. Soc. Japan* **55**, 3407 (1982); (g) L. Andrews, B. S. Friedman and N. J. Kelsall: *J. Phys. Chem.* **89**, 4016 (1985).

34. R. C. Dunbar in: *Molecular Ions: Spectroscopy, Structure and Chemistry*, T. A. Miller and V. Bondybey eds., North Holland Publ. Co, Amsterdam (1983);

35. (a) R. C. Dunbar in: *Gas Phase Ion Chemistry Vol 2*, M. T. Bowers ed., Academic Press, New York (1979); (b) R. C. Dunbar in: *Gas Phase Ion Chemistry Vol 3*, M. T. Bowers ed., Academic Press, New York (1984); (c) M. W. E. M. van Tilborg, J. van Thuijl and W. J. Van der Hart: *Int. J. Mass Spectrom. Ion Proc.* **54**, 299 (1983); (d) R. C. Dunbar, G. B. Fitzgerald and J. D. Hays: *ibid.* **66**, 313 (1985); (e) W. Van der Hart, L. J. De Koning, N. M. M. Nibbering and M. L. Gross: *ibid.* **72**, 99 (1986); (f) P. N. T. van Velzen and W. J. van der Hart: *Org. Mass Spectrom.* **19**, 190 (1984); (g) R. E: Krailler and D. H. Russell: *Anal. Chem.* **57**, 1211 (1985).

36. J. A. Syage and J. E. Wessel: *J. Chem. Phys.* **87**, 3313 (1987)

37. (a) G. S. Gudeman and T. J. Saykally: *Ann. Rev. Phys. Chem.* **35**, 387 (1984); (b) G. S. Gudeman, C. C. Martner and R. Saykally: *Chem. Phys. Lett.* **122**, 108 (1985); (c) M. C. Crofton, M.-F. Jagot, B. D: Rehfus and T. Oda: *J. Chem. Phys.* **86**, 3755 (1987).

38. For an excellen review on this subject, see A. J. Bard, A. Ledwith and H. J. Shine: *Adv. Phys. Org. Chem.* **13**, 155 (1976).

39. *Photoinduced Electron Transfer*, M. A. Fox and M. Chanon eds., Elsevier, Amsterdam (1989); for a short review, see: D. Mauzerall and S. G. Ballard: *Ann. Rev. Phys. Chem.* **33**, 377 (1982).

40. (a) A. Kira, S. Arai and M. Imamura: *J. Chem. Phys.* **54**, 498 (1971), **56**, 1777 (1972); (b) M. A. Rodgers: *Chem. Phys. Lett.* **9**, 107 (1971), *J. Chem. Soc. Farad. Trans. II* **68**, 1278 (1972). For a recent example, see: (c) Y. Chikai, Y. Yamamoto and K. Hayashi, *Bull. Chem. Soc. Japan* **61**, 2281 (1988).

41. K.-D. Asmus: *Acc. Chem. Res.* **12**, 436 (1979).

42. (a) H. Hiratsuka, K. Sekiguchi, Y. Hatano and Y. Tanizaki: *Chem. Phys. Lett.* **55**, 358 (1978); (b) H. Hiratsuka and Y. Tanizaki: *J. Phys. Chem.* **83**, 2501 (1979); (c) K. Sekiguchi, H. Hiratsuka, Y. Hatano and Y. Tanizaki: *ibid.* **84**, 452 (1980); (d) Ref [33f]; (e) T. Minegishi, H. Hiratsuka, Y. Tanizaki and Y. Mori: *Bull. Chem. Soc. Japan* **57**, 162 (1984); (f) H. Hiratsuka, Y. Hatano, Y. Tanizaki and Y. Mori: *J. Chem. Soc. Farad. Trans. II* **81**, 1653 (1985); (g) H. Hiratsuka, K. Sekiguchi, Y. Hatano, Y. Tanizaki and Y. Mori: *ibid.* **82**, 795 (1986).

43. D. F. Evans: *Nature* **176**, 777 (1955).

44. M. Kasha: *J. Opt. Soc. Am.* **38**, 1068 (1948)

45. P. Bennema, G. H. Hoijtink, J. H. Lupinski, L. J. Oosterhoff, P. Selier and J. d. W. van Voorst: *Mol. Phys.* **2**, 431 (1959).

46. J. Joussot-Dubien and R. Lesclaux: *J. Chim. Phys.* 1631 (1964); J. Joussot-Dubien and R. Lesclaux: *Compt. Rend. Acad. Sci.* **258**, 4260 (1964), *Acta Phys. Polon.* **26**, 665 (1964).

47. G. N. Lewis and D. Lipkin: *J. Am. Chem. Soc.* **64**, 2801 (1942); G. N. Lewis and J. Biegeleisen: *ibid.* **65**, 2419 (1943).

48. W. H. Hamill in: *Radical Ions*, E. T. Kaiser and L. Kevan eds, Wiley Interscience, New York 1968.

49. (a) T. Shida and W. H. Hamill: *J. Chem. Phys.* **44**, 2369, 2375 (1968); (b) T. Shida and W. H. Hamill: *ibid.* **44**, 4372 (1968), *J. Am. Chem. Soc.* **88**, 5371 (1966); (c) B. Badger and B. Brocklehurst: *Nature* **219**, 261 (1968).

50. S. Arai and M. Imamura: *J. Phys. Chem.* **83**, 337 (1979).

51. (a) C. Sandorfy: *Can. J. Spectrosc.* **10**, 85 (1965); (b) A. Grimison and G. A. Simpson: *J. Phys. Chem.* **72**, 1776 (1968).

52. (a) A. M. Barnes and H. P. Broida eds.: *Formation and Trapping of Free Radicals*, Academic Press, New York (1960); (b) B. Meyer: *Low Temperature Spectroscopy*, Elsevier, Amsterdam (1971); (c) S. Cradock & A. J. Hinchcliffe: *Matrix Isolation, A Technique for the Study of Reactive Inorganic Species*, Cambridge Univ. Press, Cambridge (1975); (c) M. Moskovits and G. A. Ozin eds.: *Cryochemistry*, Wiley, New York (1976); (d) A. J. Barnes, W. J. Orville-Thomas, A. Müller and R. Gaufrès eds.: *Matrix Isolation Spectroscopy*, Nato ASI Series C76, D. Reidel, Dordrecht (1981); (e) G. C. Pimentel: *Pure Appl. Chem.* **4**, 61 (1962); (f) G. C. Pimentel & S. W. Charles: *ibid.* **7**, 111 (1963); (g) A. J. Barnes and E. Hallam: Quart. Rev. **23**, 392 (1969); (h) J. J. Turner: *Angew. Chemie* **87**, 326 (1975); (i) I. R. Dunkin, *Chem. Soc. Rev.* **9**, 1 (1980); (j) O. L. Chapman, *Pure Appl. Chem.* **40**, 511 (1974); (k) L. B. Knight, *Acc. Chem. Res.* **19**, 313 (1986).

53. (a) L. A. Wight, B. S. Ault and L. Andrews: *J. Chem. Phys.* **60**, 81 (1974); There is some argument about the mechanism of ion formation in these experiments: While Andrews claims that ionization is effected by the 1048 and 1067 Å Ar emission lines, Jacox maintains that Penning ionization makes a significant contribution (see: M. Jacox: *Chem. Phys.* **12**, 51 (1976), *Rev. Chem. Int.* **2**, 1 (1978)).

54. (a) V. E. Bondybey, T. A. Miller and J. H. English: *J. Chem. Phys.* **72**, 2193 (1980); (b) V. E. Bondybey and T. H. Miller, *ibid.* **73**, 3053 (1980); (c) V. E. Bondybey, C. R. Vaughn, T. H. Miller and J. H. English: *ibid.* **74**, 6584 (1981); (d) V. E. Bondybey, T. H. Miller and J. H. English: *J. Chim. Phys.* **77**, 607 (1980); (e) V. E. Bondybey, J. H. English and T. H. Miller *Chem. Phys. Lett.* **90**, 394 (1982). (f) S. Leutwyler, J. P. Maier and U. Spittel: *J. Chem. Soc. Farad. Trans. II* **81**, 1565 (1985); (g) S. Leutwyler, J. P. Maier and U. Spittel: *Mol. Phys.* **51**, 437 (1984); (h) J. Fulara, S. Leutwyler, J. P. Maier and U. Spittel: *J. Phys. Chem.* **89**, 3190 (1985); (i) S. Leutwyler, J. P. Maier and U. Spittel: *Chem. Phys. Lett.* **96**, 645 (1983).

55. (a) L. Andrews, J. H. Miller and B. W. Keelan: *Chem. Phys. Lett.* **71**, 207 (1980); (b) L. Andrews and B. W. Keelan: *J. Am. Chem. Soc.* **102**, 5732 (1980); (c) J. H. Miller, L. Andrews, P. A. Lund and P. N. Schatz: *J. Chem. Phys.* **73**, 4932 (1980); (d) B. W. Keelan and L. Andrews: *J. Am. Chem. Soc.* **103**, 822 (1981); (e) B. W. Keelan and L. Andrews: *ibid.* **103**, 829 (1981); (f) R. S. Friedman, B. J. Kelsall and L. Andrews: *J. Phys. Chem.* **88**, 1944 (1984); (g) L. Andrews and T. A. Blankenship: *J. Am. Chem. Soc.* **103**, 5977 (1981); (h) L. Andrews, R. S. Friedman and B. J. Kelsall: *J. Phys. Chem.* **89**, 4550 (1985); (i) L. Andrews, B. J. Kelsall and T. A. Blankenship: *J. Phys. Chem.* **86**, 2916 (1982); (j) A. C. Puju, L. Andrews, W. A. Chupka and S. D. Colson: *J. Chem. Phys.* **76**, 3854 (1982); (k) L. Andrews, R. T. Arlinghaus abd C. K. Payne: *J. Chem. Soc. Farad. Trans. II* **79**, 885 (1983); (l) L. Andrews, B. J. Kelsall and J. A. Harvey: *J. Chem. Phys.* **77**, 2235 (1982); (m) B. J. Kelsall, R. T. Arlinghaus and L. Andrews: *High Temp. Sci.* **17**, 155 (1984); (n) L. Andrews, J. A. Harvey, B. J. Kelsall and D. C. Duffey: *J. Am. Chem. Soc.* **103**, 6415 (1981); (o) B. J. Kelsall, L. Andrews and H. Schwarz: *J. Phys. Chem.* **87**, 1295 (1983); (p) B. J. Kelsall and L. Andrews: *J. Chem. Phys.* **76**, 5005 (1982). (q) L. Andrews, B. S. Friedman and B. J. Kelsall: *J. Phys. Chem.* **89**, 4016 (1985); (r) B. J. Kelsall and L. Andrews: *ibid.* **88**, 2723 (1984); (s) I. R. Dunkin, L. Andrews, J. T. Lurito and B. J. Kelsall; *ibid.* **89**, 1701 (1985) (t) L. Andrews and J. T. Lurito: *Tetrahedron* **42**, 6343 (1986); (u) L. Andrews, I. R. Dunkin, B. J. Kelsall and J. T. Lurito: *J. Phys. Chem.* **89**, 821 (1985).

56. (a) T. Bally, S. Nitsche, K. Roth and E. Haselbach: *J. Am. Chem. Soc.* **106**, 3927 (1984); (b) T. Bally, S. Nitsche, K. Roth and E. Haselbach: *J. Phys. Chem.* **89**, 2528 (1985); (c) T. Bally, S. Nitsche and K. Roth: *J. Chem. Phys.* **84**, 2577 (1986); (d) T. Bally, E. Haselbach, S. Nitsche and K. Roth: *Tetrahedron* **42**, 6325 (1986); (e) T. Bally, D. Hasselmann and K. Loosen, *Helv. Chim. Acta.* **68**, 345 (1984).

57. J. P. Maier and D. W. Turner: *Disc. Farad. Soc.* **54**, 149 (1972).

58. All papers by Z. H. Khan: (a) Z. Naturforsch. **39A**, 668 (1984); (b) *Spectrochim. Acta* **44A**, 313 (1988); (c) *Z. Naturforsch.* **42A**, 91 (1987); (d) *Spectrochim. Acta* **45A**, 253 (1989)

59. (a) Ref. [13]; (b) E. Haselbach, T. Bally, Z. Lanyiova and P. Baertschi: *Helv. Chim. Acta* **62**, 583 (1979); (c) Y. Fujisaka, T. Shida, T. Bally and K. Roth, *manuscript in preparation*; (d) T. Bally, E. Haselbach, Z. Lanyiova and P. Baertschi: *Helv. Chim. Acta* **61**, 2488 (1978). (e) T. Bally, L. Neuhaus, S. Nitsche, E. Haselbach, J. Janssen and W. Lüttke: *Helv. Chim. Acta* **66**, 1288 (1983); (f) J.-N. Aebischer, T. Bally, K. Roth, E. Haselbach, F. Gerson and X.-Z. Qin: *J. Am. Chem. Soc.* **111**, 7909 (1989); (g) T. Bally, P. Haag, K. Roth and O. Schafer, *to be published*; (h) E. Haselbach, U. Klemm, U. Buser, R. Gschwind, E. Kloster-Jensen, J. P. Maier, O. Marthaler, H. Christen and P. Baertschi: *Helv. Chim. Acta* **64**, 823 (1981); (i) T. Bally, K. Roth and R. Straub: *J. Am. Chem. Soc.* **110**, 1639 (1988); (k) T. Bally, K. Roth and R. Straub: *Helv. Chim. Acta* **72**, 73 (1989); (l) E. Haselbach, T. Bally, R. Gschwind, U. Klemm and Z. Lanyiova: *Chimia* **33**, 405 (1979);

52

(m) T. Bally, S. Nitsche and E. Haselbach, *Helv. Chim. Acta* **67**,86 (1984); T. Bally and S. Nitsche, to be published (see [33d]) (n) K. Ushida, T. Shida and K. Shimokoshi, *J. Phys. Chem.* **93** 5388 (1989); (o) Ref [59d].

60. H. D. Roth, *Acc. Chem. Res.* **20**, 343 (1987)

61. B. F. Yates, W. J. Bouma and L. Radom: *Tetrahedron* **42**, 6225 (1986) and references cited therein.

62. (a) Ref [27b]; (b) Ref [28]; (c) S. F. Nelsen, M. F. Teasley, D. L. Kapp, C. R. Kessel and L. A. Grezzo: *J. Am. Chem. Soc.* **106**, 791 (1984); (d) S. F. Nelsen, G. T. Cunkle, D. H. Evans, K. J. Haller, M. Kaftory, B. Kirste and T. Clark: *ibid.* **107**, 3829 (1985); (e) S. F. Nelsen, W. P. Parmelee, M. Göbl, K.-O. Hilloer, D. Veltwisch and K.-D. Asmus: *ibid.* **102**, 5606 (1980).

63. (a) S. F. Nelsen, M. F. Teasley, D. F. Kapp, C. R: Kessel and L. A. Grezzo: *J. Am. Chem. Soc.* **106**, 791 (1984); (b) Ref [27a]; (c) S. F. Nelsen, M. F. Teasley and D. L. Kapp: *J. Am. Chem. Soc.* **108**, 5503 (1986).

64. (a) S. F. Nelsen, R. W. Alder, R. B: Sessions, K.-D- Asmus, K.-O. Hiller and M. Göbl: *J. Am. Chem. Soc.* **102**, 1429 (1980); (b) R. W. Alder, R. B. Sessions, J. M. Millor and M. F. Rowlin, *J. Chem. Soc. Chem. Commun.* **1977**, 747; (c) R. W. Alder and R. B. Sessions: *J. Am. Chem. Soc.* **101**, 3651 (1979); (d) R. W. Alder: Acc. Chem. Res. **16**, 321 (1983); (e) R. W. Alder, R. Gill and N. C. Goode: *J. Chem. Soc. Chem. Commun.* **1976**, 973.

65. (a) M. Göbl, M. Bonifacic and K.-D. Asmus: *J. Am. Chem. Soc.* **106**, 5984 (1984); (b) E. Anklam, K.-D. Asmus and H. Mohan: *J. Phys. Org. Chem.* **3**, 17 (1990), s.a. Ref [65e]; (c) K.-O. Hiller, B. Masloch, M. Göbl and K.-D. Asmus: *J. Am. Chem. Soc.* **103**, 2734 (1981), s.a. Ref [65i]; (d) D. Bahnemann and K.-D. Asmus: *J. Chem. Soc. Chem. Commun.* **1975**, 238; K.-D. Asmus, H. A. Gillis and G. G. Teather: *J. Phys. Chem.* **82**, 2677 (1978); (e) K.-D. Asmus, D. Bahnemann, Ch.-H. Fischer and D. Veltwisch: *J. Am. Chem. Soc.* **101**, 5322 (1979); (f) H. Mohan and K.-D. Asmus, *ibid.* **109**, 4745 (1987); (g) H. Mohan and K.-D. Asmus, *J. Chem. Soc. Perkin Trans. II* **1987**, 1795; (h) E. Anklam, H. Mohan and K.-D. Asmus: *ibid.* **1988**, 1297; (i) K.-D. Asmus, M. Göbl, K.-O. Hiller, S. Mahling and J. Mönig, *ibid.* **1985**, 641; J. Mönig, M. Göbl and K.-D. Asmus: *ibid.* **1985**, 647; (k) R. S. Glass, M. Hojjatie, G. S. Wilson, S. Mahling, M. Göbl and K.-D. Asmus: *J. Am. Chem. Soc.* **106**, 5382 (1984); (l) S. A. Chaudhri, M. Göbl, T. Freyholdt and K.-D. Asmus: *ibid.* **106**, 5988 (1984); (m) K. Kishore and K.-D. Asmus: *J. Chem. Soc. Perkin Trans. II*, **1990**, 2079.

66. K. Bobrowksi and J. Holcman: *J. Phys. Chem.* **93**,6381 (1989)

67. (a) B. Badger and B. Brocklehurst: *Chem. Phys. Lett.* **1**, 122 (1967); (b) B. Badger and B. Brocklehurst: *Trans. Farad. Soc.* **66**, 2939 (1966)

68. N. C. Baird: *J. Chem. Educ.* **54**, 291 (1977).

69. All papers by T. Clark: (a) *J. Am. Chem. Soc.* **110**, 1672 (1988); (b) *J. Comput. Chem.* **2**, 261 (1981); (c) *ibid.* **3**, 112 (1982); (d) *ibid.* **4**, 404 (1983).

70. K. Kimura, S. Katsumata, Y. Achiba, T. Yamazaki and S. Iwata: *Handbook of HeI Photoelectron Spectra of Fundamental Organic Molecules*, Halsted Press, New York 1981.

71. I. C. Lewis and L. S. Singer: *J. Chem. Phys.* **43**, 2712 (1965); s.a. O. W. Howarth and G. K. Fraenkel: *J. Am. Chem. Soc.* **88**, 4514 (1966).

72. All papers by B. Badger and B. Brocklehurst: (a) *Trans. Farad. Soc.* **65**, 2576 (1969); (b) *ibid.* **65**, 2582 (1965); (c) *ibid.* **65**, 2588 (1965).

73. (a) A. Kira, T. Nakanura and M. Imamura: *J. Phys. Chem.* **81**, 511 (1977); (b) A. Kira and M. Imamura: *ibid.* **83**, 2267 (1979); (c) A. Kira, T. Nakamura and M. Imamura: *Chem. Phys. Lett.* **54**, 582 (1978).

74. M. A. J. Rodgers: *J. Chem. Soc. Farad. II* **1972**, 1278; for more recent work on this subject, see: M. F. Desrosiers and A. D. Trifunac: *J. Phys. Chem.* **90**, 1560 (1986); A. Terahara, H. Ohiya-Nishiguchi, N. Hirota and A. Oku: *ibid.* **90**, 1565 (1986); J. M. Masnovi and J. K. Kochi: *ibid.* **91**, 1878 (1987).

75. (a) J. A. Stone and M. S. Lim: *Can. J. Chem.* **58**, 1666 (1980); (b) M. Meot-Ner: *J. Phys. Chem.* **84**, 2724 (1980); (c) M. Meot-Ner and M. S. El-Shall: *J. Am. Chem. Soc.* **108**, 4386 (1986). (d) J. R. Grover, E. A. Walters and E. T. Hui: *J. Phys. Chem.* **91**, 3233 (1987)

76. (a) T. Bally and K. Roth: *unpublished results*; (b) T. Bally, K. Roth, X.-Z. Qin and F. Gerson: *to be published*.

77. L. A. Chewter, K. Müller-Dethlefs and E. W. Schlag: *Chem. Phys. Lett.* **135**, 219 (1987).

78. see for example T. H. Morton: *Tetrahedron* **38**, 3195 (1982)

79. R. Postma, P. J. A. Ruttink, B. van Baar, J. K. Terlouw, J. C. Holmes and P. C. Burgers: *Chem. Phys. Lett.* **123**, 409 (1986).

80. J.-D. Shao, T. Baer, J. C. Morrow and M. L. Fraser-Monteiro: *J. Chem. Phys.* **87**, 5242 (1987).

81. Various schemes have been proposed for the evaluation of resonance energies. Our figures are based on that proposed by S. W. Benson [*J. Chem. Educ.* **42**, 510 (1968)] whereby the resonance energy of allyl (radical or cation) is defined as the enthalpy change for the reaction allyl+propane→n-propyl+propene. Identical results are obtained by the recently advanced procedure of Lossing [F. P. Lossing and J. L. Holmes: *J. Am. Chem. Soc.* **106**, 6917 (1984)].

82. (a) ΔH_f^0(allyl$^+$)=227±0.33 kcal/mol [J. Traeger: *Int. J. Mass Spectrom. Ion Proc.* **58**, 259 (1984)]; (b) ΔH_f^0(allyl$^.$)=39.5±0.8 kcal/mol [from (a) and I_1(allyl$^.$)=8.13±0.03 eV: F. A. Houle and J. L. Beauchamp: *J. Am. Chem. Soc.* **100**, 3290 (1978)]; (c) ΔH_f^0(n-propyl$^.$)=22.7±0.4 kcal/mol [J. L. Holmes, F. P. Lossing and A. Maccoll: *J. Am. Chem. Soc.* **110**, 7339 (1988)] (d) ΔH_f^0(n-propyl$^+$)=210.6±0.8 kcal/mol [from (c) and I_1 (propyl$^.$)=8.15±0.02 eV: J. C. Schulz F. A. Houle and J. L. Beauchamp: *J. Am. Chem. Soc.* **106**, 3917 (1984)]; (e) ΔH_f^0(propane)= -25.0±0.1 and ΔH_f^0(propene)=4.8±0.2 are taken from: J. B. Pedley, R. D. Naylor and S. P. Kirby: *Thermochemical data of Organic Compounds*, 2nd ed., Chapman & Hall, London, 1986.

83. (a) Ref [59d]; (b) H. D. Roth, M. L. M. Schilling and C. C. Wamser: *J. Am. Chem. Soc.* **106**, 5023 (1984); (c) H. D. Roth, M. L. M. Schilling and C. J. Abelt: *Tetrahedron* **42**, 6157 (1986) and references cited therein; (d) T. Momose, T. Shida and T. Kobayashi, *ibid.* **42**, 6337 (1986); see also Ref [101v]; (e) H. D. Roth and C. J. Abelt: *J. Am. Chem. Soc.* **108**, 2013 (1986); (f) S. Dai, J. T. Wang and F. Williams: *ibid.* **112**, 2837 (1990); (g) Gerson, A. De Meijere and X.-Z. Qin: *ibid.* **111**, 1135 (1989); (h) T. Miyashi, Y. Takahashi, T. Mukai, H. D. Rothand M. L. M. Schilling: *ibid.* **107**, 1079 (1985); (i) Ref [59m,n]; see also L. D. Snow and F. Williams: *Chem. Phys. Lett.* **143**, 521 (1988) and referecens cited therein. (j) A. Arnold and F. Gerson: *J. Am. Chem. Soc.* **112**, 2027 (1989); F. Williams, Q.-X. Guo and S. Nelsen: *ibid.* **112**, 2028 (1989); (k) X.-Z. Qin and F. Williams: *J. Phys. Chem.* **90**, 2292 (1986); (l) X.-Z. Qin, T. C. Pentecost, J. T. Wang and F. Williams: *J. Chem. Soc. Chem. Commun.* **1987**, 450.

84. Dimethyl- and tetramethyl allyl cations show absorptions around 300 nm [U. Pittman: *J. Chem. Soc. Chem. Commun.* **1969**, 122] but the parent species is not expected to absorb above 250 nm [see for example: A. Sabljic, N. Trinajstic, J. V. Knop, J. Koller and A. Azman: *J. Mol. Struct.* **33**, 145 (1976)].

85. G. Maier, H. P. Reisenauer, B. Rohde and K. Dehmke: *Chem. Ber.* **116**, 732 (1983) and references cited therein.

86. (a) See: L. D. Snow and F. Willliams: *Chem. Phys. Lett.* **143**, 521 (1988); M. C. R. Symons and A. W. Wren, *ibid.* **140**, 611 (1987) and references cited therein. (b) R. H. Nobes, W. J. Bouma and L. Radom, *ibid.* **135**, 78 (1987).

87. T. Bally and S. Nitsche, to be published. Several cyclic epoxides were also studied [33d] and the corresponding oxallyl cations show similar absorptions peaking between 470 (cyclohexene oxide) and 550 nm (norbornene oxide). On the other hand, polycyclic epoxides (octaline oxide [59l] or sesquinorbornene oxide [88]) which are prevented from forming oxallyl cations fail to show any optical absorptions above 250 nm after ionization. λ_{max} of ionized parent ethylene oxide in CFCl$_3$ is 460 nm [59n] whereas PD measurements indicate that in the gas phase the corresponding λ_{max} is 475 nm [F. N. T. Van Velzen and W. J. Van der Hart, *Chem. Phys. Lett.* **83**, 55 (1981)].

88. F. Williams, S. Dai, L. D. Snow, X.-Z. Qin, T. Bally, S. Nitsche, E. Haselbach, S. F. Nelsen and M. F. Teasley: J. Am. Chem. Soc. 109, 7526 (1987).

89. (a) D. Feller, E. R. Davidson and W. T. Borden: *J. Am. Chem. Soc.* **105**, 3347 (1983); (b) D. Feller, E. R. Davidson and W. T. Borden: *ibid.* **106**, 2513 (1984); (c) W. J. Bouma, J. K. MacLeod and L. Radom, *ibid.* **101**, 5540 (1979); (d) W. J. Bouma, D. Poppinger, S. Sæbø, J. MacLeod and L. Radom: *Chem. Phys. Lett.* **104**, 198 (1984).

90. X.-Z. Qin, L. D. Snow and F. Williams: *J. Phys. Chem.* **89**, 3602 (1985).

91. The largest "pure" TB interaction of two π-systems via one C-C bond was found in trans-1,2-divinylcyclobutane where it amounts to 0.68 eV: R. Gleiter: *Top. Curr. Chem.* **86**, 197 (1979).

92. For a discussion of the influence of *steric* interactions on the energies of different conformations of IV, see Ref [83c]. However, these authors did not take the stabilizing effect due to interaction of the two allyl systems into consideration. According to molecular models this should be optimal at a dihedral angle of ≈30°, as defined in Fig 4 of [83c].

54

93. C. C. Schenk, B. Diner, P. Mathis and K. Satoh: *Biochim. Biophys. Acta* **680**, 216 (1982); P. Mathis and A. W. Rutherford: *ibid.* **767**, 214 (1984).

94. D. Gust, T. A. Moore, P. A. Liddell, G. A. Nemeth, L. R. Makings, A. L. Moore, D. Barrett, P. J. Pessiki, R. V. Bensasson, M. Rougée, C. Chachati, F. C: DeSchryver, M. Van der Auweraer, A. R. Holzwarth and J. S. Conolly, *J. Am. Chem. Soc.* **109**, 846 (1987).

95. (a) K. Bobrowski and P. K. Das: *J. Phys. Chem.* **89**, 5079 (1985); (b) K. Bobrowski and P. K. Das: *J. Phys. Chem.* **90**, 927 (1986).

96. (a) E. A. Dawe and E. J. Land. *J. Chem. Soc. Farad. Trans. I* **71**, 2162 (1975); (b) J. Lafferty, A. Roach, R. S. Sinclair, T. G: Truscott and E. J. Land: *ibid.* **73**, 416 (1977); (c) R. V. Bensasson, E. J. Land and T. G. Truscott: *Flash Photolysis and Pulse Radiolysis: Contributions to the Chemistry of Biology and Medecine* (Chapter 4), Pergamon Press, Oxford, 1983.

97. J. L. Grant, V. J. Kramer, R. Ding and L. D. Kispert: *J. Am. Chem. Soc.* **110**, 2151 (1988).

98. (a) A. J. Heeger, S. Kivelson, J. R. Schrieffer and W.-P. Su: *Rev. Mod. Phys.* **60**, 781 (1988); (b) S. Roth and H. Bleier, *Adv. Phys.* **36**, 385 (1987); (c) J. L. Brédas and G. B. Street: *Acc. Chem. Res.* **18**, 309 (1985).

99. R. Schrock and T. Bally, *to be published.*

100. (a) M. Jacox: *Chem. Phys.* **12**, 51 (1976); (b) L. Andrews, C. A. Wight, F. T: Prochaska, S. A. McDonald and B. S. Ault: *J. Mol. Spectrosc.* **73**, 120 (1978)

101. (a) [55n]; (b) L. Andrews, B. J. Kelsall, C. K. Payne, O. R. Rodig and H. Schwarz: *J. Phys. Chem.* **86**, 3714 (1982); (c) [55b], B. J. Kelsall and L. Andrews: *J. Am. Chem. Soc.* **105**, 1413 (1983); (d) B. J. Kelsall, L. Andrews and G. J. McGarvey: *J. Phys. Chem* **87**, 1788 (1983); (e) [55o]; (f) B. J. Kelsall, L. Andrews and C. Trindle: *ibid.* **87**, 4898 (1983); (g) R. S. Friedman and L. Andrews: *J. Am. Chem. Soc.* **107**, 822 (1985); (h) B. J. Kelsall and L. Andrews: *J. Phys. Chem.* **88**, 5893 (1984); (i) Q.-X. Guo, X.-Z. Qin, J. T. Wang and F. Williams: *J. Am. Chem. Soc.* **110**, 1974 (1988); (j) K. Toriyama, K. Nunome, M. Iwasaki, T. Shida and K. Ushida: *Chem. Phys. Lett.* **122**, 118 (1985) and references cited therein; (k) T. Bally, K. Roth, P. Haag and O. Schafer: to be published; (l) J.-N. Aebischer, T. Bally, K. Roth, E. Haselbach, F. Gerson and X.-Z. Qin: *J. Am. Chem. Soc.* **111**, 7909 (1989); (m) S. Dai, R. S. Pappas, G.-F. Chen, Q.-X. Guo, J. T. Wang and F. Williams; *ibid.* **111**, 8759 (1989); (n) T. Shida, Y. Nosaka and T. Kato: *J. Phys. Chem* **81**, 1095 (1977); (o) [56b]; (p) B. J. Kelsall and L. Andrews: *ibid.* **88**, 2723 (1984); (q) [56d]; (r) [56c]; (s) [55s,t]; (t) K. Roth: *Dissertation Nr 886*, University of Fribourg (1989), T. Bally, K. Roth, T. Shida and Y. Fujisaka: *manuscript in preparation.* (u) [55t]; (v) T. Shida, T. Momose and N. Ono: *J. Phys. Chem.* **89**, 815 (1985).

102. T. Miyashi, K. Wakamatsu, T. Akiya, K. Kikuchi and T. Mukai: *J. Am. Chem. Soc.* **109**, 5270 (1987).

103. D. J. Belville, R. Chelsky and N. L. Bauld: *J. Comput. Chem.* **3**, 548 (1982).

104. All neutral heats of formation taken from: J. B. Pedley, R. D. Naylor and S. P. Kirby: *Thermochemical data of Organic Compounds*, 2ed., Chapman and Hall, London, 1986; Adiabatic ionization energies from: G. Bieri, F. Burger, E. Heilbronner and J. P. Maier: *Helv. Chim. Acta.* **60**, 2213 (1977).

105. I.R. Dunkin and L. Andrews: *Tetrahedron* **41**, 145 (1985).

106. See also: E. Haselbach, T. Bally and Z. Lanyiova: *Helv. Chim. Acta* **62**, 577 (1979).

107. See also: R. H. Nobes, W. J. Bouma J. K. McLeod and L. Radom: *Chem. Phys. Lett.* **135**, 78 (1987).

INFRARED AND OPTICAL ABSORPTION SPECTROSCOPY OF MOLECULAR IONS
IN SOLID ARGON

LESTER ANDREWS
Chemistry Department
University of Virginia
Charlottesville, Virginia 22901
U.S.A.

1. Introduction

The matrix isolation technique is particularly well-suited for the
study of reactive species, including free radicals, molecular ions and
molecular complexes, which react rapidly or decompose under normal
chemical conditions. Matrix studies provide a valuable complement to
gas phase work: band positions located in the matrix provide a guide
for high resolution measurements and more information can often be
obtained from a complete low resolution spectrum than from high
resolution measurements on one band system.

Many of the early matrix isolation studies of molecular ions have
been reviewed by Jacox [1] and this author [2-4]. Charged species in
noble gas matrices may be described as "isolated" or "chemically bound"
with respect to the counterion. The first ionic species characterized
in matrices, $Li^+O_2^-$, is of the latter type with Li^+ and O_2^- ions
electrostatically bound together in a C_{2v} triangular structure [5-7].
The next molecular ions identified in matrices, $B_2H_6^-$ and C_2^-, were
isolated from their counterions by the matrix [8,9]. Clearly, the
formation of "isolated" ions requires ionization energy for the subject
molecule from an external source and an acceptor atom or molecule to
trap the removed electron somewhere else in the matrix. This chapter
describes infrared spectroscopic studies of trihalomethyl cations,
polyfluoride anions, and nitrogen oxide anions, and ultraviolet-visible
spectroscopy of the naphthalene and toluene cations. Matrix studies of
other cation systems using visible and electron spin resonance
spectroscopies are described in the chapters by Bally and by Knight.

2. Experimental Methods

Molecular ions have been produced and trapped concurrently with
sample deposition using the apparatus shown in Figure 1. Sample was

A. Lund and M. Shiotani (eds.), Radical Ionic Systems, 55–72.

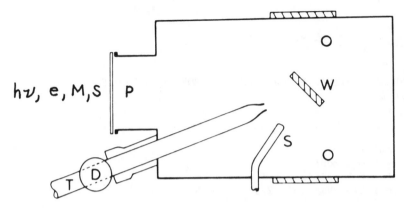

Figure 1. Vacuum vessel used for absorption spectroscopy of matrix isolated species.

deposited through line (**S**) onto a rotatable cold-window (**W**) cooled by a closed-cycle helium refrigerator to 10-12K; spectra were recorded through optical windows (**O**). The vacuum ultraviolet photoionization technique developed in this laboratory employed open-ended microwave powered discharge tubes (**T**) like the one shown in Figure 1; initial studies used He, Ne, Ar, Kr and Xe gases in the discharge in order to provide a wide range of ionization sources [10]. The argon, krypton and xenon discharge gases were subsequently collected with the matrix, but neon and helium discharge gases were evacuated through an auxiliary pumping system. Similar direct-current powered discharge resonance lamps have been used for photoelectron spectroscopy [11]. Resonance two-photon ionization of aromatic precursors has been done by irradiation with an intense high-pressure mercury arc through the photolysis port (**P**). Here the presence of an electron trapping molecule, such as CCl_4, is critical in order to "trap" the electrons removed by photoionization and prevent their neutralization of the cations formed [12]. Recently a thermionic electron source capable of codepositing 10-200μA of low energy electrons (30-400eV) with the matrix sample has been employed to produce molecular ions during sample condensation [13]. A schematic diagram of this experiment is shown in Figure 2. An important element is the ring isolated on the cold window, which is biased to attract electrons into the condensing matrix sample. A chemical ionization discharge source has also been developed to subject an argon/reagent stream to a high-voltage discharge and collect the gases on the cold window [14]. The electron or chemical ionization discharge source was mated to the matrix apparatus at **P** for the particular matrix reaction of interest.

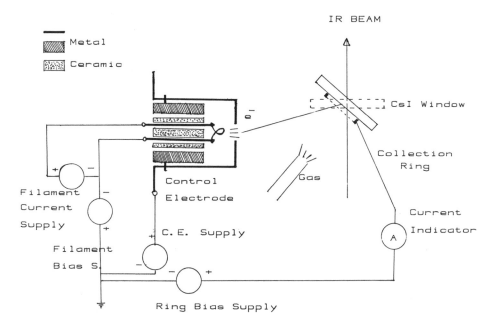

IR BEAM

Metal

Ceramic

CsI Window

Collection Ring

Control Electrode

Gas

Filament Current Supply

C.E. Supply

Current Indicator

Filament Bias S.

Ring Bias Supply

Figure 2. Thermionic electron source for matrix isolation experiments.

3. Trihalomethyl Cations

Trihalomethyl cations were prepared using the argon resonance photoionization technique. Briefly one photon produced the CX_3 radical and a second photon photoionized the radical in the matrix isolation experiments to be described below. Our goal in these studies was CF_3^+; the development of techniques for producing and trapping CF_3^+ led first to CCl_3^+ and the mixed chlorofluoromethyl cations. Jacox and Milligan performed hydrogen resonance photolysis of argon/chloroform samples and observed a large yield of the CCl_3 radical at 898 cm^{-1} and a new 1037 cm^{-1} absorption. The 1037 cm^{-1} band exhibited the 3/1 doublet splitting characteristic of three equivalent chlorine atoms and a carbon-13 shift to 1003 cm^{-1}, which supported its identification as CCl_3^+ isolated in the matrix [15]. Subsequent radiolysis studies of CCl_4, CCl_3Br, CCl_2Br_2 and CBr_4 in this laboratory produced either the 1037 cm^{-1} band, absorptions at 1019 and 957 cm^{-1} for CCl_2Br^+, absorptions at 978 and 894 cm^{-1} for $CClBr_2^+$, or a new band at 874 cm^{-1}

for CBr_3^+, from two of the above precursors, which confirmed the trichloromethyl cation identification of the 1037 cm^{-1} band [16]. Infrared absorptions for trihalomethyl cations are listed in Table 1.

TABLE 1. Carbon-halogen stretching vibrations (cm^{-1}) observed for trihalomethyl cations isolated in solid argon.

Cation	C-F	C-Cl	C-Br
CF_3^+	1665		
CF_2Cl^+	1514, 1414		
$CFCl_2^+$	1351	1142	
CCl_3^+		1037	
CCl_2Br^+		1019, 957	
$CClBr_2^+$			894
CBr_3^+			874
CF_2Br^+	1483, 1367		
CF_2I^+	1432, 1320		

A matrix argon resonance photoionization study of CCl_4 revealed different filtered mercury arc photolysis behavior for each new absorption; this was attributed to the photochemical stability of the cation in question. The photosensitive 927 and 374 cm^{-1} bands were assigned to an asymmetric CCl_4^+ species and the 502 cm^{-1} absorption to Cl_3^+. The 1037 cm^{-1} band was decreased slightly by prolonged mercury arc photolysis, which required photodetachment from chloride electron traps in the matrix, since CCl_3^+ itself probably does not dissociate in the mercury arc energy range [10]. A most important conclusion from the mercury arc photolysis studies is that isolated cations in matrices will be of two types: those which photodissociate with mercury arc light and those photochemically stable in the mercury arc range, which must exhibit a decrease in intensity from photoneutralization by detachment of electrons from counteranions in the matrix. Appearance potential data are useful for predicting the photochemical behavior of cations. Clearly, photodetachment depends on the electron trapping species, as will be discussed below for CF_3^+.

The search for CF_3^+ starting from CCl_3^+ at 1037 cm^{-1} was initiated in this laboratory by photoionization and radiolysis studies on the Freon series $CFCl_3$, CF_2Cl_2 and CF_3Cl [17,18]. Sharp new 1352 and 1142 cm^{-1} bands in $CFCl_3$ experiments exhibited appropriate carbon-13 shifts and photolysis behavior for assignment to $CFCl_2^+$. Analogous bands in CF_3Cl experiments at 1415 and 1515 cm^{-1} showed large carbon-13 shifts and slight photolysis with the full mercury arc which indicated assignment to CF_2Cl^+. These vibrations for $CFCl_2^+$ and CF_2Cl^+ were 200-300 cm^{-1} above the corresponding free radical values, which predicts that CF_3^+ may absorb above 1600 cm^{-1}.

Matrix photoionization studies have been performed on the trifluoromethyl compounds CF_3Cl, CF_3Br, CF_3I and CHF_3 [19] using the apparatus shown in Figure 1. The CF_2Cl^+ absorptions at 1514 and 1414 cm^{-1}, which shifted to 1483 and 1367 cm^{-1} for CF_2Br^+, and the CF_2I^+ absorptions were displaced to 1432 and 1320 cm^{-1}. The C-F stretching modes in the CF_2X^+ ions exhibited a pronounced heavy halogen effect. A weak 1665 cm^{-1} band in the CF_3Cl study was produced with greater intensity in the CF_3Br, CF_3I and CHF_3 experiments at the same frequency. This 1665 cm^{-1} absorption was reduced 10% by full high-pressure mercury arc photolysis for 2 h in fluoroform experiments. The production of the same 1665.2 cm^{-1} band from four different trifluoro-methyl precursors, and the photolysis behavior indicate assignment of the 1665 cm^{-1} band to CF_3^+. The trifluoromethyl cation was formed by photoionization of CF_3 radicals produced in the matrix photolysis process.

Matrix infrared spectra of CF_3^+ and $^{13}CF_3^+$ in fluoroform experiments are compared in Figure 3. The strong absorption produced from

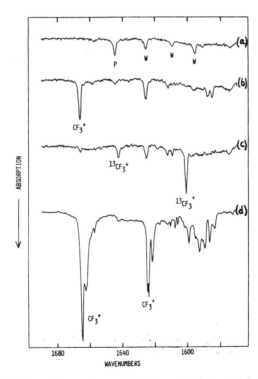

Figure 3. Infrared spectra of CHF_3 samples condensed with excess argon at 15K. (a) $Ar/CHF_3 = 200/1$, P denotes precursor and W indicates water absorptions, (b) $Ar/CHF_3 = 800/1$ with concurrent argon resonance photoionization, (c) $Ar/^{13}CHF_3 = 800/1$ with photo-ionization, (d) $Ar/CHF_3 = 700/1$ with electric discharge.

$$F — C^+ \diagdown \begin{matrix} F \\ \\ F \end{matrix}$$

<u>1</u>

$^{13}CHF_3$ at 1599.2 cm^{-1} (A = 0.20) in Figure 3(c) is appropriate for ν_3 of $^{13}CF_3^+$, and the weak $^{12}CF_3^+$ absorption (A = 0.02) at 1665.2 cm^{-1} is due to 10% ^{12}C present in the enriched fluoroform precursor. The ^{12}C-^{13}C shift, 66.0 cm^{-1}, is however, greater than expected for the antisymmetric C-F vibration, ν_3, of a planar centrosymmetric species. This unexpectedly large carbon-13 shift can be explained by Fermi resonance between ν_3 and the combination band $(\nu_1 + \nu_4)$ for the $^{13}CF_3^+$ species, since the new 1641.7 cm^{-1} product band in the $^{13}CF_3^+$ spectrum can be assigned to the combination band. The calculated position of ν_3 for $^{13}CF_3^+$, 1620 cm^{-1}, appears to coincide with the apparent position of $(\nu_1 + \nu_4)$, and these modes strongly interact and shift ν_3 down to 1599 cm^{-1} and $(\nu_1 + \nu_4)$ up to 1642 cm^{-1}. In the absence of Fermi resonance, the $(\nu_1 + \nu_4)$ combination band for $^{12}CF_3^+$ is expected to be 2-4 cm^{-1} above the $^{13}CF_3^+$ counterpart, in the region of the 1624 cm^{-1} water absorption; the spectrum in Figure 3(b) shows the 1624 cm^{-1} band, which may contain additional absorption. The markedly increased CF_3^+ yield in the electric discharge experiment, Figure 3(d), reveals a very strong 1665.2 cm^{-1} absorption (A = 0.84) and 1624 cm^{-1} absorption (A = 0.40) clearly in excess of the other water absorptions present [20]; the latter absorption is assigned to the combination band $(\nu_1 + \nu_4)$ for $^{12}CF_3^+$ provides a basis for determining the infrared inactive symmetric C-F bond stretching mode ν_1 mode of $^{12}CF_3^+$. The intensity of the ν_2 and ν_4 modes of BF_3 are approximately a factor of ten weaker than that of ν_3, and the failure to observe these weaker fundamentals of CF_3^+ in the present study is not surprising. However, ν_4 may be estimated at 500±30 cm^{-1}, which predicts ν_1 = 1125±30 cm^{-1} for $^{12}CF_3^+$.

Two similar neon resonance photoionization experiments have been performed in this laboratory codepositing Ne/CHF$_3$ = 300/1 sample at 5K with neon from a 3 mm orifice microwave discharge tube [21]. New bands were observed at 1254 cm^{-1} (A = 0.06) for CF$_3$, 1222 cm^{-1} (weak) for CF$_2$, and 1670 and 1664 cm^{-1} (A = 0.012) for CF$_3^+$. All of these bands decreased by 1/4 on annealing 5-12K. Spectra were scanned after 1-2 h periods, and product band growth rates decreased with time limiting the effective sample deposition period for good matrix isolation to about 4h. The neon condensation rate decreased rapidly as sample thickness increased thus limiting the amount of neon that can be condensed quickly enough to trap reactive species at 5K. Although neon is more inert than argon, its use for ions and small molecules in infrared experiments, <u>which require more sample</u> than in E.S.R. studies described in the chapter by Knight, is limited. Accordingly, it is possible to produce and trap an order of magnitude more CF$_3^+$ in solid argon than in solid neon. Jacox has recently employed neon resonance photoionization to produce CO$_2^+$ for observation of its infrared spectrum in solid neon [22].

The observation of CF$_3^+$ in solid neon is, however, important. The split absorptions at 1670 and 1664 cm^{-1} are characteristic of many species in this smaller matrix cavity and bracket the 1665 cm^{-1} argon value. The near agreement between neon and argon matrix absorptions for CF$_3$$^+$ and the low ionization energy for CF$_3$ radical (9.2 eV) [23] suggest that CF$_3$$^+$ does not interact strongly with these matrix environments. This agreement of neon and argon matrix values predicts the gas phase ν_3 fundamental of CF$_3$$^+$ at 1675±20 cm^{-1}.

The observed antisymmetric stretching frequency and the symmetric stretching frequency deduced from the ($\nu_1 + \nu_4$) band for CF$_3$$^+$ are large relative to the pyramidal CF$_3$ species (ν_1 = 1086 cm^{-1}, ν_3 = 1251 cm^{-1}) and the planar ^{11}BF$_3$ molecule (ν_1 = 888 cm^{-1}, ν_3 = 1454 cm^{-1}) [24,25]. From the well-known back-donation of fluorine 2p electron density to the positive carbon, it is readily apparent that CF$_3$$^+$ should exhibit extensive pi bonding, and the markedly increased ν_3 and ν_1 modes are consistent with this bonding model.

Photolysis behavior for the 1665 cm^{-1} CF$_3$$^+$ absorption in CHF$_3$, CF$_3$Cl, CF$_3$Br and CF$_3$I experiments provides evidence for the halide counterion in these studies [19]. Since dissociation of CF$_3$$^+$ to CF$_2$$^+$ + F requires at least 5 eV, which is above the mercury arc range, a photoneutralization mechanism is required for the slight decrease of CF$_3$$^+$ on mercury arc photolysis. The relative decrease in the CF$_3$$^+$ band on 290-1000 and 220-1000 nm photolysis in these experiments is more pronounced in the order I$^-$>Br$^-$>Cl$^-$>F$^-$ where these halide ions are most likely counterions in photoionization studies with CF$_3$I, CF$_3$Br, CF$_3$Cl and CHF$_3$, respectively. This photolysis behavior parallels the expected photodetachment cross section in the halide series and supports the photoneutralization model for the photolysis of CF$_3$$^+$ isolated in a matrix containing halide counteranions [2].

4. Bifluoride and Trifluoride Anions

The bifluoride ion is the best documented example of a very strong and symmetrical hydrogen bond [26,27]. Bifluoride anion has been characterized in crystalline solids [28], and as Cs$^+$HF$_2$$^-$ ion pairs in solid argon [29]. In order to examine the HF$_2$$^-$ ion free of crystal lattice and counterion effects, the isolated HF$_2$$^-$ ion was prepared by vacuum-ultraviolet irradiation of Ar/HF samples, and a strong, sharp product band was observed at 1377 cm^{-1} [30]. The analogous Ar/DF experiments gave a new 965 cm^{-1} band and the strong Ar$_n$D$^+$ band 644 cm^{-1} [31,32]. The observed HF/DF = 1377/965 = 1.427 ratio substantiated the ν_3 assignment to a linear centrosymmetric species that can have only quartic anharmonic terms in its vibrational potential function, in agreement with the solid and ion-pair studies. In order to obtain new results on polyfluoride anion systems, Ar/HF samples have been subjected to low energy electron impact during condensation at 12K using the apparatus shown in Figure 2.

The spectrum of an Ar/HF = 100/1 sample subjected to irradiation with 10 µA of electrons at 150 eV for 4h during condensation at 12K is shown in Figure 4 [33]. This sample contained strong HF cluster bands, which have been described earlier [34], and strong new product bands at

1815 and 1377 cm^{-1} labeled **T** and **B,** respectively. Sample annealing decreased **B** more than **T** and produced a broad new band at 1715 cm^{-1}. Decreasing the HF concentration decreased **T** relative to **B.** Experiments with 40%, 75% and 90% deuterium enriched hydrogen fluoride produced a 965 cm^{-1} counterpart for the 1377 cm^{-1} band and two counterparts for the 1815 cm^{-1} band, the first appearing at 1707 cm^{-1} with partial deuterium enrichment, and the second following at 1391 cm^{-1} with high deuterium enrichment.

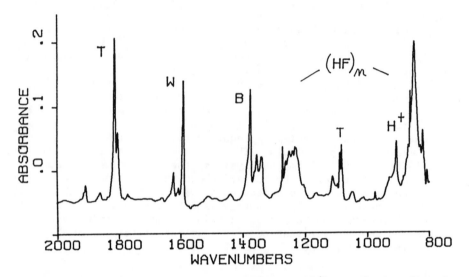

Figure 4. Infrared spectrum of Ar/HF = 100/1 sample irradiated with electrons during condensation at 12K.

Assignment of the 1377 cm^{-1} band to ν_3 of HF_2^- is reaffirmed based on the high H/D = 1377/965 = 1.427 ratio, which is higher than the 1.396 harmonic value and can arise for a linear, centrosymmetric species <u>2</u> owing to the absence of cubic and the presence of quartic

$$(\text{F--H--F})^- \qquad \underline{2}$$

$$\underline{3}$$

anharmonic terms in the vibrational potential. The results of recent high quality <u>ab initio</u> calculations [35] and diode laser studies [36] also substantiate this assignment.

The 1815 cm^{-1} band is assigned to the out-of-phase H-F stretching fundamental of $H_2F_3^-$ based on agreement with solid state spectra [37,38], the concentration dependence relative to HF_2^-, the H/D = 1815/1391 = 1.305 ratio, which is appropriate for a $\bar{C}_{2v}$ species <u>3</u> whose anharmonicity is dominated by cubic terms, and the appearance of a mixed H/D band at 1715 cm^{-1}. The $H_2F_3^-$ anion is of interest as it contains the second strongest hydrogen bond based on displacement of the H-F fundamental in (FHFDF)$^-$ to 1707 cm^{-1}; HF$_2$$^-$ contains the strongest hydrogen bond with the average H-F stretching fundamental at 1016 cm^{-1} [28]. These values characterize stronger hydrogen bonds than other strong complexes such as H$_a$F at 1870 cm^{-1} in (CH$_3$)$_3$N--H$_a$--F--H$_b$-F [38] and HF at 3041 cm^{-1} in NH$_3$--HF [39].

The broad 1715 cm^{-1} band is in excellent agreement with solid state observation for K$^+$H$_3$F$_4$$^-$ [38], and it is assigned to the isolated H$_3$F$_4$$^-$ anion in solid argon. This higher polyfluoride anion is formed at the expense of HF$_2$$^-$ and H$_2$F$_3$$^-$ on diffusion and further association of HF in the matrix.

Finally, charge balance in these experiments is maintained by the matrix solvated proton species Ar$_n$H$^+$ [31,32], which is probably formed on ionization of atomic hydrogen liberated in the dissociative capture of low energy electrons by HF molecules.

$$HF + low\ energy\ electrons \rightarrow H + F^-$$

$$nAr + H + low\ energy\ electrons \rightarrow Ar_n H^+ + 2e$$

$$HF + F^- \rightarrow HF_2^-$$

$$HF_2^- + HF \rightarrow H_2 F_3^-$$

$$H_2 F_3^- + HF \rightarrow H_3 F_4^-$$

The fluoride anion has an exceptionally high proton affinity, and it readily clusters with one, two or more HF submolecules in these matrix isolation experiments, as is found in solid polyfluorides.

5. Naphthalene Cation

The naphthalene cation is a particularly interesting case for matrix study since a wealth of data are available from photoelectron spectroscopy [40,41], absorption spectra in glassy matrices [42], and multiphoton dissociation spectra [43]. Argon discharge photo-ionization of naphthalene (hereafter N) vapor during condensation with excess argon produced new absorption band systems at 675.2 nm (A = absorbance = 0.17), 461 nm (A = 0.01) and 381 nm (A = 0.03). The spectrum of naphthalene cation from matrix two-photon ionization of naphthalene is illustrated in Figure 5: trace (a) shows the spectrum of an Ar/CCl$_4$/N = 3000/4/1 sample deposited at 20K, trace (b) illustrates the spectrum after photolysis for 15 sec with 220-1000 nm high pressure mercury arc radiation. The above absorption bands were observed with a 3-fold intensity increase over the one-photon production method, and two strong new systems were produced at 307.6 nm (A = 0.78) and 274.6 nm (A = 0.9) at the expense of N precursor absorption systems beginning at 313.3 and 272 nm, which were reduced by

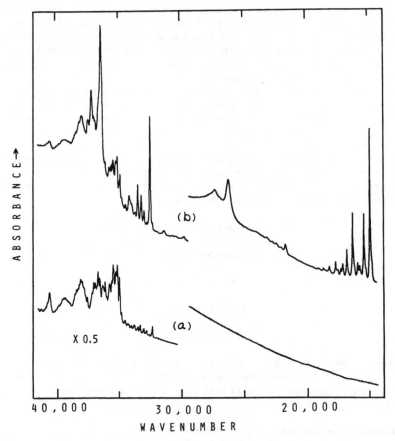

Figure 5. Visible and ultraviolet spectra of naphthalene samples in solid argon at 20K. (a) Ar/CCl$_4$/N = 3000/4/1, (b) same sample after 220-1000 nm irradiation for 15 sec.

4

70% [12,44]. The five band systems were assigned to the naphthalene radical cation (N$^+$) 4 for the following reasons:

(a) Three of these absorption band systems were observed with two different techniques capable of forming positive ions. Argon resonance radiation ionizes N molecules during matrix condensation and the N$^+$ product is trapped in solid argon; in these studies the electron removed in ionization is probably trapped by a molecular fragment or another parent molecule. Intense mercury arc radiation performed two-color resonance photoionization of N molecules isolated in solid argon with the relatively long lived S_1 state serving as an intermediate step in the photoionization process. In the latter experiments CCl_4 was added as an electron trap, and positive ion yields increased by a factor of 3 over single photon ionization experiments, giving further support to the product identification as a positive ion. The two highest energy absorption systems were obscured by precursor absorption in the first experiments, but the latter studies used one-tenth as much N, and the N^+ absorptions were produced at the expense of N bands (Figure 5) [12].

(b) The two visible band origin positions are in excellent agreement with photoelectron spectra (PES). The red absorption provides an excellent basis for comparison between PES and the sharp matrix absorption spectrum, since the difference between sharp vertical PES band origins can be measured to ± 0.02 eV. Sharp vertical ionization energies have been observed for the first (8.11 eV) and third (9.96 eV) ionic states of N^+. The difference between the first and third PES band origins is 1.85 ± 0.02 eV, which predicts an absorption band origin at 14 920 ± 160 cm^{-1} in the gas phase, assuming no change in structure between the neutral molecule and the two ionic states. The argon matrix origin at 14 810 ± 3 cm^{-1} is in agreement within measurement error, which confirms the identification of N^+ and indicates very little perturbation of N^+ by the argon matrix. The blue absorption at 21 697 ± 10 cm^{-1} is also in excellent agreement with the 21 859 ± 160 cm^{-1} difference between sharp origins of the first and fourth PES bands. The three ultraviolet bands are attributed to electron-promotion transitions with upper states not reached by PES, hence a comparison cannot be made.

(c) The first four transitions are in excellent agreement with 690, 467, 387 and 308 nm absorptions assigned to N^+ produced by radiolysis of N in a Freon glass at 77K [42]. The small red-shifts (42-400 cm^{-1}) are due to greater interaction with the more polarizable Freon medium. Vibronic structure is similar in the solid argon and solid Freon samples; the bands are sharper in solid argon which facilitates more accurate vibronic measurements.

(d) The sharp UV band origins at 32 510 and 36 417 cm^{-1} are in agreement with the broad 4-photon dissociation bands at 33 200 and 38 200 cm^{-1} for N^+; the agreement would probably be better if origins were resolved for the broad PDS bands [43].

(e) The vibronic structure in the absorption systems, particularly the red band, and the $N-d_8^+$ counterpart spacings correspond closely with N and $N-d_8$ vibrational intervals [45]. The small blue shifts in the origin bands upon deuteriation range from 75 to 145 cm^{-1}; the weak S_1 origin of N at 313.2 nm exhibits a similar 102 cm^{-1} blue shift upon deuteriation.

Two regularly repeating vibrational intervals were observed in the red N^+ absorptions. The first interval, 1422 ± 6 cm^{-1}, was unchanged upon deuterium substitution. This is in agreement with the 1420 ± 40 cm^{-1} interval in the first PES band, which is probably due to the C(9)-C(10) stretching mode $\nu_4(a_g)$. The strong fundamental at 1376 cm^{-1} in the Raman spectrum of $C_{10}H_8$ is complicated by Fermi resonance, but the strong fundamental for $C_{10}D_8$ is at 1380 cm^{-1} [44]. The second repeated interval in the spectra was approximately 505 cm^{-1} for N-h_8^+ and 485 cm^{-1} for N-d_8^+; this corresponds to the $\nu_9(a_g)$ skeletal distortion observed in Raman spectra for N-h_8 at 512 cm^{-1} and N-d_8 at 491 cm^{-1}. The 754 cm^{-1} interval corresponds to $\nu_8(a_g)$, the skeletal breathing mode observed at 758 cm^{-1} in the Raman spectrum of N. These observations show that ionization has a relatively small effect on the vibrational potential function of the large N molecule.

Matrix photoionization experiments establish a dynamic equilibrium between photoionization of a molecule M to produce M$^+$ and photodetachment of a trap T$^-$, which results in the neutralization of M$^+$. Here we depend on the relative cross-sections of these competing photo-processes. Clearly, when a good trapping molecule like CH_2Cl_2 or CCl_4 is added to the sample, the Cl$^-$ trap is more difficult to

photoionization

$$M + T \xrightarrow{} M^+ + T^-$$

photodetachment

photodetach than most molecular anions, so the yield of trapped cations M$^+$ is increased. For the 2-photon ionization method to be successful, cations must be formed more efficiently than traps are photodetached, which is clearly the case in the naphthalene system. Substituted naphthalene, condensed ring aromatic, and alkyl benzene cations have also been observed using this method of production [44,46-49].

6.   Toluene and Methylenecyclohexadiene Cations

One of the most interesting rearrangements observed in argon matrices involved alkylbenzene cations in a 1,3-hydride shift. The 2-photon ionization method described above for naphthalene has also been employed for toluene [48]. The spectrum of an Ar/CH_2Cl_2/T = 1000/2/1 sample after deposition for 4h revealed only toluene absorptions. Irradiation at 220-1000 nm for 30 sec produced broad bands at 36 200 and 23 800 cm^{-1}; longer irradiation increased the broad bands and produced sharp bands at 32 300 cm^{-1} due to benzyl radical and at 23 220 and 15 640 cm^{-1} due to a new species. A final photolysis at 290-1000 nm, which could not excite toluene, failed to increase any product absorptions and reduced the latter 2 bands by 60% [48].

Argon resonance photoionization of cycloheptatriene samples gave a strong 20 800 cm^{-1} absorption and a sharp weak 23 220 cm^{-1} band [48]. Prolonged photolysis at 470-1000 nm markedly reduced the 20 800 cm^{-1} and increased the 23 220 cm^{-1} band and an accompanying 15 640 cm^{-1} band. Irradiation at 370-460 nm reduced the latter bands by half and increased the 20 800 cm^{-1} band by 50%. A final 240-420 nm photolysis reduced all of these bands.

The 20 800 cm^{-1} band was assigned to cycloheptatriene cation, 6, in solid argon produced by 1-photon ionization [48] based on an ICR photodissociation band at 21 300 cm^{-1} [50] and a Freon glass absorption at 20 700 cm^{-1} [51]. The broad 23 800 and 36 200 cm^{-1} absorptions produced by photolysis of toluene samples, which required electron trapping molecules, were assigned to the toluene radical cation, 5, produced by sequential 2-photon ionization in solid argon [48]. These bands are in excellent agareement with gas-phase photodissociation bands of toluene cation at 24 000 and 37 400 cm^{-1} [52].

The sharp band origins at 23 220 and 15 640 cm^{-1} were assigned to the 5-methylenecyclohexadiene radical cation, 7, in solid argon based on their observation from toluene, cycloheptatriene, norbornadiene, qudricyclane, and 1-phenyl-2-butene precursors [48]. Appearance of these bands near two characteristic triene cation absorptions [51,53] confirms their assignment to a conjugated triene cation system. Support for the 7 identification has been provided by its formation from X-irradiation of 5-methylenebicyclo[2.2.0] hex-2-ene in solid argon [54].

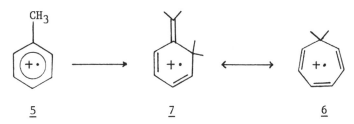

5   7   6

The photochemical rearrangements among C$_7$H$_8^+$ radical cation isomers is of particular interest here. The methylenecyclohexadiene cation, 7, was formed by rearrangement of both toluene cation, 5, and cycloheptatriene cation, 6. The former rearrangement involves a 1,3-hydrogen shift, and the latter may involve the norcaradiene cation intermediate. Similar rearrangements have been observed for alkyl substituted cation species [49,55]. Irradiation in the visible parent cation absorption initiated α-H transfer to the cation ring to give substituted methylenecyclohexadiene cations. Subsequent ultraviolet photolysis of these samples produced substituted styrene cations [49].

The 6 ↔ 7 cation rearrangement is a reversible process although the 6 → 7 direction is favored in these experiments. This dynamic rearrangement is initiated by light absorption of one structure, which activates the rearrangement; the other structure formed by rearrangement is deactivated and trapped by the cold matrix in the absence of radiation that it can absorb. The photochemical stability for the methylenecyclohexadiene cation in these experiments suggests that 7 may be important in gaseous C$_7$H$_8^+$ cation rearrangements. Analogous reversible rearrangements have been observed among octatetraene and decapentaene radical cation isomers using selective excitation by visible laser line photolysis [56]. Finally, orbital

68

symmetry conservation and frontier orbital control in these and other
organic radical cation rearrangements have been discussed by Dunkin and
Andrews [57].

7.   Nitrogen Oxide Anions

A Townsend discharge chemical ionization source has been used for
matrix infrared studies of anions in the argon-nitrous oxide system.
In addition to the neutral oxide products $(NO)_2$, N_2O_3, N_2O_4 and O_3, the
NO^-, NO_2^-, $(NO)_2^-$, NNO_2^- and O_3^- anions have also been characterized
[58]. These matrix isolation experiments provide structural
information on two different $N_2O_2^-$ anions observed in mass spectroscopy
experiments [59].

Infrared spectra in the anion region for a 2% N_2O-argon sample are
shown in Figure 6. The major product was $(NO)_2$ with a substantial
yield of NO and both symmetric and asymmetric forms of N_2O_3 owing to
atom-molecule reactions in the discharge region. In addition, new

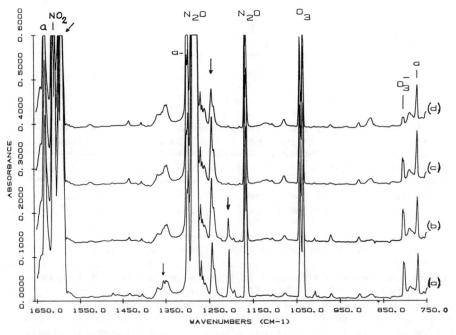

Figure 6. Infrared spectra of Ar/N_2O = 50/1 sample subjected to
chemical ionization discharge before condensation at 12K (a) and
(b) after photolysis at λ > 420 nm, (c) λ > 340 nm and (d) λ > 220
nm for 30 min periods. Arrows denote anion products, and **a**
denotes asymmetric-N_2O_3.

bands were observed at 1593, 1355, 1244, 1205 and 803 cm^{-1}, which showed different photolysis behavior (Figure 6). For $\lambda > 420$ nm, the major changes were small decreases in 1593 and 803 cm^{-1} absorptions, large decreases in the 1355 and 1205 cm^{-1} bands, and a small <u>increase</u> in the 1244 cm^{-1} band. Using $\lambda > 340$ nm, these trends continued, and the 1355 and 1205 cm^{-1} bands were virtually destroyed. For $\lambda > 220$ nm, the 803 cm^{-1} band and O_3 were reduced as expected, and the 1244 and 1593 cm^{-1} bands were markedly decreased.

Isotopic spectra provided the critical information. First, a $^{15}N_2O$ sample shifted the major anion product bands to 1573, 1218 and 1181 cm^{-1}, and a mixed O-16/O-18 sample gave triplet patterns for each band showing 2 equivalent oxygen atoms in each absorbing species.

Complementary experiments were done with NO as the reagent gas; the major difference was the very weak 1205 cm^{-1} product with a small amount of N_2O present (now as a discharge product) with a strong 1590 cm^{-1} band. In order to compare samples prepared by electron impact, a sample like that shown in Figure 6 was irradiated by electrons during deposition using the apparatus shown in Figure 2. The strongest product absorptions were 1593 and 1205 cm^{-1}; neutral products, particularly $(NO)_2$, were much weaker than in chemical ionization discharge experiments.

Ionization processes for N_2O are known from mass spectrometric studies; the major product is O^- and a minor product is NO^- [60]. The 1355 cm^{-1} band is assigned to NO^- based on agreement with gas phase and alkali metal atom matrix studies [61-63]. The other 3 products are due to O^- reactions. The 803 cm^{-1} band is due to the isolated O_3^- anion based on earlier studies [64,65]; O_3^- is presumably formed here by reaction of O^- with O_2 byproduct.

The 1244 cm^{-1} band is due to the isolated NO_2^- anion as identified by Milligan and Jacox [62]; the isotopic shifts are in agreement with this assignment to a species with 2 equivalent oxygen atoms and a 115 $\pm$ 5° valence angle. The growth of NO_2^- throughout the photolysis sequence with $\lambda > 320$ nm comes from capture by NO_2 of electrons photodetached from other anions, which attests the stability of NO_2^- and the high electron affinity of NO_2.

The most photosensitive band at 1205 cm^{-1} is also due to a species with 2 equivalent oxygen atoms and a 115 $\pm$ 5° valence angle. An earlier study produced an absorption at 1205 cm^{-1} by irradiation of an argon matrix containing N_2O and alkali metal atoms and tentatively assigned this band to the NNO_2^- anion [62]. The contrasting yields of the 1205 cm^{-1} band in N_2O and NO experiments show that N_2O is required and further substantiate the NNO_2 stoichiometry [58]. The decrease of O_3^- and 1205 cm^{-1} absorption upon irradiation at 420 nm suggests a similar photodissociation process, namely O^- elimination in the visible region [66]. Finally, the $M^+NNO_2^-$ species [62] was produced by photolysis that destroyed the isolated NNO_2^- anion, <u>8</u>, demonstrating the stabilizing effect of the coulombically bound counterion.

The sharp new product band at 1593 cm^{-1} exhibited almost identical isotopic ratios as NO and $(NO)_2$ [58]. These ratios characterize the vibration of NO diatomic oscillators, which is distinctly different from the absorption of NO_2 in the same region. The mixed oxygen isotopic pattern is a triplet with double statistical weight for the central component, which indicates 2 equivalent oxygen atoms, and hence 2 equivalent N-O diatomic oscillators, participate in this vibration. The photolysis behavior further characterizes the 1593 cm^{-1} absorber as an anion, but an anion more stable than NNO_2^-. The above evidence indicates that the 1593 cm^{-1} band is due to $(NO)_2^-$ presumably with the trans structure, 9. The structural isomers NNO_2^- and $(NO)_2^-$ observed

in mass spectrometry experiments [59] are made by reaction of O^- with the inside and outside nitrogen atoms, respectively, of N_2O. The chemical ionization and electron impact discharge techniques provide sufficient molecular anions for infrared spectroscopic study and characterization from isotopic shifts.

8. Acknowledgments

The author gratefully acknowledges financial support from the U.S. National Science Foundation for the research described in this chapter and the valuable contributions of associates whose work is described in more detail in many of the following references.

References

1. M.E. Jacox, Rev. Chem. Intermed. 2, 1 (1978).
2. L. Andrews, Ann. Rev. Phys. Chem. 30, 79 (1979).
3. L. Andrews, "Absorption Spectroscopy of Molecular Ions in Noble Gas Matrices" in **Molecular Ions**, T.A. Miller and V.E. Bondybey, eds., North-Holland Physics, Amsterdam, 1983.
4. L. Andrews, "Absorption Spectroscopy of Molecular Ions and Complexes in Noble Gas Matrices" in **Chemistry and Physics of Matrix Isolated Species**, L. Andrews and M. Moskovits, eds., North-Holland Physics, Amsterdam, 1989.
5. L. Andrews, J. Am. Chem. Soc. 90, 7368 (1968).
6. L. Andrews, J. Chem. Phys. 50, 4288 (1969).
7. L. Andrews and R.R. Smardzewski, J. Chem. Phys. 58, 2258 (1973).
8. P.H. Kasai and D. McLeod, Jr., J. Chem. Phys. 51, 1250 (1969).
9. D.E. Milligan and M.E. Jacox, J. Chem. Phys. 51, 1952 (1969).
10. F.T. Prochaska and L. Andrews, J. Chem. Phys. 67, 1091 (1977).
11. L. Andrews, J.M. Dyke, N. Jonathan, N. Keddar and A. Morris, J. Am. Chem. Soc. 106, 299 (1984).
12. B.J. Kelsall and L. Andrews, J. Chem. Phys. 76, 5005 (1982).

13. S. Suzer and L. Andrews, J. Chem. Phys. **88**, 916 (1988).
14. J. Hacaloglu and L. Andrews, Chem. Phys. Letts. **160**, 274 (1989).
15. M.E. Jacox and D.E. Milligan, J. Chem. Phys. **54**, 3935 (1971).
16. L. Andrews, J.M. Grzybowski and R.O. Allen, J. Phys. Chem. **79**, 904 (1975).
17. F.T. Prochaska and L. Andrews, J. Chem. Phys. **68**, 5568 (1978).
18. F.T. Prochaska and L. Andrews, J. Chem. Phys. **68**, 5577 (1978).
19. F.T. Prochaska and L. Andrews, J. Am. Chem. Soc. **100**, 2102 (1978).
20. B.J. Kelsall and L. Andrews, J. Phys. Chem. **85**, 2938 (1981).
21. L. Andrews and R.D. Hunt, unpublished results (1987).
22. M.E. Jacox and W.E. Thompson, J. Chem. Phys. **91**, 1410 (1989).
23. F.P. Lossing, Bull. Soc. Chim. Belges **81**, 125 (1972).
24. D.A. Dows, J. Chem. Phys. **31**, 1637 (1959).
25. D.E. Milligan and M.E. Jacox, J. Chem. Phys. **48**, 2265 (1968).
26. J.A. Ibers, J. Chem. Phys. **41**, 25 (1964).
27. J.W. Larson and T.B. McMahon, J. Am. Chem. Soc. **104**, 5848 (1982).
28. P. Dawson, M.M. Hargreave and G.R. Wilkinson, Spectrochim. Acta. **31A**, 1055 (1975).
29. B.S. Ault, J. Phys. Chem. **82**, 844 (1978).
30. S.A. McDonald and L. Andrews, J. Chem. Phys. **70**, 3134 (1979).
31. D.E. Milligan and M.E. Jacox, J. Mol. Spectrosc. **46**, 460 (1973).
32. C.A. Wight, B.S. Ault and L. Andrews, J. Chem. Phys. **65**, 1244 (1976).
33. R.D. Hunt and L. Andrews, J. Chem. Phys. **87**, 6819 (1987).
34. L. Andrews and G.L. Johnson, J. Phys. Chem. **88**, 425 (1984).
35. C.L. Janssen, W.D. Allen, H.F. Schaeffer III and J.M. Bowman, Chem. Phys. Letts. **131**, 351 (1987).
36. K. Kawaguchi and E. Hirota, J. Chem. Phys. **87**, 6838 (1987).
37. A. Azman, A. Ocvirk, D. Hadzi, P.A. Giguere and M. Schneider, Can. J. Chem. **45**, 1347 (1967).
38. L. Andrews, S.R. Davis and G.L. Johnson, J. Phys. Chem. **90**, 4273 (1986).
39. G.L. Johnson and L. Andrews, J. Am. Chem. Soc. **104**, 3043 (1982).
40. J.H.D. Eland and C.J. Danby, Z. Naturforsch., **23a**, 355 (1968).
41. P.A. Clark, F. Brogli and E. Heilbronner, Helv. Chem. Acta **55**, 1415 (1972).
42. T. Shida and S. Iwato, J. Am. Chem Soc. **95**, 3473 (1973).
43. M.S. Kim and R.C. Dunbar, J. Chem. Phys. **72**, 4405 (1980).
44. L. Andrews, B.J. Kelsall and T.A. Blankenship, J. Phys. Chem. **86**, 2916 (1982).
45. S.S. Mitra and H.J. Bernstein, Can. J. Chem. **37**, 553 (1959).
46. L. Andrews, R.S. Friedman and B.J. Kelsall, J. Phys. Chem. **89**, 4016 (1985).
47. L. Andrews, R.S. Friedman and B.J. Kelsall, J. Phys. Chem. **89**, 4550 (1985).
48. B.J. Kelsall and L. Andrews, J. Am. Chem. Soc. **105**, 1413 (1983).
49. B.J. Kelsall and L. Andrews, J. Phys. Chem. **88**, 5893 (1984).
50. R.C. Dunbar and E.W. Fu, J. Am. Chem. Soc. **95**, 2716 (1973).
51. T. Shida, T. Kato and Y. Nosaka, J. Phys. Chem. **81**, 1095 (1977).
52. H.H. Teng and R.C. Dunbar, J. Chem. Phys. **68**, 3133 (1978).

53. B.J. Kelsall and L. Andrews, J. Phys. Chem. **88**, 2723 (1984).
54. T. Bally, D. Hasselmann and K. Loosen, Helv. Chim. Acta **68**, 345 (1985).
55. B.J. Kelsall, L. Andrews and G.J. McGarvey, J. Phys. Chem. **87**, 1788 (1983).
56. L. Andrews and J.T. Lurito, Tetrahedron **42**, 6343 (1986).
57. I.R. Dunkin and L. Andrews, Tetrahedron **41**, 145 (1985).
58. J. Hacaloglu, S. Suzer and L. Andrews, J. Phys. Chem. **94**, XXXX (1990).
59. L.A. Posey and M.A. Johnson, J. Chem. Phys. **88**, 5383 (1988).
60. P.J. Chantry, J. Chem. Phys. **51**, 3369 (1969).
61. D. Spence and G.J. Schultz, Phys. Rev. A. **3**, 1968 (1971).
62. D.E. Milligan and M.E. Jacox, J. Chem. Phys. **55**, 3404 (1971).
63. D.E. Tevault and L. Andrews, J. Phys. Chem. **77**, 1646 (1973).
64. R.C. Spiker, Jr. and L. Andrews, J. Chem. Phys. **59**, 1851 (1973).
65. L. Andrews, B.S. Ault, J.M. Grzybowski and R.O. Allen, J. Chem. Phys. **62**, 2461 (1975).
66. S.E. Novick, P.C. Engelking, P.L. Jones, J.H. Futrell and W.C. Lineberger, J. Chem. Phys. **70**, 2652 (1979).

GENERATION AND STUDY OF INORGANIC CATIONS IN RARE GAS MATRICES BY ELECTRON SPIN RESONANCE

LON B. KNIGHT, JR,
Furman University
Department of Chemistry
Greenville, South Carolina 29613
USA

1. Introduction

A comprehensive treatment (through 1989) of electron spin resonance (ESR) results for inorganic cation radicals isolated in rare gas matrices will be presented in this chapter. The most recent method used to generate cations for rare gas trapping experiments will be described, namely, direct ion trapping of plasma generated by pulsed laser bombardment. [1,2] This highly versatile approach can be used alone to produce matrix spectra of isolated ions or in combination with other ionization methods during the deposition process. Examples of small inorganic cations studied by ESR in other solid hosts at low temperatures will be listed but no comprehensive treatment will be attempted in the space available for this much broader field of research which has been previously reviewed. [3-7] Several reviews of rare gas matrix ESR studies for charged and neutral radicals should be consulted for background information. [8-15] Different chapters of this monograph will discuss anion and organic cation radicals studied in the condensed phase by ESR and other spectroscopic methods. A brief section listing all gas phase cations whose magnetic hyperfine interactions have been measured will be presented for comparison with the rare gas matrix values.

The demonstrated ability of neon matrices to provide gas-like magnetic parameters and well resolved ESR spectra of small inorganic cation radicals has been the motivation in our research for interfacing this trapping method with a variety of high energy generation techniques. [9,10] Given the rich radical chemistry present in high energy environments, one could view this method as an analytical probe for characterizing and identifying the various ion and neutral intermediates that might be present. Under laser bombardment, the ~10 nsec energy pulse focused onto the sample target produces an extreme contrast in experimental conditions. The matrix deposition surface at 4 K is located only a few cm's from a spot on the sample target whose temperature suddenly increases to approximately 30,000-50,000 K. Under these trapping conditions, we have observed intense ESR signals for isolated ions and neutral

A. Lund and M. Shiotani (eds.), Radical Ionic Systems, 73–97.

radicals from a single laser pulse. Such plasma sources offer the advantage of considerably greater ion densities relative to conventional ion beams, however the ability to mass select or discriminate appears to be more feasible with conventional ion beams. [16-18]

Spectroscopic information on highly reactive ion molecules is important in monitoring and understanding the overall chemistry in numerous types of high energy environments which include: plasma sources, ion deposition and implantation processes, ion-molecule reactions, flames, upper atmospheric chemistry, electrical discharges, astrophysical problems and surface vaporization phenomena. Such matrix measurements can also assist gas phase spectroscopists in defining their searches and aid the identification of new interstellar ions and neutrals.

1.1.   TYPES OF CATIONS STUDIED

The diversity of molecular cation types that have been studied by ESR in rare gas matrices near 4 K using various generation methods include: small cluster ions (N_4^+, Cu_2^+, Si_2^+), high spin semi-conductor ions ($GaAs^+$), ion-neutral reaction products (N_2CO^+, $P_2H_6^+$), metal hydrides (CuH^+, PdH_2^+), small organic ions (CH_4^+, $C_8H_8^+$), metal carbonyls ($Cr(CO)_4^+$), and several other small cations of fundamental significance such as SiH_2^+ and C_2^+. A comprehensive listing (through 1989) of all cations studied by rare gas matrix ESR is presented for diatomics in Table I and for polyatomics in Table II.

As recently as 1983, no inorganic molecular cations had been studied by ESR in rare gas matrices and only a few such cases had been reported in other condensed solids. [3] Other cryogenic matrix materials employed for cation radical studies include halocarbons, SF_6, ionic salts, single crystals, frozen acid solutions and zeolites, just to name a few. Table 3 gives a partial listing of small inorganic cations studied by ESR in a variety of solid hosts. See specific references cited in the Table and previous reviews of these studies which have provided much valuable information. [3-7] Hydrocarbon matrices have been used successfully for numerous neutral radical ESR studies, and for a few organic cations in recent studies. [59]

2.   Importance of Nuclear Hyperfine Information

2.1.   RELATION TO THEORY

Nuclear hyperfine interactions (A tensors) provide the most direct source of experimental information for describing the atomic orbital composition of the least tightly bound electrons in molecular

Table 1. Diatomic Cation Radicals Investigated by ESR In Rare Gas Matrices

Molecular Ion	Ground State	Matrix	Nuclear hfs	Reference
C_2^+	$^4\Sigma$	Neon	^{13}C	19
CO^+	$^2\Sigma$	Neon	^{13}C; ^{17}O	20
BF^+	$^2\Sigma$	Neon	$^{10,11}B$; ^{19}F	21
N_2^+	$^2\Sigma$	Neon	^{14}N; ^{15}N	22
AlH^+	$^2\Sigma$	Argon	H; D; ^{27}Al	1
Mg_2^+	$^2\Sigma$	Argon	^{25}Mg	23
AlF^+	$^2\Sigma$	Neon	^{19}F; ^{17}Al	24
SiO^+	$^2\Sigma$	Neon	^{29}Si	25
Mn_2^+	$^{12}\Sigma$	Argon	^{55}Mn	26
CuH^+	$^2\Sigma$	Argon	H; D; $^{63,65}Cu$	23
Cu_2^+	$^2\Sigma$	Neon	$^{63,65}Cu$	27
Ag_2^+	$^2\Sigma$	Neon	$^{107,109}Ag$	27
Au_2^+	$^2\Sigma$	Neon	^{197}Au	27
GaP^+	$^4\Sigma$	Neon	$^{69,71}Ga$; ^{31}P	2
$GaAs^+$	$^4\Sigma$	Neon	$^{69,71}Ga$	28
Si_2^+	$^4\Sigma$	Neon	^{29}Si	23

radicals. A description of the highest occupied MO in terms of simple hydrogenic wavefunctions continues to be an extremely important first step in understanding the electronic structure of a small molecule in the chemically important valence region.

Since most neutral molecules are not radicals, their cations contain at least one unpaired electron and hence they become viable candidates for ESR investigation which can usually provide the important nuclear hyperfine parameters. Hence, generally applicable experimental methods for obtaining well resolved ESR spectra for a wide variety of radical cations would provide a wealth of new electronic structure information on highly reactive molecules and important chemical intermediates. Hyperfine data on small open shell ions also provide valuable information for testing the accuracy of various computational methods, thus assisting in the application of

Table 2. Polyatomic Cation Radicals Investigated by ESR in Rare
Gas Matrices

Molecular Ion	Matrix	Nuclear hfs	Reference
H_2O^+	Neon	H; D; ^{17}O	29
CH_2^+	Neon	H; ^{13}C	30
$Mg^+{}_{3-6}$	Argon	^{25}Mg	23
SiH_2^+	Neon	H; ^{29}Si	31
PdH_2^+	Neon	H; D; ^{105}Pd	32
NH_3^+	Neon; Argon	H; ^{14}N; ^{15}N	20
N_4^+	Neon	^{14}N; ^{15}N	33
O_4^+	Neon	^{17}O	34
N_2CO^+	Neon	^{13}C; ^{14}N; ^{15}N	35
$C_2O_2^+$	Neon	^{13}C; ^{17}O	36
H_2CO^+	Neon	H; ^{13}C	37
F_2CO^+	Neon	^{13}C; ^{19}F	38
CH_4^+	Neon	H; D; ^{13}C	39
CH_3CHO^+	Neon	H; D; ^{13}C	40
$P_2H_6^+$	Neon	H; ^{31}P	23
$C_8H_8^+$	Neon	H	41
$Cr(CO)_4^+$	Kr	^{13}C; ^{53}Cr	42
$Fe_2(CO)^+{}_{6,8}$	Kr	^{13}C; ^{61}Ni	43
$KrFe(CO)_5^+$	Kr	^{13}C; ^{57}Fe; ^{83}Kr	44

theory to larger chemical systems. For example, understanding and
calculating the important solid state properties of gallium arsenide
might be assisted by first applying theory to the electronic
distributions of the three unpaired electrons in $GaAs^+(X^4\sum)$ as
revealed by the Ga and As hyperfine interactions.

2.2.   IDENTIFICATION

An extremely valuable aspect of nuclear hyperfine structure,
especially for small radicals, is its ability to assist the

Table 3. Selected Examples of Inorganic Cations Studied in Various
          Solids by ESR

Molecular Ion	Matrix	Nuclear hfs	Reference
Ag_n^+	4A Zeolite	$^{107,109}Ag$	45
AgH^+	Y Zeolite: CH_3OH	H; $^{107,109}Ag$	46
Xe_2^+	SbF_5:Xe	$^{129,131}Xe$	47
PH_3^+	H_2SO_4	^{31}P	48
NH_3^+	Solid Crystals	H; ^{14}N	49
P_4^+	$CFCl_3$	^{31}P	50
SnH_4^+	$CFCl_3$	H; $^{117,119}Sn$	51
CCl_4^+	CCl_4	$^{35,37}Cl$	52
S_8^+	$CFCl_3$	^{33}S	53
$S_4N_4^+$	$CFCl_3$	--	53
$N_2O_4^+$	$CFCl_3$;SF_6	^{14}N	54
$S_3N_2^+$	D_2SO_4	$^{14,15}N$;^{33}S	55
$Se_3N_2^+$	D_2SO_4	^{14}N;^{77}Se	56
$N_2H_4^+$	$N_2H_5HC_2O_4$	H;^{14}N	57
F_3N^+	NF_4AsF_6	$^{14,15}N$; ^{19}F	58

identification process. This is usually a most difficult task when
trapping a mixture of different radical intermediates from a complex
high energy environment. Isotopic confirmation is usually a
necessity and a most valuable feature of the laser vaporization
approach is the practicality of conducting isotopic substitution with
small sample sizes. The hyperfine structure provides an internally
consistent recognition pattern that is independent of the absolute

absorption positions, thus making the identification process somewhat more direct than electronic and vibrational measurements on samples that possibly contain several different species. This advantage also applies in the identification of interstellar species.

3.   Gas Phase Hyperfine Measurements of Cations

Given the extreme reactivity and short lifetimes of open shell charged molecules, it is not usually possible to maintain sufficient gas phase concentrations for making experimental hyperfine measurements. Resolution requirements and the density of states populated in the gas phase also make such measurements and their analysis extremely difficult for most radical ions. In matrix samples, the absence of rotational levels greatly simplifies the hyperfine interpretation, but of course, other information is not obtained.

The number of inorganic and organic radical ions studied in the gas phase with sufficient resolution to obtain A tensor information is quite small - approximately nine diatomics, with H_2O^+ being the only polyatomic example, and CO^+, the only carbon containing species. A comprehensive list of these cation studies through 1989 is presented in Table 4. Experimental advances and several excellent reviews of developments in high resolution gas phase spectroscopy of ions are presented in the specific studies cited in Table 4 and their accompanying references. For example, the recent HF^+ report summarizes developments over the past 15 years. [64,70] The tremendous progress made with high resolution gas phase spectroscopy for neutral radicals has also been recently described. [71]

3.1.   GAS-MATRIX COMPARISON

The important question of what constitutes an "inert" solid for radical cation studies is, of course, the key issue when one wishes to extract information that is intrinsically that of the cation and is not severely affected by the local environment. The paucity of gas phase hyperfine data makes the determination of host interactions or matrix shifts difficult to evaluate for cation radicals. Nuclear hyperfine data for only three cations are currently available from both gas phase measurements and condensed solids, namely neon at 4 K. (The following A_{iso} and A_{dip} parameters are the standard ones defined in reference 8). The hydrogen A_{iso} parameter for H_2O^+ is -75.7(2) MHz in the gas phase as determined by laser magnetic resonance (LMR) spectroscopy; this result compares reasonably well with the neon ESR A_{iso} value of -73.4(6) MHz. Gas phase A_{iso} and A_{dip} values for ^{13}C in $^{13}CO^+$ are 1506(15) MHz and 48.2(7) MHz, compared to neon ESR results of 1573(2) MHz and 46(1) MHz, respectively. Recent neon matrix vibrational

Table 4. Molecular Cations: Gas Phase Measurement of Nuclear hfs

Molecular Ion	Ground State	Nuclear hfs	References
H_2^+	$X^2\Sigma$	H	60
NH^+	$X^2\Pi$	H; ^{14}N	61
OH^+	$X^3\Sigma$	H	62
SH^+	$X^3\Sigma$	H	63
HF^+	$X^2\Pi_{3/2}$	^{19}F	64
HCl^+	$X^2\Pi_{3/2}$	$^{35,37}Cl$	65
HBr^+	$X^2\Pi_{3/2}$	$^{79,81}Br$	66
CO^+	$X^2\Sigma$	^{13}C	67
N_2^+	$X^2\Sigma$	^{14}N	68
H_2O^+	X^2B_1	H: A_{iso}, A_{dip}	69

results for CO^+ yield a value of 2194 cm^{-1} compared to the gas phase value of 2184 cm^{-1}. [23,72,73] The gas phase A_{iso} parameter of 105(4) MHz for ^{14}N in N_2^+ is also close to the neon measurement of 104(1) MHz.

A consistency of results for the same cation in two different types of solid hosts can also help clarify whether or not inherent molecular properties are being measured. An ESR study of the acetaldehyde cation radical, CH_3CHO^+, in neon and various halocarbon matrices showed agreement for the A_{iso} value of the aldehydic hydrogen within 5-10%, despite interaction with chlorine in the halocarbon matrix. [40] However, the ESR linewidths were considerably smaller in neon making possible the resolution of the g and A tensors as well as the $-CH_3$ hyperfine structure.

Neon matrix measurements for highly reactive small neutral radicals also show close agreement with gas phase hyperfine results; A_{iso} for the hydrogen atom in the gas phase is 1420 MHz compared to 1428(1) MHz in neon [23]; for the ^{11}BO radical, the neon result is 1026(1) MHz compared to the gas phase value of 1027.4(3). [8,74] Earlier concerns that the neon matrix dipolar result for BO was substantially less than that observed in the gas phase have been clarified in recent experiments in our laboratory. Gaseous neon was pre-cooled to 20 K and deposited along with BO(g) at 3 K. This process produces a matrix sample in which the motional averaging is retarded, yielding dipolar results within a few percent of the gas phase. Ambiguity concerning motional averaging effects is not usually a problem with larger diatomics or polyatomic radicals.

An extensive comparison of gas phase and rare gas matrix vibrational frequencies and electronic band origins for approximately 200 small molecules has been compiled. [75] This analysis shows that shifts for neon matrices are typically less than 1%, with somewhat larger perturbations for the heavier rare gases and nitrogen hosts. Electronic ground state properties such as hyperfine interactions and vibrational frequencies should be less prone to matrix shifts than other properties such as g tensors and electronic transitions that also involve excited states. The more contracted cation radicals are probably less perturbed in rare gas matrices relative to anions which have a more diffuse electronic structure with the unpaired electron(s) less tightly bound.

4. Related Developments

Vibrational and electronic spectroscopic studies of charged species in rare gas matrices have been summarized in various reports. [76-80] In a few cases, laser bombardment has been used to produce neutral molecules for vibrational and electronic matrix studies. [81-83] Recent vibrational studies of certain unusual cation radicals (O_4^+, N_4^+ and $C_2O_2^+$) trapped in neon matrices corroborate nicely the assignments and structures obtained from earlier ESR experiments which employed the same type of photoionization generation method. [72]

The production of ions in most other solid hosts usually involve high energy irradiation of a frozen solution containing the neutral parent. Radiation damage in the solid, proceeding through a series of charge transfer events, ultimately produces charged radicals which are stabilized by the local environment. See other chapters for a detailed explanation and literature coverage of this important process and Table 3 for specific examples. In addition to gamma or X-irradiation methods, high quality ESR spectra of small inorganic cations have also been produced by chemical reactions, such as the formation of $S_3N_2^+$ in frozen D_2SO_4 solutions, and Xe_2^+ in crystalline SbF_5. [55,47]

The feasibility of isolating a charged species in a rare gas host for ESR study was first demonstrated for a few atomic cations in 1971. [84] Electronic excitation of a metal atom trapped in an argon matrix containing HI produced M^+, H atoms and I^- by a dissociative electron capture process. The critical observation was that M^+ and I^- were trapped in separate lattice sites and ESR measurements of the atomic cation yielded a nuclear hyperfine result close to the gas phase value. Mossbauer experiments demonstrated that Fe^+ could be isolated in rare gas matrices. [85]

No direct ion-implantation or trapping from the gas phase was reported in these early experiments. It was not until 1983 that the first molecular cation radical was observed in a rare gas lattice by ESR. [20] The technique employed was direct photoionization at 16.8 eV from a neon resonance lamp during the simultaneous deposition of the neutral parent CO(g) and neon matrix gas at 4 K. The $^{13}CO^+$ radical was carefully selected for these first trapping experiments since it was the only cation, except for H_2^+, whose nuclear hfs had been measured in the gas phase. It was also of interest since it was isoelectronic to several neutral thirteen electron radicals previously investigated in rare gases, namely CN, BO and BeF. [8] The verification that CO^+, with such a large electron affinity of 14 eV, exhibited gas-like properties in a neon matrix suggested that most any cation species could be trapped for spectroscopic study provided that an appropriate parent species could be volatilized. The trapping of small cations with large EA's in neon matrices was especially timely since negative results for this type of ion were being reported in halocarbon and other hosts, despite their success in providing a suitably inert medium for larger organic cations. [86]

5.   Properties of the Neon Matrix for Isolated Ions

The chemical inertness, high ionization energy and large conduction band gap of neon make it the only solid host capable of isolating small cations with large electron affinities. Below approximately 8 K molecular diffusion in solid neon does not occur, although electrons are apparently highly mobile even at 4 K. The high optical quality and low scattering characteristics of solid neon are important for other spectroscopic methods. Several properties of neon matrices related to ESR studies of isolated ions will be discussed in the following sections. This is not intended to be a comprehensive or quantitative treatment, but an aid to the reader's understanding of the experimental procedures and ESR results presented in later sections.

5.1.   ESR SPECTRAL CHARACTERISTICS

ESR linewidths for isolated ions in neon matrices typically vary over the range of 0.2 - 2.0 Gauss depending upon the amount of anisotropy,

motional averaging and degree of preferential orientation that might
be present. This linewidth is approximately 10-20 times less than
those usually observed for cations in halocarbon matrices. [40] Such
narrow linewidths allow a detailed analysis of the A and g tensors
provided the weaker components of the absorption envelope can be
detected. The absorption along the bond axis direction, the parallel
component, for a simple diatomic radical is usually the most
difficult to detect for a randomly oriented sample. This is
especially true for high spin radicals where the overall anisotropy
is much larger because of the D tensor or zero field splitting. A
detailed discussion of ESR lineshapes for small radicals has been
presented previously. [8] The use of spectral simulation programs is
usually essential to the analysis of such spectra. An exact
diagonalization solution coupled to either Lorentzian or Gaussian
lineshapes has been written in our laboratory that is applicable to
most any type of radical. [19,24] The program computes transition
probabilities for all possible energy level combinations and weights
the simulated spectral features accordingly. It can simultaneously
include hyperfine, quadrupole and spin-spin (D tensor) interactions
for a given radical.

The occurrence of preferential orientation for ions in neon seems
to be more common than for neutral radicals. This phenomenon,
coupled with the ability to rotate the matrix sample in the magnetic
field, can provide sufficient information for an assignment of the
various magnetic parameters along specific internal directions in the
molecule. With or without preferential orientation, simple
theoretical considerations usually allow a satisfactory assignment.
Of course, the smaller the radical and the higher its symmetry, the
more likely is this optimistic scenario to occur in practice. To
summarize, neon ESR spectra can yield almost as much information as
single crystal measurements under favorable circumstances.

5.2.   ISOLATED VS. NON-ISOLATED IONS

In these discussions, the term "isolated ion" refers to ions
surrounded only by rare gas atoms and not to cations stabilized by an
adjacent anion in the same matrix cage. The latter case would be, in
effect, an ionic neutral molecule. For such a Coulombically bound
ion pair, the actual electronic distribution about the cation center
could be severely perturbed. This distinction between isolated ions
and bound ions has been made previously. [76] Measured magnetic
hyperfine parameters reveal the extent of this perturbation in the
case of the extremely ionic molecular radical, ^{137}BaF, where the
^{137}Ba A_{iso} parameter deviates by nearly 50% from that of a free Ba$^+$
ion. [87] Recent measurements in our laboratory on $^{13}CO_2^-$ isolated in
neon matrices show significant differences in the ^{13}C A tensor

relative to earlier measurements in crystals where Na^+ was adjacent to the anion. [88]

As more matrix studies of "ions" are conducted, every attempt should be made to determine whether or not the reported species is isolated or bound. Large apparent disagreements with forthcoming gas phase ion measurements and theoretical calculations are likely to occur if this distinction is not clear. Based upon the experience to date, at least three criteria should be simultaneously satisfied to classify ions as "isolated". They are: narrow linewidths with no evidence of hyperfine interaction with a counter anion; rapid photo-bleaching with visible light (see next section); and greater sensitivity to matrix warming than comparably sized neutral radicals. If significant signal loss does not occur under these latter two conditions, the ESR signal being monitored is probably a bound not isolated ion. However, a positive photo-bleaching response is not a sufficient condition since some neutral radicals can undergo photo-chemical decomposition, although this is rare with visible light above 5000 $\mathring{A}$. One exception that we have encountered is the formyl radical, HCO, whose ESR signals rapidly decrease with photolysis. The greater sensitivity to thermal warming is caused by greater long range interactions for ions than for neutrals, hence migration and recombination would occur for ions at lower temperatures once the limited diffusion or reorganization process in the rare gas lattice begins.

5.3.   IDENTITIES OF COUNTER ANIONS

The identities of counter anions were a matter of speculation in the early ESR rare gas cation experiments. A likely candidate proposed for consideration was OH^- which can not be detected by ESR. Recent matrix vibrational studies have also indicated the presence of isolated OH^-. [89] ESR signals that we have often detected in the g_e magnetic field region and associated with "background impurities" can definitely be assigned to CO_2^- based upon unambigious ^{13}C and ^{17}O hyperfine evidence recently obtained. [23] Other ESR active anions that have been positively identified in the same neon matrix sample which also exhibits intense cation signals are CH_2^-, F_2^-, Cl_2^-, HCl^-, HBr^- and HI^-. [30,90,23] Other evidence supporting the isolated nature of cations in these neon matrices is that the A and g values obtained are independent of which counter anions are intentionally trapped. This independence could be included as a fourth criteria for an isolated ion as discussed in the previous section, although sample preparation could be difficult in some cases.

The simple model that oppositively charged ions are isolated or separated from each other by rare gas atoms and that their recombination is prevented by the rigid matrix lattice is consistent with photobleaching observations. [76] Visible light (~ 2.5 eV) is sufficiently energetic to photoionize most all isolated anions but not most bound ion pairs. The liberated electrons can readily diffuse throughout the lattice and neutralize cation radicals. By monitoring ESR signals of both anions and cations during photolysis (or photobleaching) their relative rates of decrease can be measured. [30,90] It should also be possible to measure IE's of isolated anions by varying the incident wavelengths employed. By comparison with gas phase EA's, it should be possible to obtain values for the neon matrix solvation energy of the anion.

6.   Experimental

6.1.   EQUIPMENT

A schematic diagram of the matrix isolation apparatus for conducting a wide range of ESR studies is shown in Fig. 1. The arrangement of components continues to evolve in our laboratory with new features added for increased generation flexibility, greater trapping efficiency and diagnostic probing of the ion (or plasma) trapping process. Both liquid helium and closed cycle type refrigerators are used interchangeably (APD Heli-Tran and HS-4); each one is capable of delivering about one watt of cooling at 4 K. No serious vibrational problems have been encountered with the closed cycle refrigerator which provides more stable temperature control over the limited neon range of 3-10 K.

The cryostat is mounted on a gaseous hydraulic system with appropriate vacuum seals for moving the copper matrix target smoothly and quickly between the molecular beam deposition position and the X-band microwave cavity operating in the TE_{102} mode. For controlling the electrical potential on the matrix trapping surface, the liquid helium cryostat is mounted on an insulating flange as previously described. [22] This is an important parameter to control for the direct trapping from ion beams or for electron bombardment ion generation. To reduce cavity contamination (and loss of sensitivity) during the laser vaporization of non-volatiles, a movable lid or shutter is positioned over the top cavity opening during matrix deposition. (A most difficult task is to remember to move the lid before lowering the matrix sample!).

Thermocouple and vapor pressure measurements indicate that the warmest spot on the matrix target is ~5 K during a typical neon deposition. Although it is not essential, we have found certain advantages in pre-cooling the neon matrix gas to ~18 K just prior to deposition. This is achieved by passing the matrix neon gas through

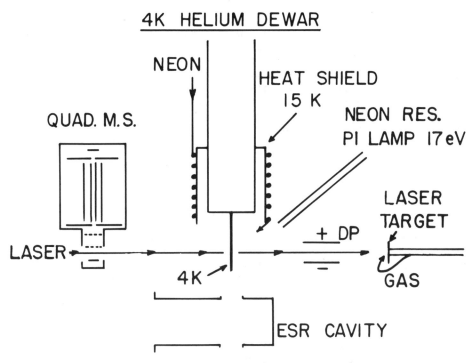

Figure 1. ESR matrix apparatus for trapping isolated ions in solid neon from the plasma produced by pulsed laser vaporization. The neon matrix gas can be pre-cooled to 15-20 K before deposition. The electrical deflection plates (DP) were used to study the deposition process. See text. Numerous ion radicals have been isolated from laser vaporization alone, without use of the photoionization source (PI).

a copper coil soldered to the lower heat shield. See Fig. 1. Neon condensation in the coil is prevented by a small regulated amount of electrical heat that has no significant warming effect on the matrix target. More efficient ion trapping is achieved by utilizing the coldest possible neon matrix gas. Also, matrix growth under these conditions seems to produce trapping sites that prevent or restrict rotational averaging which can be an important consideration in interpreting the magnetic parameters, especially for small radicals and those with low rotational or tunneling barriers.

6.2.   PHOTOIONIZATION

An open ended quartz tube (9 mm O.D.) at 45° to the matrix deposition surface serves as a neon resonance lamp which can irradiate the matrix with 16.8 eV photons and excited neutrals during deposition. Volatile compounds and neutral species vaporized from thermal effusion ovens or from laser ablation can be ionized and trapped in this manner. [21] Laser vaporization in combination with PI (photoionization) was used to trap Cu_2^+, Ag_2^+ and Au_2^+ in neon for ESR study. [27] We have found no cases where isolated ions are formed by operating the lamp after completing the matrix deposition process. A neon flow of approx 2.0 sccm through the lamp is employed with microwave discharge power levels varied from 2-60 watts. For consistent matrix results, are must be taken to reproduce the specific operating characteristics of this PI source; these include the standing wave ratio, gas flow, power, cooling rate, distance between microwave cavity and open end of lamp tube and lamp diameter. The specific operating characteristics greatly affect the relative abundances of various ion and neutral radicals actually trapped. A dramatic example of such effects have been described in a recent CH_2^-/CH_4^+ report. [30] Recent vibrational matrix studies have found certain advantages in using a much smaller opening (~1 mm) in the output end of the resonance lamp. [72]

The gas flows employed for the PI lamp and the separate matrix gas (~ 5.0 std. cc per min.) require a large background pumping speed in the vicinity of the matrix target. Otherwise, a significant fraction of radicals to be trapped from the molecular beam undergo scattering collisions and hence do not reach the trapping surface. Not including cryopumping, the pumping speed in our system is approximately 400 l/s.

6.3.   DIRECT PLASMA TRAPPING FROM PULSED LASER BOMBARDMENT

While using laser vaporization to generate new neutral radicals in our laboratory for matrix ESR studies [10,91,92], it became apparent that increased laser power levels and tighter focusing produced intense cation and anion radical signals. Such direct plasma trapping into rapidly condensing neon on a 4K surface yielded nearly identical ESR results as those obtained in earlier studies that employed photoionization or electron bombardment generation methods. A brief experimental account of this new ion-trapping technique for volatile and non-volatile applications will be presented.

Experiments conducted with electrically charged deflector plates, positioned as shown in Fig. 1, seem to clearly indicate that a plasma mechanism is responsible for trapping ions rather than a simple ion beam. The application of approximately 100 volts per cm across the plates during matrix deposition does not significantly affect the intensity of the neon matrix ESR signals for isolated ions. Screening that occurs in a plasma apparently prevents it from being severely deflected under such conditions.

Optimizing generation conditions for trapping a particular radical ion is greatly facilitated by having a real time monitoring capability rather than being dependent just upon matrix absorptions after a 30-60 minute deposit is completed. We employ a simple quadrupole mass spectrometer mounted line-of-sight and just behind the matrix deposition surface. In a slow scan or single mass mode, the MS detector response is gated to the laser firing through an appropriate delay generator. For example, the experimental conditions for producing the maximum amount of N_2^+ in the plasma can be adjusted prior to the matrix deposition process by monitoring the MS signal. The N_2^+ is generated by passing a stream of $N_2(g)$ directly over the focused laser spot on a metal surface. For most of the volatile gases tested in this manner to determine if the parent radical ion can be trapped, the specific choice of the metal is not important. Conditions that must be carefully optimized include laser power (MJ/pulse), degree of focusing, rate of beam movement across the surface and gas flow. The frequency doubled output (532 nm) from a Nd:YAG laser operating at 10 Hz is commonly used. For various types of metal vaporization/matrix deposition experiments, we find that approximately 10^{13}-10^{14} metal atoms per laser pulse are vaporized, with laser powers in the 5-15 MJ/pulse range and a spot size of approximately 0.5 mm.

In addition to generating the cation of the gas passed over the surface, it has been shown that this reactive vaporization procedure can produce new metal radicals for ESR study [10,92] such as neutral and charged metal oxides and hydrides whose production can also be optimized by MS monitoring. [32] Our attempts to trap ScN^+ in this manner have been unsuccessful. This negative evidence could mean that the ground state of ScN^+ is $^2\Pi$ not $^2\Sigma$ as predicted by a recent theoretical calculation [93], although our negative evidence is certainly not conclusive. The ScN^+ radical is isoelectronic with KO whose ground state is $^2\Pi$. Direct plasma trapping alone (no additional PI from a resonance lamp) has produced a variety of small cation radicals, including metal ions (M^+), hydride ion reaction products (PdH_2^+) and a few semi-conductor ions (GaP^+ and Si_2^+). The generation of energetic M^+ ions in this manner might be an ideal approach for conducting metal ion-neutral codeposition reactions. The passage of small reactant gases over semi-conductor surfaces undergoing laser vaporization should also yield many new radical ions for matrix ESR investigations.

7.   Semi-Conductor Cation Radicals

7.1.   SILICON CATIONS

Three small silicon containing cation radicals have been investigated by neon matrix ESR along with detailed theoretical calculations for $^{29}SiO^+$ and $^{29}SiH_2^+$. [25,31]   Direct plasma trapping from pulsed laser bombardment of single crystal silicon has produced intense ESR lines for $^{28}Si_2^+$ in neon matrices with the other isotopic species also detected with natural abundance ^{29}Si, namely $^{28,29}Si_2^+$ and $^{29}Si_2^+$. [23]   These results provide the first experimental ground state assignment for Si_2^+ which is $X^4\Sigma$ as predicted by previous theoretical calculations. [94]   Given the relatively small magnitude of the ^{29}Si hyperfine parameter observed, the three unpaired electrons seem to occupy predominantly $3p_\sigma$ and $3p_\pi$ type orbitals which are nearly degenerate.   Except for a much larger zero field parameter (D value) for Si_2^+, its electronic distribution seems similar to that characterized in detail for the isovalent C_2^+ cation which is also a $X^4\Sigma$ case. [19]   Based upon a simple comparison of atomic spin orbit parameters, the D value for Si_2^+ should be considerably larger.

From an experimental perspective, it is interesting that Si_2^+ was observed with laser vaporization alone, while C_2^+ trapping required laser vaporization to produce C_2 which had to be photoionized during neon deposition.   An electronic spectroscopic investigation of C_2^+ in neon used lower energy photolysis to produce neutral C_2 from HCCX compounds and higher energy photolysis (16.8 eV) for photoionization. [95]   The neon matrix ESR spectrum of $^{13}C_2^+$ is shown in Fig. 2.   The SiO^+ cation was originally generated by conventional high temperature vaporization of $SiO(s)$ and $SiO_2(s)$ in combination with photo-ionization. [25]   Subsequent ESR experiments have shown that passing $O_2(g)$ over crystalline silicon during laser bombardment also produces neon isolated SiO^+. [23]

The SiH_2^+ radical was discovered in attempts to produce SiH_4^+ by photoionizing $SiH_4(g)$ during neon matrix depositions. [31]   Under these conditions, no direct ESR evidence for SiH_4^+ was observed such as hfs from four hydrogen atoms.   The use of plasma techniques to deposit silicon layers in the fabrication of electronic devices and solar cells makes a fuller understanding of silicon intermediates important; SiH_2^+ is also a potential interstellar ion.

The observed ^{29}Si A_{iso} and A_{dip} values showed excellent agreement with those calculated for SiH_2^+ in its ground $^2A_1$ state which has a bond angle of $119°$ and a Si-H bond distance of 1.49 Å. [96]   The unpaired electron resides primarily on silicon in a non-bonding $3s/3p_z$ hybrid (.18/.71) with a small hydrogen spin density of 0.08, corresponding to a hydrogen A_{iso} value of 117 MHz.   The SiH_2^+ radical

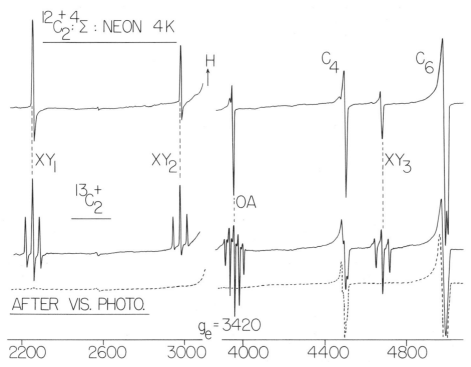

Figure 2. The pulsed laser vaporization of carbon under photo-ionization conditions produced these neon matrix $^{12}C_2^+$ and $^{13}C_2^+$ ESR spectra. See Fig. 1. The isolated C_2^+ cation species is readily distinguished from neutral radicals such as C_4 and C_6 by the spectral changes induced by visible light photobleaching. See text. The triplet nuclear hfs in $^{13}C_2^+$ is apparent and the OA group is an off-angle absorption feature with complex hfs. See ref. no. 19. The electronic ground state of C_2^+ is $X^4\Sigma$.

exhibited extreme preferential orientation in neon with two g components observable for one orientation of the trapping surface in the magnetic field; the third component became dominant when the matrix sample was rotated by 90°. In previous studies of gamma irradiated SnH4 in CFCl3 at 77 K, an ESR spectrum was observed which showed large Sn hfs and hyperfine interactions with only two hydrogen atoms. [51] A thorough analysis was conducted on this stannane radical which was assigned to SnH_4^+. Given these recent results for SiH_2^+ and the increasing instability of MH_4^+ going down this series, it is possible that the stannane hydride ion previously observed was SnH_2^+ rather than SnH_4^+. Another alternative would be H_2 loosely bound to SnH_2^+; an alternative that was considered for the SiH_2^+ case. [31]

7.2. GROUP III/V DIATOMIC CATIONS

The heavier combinations of Group III/V diatomic cations should have $^4\Sigma$ ground electronic states with large D values. Hence their ΔM_s: $1/2 \longleftarrow -1/2$ perpendicular transitions should be readily observable in the g ~ 4 ESR spectral region. ESR matrix studies for GaP^+ and $GaAs^+$ have already been reported with preliminary results obtained for $GaSb^+$ and $InAs^+$. [2,28,23] Experimental efforts to generate and trap the lighter members of this series, such as AlN^+ and AlP^+, have not yet been successful. The rich hyperfine pattern observed in the ESR spectrum for $^{69,71}GaP^+$ is shown in Fig. 3, where a quartet-of-doublets is resolved for each gallium isotope; the smaller doublet splitting results from $^{31}P(I=1/2)$ hfs.

Theoretical calculations conducted for $GaAs^+$ are consistent with the electronic structure description that is provided by the observed hyperfine parameters. [97] Theory shows that the $X^4\Sigma$ ground state electronic distribution remains essentially unchanged to the dissociation limit of $Ga^+(4s^2) + As(4s^24p^3)$. Partial donation from As(or P) into the vacant $4p_z$ orbital of Ga^+ can account for the A tensor anisotropy on gallium. This sigma type donation occurs to a smaller extent in GaP^+ relative to $GaAs^+$ based on the A tensor analysis and comparison. The difference seems to be consistent with the larger ionization energy of P relative to As, of course other factors are certainly involved. The amount of valence "s" spin density is small for both atoms in these radical cations. Recent gas phase spectroscopic studies of GaAs in pulsed supersonic beams have measured the D value of the $X^3\Sigma$ ground state. [98]

8. Cluster Cation Radicals

8.1. METAL CLUSTERS

Several small metal cluster cations have been studied by ESR in rare gas matrices, including Mn_2^+, Cu_2^+, Ag_2^+, Au_2^+ and a series of magnesium cations, Mg_{2-6}^+. A large number of such metal systems should be amenable to study by matrix ESR in future studies given the ability to form aggregates under ionizing conditions during the deposition process. The Mn_2^+ molecule is found to have eleven unpaired electrons in argon matrices with a $^{12}\Sigma$ ground state. It is especially interesting since it occurs between two uniquely

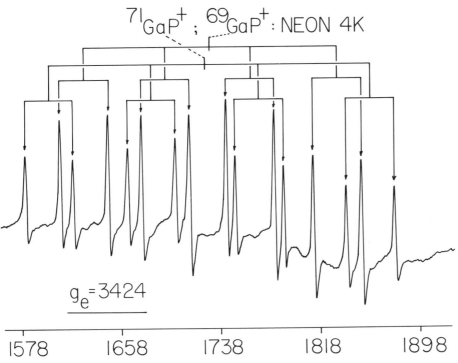

Figure 3. These quartet-of-doublets ESR hyperfine patterns for $^{69}GaP^+$ and $^{71}GaP^+$ occur in the g = 4 spectral region indicating a $X^4\Sigma$ ground electronic state. Pulsed laser vaporization of GaP(s) was used to generate this neon matrix sample of isolated cation radicals. See ref. no. 2.

different molecules: Cr_2 with multiple bonding and Mn_2 with weaker van der Waals bonding. [26] Laser vaporization combined with argon resonance photoionization was used to produce argon matrix samples of Mn_2^+.

For the Cu_2^+ molecule, the ^{63}Cu A_{iso} parameter indicates that the unpaired electron occupies, almost totally, the 4s orbital. Some A tensor anisotropy is observed for Ag_2^+ and Au_2^+ which is estimated to represent about 30 ± 20 % p character for the silver cation with a similar amount of d character for Au_2^+. [27] A most interesting comparison of magnetic parameters for Ag_2^+ in ten different solid hosts has been presented. [27] Except for benzene and neon, all A values are below 940 MHz; benzene yields $A_{||}$ = 1030(6) and

$A_\perp$ = 1014(6) with neon values of 1160(2) and 1109.2(3) MHz.

For Mg_2^+ and the higher magnesium cluster cations, the spin density resides predominantly (> 90%) in the 3s valence orbital and is equally distributed among the various atoms. [23] The ESR results alone can not establish a unique structure for the Mg_{3-6}^+ cations, but the observed ^{25}Mg hfs indicates that, at least on the ESR time scale, all atoms are equivalent. Theoretical calculations have been recently reported for these magnesium cluster cations which appear to be consistent with these preliminary ESR results. [99]

8.2.   NON-METAL CLUSTERS

Examples of non-metal cluster cations listed in Tables 1-3 include N_4^+, O_4^+, P_4^+ and S_8^+. It is apparent from these results that N_4^+ and the isovalent P_4^+ cations have quite different structures, with the former being a linear $X^2\Sigma$ radical and the latter having a Jahn-Teller distorted D_{2d} geometry. (zigzag ring structure) In P_4^+, the unpaired electron is almost equally distributed over the four atoms having a spin density of approximately 0.25 in each atom's 3p orbital. [50] In N_4^+, it is also delocalized over all four atoms with approximately 0.12 s and 0.23 p_z character on each inner nitrogen and -0.01 s and 0.15 p_z on each outer nitrogen atom. While the neutral parent of N_4^+ is bound by weak van der Waals forces, the cation itself has a bond dissociation energy of 26 kcal/mol. [100]

The ion-neutral reaction of O_2 with O_2^+ during neon deposition under photoionization conditions has enabled ESR observations of the O_4^+ cation, which is the first experimental evidence indicating a $X^4\Sigma$ electronic ground state. [34] This fundamentally important ion molecule has potential significance in upper atmospheric chemistry as a proposed intermediate in the formation of protonated water clusters. [101] Its calculated dissociation energy of 11.7 kcal/mol compares reasonably well with a recent experimental measurement of 9.2. [102,103] Its nonaxial D tensor, effective g values and small ^{17}O hfs indicate that O_4^+ is nonlinear with the three unpaired electrons occupying π type orbitals. A neon matrix IR investigation favors the planar trans configuration C_{2h}. [104]

Acknowledgments

Project support from the National Science Foundation (CHE-8508085) and student support from a Duke Endowment grant to Furman University are gratefully acknowledged. These ion trapping experiments have been conducted with the able assistance of twenty-two Furman University undergraduates over the past six years and the author is especially indebted to their participation. Appreciation is also expressed to the following colleagues for stimulating theoretical collaboration over the years; Professor E. R. Davidson, Dr. David Feller and Professor C. A. Arrington.

REFERENCES

1. L. B. Knight, Jr., S. T. Cobranchi, B. W. Gregory and E. Earl
    J. Chem. Phys. **86**, 3143 (1987).
2. L.B. Knight, Jr., and J.O. Herlong, J. Chem. Phys. **91**,69 (1989).
3. M. C. R. Symons, Chem. Soc. Rev. **13**, 393 (1984).
4. T. Shida, E. Haselbach and T. Bally, Acc. Chem. Res. **17**, 180
    (1984); J. T. Wang and F. Williams, J. Phys. Chem. **84**, 3156
    (1980).
5. X.-Z. Qin and A. D. Trifunac, J. Phys. Chem. (In Press 1990);
    M. Shiotani, Magn. Res. Rev. **12**, 333 (1987).
6. Faraday Discuss. Chem. Soc. **78** (1984); "Radicals in Condensed
    Phases".
7. Faraday Discuss. Chem. Soc. **86** (1988); "Spectroscopy at Low
    Temperatures"; D. M. Close, Magn. Res. Rev. **14**, 1 (1988).
8. W. Weltner, Jr., "Magnetic Atoms and Molecules" (Van Nostrand-
    Reinhold, New York, 1983).
9. L. B. Knight, Jr., Acc. Chem. Res. **19**, 313 (1986).
10. L. B. Knight, Jr., Chapter 7, "Chemistry and Physics of Matrix
    Isolated Species", editors, L. Andrews and M. Moskovits
    (Elsevier, Amsterdam, 1989).
11. D. M. Lindsay, M.C. R. Symons, D. R. Herschbach and A. L.
    Kwiram, J. Phys. Chem. **86**, 3789 (1982); D. M. Lindsay, G. A.
    Thompson and Y. Wang, J. Phys. Chem. **91**, 2630 (1987).
12. P. H. Kasai and P. M. Jones, J. Am. Chem. Soc. **106**, 8018 (1984);
    P. H. Kasai, J. Phys. Chem. **90**, 5034 (1986).
13. H. M. Cheung and W. R. M. Graham, J. Chem. Phys. **91**, 6664
    (1989).
14. J. M. Parnis and G. A. Ozin, J. Am. Chem. Soc. **108**, 1699 (1986).
15. R. J. Van Zee and W. Weltner, Jr. J. Am. Chem. Soc. **111**, 4519
    (1989).
16. D. M. Lindsay, F. Meyer and W. Harbich, Z. Phys. (D) **12**, 15
    (1989).
17. J. C. Rivoal, C. Grisolia, J. Lignieres, D. Kreisle, P. Fayet
    and L. Waste, Z. Phys. (D), **12**, 481 (1989).
18. D. Forney, M. Jakobi and J. P. Maier, J. Chem. Phys. **90**, (1989).
19. L. B. Knight, Jr., S. T. Cobranchi and E. Earl, J. Chem. Phys.,
    88, 7348 (1988).
20. L. B. Knight, Jr., and J. Steadman, J. Chem. Phys. **77**, 1750
    (1982); A.J. McKinley, R.F.C. Claridge and P.W. Harland Chem.
    Phys. **102**, 283 (1986).
21 L. B. Knight, Jr., A. Ligon, S. T. Cobranchi, D. P. Cobranchi,
    E. Earl, D. Feller and E. R. Davidson, J. Chem. Phys. **85**, 5437
    (1986).
22. L. B. Knight, Jr., J. M. Bostick, R. W. Woodward and J. Steadman
    J. Chem. Phys. **78**, 6415 (1983).
23. L. B. Knight, Jr., To be published.

24. L. B. Knight, Jr., E. Earl, A. R. Ligon, D. P. Cobranchi, J.R. Woodward, J. M. Bostick, E. R. Davidson and D. Feller, J. Am. Chem. Soc. **108**, 5065 (1986).

25. L. B. Knight, Jr. A. Ligon, R. W. Woodward, D. Feller and E. R. Davidson, J. Am. Chem. Soc. **107**, 2857 (1985).

26. R. J. Van Zee and W. Weltner, Jr., J. Chem. Phys. **89**, 4444 (1988).

27. R. J. Van Zee and W. Weltner, Jr. Chem. Phys. Lett. **162**, 437 (1989).

28. L.B. Knight, Jr. and J.T. Petty, J. Chem. Phys. **88**, 481 (1988).

29. L. B. Knight, Jr. and J. Steadman, J. Chem. Phys. **78**, 5940 (1983).

30. L. B. Knight, Jr., M. Winiski, P. Miller, C. A. Arrington and D. Feller, J. Chem. Phys. **91**, 4468 (1989).

31. L. B. Knight, Jr., M. Winiski, P. Kudelko and C. A. Arrington, J. Chem. Phys. **91**, 3368 (1989).

32. L. B. Knight, Jr. S. T. Cobranchi, J. Herlong, T. Kirk, K. Balasubramanian and K. K. Das, J. Chem. Phys. **92**, 2721 (1990).

33. L. B. Knight, Jr., K. D. Johannessen, D. C. Cobranchi, E. A. Earl, D. Feller and E. R. Davidson, J. Chem. Phys. **87**, 885 (1987).

34. L. B. Knight, Jr., S. T. Cobranchi and Jeff Petty, J. Chem. Phys. **91**, 4423 (1989).

35. L. B. Knight, Jr., J. Steadman, P. K. Miller and J. A. Cleveland, Jr., J. Chem. Phys. **88**, 2226 (1988).

36. L. B. Knight, Jr., J. Steadman, P. K. Miller, D. E. Bowman, D. Feller and E. R. Davidson, J. Chem. Phys. **80**, 4593 (1984).

37. L. B. Knight, Jr. and J. Steadman, J. Chem. Phys. **80**, 1018 (1984).

38. L. B. Knight, Jr. and J. Ott, Faraday Discuss. Chem. Soc. **86**, 71 (1989).

39. L. B. Knight, Jr., J. Steadman, D. Feller and E. R. Davidson, J. Am. Chem. Soc. **106**, 3700 (1984).

40. L. B. Knight, Jr., B. W. Gregory, S. T. Cobranchi, F. Williams and X. Z. Qin, J. Am. Chem. Soc. **110**, 327 (1988).

41. L. B. Knight, Jr., C. A. Arrington, B. W. Gregory, S. T. Cobranchi, S. Liang and L. Paquette, J. Am. Chem. Soc. **109**, 5521 (1987).

42. S. A. Fairhurst, J. R. Morton and K. F. Preston, Chem. Phys. Letters **104**, 112 (1984).

43. J. R. Morton and K. F. Preston, Inorg. Chem. **24**, 3317 (1985); M. C. R. Symons, J. R. Morton and K. F. Preston, ACS Symposium Series No. 333, "High Energy Processes in Organometallic Chemistry" editor K.S. Suslick, American Chemical Society; 1987.

44. S. A. Fairhurst, J. R. Morton, R. N. Perutz and K. F. Preston, Organometallics **3**, 1389 (1984).

45. J. R. Morton and K. F. Preston, J. Magn. Reson. **68**, 121 (1986);
    See also Landolt-Barnstein, New Series Vol **9a**, Springer-Verlag,
    New York 1977.
46. M. C. R. Symons, R. Janes, and A. D. Stevens, Chem. Phys.
    Letters, **160**, 386 (1989); A. Abou-Kais, J. C. Vedrine and C.
    Naccache, J. Chem. Soc., Faraday Trans. II, **74**, 959 (1978).
47. L. Stein, J. R. Norris, A. J. Downs and A. R. Minichan, J. Chem.
    Soc. Chem. Comm., p. 502 (1978).
48. A. Begum, A. R. Lyons and M. C. R. Symons, J. Chem. Soc. (A)
    2290 (1971); A. Hasegawa, G. D. G. McConnachie and M. C. R.
    Symons, J. Chem. Soc. Faraday Trans. **1** **80**, 1005 (1984).
49. "Magnetic Properties of Free Radicals", editors J. R. Morton and
    K. F. Preston, Landolt-Bornstein New Series II/17a, p. 42-43
    (Springer-Verlag New York 1986); K. Murthy and S. V. Bhat, Chem.
    Phys. Letters **133**, 455 (1987).
50. A. Hasegawa, Jane L. Wyatt and M. C. R. Symons, J. Chem. Soc.
    Chem. Comm. (1989, in press).
51 A. Hasegawa, S. Kaminaka, T. Wakabayashi. M. Hayashi, M. C. R.
    Symons and J. Rideout, J. Chem. Soc. Dalton Trans. 1667 (1984).
52. H.Muto, K.Nunome and M. Iwasaki, J. Chem. Phys. **90**, 6827 (1989).
53. H. Chandra, D. N. Rao and M. C. R. Symons, J. Chem. Soc. Dalton
    Trans. 729 (1987).
54. D. N. Rao and M. C. R. Symons, J. Chem. Soc. Dalton Trans. 2533
    (1983); J. R. Morton , K. F. Preston and S. J. Strach, J. Phys.
    Chem. **83**, 533 (1979).
55. K. M. Johnson, K. F. Preston and L. H. Sutcliffe, Mag. Reson. in
    Chem. **26**, 1015 (1988); S. A. Fairhurst, K. F. Preston and L. H.
    Sutcliffe, Can. J. Chem. **62**, 1124 (1984); K. F. Preston, J.-P.
    Charland and L. H. Sutcliffe, Can. J. Chem. **66**, 1299 (1988).
56. E. Awere, J. Passmore, K. F. Preston and L. H. Sutcliffe, Can.
    J. Chem. **66**, 1776 (1988).
57. E. Sagstuen, O. Awadelkarim, A. Lund and J. Masiakowski, J.
    Chem. Phys. **85**, 3223 (1986).
58. I. B. Goldberg, H. R. Crowe and K. O. Christie Inorg. Chem. **17**,
    3189 (1978).
59. J. A. Howard and B. Mile, Chapter 5 in "Electron Spin Resonance"
    Vol. 11B, M. C. R. Symons, editor; (The Royal Society of
    Chemistry, Burlington House, London 1989); T. Ichikawa, M.
    Shiotani, N. Ohta and S. Katsumata, J. Phys. Chem. **93**, 3826
    (1989).
60. K. B. Jefferts, Phys. Rev. Letters **23**, 1476 (1969).
61. P. Verhoeve, J. J. Ter Meulen, W. L. Meerts and A. Dymanus,
    Chem. Phys. Letters **132**, 213 (1986).
62. M. H. W. Gruebele, R. P. Muller and R. J. Saykally, J. Chem.
    Phys. **84**, 2489 (1986); J. P. Bekooy, P. Verhoeve, W. L. Meerts
    and A. Dynamus, J. Chem. Phys. **82**, 3868 (1985).

63. D. C. Hovde and R. J. Saykally, J. Chem. Phys. 87, 4332 (1987);
    C. P. Edwards, C. S. Maclean and P. J. Sarre, Mol. Phys. 52,
    1453 (1984).

64. J. V. Coe, J. C. Owrutsky, E. R. Keim, N. V. Agman, D. C. Hovde
    and R. J. Saykally, J. Chem. Phys. 90, 3893 (1989).

65. K. G. Lubic, D. Ray, D. C. Hovde, L. Veseth and R. J. Saykally,
    J. Mol. Spectrosc. 134, 21 (1989); R. J. Saykally and K. M.
    Evenson, Phys. Rev. Letters 43, 515 (1979).

66. K. G. Lubic, D. Ray, D. C. Hovde, L. Veseth and R. J. Saykally,
    J. Mol. Spectrosc. 134, 1 (1989); D. Ray, K. G. Lubic and R. J.
    Saykally, Mol. Phys. 46, 217 (1982).

67. N. D. Piltch, P. G. Szanto, T. G. Anderson, C. S. Gudeman, T. A.
    Dixon and R. C. Woods, J. Chem. Phys. 76, 3385 (1982); A.
    Carrington, D. R. J. Milverton and P. J. Sarre, Mol. Phys. 35,
    1505 (1978).

68. S. D. Rosner, T. D. Gaily and R. A. Holt, J. Mol. Spectrosc.
    109, 73 (1985); Linda Young, Private Communcation.

69. S. E. Strahan, R. P. Mueller and R. J. Saykally, J. Chem. Phys.
    85, 1252 (1986).

70. "Molecular Ions: Geometric and Electronic Structure" edited by
    J. Berkowitz and K. Groeneveld (Plenum, New York, 1983); "Ion
    and Ion Cluster Spectroscopy and Structure" edited by J. P.
    Maier (Elsevier, Amsterdam, 1989).

71. T. A. Miller, J. Phys. Chem. 93, 5986 (1989).

72. M. E. Jacox and W. E. Thompson, Res. Chem. Intermed. 12, 33
    (1989).

73. K. P. Huber and G. Herzberg, "Constants of Diatomic Molecules"
    (Van Nostrand Reinhold, New York, 1979).

74. M. Tanimoto, S. Saito and E. Hirota, J. Chem. Phys. 84, 1210
    (1986).

75. M. E. Jacox, J. Mol. Str. 157, 43 (1987).

76. L. Andrews, Annu. Rev. Phys. Chem. 30, 89 (1979).

77. M. E. Jacox and W. E. Thompson, J. Chem. Phys. 91, 1410 (1989).

78. L. Andrews in "Chemistry and Physics of Matrix-Isolated Species"
    Chapter 2, edited by L. Andrews and M. Moskovits (Elsevier,
    1989).

79. B. S. Ault, Rev. Chem. Intermed. 9, 233 (1988).

80. T. A. Miller, V. E. Bondybey and J. E. English, J. Chem. Phys.
    70, 2919 (1979).

81. L. A. Heimbrook, M. Rasanen and V. E. Bondybey, J. Phys. Chem.
    91, 2468 (1987).

82. N. P. Machara and B. S. Ault, Chem. Phys. Lett. 140, 411 (1987);
    J. Chem. Phys. 88, 2845 (1988).

83. R. Withnall, M. McCluskey and L. Andrews, J. Phys. Chem. 93, 126
    (1989).

84. P. H. Kasai, Acc. Chem. Res. 4, 329 (1971).

85. P. A. Montano, P. H. Barrett, H. Micklitz, A. J. Freeman, and J. V. Mallow, Phys. Rev. B 17, 6 (1978).

86. M. C. R. Symons, T. Chen, and C. J. Glidewell, J. Chem. Soc. Chem. Comm. 326 (1983).

87. L. B. Knight, Jr., W. C. Easley, W. Weltner, Jr. and M. Wilson, J. Chem. Phys. 54, 322 (1971).

88. A. Carrington and A. D. McLachlan, "Introduction to Magnetic Resonance" p. 138 (Harper and Row, New York 1967); J. Westerling and A. Lund, Chem. Phys. Letters 147, 111 (1988).

89. S. Suzer and L. Andrews, J. Chem. Phys. 88, 916 (1988).

90. L. B. Knight, Jr., E. Earl, A. R. Ligon and D. P. Cobranchi, J. Chem. Phys. 85, 1228 (1986).

91. L. B. Knight, Jr. S. T. Cobranchi, J. Petty and D. P. Cobranchi, J. Chem. Phys. 91, 4587 (1989).

92. L. B. Knight, Jr., S. T. Cobranchi, J. T. Petty, E. Earl, D. Feller, and E. R. Davidson, J. Chem. Phys. 90, 690 (1989); L. B. Knight, Jr., S. T. Cobranchi, B. W. Gregory and G. C. Jones, Jr. J. Chem. Phys. 88, 524 (1988).

93. K. L. Kunze and J. F. Harrison, J. Phys. Chem. 93, 2983 (1989).

94. K. Raghavachari, Private Communication.

95. D. Forney, H. Althaus and J. P. Maier, J. Phys. Chem. 91, 6458 (1987); J. P. Maier and M. Rosslein, J. Chem. Phys. 88, 4614 (1988).

96. M. C. Curtis, P. A. Jackson, P. J. Sarre and C. J. Whitham, Mol. Phys. 56, 485 (1985).; M. N. Paddon-Row and S. S. Wong, J. Chem. Soc. Chem. Comm. 1585 (1987); J. A. Pople and L. A. Curtiss, J. Phys. Chem. 91, 155 (1987); R. F. Frey and E. R. Davidson, J. Chem. Phys. 89, 4227 (1988).

97. K. Balasubramanian, J. Chem. Phys. 86, 3410 (1987); Chem. Phys. Letters 150, 71 (1988).

98. G. W. Lemire, G. A. Bishea, S. A. Heidecke and M. D. Morse, J. Chem. Phys. (1989, in press).

99. G. Durand, J. Chem. Phys. 91, 6225 (1989).

100. P. Kebarle, Annu. Rev. Phys. Chem. 28, 445 (1977); J. D. Payzant and P. Kebarle, J. Chem. Phys. 53, 4723 (1970); H. H. Teng and D. C. Conway, ibid, 59, 2316 (1973); R. N. Varney, ibid, 33, 1709 (1960); C. Y. Ng, Adv. Chem. Phys. 52, 263 (1983).

101. G. R. Reid, Adv. At. Mol. Phys. 12, 375 (1976); E. E. Ferguson, F. C. Fehsenfeld and D. L. Albritton in "Gas Phase Ion Chemistry", edited by M. T. Bowers (Academic, New York, 1979).

102. D. C. Conway, J. Chem. Phys. 50, 3864 (1969).

103. K. Hiraoka, J. Chem. Phys. 89, 3190 (1988).

104. W. E. Thompson and M. E. Jacox, J. Chem. Phys., 91, 3286 (1989).

ESR STUDIES ON CATION RADICALS OF SATURATED HYDROCARBONS
Structure, Orbital Degeneracy, Dynamics, and Reactions

KAZUMI TORIYAMA
Government Industrial Research Institute, Nagoya, Hirate, Kita, Nagoya 462, Japan

1. Introduction

The most prominent development in the study of cation radicals by electron spin resonance (esr) during the past 20 years is the detection and the structural study of alkane cation radicals[1,2]. Although an olefin cation was detected by Ichikawa and Ludwig by the radiolysis of tetramethylethylene(TME) in an inert alkane with an electron scavenger[3], detection of alkane cations had been unsuccessful despite many efforts such as irradiation of alkanes at 4 K with or without an electron scavenger[4]. As a consequence, alkane cations had been thought to be unstable and decomposed by the instantaneous deprotonation. Nauwelaerts and Ceulemans applied the technique similar to that of Ichikawa to an alkane system and reported the spectrum of "n-$C_8H_{18}^+$". However, it was too obscure to identify the radical as an alkane cation[5]. In 1979, Smith and Symons obtained the esr spectrum of the hexamethylethane cation radical (HME$^+$) as the first example of an alkane cation radical, by the radiolysis of HME in CBr_4 or CCl_4[6]. This result was, however, not sufficient to be accepted by the general researcher, although its assignment was supported by the subsequent work by Wang and Williams[7] and later by Shida et al. using Freons as the matrices[8]. It was the work by Iwasaki and his coworkers on $C_2H_6^+$ in SF_6 at 4 K, published in 1981, that gave a definite proof for the detection of the alkane cation radical with clear information on the structure of it[9]. This brought about the bursting development in the esr studies on the cation radicals, not only for alkanes but also for the basic organic compounds, performed in the last ten years[2].

In this chapter, we are concerned with cation radicals of saturated hydrocarbons which are generated by the radiolysis of alkanes in low temperature matrices. The esr spectra of alkane cation radicals are characterized by large and essentially isotropic hyperfine coupling constant (hfcc) due to the unpaired electron in H_{1s} orbital[9,10]. This feature is brought by the fact that the unpaired electron is delocalized in a σ-orbital, in contrast with aromatic hydrocarbon cations in which the unpaired electron is delocalized in a π-orbital[1]. σ-Delocalization of the unpaired electron is the most important concept which has been demonstrated clearly through these studies, because it is closely

99

A. Lund and M. Shiotani (eds.), Radical Ionic Systems, 99–124.
© 1991 *Kluwer Academic Publishers. Printed in the Netherlands.*

related with the reactivity of the non conjugated organic ions[11].

Another important result is the direct detection of static Jahn-Teller (J-T) distortion[12], which usually had been detected indirectly through the temperature dependence or isotope effect on hfcc[12,13], except for a few cases[14]. For these latter cases, the effect of the surroundings had been suggested for the distortion because of the unexpected stability of the specific distorted form. However, in the case of alkane-cation radicals, the coordinate of distortion and its direction is not altered with the matrices, so that it is regarded that the mode of distortion is intrinsic to the cation radical, and the role of the matrix is just to stabilize it[15]. It was also found that the dynamic J-T process involves site-jumping (or tunneling) and cannot be interpreted as the change in Boltzmann populations with temperature of the two near degenerate states which has been often assumed in the solution esr of J-T active species[12,13].

2. Method

Esr detection of the alkane-cation radical was made possible by the radiolysis of the alkane in a frozen solution of SF_6, Freons, or other halocarbons with a high ionization potential (IP). In these matrices, the cation radical is generated through the positive charge transfer from the matrix cation to the solute alkanes[16]. The counter electron is captured by a matrix molecule to form an anion due to the positive electron affinity (EA). In addition, the subsequent ion-molecule reaction, which may be the most probable cause of the instability of these cations[16], is prevented by the isolation of alkane cations from other alkane molecules.

$$RX \quad \rightsquigarrow \quad RX^+ \; + \; e^- \tag{1}$$
$$RX^+ \; + \; RH \; \longrightarrow \; RX \; + \; RH^+ \tag{2}$$
$$RX \; + \; e^- \; \longrightarrow \; RX^- \; (or \; R\cdot + X^-) \tag{3}$$

The halogenated compounds which have been used frequently as the matrix are, SF_6, $CFCl_3$ (F11), $CF_2ClCFCl_2$ (F113), CF_3CCl_3, perfluoro-methylcyclohexane (PFMCHx) and the derivatives of them. The perfluoro-n-alkanes, CBr_4, and CCl_4 are also used, although CBr_4 and CCl_4 give fairly strong background signals. There may be some factor other than IP for stabilizing alkane cation radicals. For example, the use of SF_6 or C_2F_6 was needed to stabilize the cation radical of ethane[9], although the IP of F113 is higher than that of ethane by ca. 0.5 eV. Similarly, SF_6 (IP = 15.6 eV) was still insufficient to generate the cation radical of methane (IP = 13.5 eV) at 4 K[18], and the photolysis by Ne discharge was needed for esr detection of methane cation[19].

The advantage of using fluorine containing compounds as the matrices is that the background esr signals of fluorine containing radicals are broadened due to the large anisotropy of α-fluorine hyperfine couplings[20]. The usefulness of $CFCl_3$ (F11) for the matrix was found by Shida and Kato[21], as an extension of the use of Freon mixture (FM) glass to observe the optical spectra of cationic species[22]. This is the most widely used matrix since the cation radical trapped in this matrix shows well resolved hyperfine structures. However, it has been reported that upon freezing the solution the solute as well as F11 molecules are sometimes oriented partially along an axis so that the undistorted powder spectrum can not be obtained. An example of this partial orientation was demonstrated with benzene cation[23].

Toriyama et al. have shown that the synthetic zeolite, which had been known to work as a positive charge donating matrix to form and stabilize the cation of conjugated hydrocarbons[24], can also be used to stabilize alkane cation radicals[25]. The advantage of using zeolite for the investigation on the reaction mechanism of alkane cation is that the conformation as well as the mobility of the solute molecule can be controlled through the selection of the pore size.

Recently, the successful detection of the alkane cation radicals by fluorescence detected esr (FDMR or ODESR) has been reported[26,27]. This technique is powerful to investigate the reaction and the dynamics of cation radical. The details of these results will be described in other chapters.

3. Cation Radicals of Linear Alkanes

3.1. ETHANE

Ethane has D_{3d} symmetry so that the highest occupied orbitals (HOMO) are a_{1g} and e_g. A number of studies were reported to predict the singly occupied molecular orbital (SOMO) of $C_2H_6^+$. Most of the ab initio calculations suggested that the ground state of $C_2H_6^+$ was $^2A_{1g}$ at large C - C distance but was ^2E_g or a mixture of $^2A_{1g}$ and ^2E_g at the shorter C - C distance[28]. On the other hand, results of photoelectron spectroscopy (PE) indicated that SOMO was e_g, which would split to a_g and b_g by the J-T distortion to C_{2h} symmetry[29](Fig. 1). In 1981, Iwasaki and his coworkers succeeded in the esr detection of $C_2H_6^+$[9,10]. As is shown in Fig. 2, the esr spectrum of $C_2H_6^+$ observed at 4 K in SF_6 consists of a triplet of 1 : 2 : 1 intensity with a large splitting of 15.25 mT, which corresponds to the spin density of $\approx$ 0.3 on H_{1s}. The hyperfine coupling constant(hfcc) of the rest four protons are less than the line width ($\leq$ 1.0 mT). Therefore, it was concluded that the $C_2H_6^+$ is distorted to C_{2h} and the SOMO is $4a_g$ (Fig. 1). The mode of deformation was associated with the J-T active e_g methyl rocking vibration (v_{12})[30]. From the molecular symmetry and INDO MO calculations, the geometry as in inset I was proposed, in which the methyl groups tilt slightly ($\sim$10°) so that two in-plane C - H bonds come to the upright position with respect to the C - C bond. This structure resembles that of diborane (B_2H_6)[31] rather than that of its anion, $B_2H_6^-$, which is isoelectronic with $C_2H_6^+$ and has been thought to have D_{3d} symmetry[32].

Upon raising the temperature to 77 K, the three-line spectrum changed reversibly into an equally spaced seven-line one with a splitting of 5.04 mT, which is 1/3 of that at 4 K (Fig. 2). This spectral change was attributed to the dynamic J-T effect, i.e., rapid exchange (or tunneling) among the three equivalent distorted forms with one of the three methyl protons at the upright position in the plane[10,33]. Such averaging of the hfcc cannot be achieved by the intramolecular rotation around the C - C bond. It was also confirmed from the line broadening observed at temperatures between 4 to 77 K, that the spectral change could neither be interpreted as the change in the Boltzmann populations with temperature of the two nearly degenerate levels, which has been often assumed in the solution esr of J-T active species[12]. The apparent activation energy(E_a) of 250 cal/mole was obtained, through the Bloch treatment of the spectra[33]. E_a is smaller by one order of magnitude than that for intramolecular rotation of the methyl group of ethane molecule in the gas phase (2.8 kcal/mole).

A deuterium isotope effect was found in the J-T distortion as well as in its dynamics for $CH_2DCH_3^+$, $CHD_2CH_3^+$, $CH_2DCH_2D^+$, $CH_2DCD_3^+$, and $CH_3CD_3^+$[33]. In partially deuteriated methyl groups in these cation radicals a C - H bond preferentially occupies the in plane upright position, whereas a C - D bond occupies essentially the same upright position when CH_3 is replaced by CD_3 (a minor difference was found; see table 1). A larger E_a ($\geq$ 500 cal/mole) is needed for averaging the hfcc when methyl group is partially deuteriated.

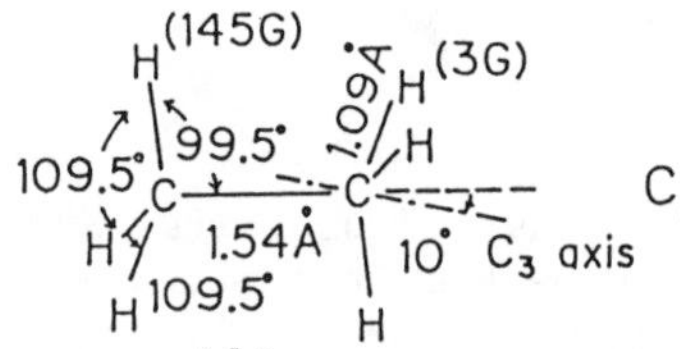

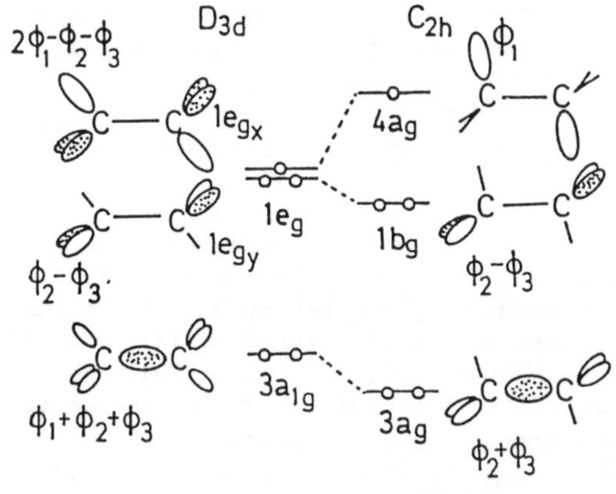

Fig. 1. Schematic representation of the Jahn-Teller split orbitals associated with the C_{2h} distortion along the e_g rocking mode (v_{12}) of $C_2H_6^+$.

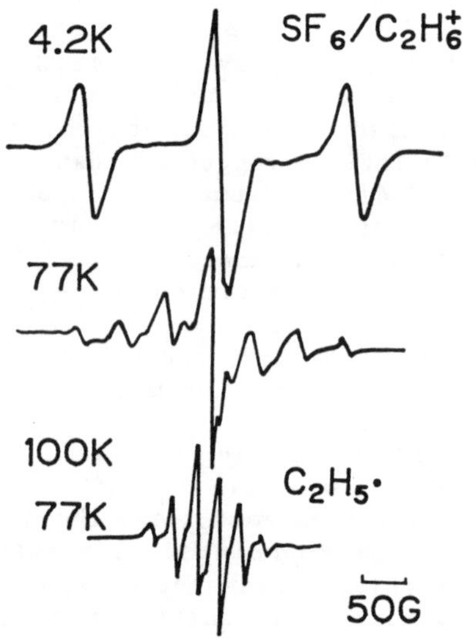

Fig. 2. ESR spectra of (a) $C_2H_6^+$ observed at 4.2 K, (b) $C_2H_6^+$ at 77 K, and (c) C_2H_5 observed at 77 K after annealing at 100 K. Matrix is SF_6.

Table 1. The Observed Hyperfine Coupling Constants of Alkane Cations[10]

cations	symm.	SOMO	matrix	temp. /K	hfcc/ mT
$CH_4{}^+$	(C_{2v})	b_1	Ne	4	5.48(4H)[a]
$CH_2D_2{}^+$	C_{2v}	b_1	Ne	4	12.17(2H); 0.22(2D)[a]
$C_2H_6{}^+$	C_{2h}	$4a_g$	SF_6	4	15.2(2H)
				77	5.03(6H)
$CH_3CD_3{}^+$		$4a_g$	SF_6	4	14.05(1H); 2.19(1D); .9(4H)
$C_3H_8{}^+$	C_{2v}	$4b_1$	SF_6	4	9.8(2H); -0.3(4H)[b]
$C_3H_8{}^+$	C_{2v}	$2b_2$	F113	4	10.5(2H); 5.25(4H)
$CD_3CH_2CD_3{}^+$		$4b_1$	SF^6	4	1.48(2D)
$n\text{-}C_4H_{10}{}^+$	C_{2h}	$7a_g$	F113	77	6.13(2H)
$n\text{-}C_6H_{14}{}^+$	C_{2h}	10_{ag}	F11	77	4.10(2H)
$(CH_3)_3CH^+$	C_s	$11a_1$	SF_6	4	5.8 (2H)
$(CH_3)_3CH^+$	C_s	$11a_1$	F113	4	5.25(2H)
$(CH_3)_3CH^+$	C_{3v}	$6a_1$	F113	77	25.0(1H); 4.75(3H)
$(CH_3)_3CD^+$	C_s	$11a_1$	F11	4	5.42(2H)
$(CD_3)_3CH^+$	C_s	$11a_1$	F11	4	0.85(2D); 1.8(1H)
$(CH_3)_4C^+$	C_{3v}	$8a_1$	SF_6	4	4.00(3H)
$(CH_3)_4C^+$	C_{3v}	$8a_1$	F113	4	3.98(3H)
$(CH_3)_4C^+$	C_{3v}	$8a_1$	F11	4	4.07(3H)
$(CH_3)_3CC(CH_3)_3{}^+$	D_{3v}	$6a_{1g}$	F11	77	2.8(6H); 0.38(12H)

[a] Reference [19].
[b] Reference [35].

3.2. PROPANE

Since the highest three occupied levels of propane, i.e., $2b_2$ (II), $4b_1$ (III), and $6a_1$ (IV), have energies close to each other, it is of interest which of these is the SOMO of its cation. A number of MO calculations as well as photoelectron spectroscopic works have been reported with controversial results[34]. Experimentally, two types of cations, II and III, were observed depending on the matrices[10].

The spectrum observed in F113 or in F11 was 2 x 5 lines and was attributed to the cation radical with $2b_2$ SOMO, which mainly consists of the three out-of-plane pseudo π_{CH2} orbitals with antibonding combination: [$\sqrt{2}\ \phi_2 - \phi_1 - \phi_3$]. This cation is characterized by its pseudo π orbital delocalized over the molecule and is unique with its large hfcc; i.e., 10.5 mT assigned to the two protons on the central carbon and 5.25

mT to the four out of plane methyl protons. The relative magnitude between these is about 2 : 1 in accord with the coefficients of the molecular orbitals.

On the other hand, a three line spectrum with a^H = 9.8 mT(2H) was observed in SF_6 at 4 K and assigned to the cation with $4b_1$ SOMO, which mainly consists of two in-plane C - H and two C - C σ-orbitals (III). This is a delocalized σ-radical, and the protons which give a large splitting of 9.8 mT are two in-plane end protons. The hfcc for the out-of-plane protons were not resolved explicitly but were obtained through the analysis of the line shape observed at higher temperatures[35]. They are - 0.3 mT for the protons on C_1 and smaller than 0.3 mT for those on the central carbon. The hfcc's of out-of-plane protons will be discussed in the next section.

In an INDO MO calculation, the cation with the $2b_2$ SOMO was obtained assuming the standard geometry, while that with the $4b_1$ SOMO is obtained if two CH_3 axis were tilted slightly (≥ 7°) toward the central CH_2 group in the molecular plane, maintaining the C_{2v} symmetry[10].

3.3. OTHER n-ALKANES

3.3.1. $[H(CH_2)_nH]^+$. Similar to the case of $C_2H_6^+$ and $C_3H_8^+$ with $4b_1$ SOMO, the cation radical of larger n-alkanes also give a nearly isotropic three-line esr spectrum with the intensity ratio of 1 : 2 : 1, due to the two equivalent protons[10,25,37,38]. The splitting decreases sharply with increasing the carbon number, from 15.0 mT for $C_2H_6^+$ to 1.0 mT for $C_{10}H_{22}^+$ (Fig. 3). By analogy from the results for $C_2H_6^+$ and $C_3H_8^+$ with $4b_1$ SOMO and with the aid of MO calculations, SOMO of the linear alkane cation radical with the fully extended conformation (all trans) was attributed to a σ-orbital which spreads all over the molecular skeleton. The symmetry is a_g or b_1 depending on whether the carbon number is even or odd (Fig. 4). The protons which give the largest hfcc are chain-end in-plane ones. Therefore, Toriyama et al. proposed an expression as $[H(CH_2)_nH]^+$ for the linear alkane-cation radicals with the fully extended form[37]. The sharp decrease of the hfcc with the increase of carbon number is brought from the spin delocalization throughout the molecules, which reduces the spin density in the chain-end in-plane C - H bonds.

The hfcc's of inner protons and out-of-plane methyl protons are usually small and do not give a resolvable hyperfine structure except for $n-C_4H_{10}^+$ and $n-C_6H_{14}^+$ (see Table 2). For $C_3H_8^+$ and $n-C_4H_{10}^+$, however, the hfcc's of the out-of-plane methyl protons were estimated including the signs through the analysis of the spectral change with the observing temperature[35,46] (see 3.3.2). The interesting results are small absolute values for propane cations and positive hfcc for the out-of-plane methyl protons of butane cation. Since these protons are located close to nodal plane of the unpaired electron orbital, negative hfcc's due to spin polarization is expected as is the case of α-proton in a neutral π-radical[36]. The cause of the unexpected positive and fairly large hfcc of n-$C_4H_{10}^+$ was attributed to the elongation of C_2 - C_3 bond, which brings a higher spin density on C_2 and C_3. As a result the negative spin induced on the methyl proton through the spin polarization by the unpaired electron on the methyl carbon is overcome by the positive spin raised through hyperconjugation from the spin on C_2 and C_3. Thus, a positive coupling is given to the out of plane methyl protons as is the case of a "β-proton" of localized π-radical[36]. The unexpectedly small hfcc's of out-of-plane protons of $C_3H_8^+$ (less than 1/2 of those for n-$C_4H_{10}^+$) can be understood in the same line. That is, the interaction with the unpaired electron on the adjoining carbon is offset by the interaction with that on the neighbor carbon.

In the case of n-$C_4H_{10}^+$ in CF_3CCl_3, non-planar geometry of the main frame is proposed, since the four methylene protons in this matrix are not equivalent but the existence of two pairs of protons is indicated[39-41,45].

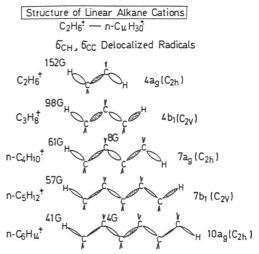

Structure of Linear Alkane Cations
$C_2H_6^+$ — n-$C_4H_{30}^+$

δ_{CH}, δ_{CC} Delocalized Radicals

Fig. 4. Schematic unpaired electron orbitals of n-alkane cations from $C_2H_6^+$ to n-$C_6H_{12}^+$ and their symmetry.

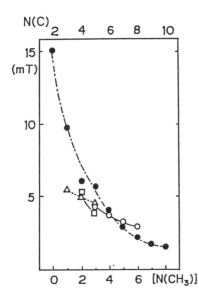

Fig. 3. Carbon number dependence of the hfcc of the in-plane chain end protons of the linear alkane cation radicals($\bullet$), and C - H_β proton coupling of methyl-substituted butane cation (O), methyl-substitute propane cation ($\square$), and of neutral π radicals ($\triangle$).

3.3.2. *Gauche Conformers* . All of the above descriptions are valid only for the linear alkanes with fully extended conformation. Gauche conformers were also detected in the frozen matrices[10,39-43,45]. In the gauche form, the spin delocalization along the carbon skeleton is terminated at the gauche carbon, and most of the unpaired electron is confined in the longer half. As a result, the hfcc of the in-plane protons become larger than those for the extended conformers. Those giving the main hfcc's are the chain-end in-plane proton and one of the protons bonded to the gauche carbon (Fig. 5). In this cation, however, the esr spectrum is not a three-line one of 1 : 2 : 1 intensity because these two in plane protons are not equivalent. From the INDO MO calculations[10,40] and the experiments with specifically deuteriated alkanes[40], the larger coupling was attributed to the proton on the gauche carbon.

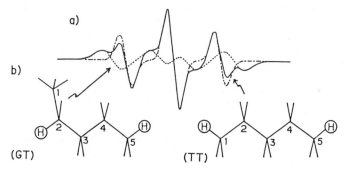

Fig. 5. (a)Esr spectrum of n-$C_5H_{12}^+$ in F113 observed at 77 K and (b)the schematic representation of *TT* and *TG* conformers. In (a), Dotted line shows TG conformer and the broken line TT conformer.

The probability that the alkane cation takes the gauche form and the position where the chain folding takes place depend on the chain length of the alkanes as well as the matrix adopted. In CF_3CCl_3 many types of gauche conformer (*GG* as well as *TG*) have been identified[39-43,45]. It is also reported that partially deuteriated isomers take folded conformation more frequently than the protiated molecules[45]. In addition, they sometimes have intermediate rotational angle around the C - C bond rather than ± 120°[39-43,45]. An isotope effect that C - H bond preferentially takes the in-plane position was found for the partially deuteriated n-butane cation radicals[45]. Isotope effects are described in another chapter.

3.3.3. *Intramolecular Motion of a Methyl Group.* The three line spectrum of the linear alkane cation radical indicates that the intramolecular CH_3 rotation is restricted. The rigidity of the methyl group at around 77 K is usual in linear and branched alkane-cation radicals, in contrast with those in olefinic or aromatic π-type radicals, in which the methyl group usually rotates freely in the same matrices at 77 K[1,2,47,48]. This is mainly due to the difference in the symmetry of intramolecular potentials, i.e. V_6 for the latter and V_3 for alkane cations with sp^3 configuration[48]. At a temperature higher than 77 K, however, the esr spectrum is sometimes broadened due to the restricted

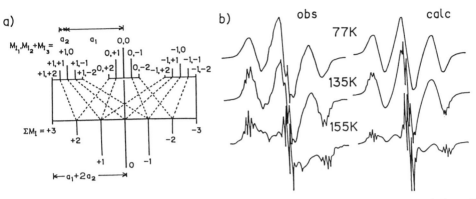

Fig. 6. (a)Illustration of the alternation of the line position by reorientation of the methyl groups and the resulting line broadening of the spectra for the system with two methyl groups having a set of one large (a_1) and two small (a_2) hfcc with the same sign. (b)Observed temperature change of the esr spectra of n-$C_4H_{10}^+$ in F11 and their simulation by the modified Bloch treatment.

Table 2. Linear alkane-cation radicals

n(C)	matrix	SOMO	temp. /K	hyperfine couplings/mT in-plane H	out-of-plane H methylene	methyl	[ref.] and other matrices examined
2	SF_6	a_g	4	15.2(2)[a]		≤ 1.0(4)	[10], C_2F_6
3	SF_6	b_1	4	9.8(2)	≤ 0.3(2)	-0.3(4)	[10],[35]
4	F113	a_g	77	6.1(2)			[37],F11[38]
	F11	a_g	130	6.0(2)	(-)0.55(4)	+0.78(4)	C_4F_{10}[35],SF_6[44]
	F113a		77	5.9(2)	0.74(2),0.5(2),0.04(2)		[40]
5	F113	b_1	77	5.7(2)			[37,45],F11[39]
	G			5.8(1) 9.9(1)			SF_6[44]
	F113a	GG	100	8.49(2)	(-)1.7(6)		[39]
6	F11	a_g	77	4.4(2)	(-)0.4(8)		[37],SF_6[44]
	F113	G	77	5.3(1) 7.4(1)			[10]
	F113a		141		(-)0.67(8)		[39]
7	F113	b_1	77	3.0(2)			[37],SF_6[44],CCl_4[
8	F113	a_g	77	2.2(2)			[37],C_8F_{20}[37]
	F11	G	77	2.2(1) 4.1(1)			[10],ZSM-5[25]
9	F113	b_1	77	1.7(2)			[37],F113a[39]
10	F113	a_g	77	1.6(2)			[37],F113a[39]
11-14	F113a		77				[39]

[a] Number of the equivalent proton.
[b] F113a stands for $CFCl_3CCl_3$.

methyl rotation. A typical case (n-$C_4H_{10}^+$ in F11) is shown in Fig. 6[46]. Through the spectral simulation with the modified Bloch equation, the activation energies for internal rotation of a methyl group around an electron deficient C - C σ-bond was estimated for $CH_3CD_2CH_3^+$ with $4b_1$ SOMO and n-$C_4H_{10}^+$[35,46]. They were 2.5 Kcal/mole for the former in SF_6, and 2.3 to 2.4 Kcal/mole for n-$C_4H_{10}^+$ in F11 and n-C_4F_{10}. Assuming the zero-point energies of 0.3 - 0.4 kcal/mole, these correspond to the potential barriers of 2.9 - 2.7 kcal/mole[50], which are not very much reduced from those of the corresponding mother molecules. This fact indicates that the electron deficiency in those C - C σ-bonds dose not reduce the barrier not so much.

In addition to the barrier height, as is seen from Fig. 6, hfcc of the out-of-plane methyl protons and its sign can be estimated from the temperature change of the over-all spectral width[35,46].

3.4. METHANE AND ITS HOMOLOGUES WITH T_d SYMMETRY

Esr observations of the cation radical of methane (IP = 13.5eV) in frozen matrices had been unsuccessful for a long time. The radiolysis in SF_6 (IP = 15.6 eV) at 4 K gave only the complex of methyl radical with a fluorinated compound [$CH_3 \cdots FX$] as the precursor of methyl radical[18]. In 1984, however, Knight et al. succeeded in esr detection of the methane cation radical by Ne discharge photoionization at 4 K[19]. They observed a quintet spectrum with $|a^H|$ = 5.48 mT for CH_4^+ and triple-quintet with $|a^H|$ = 12.17 mT(2H) and $|a^D|$ = .222 mT(2D) (corresponds to 1.4 mT for H) for $CH_2D_2^+$.

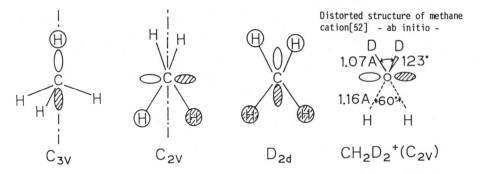

Distorted structure of methane cation[52] - ab initio -

For the J-T distortion of the cation radical of a tetrahedral molecule (T_d symmetry), C_{3v}, C_{2v}, and D_{2d} are expected. Since hfcc's of the four protons are equal, the distorted geometry of CH_4^+ might be attributed to D_{2d} symmetry. However, the result for $CH_2D_2^+$ indicates that the protons in methane cation are classified into two non-equivalent pairs. Thus, they assigned the SOMO of CH_4^+ as well as that of $CH_2D_2^+$ to be b_1 orbital of C_{2v} symmetry. In the case of CH_4^+, only the averaged state of J-T distortion can be detected, due to the lower activation energy of dynamic J - T as compared to that of $CH_2D_2^+$. The C_{2v} symmetry is the same as that of BH_4[51], which is isoelectronic with CH_4^+. The distorted geometry which gives the observed hfcc (<HCH = 59°,

C - H = 1.164 Å, <H'CH' = 123 °, and C - H' = 1.075 Å : inset V) was shown by using ab initio MO calculation[19]. In $CH_2D_2^+$, deuterons occupy the shorter bonds preferentially. Paddon-Row and his coworkers attributed the cause of this isotope effect on the site preference to the difference in the zero-point energies of the normal vibrations[52]. It is noteworthy that the trend that the C - H bonds prefer the position which is distorted more than the alternative position is similar with the case of partially deuterated ethane cation radicals[33].

In Table 3, esr results on the cation radicals of methane and its homologues with T_d symmetry are listed. No cation with D_{2d} has been detected so far. Among these, SnH_4^+ is unique since it takes two forms with different SOMO, C_{3v} and C_{2v}, though the latter is less stable and converts itself to the C_{3v} cation during the storage at 77 K[53]. In addition, the SOMO of SnH_4^+ with C_{2v} symmetry is a_1 in contrast with the other C_{2v} radicals, CH_4^+ and BH_4, which have b_1 SOMO.

Neopentane cation radical is another J-T active alkane-cation produced from a molecule of T_d symmetry. For this cation, C_{3v} SOMO was suggested from the equally spaced four line spectrum with the coupling of 4.0 mT[9,10]. From the INDO MO calculation, a distortion with flattened $C(CH_3)_3$ group was suggested[10]. Similarly, the loss of 5s character in the SOMO of $Sn(CH_3)_4^+$ with C_{3v} symmetry is suggested[53].

Table 3. Cation Radicals with T_d Symmetry

cations	symm.	SOMO	g-value	hyperfine coupling (mT)			ref.
CH_4^+	C_{2v}	b_1	2.0029	5.48(4H)			[22]
$CH_2D_2^+$	C_{2v}	b_1	2.0029	12.17(2H)	0.222(2D)		[22]
SnH_4^+	C_{2v}	a_1	2.0100	8.5(2H)	-261.0(^{119}Sn)		[53]
	C_{3v}		2.005	17.5(1H)	-333.6(^{119}Sn)		
BH_4 [a]	C_{2v}	b_1		10.71(2H)	0.68(2H)	1.45(B)	[51]
$C(CH_3)_4^+$	C_{3v}	a_1	2.0025	4.2(3H)			[10]
$Sn(CH_3)_4^+$	C_{3v}	a_1	2.029	-1.35(3H)	-12.2(Sn)	7.53(^{13}C)	[53]
$Si(CH_3)_4^+$	C_{2v}	(a_1)	2.0050	1.36(6H)	0.48(6H)		[54]
$Ge(CH_3)_4^+$	C_{2v}	(a_1)	2.0196	1.47(6H)	0.42(6H)		[54]

[a] A neutral radical isoelectronic with CH_4^+.

4. Cation Radicals of Branched Alkanes

4.1. ISOBUTANE

The simplest branched alkane is 2-methylpropane (isobutane). Again, two kinds of cation radicals were detected[10] due to the small energy difference between the two highest occupied levels in the isobutane molecule with C_{3v} symmetry, that is, $6a_1$ (VII) and $5e$ (VIII)[55] (see Table 1).

C_{3v} C_S

$6a_1$ ———•— ϕ_{CH} $2\phi_1\text{-}\phi_2\text{-}\phi_3$ $11a'$ ———•— ϕ_1

$5e$ ═══ $2\phi_1\text{-}\phi_2\text{-}\phi_3$ $\phi_2\text{-}\phi_3$ $5e$ ——•— $\phi_2\text{-}\phi_3$

             $\phi_2\text{-}\phi_3$ $6a_1$ $6a''$

                      ϕ_{CH} —•—•—

(VII)   (VIII)

The cation radical formed in SF_6 at 4 K gives an esr spectrum of a triplet with the splitting of 5.8 mT, and was attributed to a C_s cation[9,10]. The SOMO is 11a', which is derived from 5e orbital by J-T distortion. In this cation, the unpaired electron is confined mainly in one of the C - C bonds, and the two trans β-protons, one from each methyl group, have a large hfcc. The cation radical with C_s symmetry was also generated both in F11 and F113 if it was irradiated at 4 K[10]. In these matrices, however, the symmetry of the cation changed to C_{3v} (SOMO is $6a_1$) irreversibly when the sample is warmed up to 77 K, suggesting that the C_{3v} cation is more stable than that with C_s symmetry. In the C_{3v} cation, the unpaired electron is mainly confined in the tertiary C - H bond and this proton shows a large hfcc of 25.0 mT[10].

4.2. METHYL SUBSTITUTED BUTANES

4.2.1. *Hexamethylethane (HME)* . The hexamethylethane cation radical (HME^+) is the first alkane cation radical observed by esr. From the seven-line esr spectrum with the splitting of 2.9 mT, it was suggested that the unpaired electron is mostly confined in the central C - C bond and the six protons (one from each six methyl groups) located in the trans position with respect to the central C - C bond couple with the unpaired electron via hyperconjugation[7]. Shida et al. observed the hyperfine structure for the rest 12 protons (gauche C - H_β : a = 0.42 mT) and confirmed that the methyl groups are fixed at 77 K, in contrast with the case of hexamethyldisilane cation[8]. They attributed the difficulty of the methyl group rotation to the shorter C - CH_3 bond compared with the Si - C bond.

4.2.2. *Other Methyl-Substituted Butanes* . Five kinds of methyl substituted butane cations, including HME^+, were investigated as a series to study the origin of the β-proton coupling in a cation with a σ-localized SOMO[56]. The methyl substitutions at C_2 and C_3 result in a regular increase of the hyperfine lines by one per one methyl substitution, with the concomitant decrease in each proton coupling (Table 4). From the fact that the decrease of proton coupling with increasing the number of methyl substitution is not so sharp as compared with the sharp carbon-number dependence in n-alkane cations (Fig. 3), Iwasaki and his coworkers deduced that the unpaired electron in a branched alkane cation tends to localize in one of the σ-bonds as to maximize the hyperconjugation effect. They also estimated B_2 values in the $\cos^2\theta$ rule for the β-proton coupling of cation radicals assuming:

$$a_\beta = P_c \, B_2 \cos^2\theta \quad : \quad \text{where} \quad P_c = 2P_{cc}. \tag{4}$$

They are 11.8 mT for the trans C - H_β and 1/2 of it for the gauche C - H_β. A large difference in the B_2 values for the front and the back lobe interaction is a characteristic feature of a bent radical, common to cation and neutral radical. In the case of neutral radical, it has been confirmed experimentally[57] as well as theoretically[58].

σ-cation radical   neutral π-radical

2.85 mT   5.37 mT

0.38 mT   1.34 mT

$a(H_g)/a(H_t) \approx 1/8$ $a(H_g)/a(H_t) \approx 1/4$

It should be noticed here that the magnitude of B_2 value for the alkane cation radical (11.8 mT) is fairly large compared with that for a neutral π radical (5.8 mT). This is common to olefinic cations (12.1 mT) and is attributed to the positive charge effect proposed for the aromatic cation radical[59].

4.2.3. *Other Branched Alkanes* . Radical cations of 3-methylpentane (3MP$^+$)[60,61], 2-methylbutane (2MB$^+$)[61], and 3-methylheptane (3MH$^+$)[62], have been observed by esr. Since the spectrum of 3-MP$^+$ was almost the same as that of 2-MB$^+$ (4-line spectrum with the separation of ca. 4.5 mT), Toriyama et al. concluded that 3MP$^+$ takes the GT conformation in F113 and the unpaired electron is confined mainly in C_3 - C_4 bond. As the result, the large hfcc's are arisen from one proton on each of three carbons, C_2, C_5, and C_6, which corresponds to trans C - H_β with respect to C_3 - C_4 bond[60]. By using some specifically deuteriated isomers in the experiment as well as by the spectral simulations, Shiotani et al. refined the hfcc's for 3-MP$^+$ and 2-MB$^+$ and proposed a slightly different geometry to both of them[61]. Ichikawa and Ohta, reported that esr detection of the cation radical of some methyl-substituted butanes and heptanes as well as n-alkanes (n = 6 - 11) is possible even in the saturated hydrocarbon matrices (3-MP, 4-methyl-nonane etc.), if an electron scavenger like a Freon is added[62]. They extracted those spectra with the same method adopted by Nauwelaerts and coworkers[5].

Conformation of 2-MB$^+$ (IX) and 3-MP$^+$ (X)

(IX)   (X)

Table 4. Branched Alkane Cation Radicals

cations	matrices	a^H(trans)/mT	a^H(gauche)/mT	ref
2MB	F113	4.30(3H)		[56]
		5.05(1H) 4.33(2H)		[61]
2,2 DMB	F113	3.70(4H)		[56]
2,3 DMB	F113	3.75(4H)		[56]
	CFCl$_3$	4.10(4H)		[38]
2,2,3 TMB	F113	3.20(5H)		[56]
HME	F113	2.90(6H)		[56]
	CBr$_4$	3.2(6H)	0.45(12H)	[6]
	FM	2.90(6H)		[7]
	CFCl$_3$	2.90(6H)	0.42(12H)	[8]
		2.88(6H)	0.38(12H)	[37]
3MP	F113	4.5(3H)		[60]
		5.29(1H) 49.9(1H) 3.93(1H)		[61]

5. Cation Radicals of Cycloalkanes

Cation radicals of basic mono-cyclic alkanes with the carbon number N_C = 3 to 8 have been studied[12,63-75]. All of them except c-$C_5H_{10}^+$ and c-$C_7H_{14}^+$ are J-T active, and the mode of J-T distortion was the primary concern. The observed hfcc's and the symmetry of SOMO deduced are listed in Table 5 and Fig. 7. The characteristic features found commonly in the cycloalkane cation radicals are: a) delocalization of the unpaired electron all over the molecular frame, and b) a fairly large positive g-shift in comparison with the usual carbon centered radicals or aromatic J-T cations. The large g-shift was attributed mainly to the excitation of an electron from σ_x to the half filled σ_y, which is expected to be effective in these J-T active σ-cation radicals.

$$\text{J-T } \sigma\text{-cation} \quad (\sigma_y)^2(\sigma_x)^1 \xrightarrow{L_z} (\sigma_y)^1(\sigma_x)^2 \quad (5)$$

$$\text{J-T } \pi\text{-cation} \quad (\pi)^2(\pi')^1 \xrightarrow{\;\;L_z\;\;}\!\!\!\!\!\!\!\times (\pi)^1(\pi')^2 \quad (6)$$

The overall spectral width (or the sum of the hfcc's of the protons) of cycloalkane-cation radicals increases with the increase of the number of carbons comprising the ring. This is in marked contrast with the carbon number dependence of the hfcc's for linear alkane cation radicals. This is caused by the fact that in a σ-delocalized radical, only the protons located in the molecular plane can participate in SOMO to have large hfcc's. Only two protons satisfy this condition in n-alkanes regardless of the length, while the number of in plane protons increases directly with the length in cyclo-alkanes.

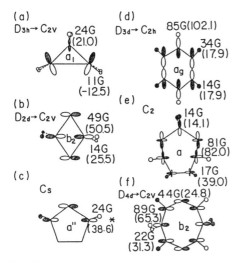

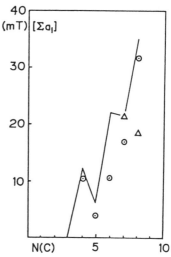

Fig. 7. The static distortions and the unpaired electron orbitals of a series of cycloalkane cation radicals. The assignment of the observed hfcc and the comparison with the INDO values (in parenthesis) are also given.

Fig. 8. Carbon number dependence of the sum of the all hfcc's of c-alkane cation radicals. N(c) is the number of carbon constructing the ring. (o): cyclo alkane cations, (△): bicycloalkane cations.

Table 5. Cycloalkane-cation radicals[15]

cations	matrices	symm.	SOMO	g (4 K)	hfcc (4 K/mT)	position of H	Ref.[a]
c-$C_3H_6^+$	F113	D_{3h} - C_{2v}	a_1	2.0040	2.40(2H)	C_1	63
					-1.10(2H)	C_2	
c-$C_4H_8^+$	F11	D_{2d} - C_{2v}	b_2	2.0045	4.9(2H)	C_2(e)	
					1.4(2H)	C_2(a)	
c-$C_5H_{10}^+$	F113	C_s	a''	2.0036	2.5(2H)	C_2(e)	63,75,76
c-$C_6H_{12}^+$	F11	D_{3d} C_{2h}	a_g	2.0066	8.5(2H)	C_1(e)	66-70
					3.4(2H)	C_2(e)	
					1.4(2H)	C_3(e)	
c-$C_7H_{14}^+$	F113		a	2.0042	8.1(2H)	C_3(e)	
					1.7(2H)	C_4(e)	
					1.4(2H)	C_1	
c-$C_8H_{16}^+$	F11	D_{4d} - C_{2v}	b_2	2.0059	8.9(2H)	C_3(e)	
					4.4(4H)	C_2(e)	
					4.2(2H)	C_3(a)	
BCH$^+$	F113		a_2	2.0052	6.51(4H)	C(exo)	78,79
					3.5(2H)	C_1	
BCO$^+$	F11	D_{3h} - C_{2v}	b_2	2.0098	3.85(4H)	C(end)	79
					1.58(4H)	C7,8	
					8.2(4H)	C(exo)	
cis-DEC$^+$	C_6F_6/cis-DEC				5.0(4H)		77

[a] References other than [15].

5.1. CYCLOPROPANE

Cyclopropane (c-C_3H_6) with D_{3h} symmetry possesses a degenerate highest occupied orbital ($3e'$). For c-$C_3H_6^+$, two deformed structures with C_{2v} symmetry can be expected through the J-T distortion. They are the $^2A_1$ cation with $6a_1$ SOMO and that of the $^2B_1$ with the SOMO of $3b_1$[30]. Shida and his coworkers observed the esr spectrum of its cation at 77 K, and obtained only a single line (peak to peak width was $\approx$ 1.5 mT). Therefore, they calculated the electronic state with an ab initio method and obtained $^2B_1$ ground state[63].

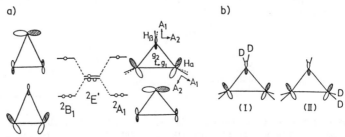

Fig. 9. (a)The unpaired electron orbitals of c-propane cation radical in the $^2A_1$ and $^2B_1$ states. (b) Isotope site preference in the distorted c-propane cation in $^2A_1$ state.

On the other hand, Iwasaki et al. obtained a well resolved and widely spread esr spectrum at 4.2 K in F11 and F113, and interpreted these with four protons with equivalent hf tensor (A = -0.28, -1.23, -1.79, a = -1.17 mT) and two protons with less anisotropic hfcc (A = 2.7, 2.3, 2.18, a = 2.04 mT)[64]. The anisotropy in these two hf tensors are characteristic of those of α and β proton, respectively. Thus, it was attributed to the radical with $6a_1$ SOMO, in which the unpaired electron is confined mainly in the basal C - C bond of the obtuse triangle. The distortion is along the normal vibration mode of e' ring deformation.

At a higher temperature rapid exchange among the three equivalent distorted forms takes place[15,64]. The dynamic J-T effect in this cation shows an interesting phenomenon, in which three hfcc's are averaged out to approximately zero at the high exchange limit ($[a_1 + a_2 + a_3]/3$ = [2.4 - 1.17 x 2]/3 = .02 mT). The activation energy of the dynamics is dependent on matrix (E_a = 34 - 680 cal/mole, A = 0.6 - 3.4 x 10^9 /s) and is especially low in SF_6.

Isotope effects on the J-T distortion and on its dynamics were also found in cyclopropane-1,1-d_2^+[65]. The SOMO is $6a_1$ of C_{2v} symmetry as shown above for c-$C_3H_6^+$, and CD_2 preferentially occupies the top of the obtuse triangle. In contrast with the partially deuteriated methane and ethane cation radicals, the unstable form with CD_2 on one of the basal positions is also detected at temperatures below 20 K. The potential barrier of the dynamic J-T increases with this partial deuteriation.

5.2. CYCLOHEXANE AND THEIR DERIVATIVES

The cation radical of c-C_6H_{12} has been investigated by many groups from the earliest period[66-70,15], since this is a representative system in radiation chemistry. The spectrum at 77 K as well as at 4 K were broad and difficult to analyze, but it becomes well resolved at T ≥ 140 K to give 7 lines (4.3 mT) due to 6-equivalent equatorial protons[68]. The spectrum at 4 K was interpreted with three pairs of protons with the hfcc's of 8.5, 3.4, and 1.4 mT, respectively, and these hfcc's were assigned to the equatorial protons of the cation in ^2A_g state[15,69]. The distortion to elongate two parallel C - C bonds was suggested.

The highest occupied orbital of c-C_6H_{12} with D_{3d} symmetry is the degenerate e_g, which splits into a_g and b_g orbitals by the distortion to C_{2v}. This distortion may be arisen from the vibronic coupling with the J-T active e_g vibration mode. From the symmetry of these orbitals, two protons with a large hfcc are expected for a_g SOMO, and four protons for b_g SOMO[30]. The observed three pairs of coupling protons do not satisfy either of these demands. Toriyama and Iwasaki regarded the two smaller couplings as basically in one group and explained the small difference between these two couplings by a deviation from the C_{2v} symmetry. In addition, this deviation was attributed to the restricted dynamics, which takes place even at 4 K, along the potential surface with three nonequivalent energy-minima (Fig. 9). The activation energy for this dynamics was estimated as low as 170 and 240 cal/mole through the modified Bloch analysis[15]. Lunell and Lund proposed another interpretation based on an ab initio MO calculation, that the cation is distorted statically to a C_s symmetry at 4 K, which gives three pairs of protons (7.0, 2.8, -1.9 mT)[70]. In this model, however, a pair of protons have a negative hfcc so that it is hard to explain the spectral change at temperatures from 4 K to 77 K, during which the spectral width was kept unchanged.

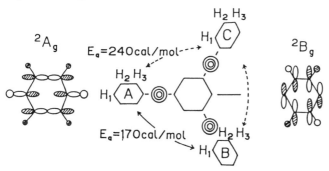

Fig. 10. The unpaired electron orbitals of c-hexane cation radical and the illustration of the dynamic J-T on the potential surface with non-equivalent three minima.

Shiotani et al. investigated a series of methyl and ethyl substituted cyclohexane cation radicals[71-72]. They asserted that the SOMO of the mono-substituted cation is a modified a_g. Details will be described in a separate chapter. The present author, however, considers that the effect of alkyl substitution in a σ-radical may not be merely an electrostatic perturbation to the SOMO as in the case of a π-cation, and that the molecular frame for MO consideration should be expanded to the substituents[73].

5.3. OTHER CYCLOALKANES

c-C_4H_8 is known to have a puckered ring with D_{2d} symmetry, and two kinds of ring deformation are expected for the static J-T effect: D_2 or C_{2v} symmetry[74]. From the esr spectrum observed in $CFCl_3$ at 4 K (a triple-triplet with hfcc's of 4.9 mT(2H) and 1.4 mT(2H)), the distorted form of this cation was determined to be C_{2v} with b_2 SOMO[75]. Once the sample was annealed at 77 K, reduction of the barrier for the ring puckering and/or the flattening of the ring is induced irreversibly to give a 5-line spectrum at 4 K. At a temperature higher than 77 K, a 9-line spectrum from 8-equivalent protons (a = 1.33 mT) was observed as the result of the "pseudo-rotation" by the dynamic J-T effect.

The esr spectrum of c-$C_5H_{10}^+$ is a simple three line with a hfcc of 2.5 mT at a temperature below 77 K, and is easily attributed to the a" SOMO with the C_s symmetry, where the unpaired electron is located in the W-skeleton made of C_5 - C_1 - C_2 carbons and two equatorial protons on C_5 and C_2[63,76,77]. There may be some ambiguity in the assumed geometry, however, because no MO calculation could reproduce the observed hfcc of this cation well. (The hfcc's of most of the other alkane cations investigated so far are well reproduced by INDO or ab initio MO calculations.) At a temperature higher than 77 K, all the protons in the ring become equivalent as the result of ring inversion as well as the pseudo rotation around the C_5 axis through the ring puckering[77]. The activation energies for these processes were estimated through the spectral simulation to be 1.2 and 3.6 kcal/mole, respectively[76].

The c-$C_7H_{14}^+$ formed in F113 was interpreted as a cation with twisted-chair form of C_2 symmetry[15]. c-$C_8H_{16}^+$ showed an esr spectrum which is interpreted with three sets of non-equivalent coupling protons. The molecular symmetry was determined to be C_{2v} and the SOMO is b_2[15].

5.4. BICYCLOALKANES

The cation radicals of norbornane (or bicyclo[2,2,1]heptane: BCH) and bicyclo[2,2,2]octane (BCO) were investigated using F113 and F11 matrices. The SOMO of BCH cation is $4a_2$ (XI), in which the unpaired electron delocalizes over the four exo-C - H bonds, so that those four protons give a large hfcc of 6.51 mT[78,79]. BCO cation is J-T active and exhibits a static distortion from D_{3h} to C_{2v} at 4 K in F11. The SOMO was tentatively assigned to $6b_2$ (XII), in which the unpaired electron delocalizes over the four endo-C - H bonds, giving the largest hfcc of 3.8 mT. At 77 K in c-C_6F_{12} or F113 a dynamically averaged esr spectrum with 12 equivalent protons was observed[79].

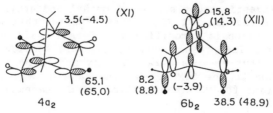

The esr detection of the cation of quadricyclane, a highly strained bicycloalkane, was unsuccessful, instead the esr spectrum of norbornadiene cation was observed even at 4 K[79]. This result indicates that the quadricyclane cation, once formed, is converted to the norbornadiene cation during the irradiation.

The cation radical of cis-decalin was first detected by ODESR in solution by Molin et al.[80], and the hfcc of 5.0 mT was assigned to the four protons in trans positions with respect to the central C - C bond, in which most of the unpaired electron is confined[56].

6. Reactions of Alkane Cation Radicals

Reactions of alkane-cation radicals by thermal- and photo-excitation have been investigated in various matrices for almost all the cations detected by esr. The major products of these identified reactions are listed in Table 6. In general, thermal excitation induces deprotonation to form the corresponding alkyl radicals, while photo-excitation causes elimination of H_2 or CH_4 to form olefin cations[10,15,81,82].

6.1. THERMAL REACTIONS: SELECTIVE DEPROTONATION

Thermal excitation of alkane-cation radicals usually induces deprotonation reaction to give a neutral alkyl radical. This reaction takes place in F113, SF_6, and PFMCHx for normal-, branched-, and cyclo-alkane cation radicals[81,15].

$$RH_2^+ \longrightarrow RH\cdot + H^+ \tag{7}$$

It has been deduced from a series of experiments that deprotonation takes place preferentially from the C - H bond which has the highest unpaired electron density in the cation radical[10]. This mechanism was derived from the following observations.

A different alkyl radical was formed if the SOMO of the alkane cation was changed in a different matrix. That is: 1-propyl radical is mainly formed from $C_3H_8^+$ with $2b_1$ SOMO, in which protons having the highest spin density are in-plane chain-end protons, while 2-propyl radical is generated from the $C_3H_8^+$ with $2b_2$ SOMO. The latter cation is a π-type radical and the higher spin density appears in the C_2 - H bond. Similarly, iso-$C_4H_{10}^+$ with C_{3v} symmetry is converted to t-C_4H_9 radical, and the cation with C_s SOMO (C-C σ-radical) is converted to iso-C_4H_9 by loosing one of the methyl protons[10]. A highly selective reaction was also observed in SF_6 for a series of n-alkane cations with n(C) = 4 to 7. In this case, 1-alkyl radicals were formed selectively from the corresponding n-alkane cations with the extended conformation by the deprotonation from one of the methyl groups[44]. In the methyl substituted butane cation, deprotonation from the primary C - H with a high spin density was found as the initial alkyl radical, although it was converted further to a tertiary alkyl radical by the successive intramolecular proton transfer[82].

Table 6. Products of the Reaction of Alkane-Cation Radicals

cation radicals	matrices	character of SOMO	product thermal	product photo-excitation	ref
$C_2H_6^+$		σ_{CH} del	CH_2CH_3		10
$CH_3CD_3^+$			CH_2CD_3, CH_3CD_2		10
$C_3H_8^+$	SF_6	$\sigma_{CH,CC}$ del	$CH_2CH_2CH_3(0.7)$		10,79
			$CH_3CHCH_3(0.3)$		
	F113	π_{CH} del	CH_3CHCH_3		10
	F11		$CH_2CH=CH_2^+$		
i-$C_4H_{10}^+$	SF_6	σ_{CC} loc	$CH_2CH(CH_3)_2$		10
	F113	σ_{CH} loc	$C(CH_3)_3$	$C(CH_3)_3$	
	F11	σ_{CH} loc	$(CH_3)_2C=CH_2^+$	$(CH_3)_2C=CH_2^+$	10
neo-$C_5H_{12}^+$	SF_6	σ_{CC} loc	$CH_2C(CH_3)_3$		10
	F113	σ_{CC} loc	$CH_2C(CH_3)_3$	$(CH_3)_2C=CH_2^+$	79
	F11	σ_{CC} loc	$(CH_3)_2C=CH_2^+$	$(CH_3)_2C=CH_2^+$	79
n-$C_4H_{10}^+$	SF_6	$\sigma_{CH,CC}$ del	$CH_2CH_2CH_2CH_3$		44
	F113		$CH_3CHCH_2CH_3$		10,37
	CF_3CCl_3			$CH_3CH=CHCH_3^+$	39
n-$C_mH_{2m+2}^+$	SF_6	$\sigma_{CH,CC}$ del	$CH_2(CH_2)_{m-2}CH_3$		44
m = 5 - 7	CF_3CCl_3			$CH_3CH=CHCH_3^+$	39
2-MB	F113	σ_{CC} loc	$CH_3CH(CH_3)CH_2CH_2$		79
			$CH_3CH(CH_3)CHCH_3$		
2,2DMB		σ_{CC} loc	$(CH_3)_3CCH_2CH_2$	$(CH_3)_2C=CH(CH_3)^+$	79
2,3DMB		σ_{CC} loc		$(CH_3)_2C=CH(CH_3)^+$	79
2,2,3TMB		σ_{CC} loc	$(CH_3)_3CCH(CH_3)CH_2$	$(CH_3)_2C=C(CH_3)_2^+$	79
			$(CH_3)_3CC(CH_3)_2$		
c-C_3	SF_6	σ_{CC} del	c-C_3H_5, $CH_2CH_2CH_2^+$		15,92
	F113		$CH_2CH_2CH_2^+$		84
c-C_4	SF_6	$\sigma_{CH,CC}$ del	c-C_4H_7		89
	F11			$CH_2=CHCH_2CH_3^+$	89
				$CH_3CH=CHCH_3^+$	89
c-C_5 - C_8	F113	$\sigma_{CH,CC}$ del	c-C_mH_{2m-1}		15
c-C_6	CF_3CCl_3	$\sigma_{CH,CC}$ del		c-$C_6H_{10}^+$	39
BCH	F113	$\sigma_{CH,CC}$ del	(- H^+)		82
BCO	F113	$\sigma_{CH,CC}$ del	(- H^+)		82

[a] abbreviations: loc: localized, del: delocalized

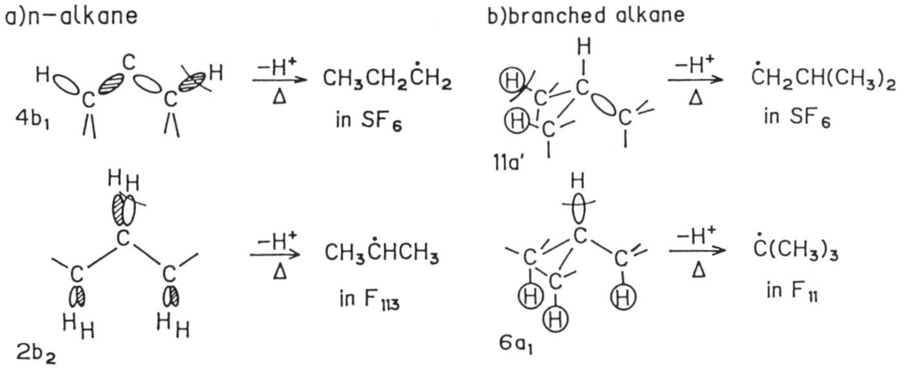

a)n−alkane

b)branched alkane

$4b_1$ $\xrightarrow[\Delta]{-H^+}$ $CH_3CH_2\dot{C}H_2$ in SF_6

$11a'$ $\xrightarrow[\Delta]{-H^+}$ $\dot{C}H_2CH(CH_3)_2$ in SF_6

$2b_2$ $\xrightarrow[\Delta]{-H^+}$ $CH_3\dot{C}HCH_3$ in F_{113}

$6a_1$ $\xrightarrow[\Delta]{-H^+}$ $\dot{C}(CH_3)_3$ in F_{11}

Fig. 11. Selective deprotonation to form alkyl radicals.

The observations which conflict with this rule were reported in the reactions of n-butane and n-pentane cations in F113, in which only 2-butyl and 2-pentyl radical, respectively, were formed[37,10,39,42]. Toriyama et al. suggested two possible causes for the formation of 2-alkyl radical in F113: one is the conformation change of the alkane cation radical to the gauche form in advance to the deprotonation reaction, and the other is the isomerization of originally generated 1-alkyl radical into 2-alkyl radical. Both of them may be possible at around the reaction temperatures (above 105 K) in a soft matrix like F113.

Although the isotope effect in the site selection for the distortion of alkane cation, which is a preliminary step for deprotonation, is considerable, that in the deprotonation step itself is not so large. This is deduced from the observation that the isotope effect in the yield of deprotonation reaction is only ca. 1.3 for $CH_3CD_3^+$[10], while that for $CHD_2CHD_2^+$ is as large as 4.0[83]. Decrease of the alkyl-radical yield by deuteriation was also observed in other system[10,40].

A ring opening was reported for the thermal reaction of c-propane cation radical in F113[84].

$$(8)$$

The product suggested in the above mechanism had attracted a strong interest of many researchers because it has a structure with the positive charge and the spin separated from each other. In order to explain the spin localization, the "orthogonal" structure, in which the planes of the two outer CH_2- groups are at right angles, was postulated. Further studies on methyl substituted c-propanes and other related compounds have been made to confirm this reaction[85,86]. However, with all these efforts, there still remains some ambiguity if the product really has a positive charge or not. The formation of a complex with the matrix or a product from it has been suggested to stabilize this cation[87,88]. From the esr spectrum itself, this radical cannot be distinguished from 1-propyl radical or its chlorine derivatives. In fact, this reaction takes place efficiently only in the matrices, F113, SF_6, and PFMCHx, in which efficient alkyl radical formation from the cation, probably by a bimolecular reaction, has been observed for the other alkane cations.

6.2. PHOTOLYSIS OF CATIONS: DEHYDROGENATION

Most of the alkane cations have absorption bands in the visible or near UV region. The elimination of H_2 or CH_4 to generate corresponding olefinic cations by visible light illumination was observed for i-$C_4H_{10}^+$, neo-$C_5H_{12}^+$[10,81], and methyl-substituted alkane cations[81] produced in F113 or F11.

$$RH_2^+ \quad \longrightarrow \quad R^+ + H_2 \quad (\text{ or } CH_4 \quad) \qquad (9)$$

Photolysis of linear and cycloalkane cations in these matrices does not produce any new radicals[41,82]. In CF_3CCl_3, however, the alkane cations behave differently with those in other matrices. For example, the elimination of H_2 from c-hexane cation to form c-hexene cation was observed[67].

$$c\text{-}C_6H_{12}^+ \quad \longrightarrow \quad c\text{-}C_6H_{10}^+ + H_2 \qquad (10)$$

In addition, C - C bond cleavage to form 2-butane cation was induced by visible light ($\lambda \geq 390$ nm) for n-alkane cation radicals with n(C) = 4 to 7 in CF_3CCl_3[39]. This reaction was greatly suppressed by partial deuteriation of the alkanes[41].

$$C_nH_{2n+2}^+ \quad \overset{h\nu}{\longrightarrow} \quad CH_3CH=CHCH^+ + C_{n-4}H_{2(n-4)+2} \qquad (11)$$

A photo-induced ring opening was reported for c-$C_4H_8^+$ in F11. This cation is converted to 1-butene cation by C - C cleavage followed by the 2 - 4 proton transfer[89]. Further illumination by visible light induces another proton transfer from 3 to 1 to produce 2-butene$^+$, as was already reported by Shida and coworkers for the 1-butene cation formed from the radiolysis of 1-butene[90].

$$
\begin{array}{ccc}
\begin{array}{ccc} CH_2 & - & CH_2 \\ | & + & | \\ CH_2 & - & CH_2 \end{array} &
\longrightarrow &
\begin{array}{ccc} CH_2 & & CH_3 \\ \| & + & | \\ CH & - & CH_2 \end{array} \quad
\longrightarrow \quad
\begin{array}{ccc} CH_3 & & CH_3 \\ | & + & | \\ CH & = & CH \end{array}
\end{array} \qquad (12)
$$

Visible light induced hole transfer from the solute alkane (n-heptane) cation to the matrix (CCl_4) was found[68].

$$CCl_4^+ + n\text{-}C_7H_{16} \quad \underset{\text{Visible light } (\lambda \geq 600nm)}{\overset{\text{thermal } (T \geq 130 \text{ K})}{\rightleftharpoons}} \quad CCl_4 + n\text{-}C_7H_{16}^+ \qquad (13)$$

The solute cation, n-$C_7H_{16}^+$, was only observed upon limited warming of the solid solution of n-C_7H_{16}(IP = 9.9 eV) in CCl_4(IP = 11.4 eV) irradiated at 77 K. Successive photobleaching at 77 K recovers CCl_4^+

semi-quantitatively. This result may suggest that some activation energy is needed for the hole transfer to a solute in a matrix such as CCl_4.

6.3. MATRIX DEPENDENCE IN CATION RADICAL REACTION

The deprotonation reaction of an alkane-cation radical is strongly dependent on the matrix. This reaction proceeds with a considerable efficiency in SF_6, F113, and PFMCHx, while in F11 and CF_3CCl_3 no positive evidence for this has been reported except an experiment at extremely high concentration[68]. On the other hand, photo-induced elimination of H_2 or CH_4 takes place in both types of these matrices.

One possible cause of this matrix dependence may be the difference in the "softness" of them at around 100 K, at which temperature most of the reaction of cation radicals take place. That is: F11 and CF_3CCl_3 are "hard" matrices, in which the temperature of the phase transition to the plastic phase is higher than 130 K, so that the migration of the solute at the lower temperature is inhibited. On the other hand, matrices such as F113, SF_6, and PFMCHx are "soft" (T_t of SF_6 is ca. 93 K), thus reactions which need molecular diffusion are possible in them. In fact, the reactions observed below 130 K in the former matrices so far are uni-molecular decomposition reactions.

6.4. THE MECHANISM OF THE DEPROTONATION REACTION

The following mechanisms are considered for the conversion of the cation radicals to alkyl radicals upon warming. They are:
I) ion-molecule reactions,

$$RH^+ + RH \longrightarrow R\cdot + RH_2^+, \tag{14}$$

II) charge neutralization reactions,

$$RH^+ + SX^- \longrightarrow R\cdot + S\cdot + HX \tag{15}$$

$$\text{or} \quad RH^+ + X^- \longrightarrow R\cdot + HX, \tag{16}$$

III) positive charge migration and the subsequent H abstraction by an X atom,

$$RH^+ + X^- \longrightarrow RH + X\cdot \tag{17}$$

$$X\cdot + RH \longrightarrow R\cdot + HX. \tag{18}$$

Employing the mechanism (III), it seems difficult to explain the selective l-alkyl radical formation observed for n-alkane cations in SF_6[81], since the bond energy of primary C - H is larger than that of secondary C - H. Mechanism (II) explains the site selectivity described above (section 6.1), since X^- may attack the C - H bond with a higher positive charge. On the other hand, the ion-molecule reaction (I) is exothermic only for methane and ethane, in the gas phase, and is impossible energetically for the alkanes larger than or equal to propane, once they are thermalized. In solid matrices, however, the solvation energy may contribute to overcome this difficulty. The matrix dependence in the reaction efficiency is acceptable taking into account the mobility of cation or neutral alkane molecules in the matrices.

Whether the ion-molecule reaction is possible or not is one of the most interesting problems in radiation chemistry[17]. Recently, an evidence which supports the feasibility of the ion-molecule reaction of a thermalized alkane cation of fairly large molecules was reported[25]. That is, the alkane cation radical (n-hexane, n-octane) generated by the radiolysis at 4 K in ZSM-5, at a very low solute concentration, decayed without forming the alkyl radical upon warming. At a higher alkane concentration, the alkane cation was not stabilized even at 4 K, but it was converted to an alkyl radical during the irradiation. This suggests that deprotonation from an alkane cation occurs only if it encounters a neutral alkane, since another molecule may not be trapped in the cavity, at low concentration. On the other hand if there is another alkane in the cavity at a high concentration, the cation radical can react.

7. Summary

The combination of the cryogenic techniques, ionizing radiation, and esr measurement using matrices having a high IP and a positive EA, has disposed the longstanding problem of esr detection of alkane cation radicals. Thus the out-line of the electronic structure of alkane cation radicals has been established. The results obtained throughout these studies are also useful for the testing and improvement of molecular orbital theory for cationic radical species[90]. Then, it will in turn feed back to the σ-electron system precise information on the electronic and geometric structures of such species.

As for the reactions of cation radicals, the data observed so far are strongly dependent on the matrices. Further studies on the reaction as well as the cause of this matrix dependence itself would bring a new incite on the reactivity of cation radicals.

Finally the author wishes to express sincere thanks to Dr. M. Okazaki for the valuable discussion throughout the preparation of this review.

References

1. G. Vincow: "Radical Ions", Ed. E.T. Kaiser & L. Kevan, Interscience pub., Chap. 4, (1968).
2. M.C.R. Symons: Chem. Soc. Rev., 13, 393 (1984); M. Shiotani: Mag. Res. Rev., 12, 333 (1987); T. Shida: Acc. Chem. Res., 17, 186 (1984); L.B. Knight: ibid. 19, 313 (1986).
3. M. Iwasaki, K. Toriyama, H. Muto, and K. Nunome: J. Chem. Phys., 65, 596 (1968).
4. T. Ichikawa and P.K. Ludwig: J. Am. Chem. Soc., 91, 1023 (1969).
5. F. Nauwelaerts and Ceulemans: Chem. Phys. Lett., 38, 354 (1976); F. Nauwelaerts, M. Lemahieu and J. Ceulemans: J. Chem. Phys., 66, 140 (1977).

6. I.G. Smith and M.C.R. Symons: *J. Chem. Res. (S)*, 382 (1989); M.C.R. Symons: *Chem. Phys. Lett.*, **69**, 198 (1989).
7. J.T. Wang and F. Williams: *J. Phys. Chem.*, **84**, 3156 (1980).
8. T. Shida, H. Kubodera, and Y. Egawa: *Chem. Phys. Lett.*, **79**, 179 (1981).
9. M. Iwasaki, K. Toriyama, and K. Nunome: *J. Am. Chem. Soc.*, **103**, 3591 (1981).
10. K. Toriyama, K. Nunome, and M. Iwasaki: *J. Chem. Phys.*, **77**, 5891 (1982).
11. J.W. Verhoeven and P. Pasman: *Tetrahedron*, **37**, 943 (1981).
12. M. K. Carter and G. Vincow: *J. Chem. Phys.*, **47**, 292 (1967), and papers cited therein.
13. M. Kira, M. Watanabe, and H. Sakurai: *Chem. Letters*, 973 (1979); T. Clark, J. Chandrasekhar Jr., and P.V.R. Schleyer: *J. C. S. Chem. Comm.*, 26 (1980).
14. G. R. Liebling and H. M. McConnell: *J. Chem. Phys.*, **42**, 3931 (1965);G. C. Closs and O. D. Redwine: *J. Am. Chem. Soc.*, **108**, 506 (1986).
15. M. Iwasaki, K. Toriyama, and K. Nunome: *Faraday Discuss. Chem. Soc.*, **78**, 19 (1984).
16. P.W.F. Louwrier and W. Hamill: *J. Phys. Chem.*, **74**, 1418 (1970).
17. F.P. Schwarz, D. Smith, S.G. Lias, and P. Ausloos: *J. Chem. Phys.*, **75**, 3800 (1981), and papers cited therein.
18. K. Toriyama, K. Nunome, and M. Iwasaki: *J. Phys. Chem.*, **92**, 5097 (1988).
19. L. B. Knight, Jr. and J. Steadman: *J. Am. Chem. Soc.*, **106**, 3700 (1984).
20. M. Iwasaki, K. Toriyama, and B. Eda: *J, Chem. Phys.*, **42**, 63 (1965); M. Iwasaki, *J. Chem. Phys.*, **45**, 990 (1966).
21. T. Shida and T. Kato: *Chem. Phys. Lett.*, **68**, 106 (1979).
22. A. Grimison and G.A. Simpson: *J. Phys. Chem.*, **72**, 1776 (1968).
23. M. Iwasaki, K. Toriyama, and K. Nunome: *J. Chem. Soc. Chem. Comm.* , 320 (1983).
24. J. C. Verdine, A. Suroux, V. Bolis, P. Dajaifve, C. Naccache, P. Wierzhowski, E.G. Derouane, J.B. Nagiy, J. Gilson, J.H.C. VanHooff, J.P. vn den Berg, and J. Wolthuizen: *J. Catal.*, **59**, 248 (1979).
25. K. Toriyama, K. Nunome, and M. Iwasaki: *J. Am. Chem. Soc.*, **109**, 4496 (1987).
26. B.M. Tadjikov, V.I. Melechov, O.A. Anisimov, and Yu.N. Molin: *Rad. Phys. Chem.*, **34**, 353 (1989), and papers cited therein.
27. D.W. Werst, M. G. Bakker, and A. D. Trifunac: *J. Am. Chem. Soc.*, **112**, 40 (1990), and papers cited therein.
28. A. Richarts, R.J. Buenker, P.J. Bruna, and S.D. Peyerimhoff: *Mol. Phys.*, **33**, 1345 (1977), and papers cited therein.
29. J.W. Rabalais and A.Katrib: *Mol. Phys.*, **27**, 923 (1974).
30. G. Herzberg: *Electronic Spectra and Electronic Structure of Polyatomic Molecules* (Van Nostrand Reinhold, New York, 1966), p. 50.
31. B. M. Gimarc: *J. Am. Chem. Soc.*, **95**, 1417 (1973).
32. T.A. Claxton, R.E. Overill and M.C.R. Symons: *Mol. Phys.*, **27**, 701 (1974).
33. M. Iwasaki, K. Toriyama, and K. Nunome: *Chem. Phys. Lett.*, **111**, 309 (1984).
34. W.C. Herndon, M.L. Ellzey,Jr., and K.S. Raghuveer: *J. Am. Chem. Soc.*, **100**, 2645 (1978); R.G. Dromey and J.B. Peel: *J. Mol. Struct.*, **23**, 53 (1974), and papers cited therein.
35. K. Matsuura, K. Nunome, K. Toriyama, and M. Iwasaki: *J. Phys. Chem.*, **93**, 149 (1988).
36. H. M. McConnell: *J. Chem. Phys.*, **24**, 632 (1956).
37. K. Toriyama, K. Nunome, and M. Iwasaki: *J. Phys. Chem.*, **85**, 2149 (1981).
38. J. T. Wang and F. Williams: *Chem. Phys. Lett.*, **82**, 177 (1981).
39. M. Tabata and A. Lund: *Rad. Phys. Chem.*, **23**, 545 (1984).
40. M. Lindgren, A. Lund, and G. Dolivo: *Chem. Phys.*, **99**, 103 (1985).
41. A. Lund, M. Lindgren, G. Dolivo, and M. Tabata: *Rad. Phys. Chem.*, **26**, 491 (1985).
42. G. Dolivo and A. Lund: *J. Phys. Chem.*, **89**, 3977 (1985).
43. G. Dolivo A. Lund: *Z. Naturforsch.*, **40a**, 52 (1985).
44. K. Toriyama, K. Nunome, and M. Iwasaki: *J. Phys. Chem.*, **90**, 6836 (1986).
45. A. Lund, M. Lindgren, M. Tabata, S. Lunell, G. Dolivo, and T. Gaumann: "*Elect. Mag. Res. of Solid state*",*Ed. J.A. Weil*, Can. Soc. Chem., **20** (1987).
46. M. Iwasaki and K. Toriyama: *J. Am. Chem. Soc.*, **108**, 6441 (1986).
47. E. Tannenbaum, R.D. Johnson, R.J. Myers, W.D.J. Gwinn: *J. Chem. Phys.*, **22**, 949 (1954).
48. K. S. Pitzer: *Discuss. Faraday Soc.*,. **10**, 66 (1951).
49. D.N.Rao and M.C.R. Symons: *J. Chem. Soc. Perkin Trans.*. **2**, 991 (1985).
50. (for example E_a = 3.4 - E(zero point energy, ca. 0.4) kcal/mole for propane [48])

124

51. M. C. R. Symons, T. Chen, and C. Glidewell: *J. C. S. Chem. Commun.*, **326** (1983); T. A. Claxton, T. Chen and M. C. R. Symons: *Faraday Discuss. Chem. Soc.*, **78**, 1 (1984).

52. M. N. Paddon-Row, D. J. Fox, J. A. Pole, K. N. Houk, and D. W. Pratt: *J. Am. Chem. Soc.* **107**, 7696 (1985).

53. M.C.R. Symons: *J. Chem. Soc. Chem. Commun.* , 270 (1982); A. Hasegawa, S. Kaminaka, T. Wakabayashi, M. Hayashi, and M.C.R. Symons: *J. Chem. Soc., Chem. Comm.*, 1199 (1983); A. Hasegawa, S. Kaminaka, T. Wakabayashi, and M. Hayashi: *J. Chem. Soc.*, 1667 (1984).

54. B. W. Walter and F. Williams: *J. Chem. Soc. Chem. Commun.* , 869 (1982).

55. J. N. Murrel and W. Schmidt: *J.C.S. Faraday Trans.*, **2**, 1709 (1972).

56. K. Nunome, K. Toriyama, and M. Iwasaki: *J. Chem. Phys.*, **79**, 2499 (1983).

57. E. L. Cochran, F.J. Adrian, and Boweres: *J. Chem. Phys.*, **40**, 213 (1964); H. Muto, M. Iwasaki, and Y. Takahashi: *J. Chem. Phys.*, **66**, 1943 (1977).

58. R.S. Mulliken, C.A.Rieke, D. Orloff, and H. Orloff: *J. Chem. Phys.*, **17**, 1248 (1949).

59. R. Hulme and M.C.R. Symons: *J. Chem. Soc.*, 1120 (1965).

60. K. Toriyama, K. Nunome, and M. Iwasaki: *Chem. Phys. Lett.*, **132**, 456 (1986).

61. M. Shiotani, A. Yano, N. Ohta, and T. Ichikawa: *Chem. Phys. Lett.*, **147**, 38 (1988).

62. T. Ichikawa and N. Ohta: *J. Phys. Chem.*, **91**, 3244 (1987).

63. K. Ohta, H. Nakatsuji, H. Kubodera, and T. Shida: *Chem. Phys.*, **76**, 271 (1983).

64. M. Iwasaki, K. Toriyama, and K. Nunome: *J. Chem. Soc. Chem. Comm.*, 202 (1983).

65. K. Matsuura, K. Nunome, M. Okazaki, K. Toriyama and M. Iwasaki: *J. Phys. Chem.*, **93**, 6642 (1988).

66. H. Kubodera and T. Shida: *23rd Symposium of Radiation Chemistry*, B202, Kyoto, Oct. 1980.

67. M. Tabata and A. Lund: *Chem. Phys.*, **75**, 379 (1983).

68. T. Shida and Y. Takemura: *Rad. Phys. Chem.*, **21**, 157 (1983).

69. K. Toriyama, K. Nunome, and M. Iwasaki: *J. Chem. Soc. Chem. Comm.*, 143 (1984).

70. S. Lunell, M.B. Huang, O. Claesson, and A. Lund: *J. Chem. Phys.*, **82**, 5121 (1985).

71. M. Shiotani, N. Ohta, and T. Ichikawa: *Chem. Phys. Lett.*, **149**, 185 (1988).

72. M. Lindgren, M. Shiotani, N. Ohta, T. Ichikawa, and L. Sjoqvist: *Chem. Phys. Lett.*, **161**, 127 (1989).

73. K. Toriyama, K. Matsuura, and K. Nunome: *Chem. Phys. Lett.*, ... (1990).

74. A. Almenningen, O. Bastiansen, P.N. Skancke: *Acta. Chemica. Scand.*, **15**, 711 (1961).

75. K. Ushida, T. Shida, M. Iwasaki, K. Toriyama, and K. Nunome: *J. Am. Chem. Soc.*, **105**, 5496 (1983).

76. L. Sjoqvist, A. Lund, and J. Maruani: *Chem. Phys.*, **125**, 293 (1988).

77. M.B. Huang, S. Lunell, and A. Lund: *Chem. Phys. Lett.*, **99**, 201 (1983).

78. K. Toriyama, K. Nunome, and M. Iwasaki: *J. Chem. Soc. Chem. Comm.*, 1346, (1983).

79. K. Nunome, K. Toriyama, and M. Iwasaki: *Tetrahedron*, **42**, 6315 (1986).

80. V.I. Melekhov, O.A. Anisimov, A.V. Veselov, and Yu.N. Molin: *Chem. Phys. Lett.*, **127**, 97 (1986).

81. M. Iwasaki, K. Toriyama, and K. Nunome: *Rad. Phys. Chem.*, **21**, 147 (1983).

82. K. Nunome, K. Toriyama, and M. Iwasaki: *Chem. Phys. Lett.*, **105**, 414 (1984).

83. K. Toriyama, K. Nunome, and M. Iwasaki: *52nd. Ann. Meet. Japan. Chem. Soc.*, **3F18**, Kyoto, Apl. (1986).

84. X.Z. Quin and F. Williams: *Chem. Phys. Lett.*, **112**, 79 (1984).

85. X.Z. Quin and F. Williams: *Tetrahedron* **42**, 6301 (1986).

86. X.Z. Quin and F. Williams: *J. Phys. Chem.*, **90**, 2299 (1986); X.Z. Quin, T.C. Pentecost, J.T. Wang, and F. Williams: *J. Chem. Soc. Chem. Comm.* , 450 (1987); F.W. McLafferty, M.P. Barba las, and F. Turecek: *J. Am. Chem. Soc.*, 2 (1983); B.F. Yates, W.J. Bouma, and L. Radom: *Tetrahedron*, **42**, 6225 (1986).

87. M.C.R. Symons: *Chem. Phys. Lett*, **117**, 381 (1985).

88. X.Z. Quin, L.D. Snow, and F. Williams: *Chem. Phys. Lett.*, **117**, 383 (1984)

89. K. Toriyama, K. Nunome, and M. Iwasaki, T. Shida, and K. Ushida: *Chem. Phys. Lett.*, **122**, 118 (1985).

90. T. Shida, Y. Egawa, H. Kubodera, and T. Kato: *J. Phys. Chem.*, **73**, 5963 (1980).

91. W.J. Bowma, D. Poppinger, and L. Radom: *Isr. J. Chem.*, **23**, 21 (1983).

92. K. Matsuura, K. Nunome, and K. Toriyama: unpublished.

ESR STUDIES OF RADICAL CATIONS OF CYCLOALKANES AND SATURATED HETEROCYCLES.

MIKAEL LINDGREN
Department of Physics and Measurement Technology, Chemical Physics, Linköping Institute of Technology, S-581 83 Linköping, Sweden

MASARU SHIOTANI
Department of Applied Physics and Chemistry, Faculty of Engineering, Hiroshima University, 0724 Higashi-Hiroshima, Japan

1. Introduction

Since ESR spectroscopists started to use the halocarbon matrix isolation technique about a decade ago to study radical cations, many unstable intermediates have been characterised [1]. Most notable in this context are perhaps the radical cations of saturated organic compounds. The radical cations of alkanes have their semi occupied molecular orbital (SOMO) confined into σ-bonds which can be extended over several carbons in the molecular framework. As a result, such bonds are weakened and elongated. The presence of a heteroatom in a saturated system gives a SOMO which involves the lone-pair orbital(s). In such cases, the radical cations usually have a π-type character, and the unpaired electron delocalises to hydrogens on the adjacent carbons *via* hyperconjugation.

The electronic ground states and dynamical features of radical cations can be deduced from ESR studies, and the aim of this chapter is to give an overview of the experimental ESR results of radical cations of cycloalkanes, bicycloalkanes and saturated heterocycles. In addition to structural and dynamical features of the cations, some of the fragmentation and rearrangement reactions will be discussed, such as the ring-opening reactions of three-membered rings, intramolecular hydrogen transfer reactions, and the ion molecule reactions giving rise to neutral radicals which usually occur at phase transitions of the solid matrices.

All of the cations discussed here have been stabilised in solid (frozen) halocarbon matrices. The basic chemical technique to generate cations in halocarbon matrices was described in an earlier chapter by Toriyama, and we will start directly with the presentation of the cation structures.

2. Radical Cations of Three- and Four-Membered Rings

Three- and four-membered saturated ring systems with and without a heteroatom are all relatively stable. However, because of the strain within the ring they tend to undergo ring-

A. Lund and M. Shiotani (eds.), Radical Ionic Systems, 125–150.

opening reactions. The ring-opened isomers of cations of the three-membered rings containing a heteroatom have structural resemblance with the allyl radical, since the lone-pair orbitals on the heteroatom can overlap with the π-orbitals on the methylene groups. A thorough discussion of the electronic structures of such cations deduced from optical absorption spectroscopy is found in the chapter by Bally. ESR parameters found for the radical cations of cyclopropane and its derivatives are summarised in Table 1. The ESR parameters for cations of three-membered rings with heteroatoms and the four-membered rings, are summarised in Tables 2 and 3, respectively. Details are discussed below.

Table 1. ESR parameters of the cyclopropane cation and its methylsubstituted derivatives being ring-closed and ring-opened (methyl-substituted trimethylene cations).

Cation	Matrix	T(K)	ESR-parameters[a]	Ref
cyclopropane$^+$				
closed	$CFCl_2CF_2Cl$	4.2	g=2.0042, a(4H)=-12.5, a(2H)=21	[3]
open localised	$CFCl_2CF_2Cl$	108	g=2.0028, a($2H_\alpha$)=22.4, a($2H_\beta$)=30.2	[8]
cis1,2-Me$_2$-cC3$^+$				
closed	$CFCl_3$	150	g=2.0041, a(2CH$_3$)=20.5	[7]
			a($2H_\alpha$)=10.4, a($2H_\beta$)=20.5	
open localised	$CFCl_2CF_2Cl$	115	g=2.0028, a(CH$_3$)=24.7, a($1H_\alpha$)=24.7,	[7]
			a($1H_\beta$)=32.4 or a($2H_\beta$)=16.2[b]	
trans1,2-Me$_2$-cC3$^+$				
closed	$CFCl_3$	150	g=2.0041, a(2CH$_3$)=21.8	[7]
			a($2H_\alpha$)=11.9, a($2H_\beta$)=21.8	
open localised	$CFCl_2CF_2Cl$	115	g=2.0028, a(CH$_3$)=24.7, a($1H_\alpha$)=24.7,	[7]
			[b]a($1H_\beta$)=32.4 or a($2H_\beta$)=16.2	
1,1,2,-Me$_3$-cC3$^+$				
closed	$CFCl_3$	150	g=2.0040, a(2CH$_3$)=20.6, a(CH$_3$)=14.5,	[7]
			a($1H_\alpha$)=9.8, a($2H_\beta$)=17.9	
open localised	$CFCl_2CF_2Cl$	115	g=2.0029, a(2CH$_3$)=23.6, a($2H_\beta$)=11.8	[7]
1,1,2,2-Me$_4$-cC3$^+$				
closed	$CFCl_3$	145	g=2.0033, a(4CH$_3$)=15.0, a($2H_\beta$)=18.7	[7]
closed	CF_2ClCCl_3	155	g=2.0033, a(4CH$_3$)=15.0, a($2H_\beta$)=18.7	[7]
open localised	$CFCl_2CF_2Cl$	117	g=2.0029, a(2CH$_3$)=23.3, a($2H_\beta$)=11.7	[7]

[a] $a=a_{iso}$, splittings are given in Gauss (10G=1mT). [b] There are two ways to interpret the H$_\beta$ hf splitting, see [7] for details.

2.1. CYCLOPROPANE AND ITS METHYL-SUBSTITUTED DERIVATIVES

2.1.1. The Ring-closed Cyclopropane Cation. The radical cation of cyclopropane has been studied using several matrices [2-7] and by *ab initio* MO calculations [4,5]. The low temperature ESR spectrum resolves a triplet of quintets having the isotropic hf (hyperfine) splittings, 12.5 G (2H) and 21 G (4H), respectively. The values can be explained in terms of an a_1 SOMO in C_{2v} symmetry which is expected from a Jahn-Teller distortion of the $^2E'$ state of the neutral molecule (D_{3h} symmetry). The a_1 orbital is bonding between two carbons, and as a consequence of being singly occupied, the bond becomes weakened and elongated. More details on the structure and dynamics are given in other chapters (Toriyama, Shiotani/Lund).

2.1.2. *1,2-dimethyl-, 1,1,2-trimethyl-, and 1,1,2,2-tetramethyl-cyclopropane.* It is well known that the weakest bond in an alkyl-substituted cyclopropane ring is the one connecting to the most substituted carbon atom [7,9]. In accordance with this, it was found that the main part of the unpaired electron in the cation of various methyl-substituted cyclopropanes, such as *cis* and *trans* 1,2-dimethyl-, 1,1,2-trimethyl-, and 1,1,2,2-tetramethyl-cyclopropane, is confined into a σ-bond at the most substituted ring carbon [7]. The distribution of spin density in alkyl-substituted cyclohexane cations, which will be discussed in a following section, has been concluded to follow a similar rule.

A hf splitting of *ca.* 15-20 G due to all of the methyl hydrogens was resolved in the ESR spectra of methyl-substituted cations in $CFCl_3$ at 150K [7]. It means that, for a rigid structure, one of the C-H bonds of each methyl group is in a position favourable to obtain spin density *via* hyperconjugation from the C-C bond in the ring structure of high spin density. The ESR parameters are summarised in Table 1. The methylsubstituted cyclopropanes were found to undergo ring-opening reactions in $CFCl_2CF_2Cl$ at elevated temperatures [7,9], to be discussed below.

2.1.3. *Derivatives of Trimethylene.* In certain matrices, $CFCl_2CF_2Cl$ and CF_2ClCF_2Cl, the ESR signal of the cyclopropane cation [7,8] and its methyl-substituted derivatives [7,9] was found to change irreversibly on increasing the temperature. Upon warming samples containing the cyclopropane cation to 80-100 K the ESR hf pattern of the ring-closed structure gradually changes into a triplet of triplets having the splittings 30.2 and 22.4 G, respectively. Williams and coworkers [8,9] assigned the hf splittings to the trimethylene cation, i.e., the "ring-opened" isomer of cyclopropane. The ESR data clearly demonstrated that the unpaired electron was localised to one of the terminating methylene groups: one pair of splittings could be assigned to two equivalent α-hydrogens (22.4 G) and the other triplet to two identical β-hydrogens (30.2 G). It was proposed that the trimethylene cation took a structure with the two terminal methylene groups orthogonal, and as a consequence, the charge and unpaired electron were separated from each other to give a *distonic* radical cation having the unpaired electron localised to one of the methylene groups, see structure I below.

I   II

Symons argued that the methylene group housing the charge should be susceptible to bind a chlorine of a matrix molecule [10], see structure II above, and that such a *weak* complex can explain the asymmetrical distribution of the charge and spin density. However, no further experimental evidence has been given in support for such an adduct.

The changes occurring for the ESR signal of the methyl-substituted analogues could be explained in a similar way [7]. Details on the ESR parameters are collected in Table 1.

Toriyama *et. al.* [2] pointed out the possibility that the signal of the *distonic* cation actually might be due to the 1-propyl radical, formed by an intermolecular hydrid ion transfer from a neutral cyclopropane molecule to the trimethylene cation. The results on methyl-substituted cyclopropanes, however, did not give support for this model.

2.2. CATIONS OF ETHYLENE OXIDE AND ITS DERIVATIVES

2.2.1. *Ethylene Oxide*. Ethylene oxide (or oxirane) is a highly reactive molecule. It was assumed to be the cause of the fire that destroyed the Apollo rocket in 1967 [12]. The problem of assigning the strucure of the oxirane cation stabilised in halocarbon matrices has been considered in a relatively large number of studies [13-20]. There is a continuing disagreement between the groups of Williams and Symons over: i) the interpretation of experimental data (^{13}C hf interactions) [16,18,20], ii) the question of having a ring-closed or open delocalised (allylic) structure [15-18], iii) the structure of both the opened and closed forms [17-20]. Here a survey of the results and the discussions is given. The structures which have been discussed (**III-VI**) are depicted below.

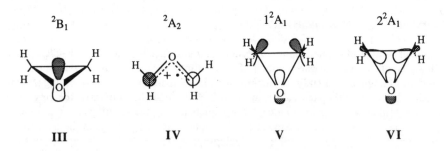

The low temperature ESR spectra of the oxirane cation in various matrices comprise a hf quintet with a line separation of about 16 G. This rules out the $^2B_1$ (**III**) state since the unpaired electron in this case would localize mainly in the oxygen lone-pairs, giving rise to large hf interactions (at least 35 G [20]) to the methylene hydrogens via hyperconjugation [13,14]. Using $CFCl_3$ as matrix, Williams and coworkers assigned the spectrum recorded at 77 K to a ring-opened oxallyl structure (**IV**) from the resemblance of the associated hf parameters to the isoelectronic allyl radical. In an independent study, Symons and Wren proposed a ring-closed $^2A_1$ state (**V**) [14].

Using $CFCl_2CF_2Cl$ as matrix, Williams and coworkers observed localised forms of the ring-opened oxirane cation and various methyl-substituted derivatives [15]. They proposed an orthogonal structure analogous with the *distonic* radical cations observed for the methylsubstituted cyclopropanes (see section 2.1.2 and 2.1.3.). This structure is referred to as *open localised* in Table 2.

The optical absorption spectrum of ionised oxirane shows a strong absorption band in the visible, 465-550 nm [21] which has been assigned to the (ring-opened delocalised) ox-allyl structure. Thus, Symons and coworkers [17] argued that samples containing the ox-allyl cation should be pink or reddish, and using $CFCl_3$ as matrix, they deduced a ring-closed structure opposing the conclusions of Qin and Williams [13]. The cations of both 2-methyl-, and *cis*2,3-dimethyl-oxirane were found to be without color at low temperature (4-40K), but the color as well as the associated ESR hf signatures changed on annealing to 77K or above [17]. The color change towards red/purple was taken as evidence of a ring-opening reaction. Since the oxirane cation was found colorless in $CFCl_3$ at all temperatures, it was argued that the cation had a closed structure in this matrix. From the same considerations, the cation structure was concluded to be open in SF_6. The color of the sample was here red, and the ESR spectrum showed the same quintet hf pattern, however, with a slightly smaller splitting. Details about the ESR parameters are summarised in Table 2.

The findings of Symons and coworkers [17] are incomplete since an analysis of the low-temperature ESR data of the ring-closed structures of the methyl-derivatives was not offered. In addition, some confusion arose when Rhodes and Symons [18] later proposed a different electronic state for the closed structure (VI), an interpretation which was called in question by Snow and Williams [20]. There is a also a continuing dispute over the analysis of the ^{13}C hf splittings in ^{13}C-labelled samples [16,18,20]. Therefore, we are not willing to pursue the discussion further here but merely want to point out that the study by Symons and coworkers [17] might have shown the difficulty to deduce the ring-opened structure solely from ESR data. It seems desirable to investigate the oxirane cation also with other spectroscopic methods, such as IR or Raman, under the same conditions as in the ESR studies, which hopefully would provide more unambiguous arguments.

2.2.2. *Substituted Oxirane Cations Forced to be Ring-closed.* The cations of cyclopentane oxide and cyclohexane oxide were found to undergo ring-opening reactions [17]. The irradiated solutes/$CFCl_3$ were found to be pink with the associated ESR signals supporting ring-opened oxallyl structures.

In nicely designed experiments Williams and coworkers characterised cations of two oxirane cations forced to be ring-closed, namely, the cations of 9,10-octalinoxide and *syn*-sesquinorbornene oxide [19]. Owing to the sterical constraints, the oxirane ring cannot easily open after ionisation, and information of a ring-closed oxirane cation was obtained. An analysis of the temperature dependent ESR line shape of the 9,10-octaline oxide cation (which was due to molecular motion) gave a unique assignment of the hf splittings of the rigid structure. By correlating the results with semi-empirical calculations it was concluded that the unpaired electron is mainly confined to the σ-bond between the two carbons in these structures.

2.3. AZIRIDINE, CYCLOPROPYLAMINE AND THIIRANE

2.3.1. *Aziridine.* The ESR signal of the aziridine cation in $CFCl_3$ shows the following hf splittings: 16.1 G (4H), 7.7 G (1N) and 4.3 G (N-H). The hydrogen splittings resemble those of the allyl radical and the spectrum was assigned to a ring-opened allyl-type radical structure [22]. ESR spectra of the cation in $CFCl_2CF_2Cl$ or CF_3CCl_3 showed signals from the neutral 1-aziridinyl radical [22]. It was suggested that the 1-aziridinyl radical was formed by hydrogen abstraction from a neutral solute molecule by the cation.

2.3.2. *Cyclopropylamine.* The cation radical of cyclopropylamine was found in a ring opened distonic form [29], with the charge localised at the amino group, and the unpaired electron residing in the terminating methylene group. The cation decomposed by H^+ abstraction at the C-N bond in $CFCl_2CF_2Cl$ [29].

Table 2. ESR parameters of the cations of three-membered heterocycles with ring-closed, allylic (open delocalised) or *distonic* (open localised) structures.

Cation	Matrix	T(K)	ESR-parameters[a]	Ref
[b] oxirane[+]	CFCl$_3$	77-150	g=2.0024, a(2CH$_2$)=16.6	[13,17]
open delocalised	SF$_6$	77	a(2CH$_2$)=15.7	[17]
open localised	CFCl$_2$CF$_2$Cl	90	g=2.0035, a(CH$_2$)=21	[15]
Me-oxirane[+],				
closed	CFCl$_3$	4	not clarified	[17]
opened[c]	CFCl$_3$	-	I a(CH$_2$)=21, a(CH)=12, a(CH$_3$)=12	[17]
		ca. 130	II a(CH$_2$)=12, a(CH,CH$_3$)=21	
open localised	CFCl$_2$CF$_2$Cl	90	g=2.0035, a(CH$_2$)=21	[15]
*cis*Me$_2$-oxirane[+]				
closed	CFCl$_3$	4	not clarified	[17]
open delocalised	CFCl$_3$	130	a(8H)=16.5	[17]
open localised	CFCl$_2$CF$_2$Cl	95	g=2.0033, a(CH,CH$_3$)=22.5	[15]
*trans*Me$_2$-oxirane[+]				
open localised	CFCl$_2$CF$_2$Cl	100	g=2.0038, a(CH,CH$_3$)=23	[15]
Me$_3$-oxirane[+]				
open localised	CFCl$_2$CF$_2$Cl	105	g=2.0033, a(CH,CH$_3$)=22	[15]
Me$_4$-oxirane[+]				
open delocalised	CFCl$_3$	77	a(4CH$_3$)=15.2	[17]
open delocalised	CFCl$_2$CF$_2$Cl	95	g=2.0023, a(4CH$_3$)=15.1	[15]
open localised	CFCl$_2$CF$_2$Cl	115	g=2.0026, a(2CH$_3$)=21	[15]
aziridine[+]				
open delocalised	CFCl$_3$	150	a(2CH$_2$)=16.1, a$_{NH}$=4.3, a(^{14}N)=7.7	[22]
thiirane[+]				
closed	CFCl$_3$	90	g$_{//}$=2.002, g$_\perp$=2.028, a(2CH$_2$)=16.1	[23,28]
dimer	CFCl$_3$	90	g$_{iso}$=2.012, a(8H)=5.8	[23]
c-propylamine[+]				
open *distonic*	CFCl$_3$	80-160	g$_{iso}$=2.0027, a(2H$_\alpha$)=22.5, a(2H$_\beta$)=17.0	[29]

[a] a=a$_{iso}$, splittings are given in Gauss (10G=1mT). [b] See the text for details about open or closed structure, ^{13}C splittings are discussed in [16,18,20]. [c] The analysis of the ESR hf structure given in [17] for the 2-methyloxirane cation is not clearly described. Interpretation **I** is given in the stick plot of Fig. 5 and in Table 1 in [17] whereas **II** is proposed in the text.

2.3.3. *Thiirane*. In contrast to the cations of aziridine and oxirane, the sulphur containing analogue, thiirane, was concluded to take a ring-closed structure [23]. Using CFCl$_3$ as matrix, an ESR signal due to the ring-closed cation was found in addition to a cation dimer. The ESR spectrum of the monomer resolved a hf splitting of 16.1 G (4H), due to the four methylene hydrogens [23,28]. Both decayed out above 140 K without giving any successor radicals. In $CFCl_2CF_2Cl$ and CF_3CCl_3 exclusively the dimer cation was detected. On increasing the temperature of the samples above phase transitions of the matrices, about 150K in CF_3CCl_3 and 100K in $CFCl_2CF_2Cl$, a variety of radical reactions were found to occur, involving ethylene extrusion and formation of matrix radicals. The details concerning ion-molecule reactions which occur in $CFCl_2CF_2Cl$ involving various ether, thioether and olefin radical cations have been discussed in the papers by Williams and coworkers [23-25].

2.4. FOUR-MEMBERED RINGS

The ESR parameters of the radical cations of the four-membered rings are summarised in Table 3. Details are discussed below.

Table 3. ESR parameters of the cations of four-membered rings.

Cation	Matrix	T(K)	ESR-parameters[a]	Ref
cyclobutane$^+$	$CFCl_3$	4.2	a(2H)=49, a(2H)=14	[26]
-"- annealed at 77 K		4.2	a(2H)=44, a(2H)=22	[26]
oxetane$^+$	$CFCl_3$	152	$g_{//}$=2.0046, $g_\perp$=2.0135 $A_{//}$=65.5 $A_\perp$=65.7 (4H) $A_{//}$=10.5, $A_\perp$=11.1 (2H)	[13]
thietane$^+$	$CFCl_3$ ($CFCl_2CF_2Cl$)	90	$g_{//}$=2.008, $g_\perp$=2.027 a(4H)=31.1	[28]
thietane$^+$	$CFCl_3$	120	$g_{//}$=2.002, $g_\perp$=2.023 a(4H)=31	[27]
azetidine$^+$	$CFCl_3$	90	g_{iso}=2.0037, a(4H)=54.2, a_{N-H}(1H)=22.6 $A_{//}(^{14}N)$=41.4, $A_\perp(^{14}N)$=8.0,	[22]
	$CFCl_3$	140	g_{iso}=2.0038, a(4H)=54.1, a_{N-H}(1H)=22.7 $A_{//}(^{14}N)$=38.2, $A_\perp(^{14}N)$=9.5	[22]

[a] $a=a_{iso}$, splittings are given in Gauss (10G=1mT).

2.4.1. The Ring-closed and Ring-opened Cyclobutane Cation. The neutral cyclobutane molecule has a puckered ring structure with a doubly degenerate e HOMO. The hf parameters of the cation, see Table 3, were explained in terms of a C_{2v} structure [2, 26]. It was suggested that the SOMO is b_2, following from a Jahn-Teller distortion of the D_{2d} structure. By illuminating the cyclobutane cation in $CFCl_3$ at 77 K by UV light, the ESR signal of the cyclobutane cation disappeared while a signal belonging to the 1-butene cation was growing in the spectra. It was suggested that ring-opening occurred, followed by immediate hybrid ion transfer from C2 to C4. The 1-butene cation was shown to isomerise further to the 2-butene cation by illuminating with visible light, and it was proposed that a 1,3-shift of a hydride ion occurred. More details are found in the chapter by Toriyama.

2.4.2. *Heterocyclic Four-Membered Rings*. All of the four-membered heterocyclic cations have been found to take an electronic structure having the largest part of the unpaired spin density in lone-pair orbitals of the heteroatom. This gives rise to large hf splittings for the β-hydrogens on the adjacent methylene groups, and in the case of trimethylene oxide (oxetane) and thietane, a considerable g-anisotropy, see Table 3.

The effect of ring-size on the magnitude of the 4 equivalent hf-splittings of the β-hydrogens in the ring-closed cations of thiirane and thietane was discussed by Qin and Williams [28]. For the thietane cation the splitting was found to be 31.1 G (Table 3) compared to 16.1 G for the cation of the three-membered ring, thiirane (see section 2.3.3). The lower value was explained by a smaller admixture of the CH_2 group orbital into the b_1 SOMO of the heterocycle.

The cations of four-membered rings containing a heteroatom do not undergo ring-opening reactions. The cations were found to decay at elevated temperatures with concomitant formation of neutral radicals. Trimethylene oxide (oxetane) and thietane yielded the oxetan-2-yl [24] and thietan-2-yl [23] radicals, respectively, whereas azetidine gave the 1-azetidinyl radical [20]. It was suggested that these radicals were formed in ion-molecule reactions [20,23,24].

3. Cations of Five-Membered Rings

Rigid structures of saturated five-membered-ring molecules are generally puckered, and the forces which tend to retain tetrahedral bond angles are almost balancing the torsional strain. As a consequence, the energy barriers for interconversion between different puckered structures or *pseudo* rotation of the puckering distortion are small, and much theoretical and experimental efforts have been put on the analysis of structure and dynamics of five-membered rings [30].

The hf-splittings associated with the radical cations, as well as the temperature-dependent ESR line-shape, are indirectly a measure of the molecular structure around the radical center and molecular motion (chemical exchange). In this section we will review how information about structure and dynamics has been obtained from ESR studies of radical cations of some saturated five-membered-ring molecules. The reactions to form neutral radicals from the cations are similar to those found for the four-membered rings, see references given in section 2.3, and will not be considered here. The ESR parameters are summarised in Table 4. The activation energies of the dynamical exchange are summarised in Table 5 in a following section (4.2.2).

3.1. CYCLOPENTANE

It is well known that a rigid form of cyclopentane can exist as a puckered C_2 envelope or a C_s half chair conformation, see Fig. 1a. These two structures are actually very similar, and the reader can easily check this by inspecting a molecular model. By choosing one of them and defining the unique carbon atom as C1, a structure similar to the *other* conformation is recognized by looking along the line set by C3 (or C4) and the center of the bond C5-C1 (or C1-C2). The structure and dynamics of the neutral molecule have attracted much attention [31], and the resemblance between the C_s and C_2 structures also makes the conclusions of the cation structure ambiguous, as will be discussed below.

3.1.1. *The Rigid Structure of Cyclopentane*⁺. The ESR spectrum of the cyclopentane cation comprises a 1:2:1 hf triplet in the rigid state, with a splitting of *ca* 25 G due to two equivalent hydrogens. This is consistent with both the C_s and C_2 structure, see Fig. 1a. From *ab initio* calculations a $^2A''$ state has been suggested to be the ground state structure [33,34], with a σ-type "W"-shaped SOMO delocalised over three carbon atoms and two equatorial hydrogens, centred on the unique carbon in the C_s conformation (see Fig. 1a,b).

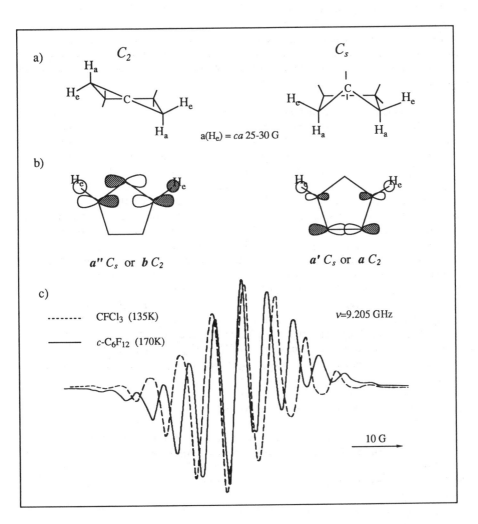

Fig. 1. (a) The rigid C_2 and C_s conformations of c-pentane. (b) Representation of possible SOMOs. (c) Experimental spectra of the cyclopentane cation in the fast exchange limit in two different matrices. All ten hydrogens are equivalent on the ESR time scale owing to rapid ring inversion and *pseudo* rotation.

However, the agreement between the calculated and experimental hf splittings is poor, 39-44 G (calc) *vs.* 22-25 G (exp). It has been speculated upon another SOMO in the ring structure [35] (depicted to the right in Fig. 1b), which can be viewed as a localisation of the unpaired electron mainly to one C-C bond. Preliminary MO calculations [35] for this structure in a C_2 conformation gave a more reasonable hf splitting, *ca* 25-30 G for the two hydrogens H_e. The two hypothetical SOMOs projected in the ring are depicted in Fig. 1b.

The C_2 conformation has never been considered in the analysis of experimental data. It is depicted in Fig. 1a together with the C_s conformation. Because of the σ-character of the SOMOs it is not easy to justify that one of the conformations (C_s or C_2) should be in favour of the other, at least not until closely examining all possibilities. Thus, further theoretical work seems necessary. The hf splittings resolved in the ESR spectra of the cations in a rigid conformation would in all cases be assigned to two equatorial hydrogens, H_e in Fig. 1.

3.1.2. *The Dynamics of Cyclopentane*+. A study of the dynamical behaviour of the cation in CCl_3CF_3 [36] revealed two characteristic motions: ring inversion (E_a=1.2 kcal/mole) and *pseudo* rotation (E_a=3.6 kcal/mole). Here, ring inversion means that the molecule changes into its mirror image and it causes the two equatorial hydrogens giving the large hyperfine splitting (H_e in Fig. 1) to interchange with the axial hydrogens, H_a, on the same methylene group. It is not possible to discriminate between the two structures, C_s or C_2, only considering the rate constants of the exchange. At temperatures above *ca.* 100 K both processes, inversion and *pseudo* rotation, can be considered as being in the fast exchange limit, and the ESR signal is characterised by an 11 line spectrum due to 10 equivalent hydrogens (*i.e.*, the rate constants of inversion and *pseudo* rotation being higher than *ca* 1.4 x 10^9 s^{-1} and 2.8 x 10^8 s^{-1}, respectively). Interestingly, the magnitude of the hf splitting associated with the 10 equivalent hydrogens depends on the choice of matrix, for example; 5.9 G in c-C_6F_{12} and 7.7 G in $CFCl_3$, as do the line shapes recorded with the sample at 77 K [37], see Fig. 1c. The latter indicates that the rate constants for the two dynamical processes depend on the solvent as well as on the temperature. Altogether, these facts imply that the electronic ground state might be different, depending on the matrix. Conclusively, all combinations of the conformations and SOMOs in Fig. 1 are plausible ground states for the cyclopentane cation until further studies have been made. It seems desirable to disclose in detail particularly the matrix effects.

3.2. FIVE-MEMBERED RINGS WITH HETEROATOMS

3.2.1. *Tetrahydrofuran.* Radical cations of tetrahydrofuran (THF) and certain methyl-derivatives; 2-methyl-THF, and, *cis*- and *trans*-2,4-dimethyl-THF, have been studied using $CFCl_3$ as matrix [38]. The results are summarised in Table 4. The ESR spectrum of THF+ at 77 K corresponds to a rigid structure, and the hf structure comprises a triplet of triplets with the splittings 89 G (2H) and 40 G (2H), respectively. The results of the methyl-derivatives are consistent with those of THF. It was concluded that all of the cations took a twisted C_2 (or pseudo C_2) structure with the oxygen located in the unique ring position, *cf.* Fig. 1a where the C_2 structure of cyclopentane is depicted. The proposed SOMO consists of the oxygen lone-pairs, directed approximately perpendicular to the C-O-C fragment, and consequently, the large hf splitting (89 G for THF+) are assigned to the axial hydrogens on C2 and C4. The case is analogous with the oxetane cation (section 2.4.2), however, the large splittings due to the methylene hydrogens in THF+ are not equivalent since the C_2 conformation discriminates between the axial and equatorial hydrogens.

The ESR line-shape of THF$^+$ has a considerable temperature dependence. Increasing the temperature from 77K, the triplet of triplets gradually changed into a quintet having an approximate binomial intensity distribution. This was caused by an exchange between the hydrogens in axial and equatorial positions. Computer simulations of the line-width alternated spectra recorded at intermediate temperatures gave an activation energy of 1.65 kcal/mole for a puckering motion. Results of cation dynamics are summarised in Table 5, and discussed in section 4.2. A similar temperature-effect on the ESR line-shape was observed for 2-methyl-THF$^+$, however, no detailed analysis was offered. Further discussions of aliphatic ether cations are given in the chapter by Shida/Momose/Matsushita.

Table 4. ESR parameters of cations of five-membered rings.

Cation	Matrix	T(K)	ESR-parameters[a]	Ref
cyclopentane$^+$	CF_3CCl_3	5-108	a(2H)=24, a(2H)=3.5, a(2H)=1.5, a(2H)=2.5[b]	[36]
	$CFCl_3$	ca.120	a(10H)=7.7	[37]
	c-C_6F_{12}	ca.120	a(10H)=5.9	[37]
THF$^+$	$CFCl_3$	77	a(2H)=89, a(2H)=40	[38]
2-Me-THF$^+$	$CFCl_3$	77	a(1H)=83, a(2H)=42	[38]
cis-2,2'-Me-THF$^+$	$CFCl_3$	77	a(1H)=95, a(2H)=35	[38]
trans-2,2'-Me-THF$^+$	$CFCl_3$	77	a(2H)=97	[38]
pyrrolidine$^+$	$CFCl_3$	100	g=2.003, $A_{//}$=44.0 $A_\perp$=8.0 (^{14}N) a(2H)=70.5, a(2H)=34, a_{N-H}=24.5	[40]
	CF_3CCl_3	4	$g_{//}$=2.006, $g_\perp$=2.0045, a(2H)=34.5, $A_{//}$=38.0 $A_\perp$=10 (^{14}N), a(2H)=69.1, a_{N-H}=24.5	[40]
	$CF_2ClCFCl_2$	4	$g_{//}$=2.0065, $g_\perp$=2.0035, a(2H)=32.3, $A_{//}$=42.5 $A_\perp$=8.0 (^{14}N), a(2H)=73.5, a_{N-H}=25.5	[40]
N-Me-pyrrolidine$^+$	$CFCl_3$	4	g=2.003, $A_{//}$=42.5 $A_\perp$=10 (^{14}N), a(2H)=57.8, a(2H)=29, a(CH$_3$): 14.3 (2H), 57.0 (1H)	[40]
		77	a(CH$_3$)=28.5, $A_{//}$=42.5 $A_\perp$=10 (^{14}N), a(2H)=57.8, a(2H)=28.5	[40]
tetrahydrothiophene$^+$	$CFCl_3$	77	g_x=2.027, g_y=2.014, g_z=2.002 a(2H$_{\beta1}$)=40, a(2H$_{\beta1}$)=20	[27]
	$CF_2ClCFCl_2$	81	g_x=2.027, a(2H$_{\beta1}$)=40, a(2H$_{\beta1}$)=20	[24]

[a] a=a_{iso}, splittings are given in Gauss (10G=1mT). [b] From a study of exchange broadened spectra.

3.2.2. *Pyrrolidines.*The structure, reaction and dynamics of the pyrrolidine cation, $C_4H_8NH^+$, have been investigated by ESR [40]. The magnetic parameters of the cation in various halocarbon matrices are summarised in Table 4. From the anisotropic ^{14}N hf splitting ($A_{//}$=38.0 G in CF_3CCl_3 at 4 K), *ca* 55% of the unpaired electron is estimated to reside in the non-bonding π-type orbital of the nitrogen atom. The two triplets, 69.1 G (2H) and 34.5 G (2H) in CF_3CCl_3, respectively, are attributed to the axial and the equatorial hydrogens on the methylene group adjacent to the nitrogen, the couplings arising from the hyperconjugative mechanism. The hf splitting of the N-H proton was measured to 24.5 G in CF_3CCl_3. As in the case of the cyclopentane cation $C_4H_8NH^+$ can take either a C_s or C_2 conformation. From *ab initio* UHF, MP2 and SDCI calculations a ^2B ground state in a twisted C_2 structure was proposed, having the calculated hf splittings in qualitative agreeement with the experimental ones [40]. A twist angle of 22° was calculated on the UHF/6-21* level. Although the magnitude of the hf splittings slighly depends on the matrix (see Table 4), the matrix effects are too small to alter the electronic ground state.

Upon warming the sample above *ca* 100 K in the $CF_2ClCFCl_2$ and CF_2ClCF_2Cl matrices the ESR signal of $C_4H_8NH^+$ was lost and signals due to that of a neutral radical appeared, $C_4H_8N\cdot$, formed by a selective deprotonation of the N-H bond of the cation.

The ESR line-shape of $C_4H_8NH^+$ was temperature-dependent, Fig. 2 left column; a-d.

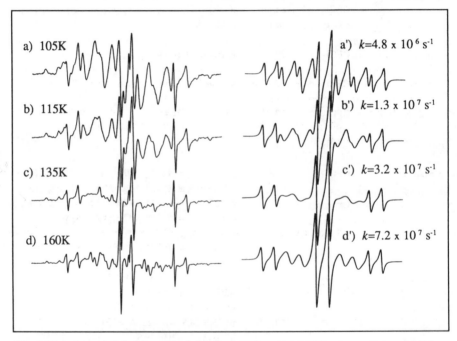

Fig. 2. Experimental (left column) and simulated (right column) ESR spectra of the pyrrolidine cation in the temperature range between 100K and 145K. $CFCl_3$ was used as matrix. The values "k" are the rate constant of the ring inversion at appropriate temperatures.

This process can be analyzed in terms of a ring inversion between two equivalent twisted C_2 structures which brings about an averaging of the two pairs of non-eqivalent hf splittings, H5 (H8) and H6 (H7) in the inset below (next page).

The simulations obtained by adopting this two site model are shown in Fig. 2, right column; a'-d'. Even though the g-anisotropy and the anisotropic ^{14}N hf splittings are neglected, the simulated line shapes reproduce qualitatively the experimental ones. The Arrhenius plot yielded an activation energy of 1.6 kcal/mole, which is of the same magnitude as those deduced for the inversion of the cations of cyclopentane and THF, see Table 5.

Also the cation of N-methylpyrrolidine was proposed to take a ^2B electronic ground state in a twisted C_2 conformation [40]. The rigid structure (at 4K) was found to have one of the C-H bonds of the methyl group oriented parallel to the $2p_z$ orbital of the nitrogen. Warming the sample to *ca* 20 K in $CFCl_3$ the splitting associated with this hydrogen was averaged into a hf quartet indicating methyl group rotation. In contrast to $C_4H_8NH^+$, no ring inversion was observed upon further warming up to ca. 140K. Thus, the presence of a methyl group on the nitrogen might have a crucial influence on the barrier for inversion.

4. Cations of Six-membered Rings

Cations of saturated six-membered rings with heteroatoms have attracted little attention. In view of this and the space limitations of the chapter we have chosen to exclude the topic here, and the reader is referred to a review by Symons [1a] for a collection of the associated ESR parameters and a discussion of the data. This section will focus on our recent studies of cations of cyclohexane and its alkyl-substituted derivatives.

4.1. CYCLOHEXANE AND ITS ALKYL-SUBSTITUTED DERIVATIVES

The highest occupied MOs in alkanes are usually close in energy. There are in most cases at least two orbitals (doubly occupied) within one electron volt [32]. This is particularly evident in molecules having a symmetry condition which makes some of the MOs degenerate. The ionisation of a non-linear *free* molecule will then be followed by a geometrical distortion to structures of lower symmetry due to the Jahn-Teller effect. The introduction of substituents is another way to remove orbital degeneracy, and from such molecules it is possible to extract information of electronic states similar to those following a Jahn-Teller distortion.

The cation of cyclohexane has been presented in the chapter by Toriyama, see also [41-44]. Since the cyclohexane system is of fundamental importance it is of interest to collect information on structure, dynamics and reactions of cations of its derivatives. We start with a brief introduction of the electronic structures possible for the unsubstituted cation. This is followed by a presentation of cyclohexane cations substituted on one carbon, which exclusively stabilise an a_g like SOMO in the ring structure. Thereafter it is shown how various modes of methyl substitution stabilises electronic structures in the ring similar to the

^2A_g, ^2B_g and $^2A''$ states of the cyclohexane cation.

4.1.1. *The Jahn-Teller Split HOMO of Cyclohexane*. The ionisation of the cyclohexane chair conformation can give rise to two possible electronic states, ^2A_g or ^2B_g (in C_{2h}), originating from a Jahn-Teller split of the degenerate E_g (D_{3d} chair) structure [41-44]. It has been suggested that the ^2B_g state was further distorted into a $^2A''$ state (in C_s) of lower energy according to *ab initio* calculations [43,44]. The SOMOs of the three structures are depicted below together with the associated theoretical hf splittings (INDO) due to certain equatorial ring hydrogens in each case [45].

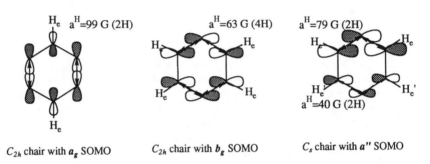

C_{2h} chair with a_g SOMO C_{2h} chair with b_g SOMO C_s chair with a'' SOMO

The arrows indicate distortions in the ring structure in terms of elongated bonds, usually *ca* 1.55-1.60 Å [43-45].

4.1.2. ^2A_g *like Structures for 1-alkyl- and 1,1-dialkyl-substitution*. The cations of 1-alkyl-, or 1,1-dialkyl-cyclohexanes possess a common SOMO within the ring structure which resembles the a_g orbital of the unsubstituted cyclohexane cation [45-47]. The various cations can be grouped into two classes depending on how the alkyl group(s) is (are) located with respect to an a_g like SOMO in the ring structure: asymmetrically, or symmetrically, see Fig. 3.

The asymmetrical case occurs for cations having alkyl groups such as methyl, ethyl, *n*-propyl and *iso*-butyl. To this class belong also the cations of 1,1-dimethylcyclohexane and 1-methyl,1-ethylcyclohexane. The hf pattern is dominated by an 1:2:1 triplet with a splitting being less in magnitude than *ca* 75 G, depending on the number, size and conformation of the substituent(s). It decreases as the size or the number of substituents increases. An additional hf splitting(s) smaller than 34 G appears in certain cases. In such cations the substituent(s) take a conformation having one C-H bond coaxial with respect to the elongated bonds in the ring structure. With substituents larger than ethyl, drastic matrix effects on the hf interaction due to hydrogens on the substituents were observed, implying that alkyl substituents with longer tails can take various conformations. The hf data obtained at 4K using the CF_3-c-C_6F_{11} matrix, are summarised in Fig. 3a. Further details regarding matrix effects and the conformations of the larger substituents are found in [47]. A dynamical effect observed for the methyl- and 1,1-dimethylcyclohexane cations will be discussed in a following section.

The symmetrical case occurs in cations having an alkyl group such as *iso*-propyl and *tert*-butyl, see Fig. 3b. The hf structure resolved in the ESR spectra are due to the methyl hydrogens on the substituent (one for each methyl group) being axial with respect to the σ-bond connecting to the cyclohexyl group. These are of the same magnitude as in the cation of 2,2,3,3-tetramethylbutane [85]. It implies a similar structure over the *iso*-propyl and *tert*-butyl fragments and that a large fraction of the unpaired electron is confined into the bond connecting to the cyclohexyl group, see Fig. 3b.

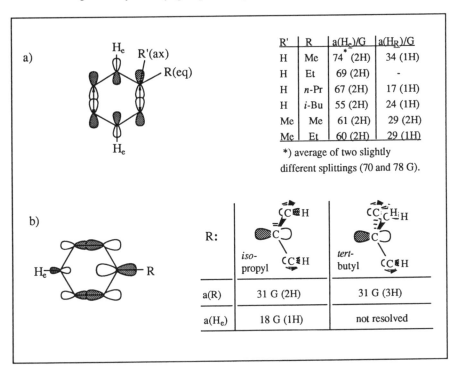

Fig. 3. The cations of 1-alkyl and 1,1-dialkylsubstituted cyclohexanes. (a) The case with an a_g like SOMO in the ring asymmetrically with respect to the position of the substituent. (b) The case with the a_g like SOMO symmetrical.

4.1.3. ^2B_g and ^2A_g like Structures Following Two Modes of Dimethyl-Substitutions.

The results obtained for two modes of dimethylsubstitution is summarised in Fig. 4, left column. Each electronic structure is depicted as a schematic representation of the SOMO in the ring, and the associated experimental hf splittings are indicated. The splittings are assigned to equatorial ring hydrogens, H_e, and in some cases the hydrogens coaxial with a bond in the ring structure having a large fraction of the unpaired spin density, here denoted *trans* methyl hydrogens, H_t. The mechanism for spin transfer is thus analogous to the cases discussed in earlier sections. Because of space limitations we cannot discuss all of the cases in detail, however, a few comments are made [45].

It is well known that cyclohexane can exist as two conformations, the chair and the twisted boat, with the energy of the former being ca. 5-6 kcal/mole lower [48-52]. There are always two chair conformations for each mode of 1,X-dimethylsubstitution (X:2,3,4) [51-53]. In some of the cases both of the methyl groups are equatorial in one chair conformation and axial in the other. In such cases there are no doubts that the former is of lowest energy and abundant at low temperatures. In the remaining cases both chair forms are sterically equivalent having one axial and one equatorial methyl group. The sterical interaction between the axial methyl group and the axial hydrogens on C3 and C5 gives rise to additional strain in the ring structure, and it follows that the twisted boat conformation becomes closer to the chair structure in energy since it can have both methyl groups equatorial. The energy difference in the case of cis1,4-dimethylcyclohexane+ is 2-3 kcal/mole [54a], and thus, it is not obvious that the twisted boat should be excluded in every case.

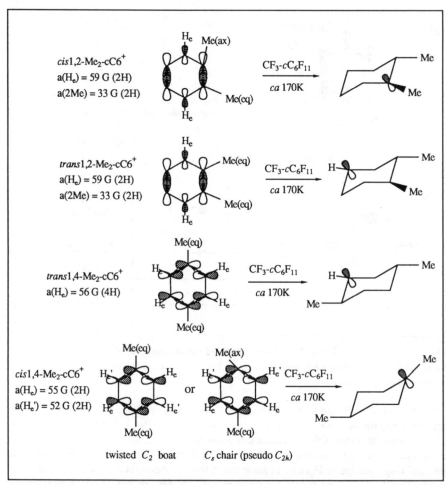

Fig. 4. Left column: The electronic structure and hf splittings of the cations of cis and $trans$ isomers of 1,4- and 1,2-dimethylcyclohexane in CF_3-c-C_6F_{11}. Right column: The neutral alkyl radicals formed during the thermal decay of the cations (temperature ca 170 K).

The ESR spectrum of the $trans$ isomer of the cation of 1,4-dimethylcyclohexane comprised a hf quintet, a^H=56G (4H), of binomial intensity distribution. INDO calculations on a MNDO optimised geometry gave essentially the same hf splitting [45]. The electronic structure and geometry were found similar to the analogous structure calculated for the unsubstituted cyclohexane cation (^2B_g) discussed in an earlier section. The ESR spectrum of the cis isomer is similar to that of the $trans$, however, the relative intensities of the hf lines deviate from the binomial distribution expected from an assumption of four equivalent splittings. The spectrum can be simulated using two pairs of slightly different splittings, a^H=52G (2H) and a^H=55G (2H) [54]. An ambiguity in the determination of cation conformation was found since the hf splittings might be explained by a C_s chair conformation or the C_2 twisted boat. The former has an electronic structure analogous with the *trans* isomer, slightly distorted owing to the lower symmetry. The two cases are depicted in Fig. 4, left column. Note that both structures have a very similar SOMO within the σ-bonds of the ring and associated distribution of the unpaired electron into analogous equatorial C-H bonds.

Both isomer cations of 1,2-dimethylcyclohexane [45] were found to take a chair conformation. Their SOMO and associated hf splittings are summarised in Fig. 4. The results are consistent with a SOMO in the ring structure which resembles the a_g SOMO of the unsubstituted cyclohexane cation. One of the elongated bonds is between the ring carbons having the methyl substituents (C1-C2), and thus, the large hf splittings (59 G) are due to two equatorial hydrogens in the ring, H_e in Fig. 4. The splitting of 33 G can be attributed to one hydrogen on each of the methyl groups - the hydrogens which are coaxial with respect to the C1-C2 bond, in consistency with the a_g-like structures discussed in section 4.1.2.

4.1.4. *The Cis/Trans Isomer Effect For 1,3,5-Trimethyl-Substitution.*

The electronic structures of the *cis* and *trans* 1,3,5-trimethylcyclohexane cation were clarified recently [45]. The most stable conformation of the *cis* isomer has all of the methyl groups equatorial. The trans isomer was found to take a chair conformation having one axial and two equatorial methyl groups. The magnitude and assignment of experimental hf splittings are given in Fig. 5, together with schematics of the electronic structures. The $^2A''$ and $^2A'$ states (in C_s) found for the *cis* and *trans* isomer were deduced by comparing the experimental hf splittings with those calculated with the MNDO and INDO methods [45].

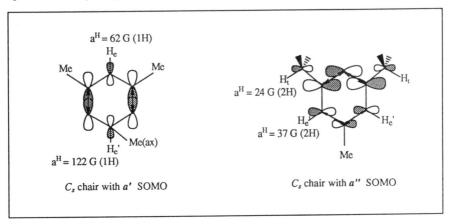

Fig. 5. A schematic representation of the SOMO in the cations of *cis* and *trans* isomers of 1,3,5-trimethylcyclohexane, together with the magnitude and assignment of the hf splittings (in Gauss). Matrix; CF_3-c-C_6F_{11}. Sample temperature; 4.2K.

The presence of an axial methyl group on a cyclohexane chair structure (neutral) is known to increase the total energy with *ca* 1.8 kcal/mole compared with the conformation having one equatorial methyl group [52,53]. This gives an upper limit of the energy difference between the two electronic states, $^2A''$ and $^2A'$, in the *cis* conformation (the *cis* conformation being of lower energy). It implies that the two corresponding states of the cation of *cis*1,3-dimethylcyclohexane should be even closer in energy since the substitutent decisive for the stabilisation of a particular ground state is not present. Furthermore, assuming that the a'' and a' levels are good approximations of the a_g and b_g levels of the cyclohexane cation, the upper limit of the stabilisation energy of the Jahn-Teller distortion can be estimated to *ca* 1.8 kcal/mole.

4.2. DYNAMICAL MIRROR INVERSION OF TWO METHYLCYCLOHEXANE CATIONS

Methane, ethane and several cyclic alkanes have a degenerate ground state and the ionisation must therefore be followed by a geometrical distortion due to the Jahn-Teller effect. These cations show dynamical features described as "site jumping" of the electronic structures between different but equivalent potential minima [41,42]. The activation energy of such site jumping is usually low, typically 0.2 - 0.3 kcal/mole, which is about one order of magnitude lower than dynamical processes involving "true" physical reorientation or conformational changes of molecular fragments (see Table 5).

In this section dynamics of the methyl- and 1,1-dimethylcyclohexane cations is discussed. The neutral molecules have low symmetry but even in these cases the dynamics of the cations, site jumping between equivalent mirror structures, has an activation energy which is comparable to that for the Jahn-Teller distorted alkane cations [41,42].

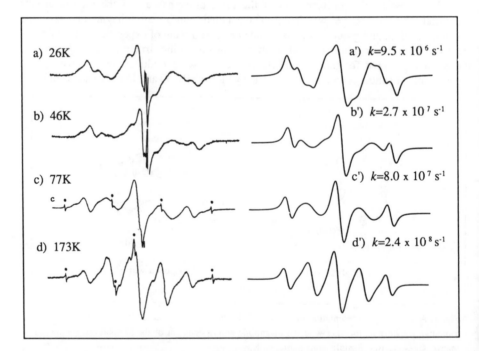

Fig. 6. Experimental (left column) and simulated (right column) ESR spectra of methyl-d_3-cyclohexane$^+$ in the temperature region 4.2K to 173K. The cation was stabilised in CF_3-c-C_6F_{11}.

4.2.1. *Dynamics of the Methylcyclohexane cation.* The electronic structure of the methyl- and 1,1-dimethyl-cyclohexane cations was discussed in section 4.1.2. It was demonstrated how the cation takes a SOMO in the ring structure resembling the a_g SOMO of the cyclohexane cation with one of the elongated bonds in the ring (C1-C2) involving the tertiary or quaternary carbon.

The ESR spectra of the methyl-d_3-cyclohexane cation are simple and therefore instructive to discuss in detail, see Fig. 6. The ESR spectrum recorded at 4.2K consists of an approximate 1:2:1 hf triplet with the splitting being *ca* 74 G [46,55-57]. At the high temperature limit (173K, near a phase transition in the CF_3-c-C_6F_{11} matrix) the ESR signal can be reproduced in simulations with the following hf parameters: $a_1 = 48.8$ G (2H) and $a_2 = 42.7$ G (2H). The reversible changes in the ESR line shape between 4.2 and 173 K and the line width alternation can be explained in terms of a dynamical effect. The various line shapes could be reproduced in simulations using a program to calculate exchange broadened isotropic ESR spectra [57], see Fig. 6, right column.

The proposed model is depicted below for methylcyclohexane$^+$.

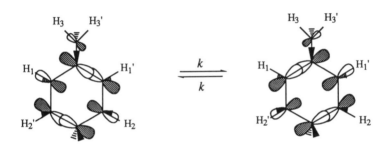

In the dynamical model a *distorted* a_g structure transforms to its own mirror image. The hydrogens H_1 and H_2 are equatorial hydrogens and the hydrogens H_3 on the methyl group are coaxial with the elongated C-C bonds in each of the mirror structures, respectively (cf. sections 4.1.2 and 4.1.3). The following hf splittings could be deduced for the protiated cation in the rigid state: $a(H_1) = 70$ G, $a(H_1') = 15.4$ G, $a(H_2) = 78$ G, $a(H_2') = 19.6$ G, $a(H_3) = 34$ G, and $a(H_3') = 6.2$ G. The numbering refers to the molecule to the left above. The splittings used to simulate ESR spectra of methyl-d_3-cyclohexane$^+$ are obtained from the splittings of the methyl group, $a(H_3)$ and $a(H_3')$, and the gyromagnetic ratio of deuterium atoms. Note that the methyl group is rigid at 173 K.

The temperature-dependent ESR line shape observed for the 1,1-dimethylcyclohexane cation were simulated using the same dynamical model [57]. The methyl groups are rigid up to *ca* 170K also in this cation. The exchange rate constants found in the simulations as a function of the temperature of the associated experimenetal spectra give the activation energy of the process. The values of the two cations are given in Table 5 together with those of other matrix stabilised alkane cations.

4.2.2. *Activation Energies of Molecular Dynamics in Halocarbons.* Some of the parameters characteristic for the dynamics of radical cations and neutral radicals of saturated compounds in halocarbons are summarised in Table 5. The activation energies fall into two groups; those in the order of *ca* 1-3 kcal/mole, and those between 0.05 and 0.7 kcal/mole. The former cases are for cations or radicals in which the dynamical process involves a reorganisation of the molecular structure, as, *e.g.*, the ring inversion of the five membered rings. Low activation energies are found particularly among the Jahn-Teller active molecular cations of c-propane and c-hexane. Matrix effects have been discussed in [42]. The activation energies of the mirror inversion of the electronic state in methyl- and 1,1-dimethylcyclohexane cations are close to the values reported for the cyclohexane cation, and conclusively, both the electronic states and exchange processes seem to be similar. Furthermore, the results of the former cations show that high symmetry and a distortion of a particular Jahn-Teller active molecule is not a necessary condition for having dynamical site jumping of low activation energy.

Table 5. Dynamical motion and activation energies for cyclic saturated radical cations.

Cation	Matrix	T(K)	Comments	E_a(kcal/mole)	Ref
$c\text{-}C_3H_6^+$	SF_6	4-102	three site jump	0.034	[42]
$c\text{-}C_3H_6^+$	$CFCl_3$	4-102	three site jump	0.05	[42]
$c\text{-}C_3H_6^+$	CF_3CCl_3	4-113	three site jump	0.68	[42]
$c\text{-}C_6H_{12}^+$	$CFCl_3$	18-37	two site jump	0.17	[41]
$c\text{-}C_6H_{12}^+$	$CFCl_3$	30-140	three site jump	0.24	[41]
$Me\text{-}cC_6H_{11}^+$	$CF_3\text{-}cC_6F_{11}$	4-173	two site jump	0.20	[57]
$1,1\text{-}Me_2\text{-}cC_6H_{10}^+$	$CF_3\text{-}cC_6F_{11}$	4-173	two site jump	0.30	[57]
$c\text{-}C_5H_{10}^+$	CF_3CCl_3	4-110	inversion/pseudo rot.	1.2/3.6	[36]
THF^+	$CFCl_3$	77-155	puckering	1.65	[38]
pyrrolidine$^+$	$CFCl_3$	75-165	two site	1.6	[40]
pyrrolidine$^+$	$CFCl_2CF_2Cl$	74-111	two site	2.0	[40]

4.3. THE ESEM TECHNIQUE TO MEASURE MATRIX INTERACTIONS

Using the halocarbon matrix isolation technique, strong interactions between the solvent molecules and cations have been found in several cases, particularly for cations having a rather localised SOMO with an ionisation potential of the corresponding neutral molecule being near that of the solvent. Two kinds of σ^*-complexes have been reported. Complexes changing irreversibly with temperature into the solute cations were found for methyl formate/$CFCl_3$ [58] and analogous ether/halocarbon mixtures [59,60]. A reversible temperature effect on the ESR hf pattern were found in several acetaldehyde/Freon systems [61]. The low temperature spectra showed additional resolved anisotropic contributions from matrix chlorine atoms which disappeared upon moderate warming. A dynamical effect which gave rise to a motional averaging of the anisotropic hf tensors was suggested. Reversible temperature effects have been reported also for dimethyl ketene/freon [62] and alkene/freon [63] systems. In the latter two cases, the cation interacts with matrix fluorine - an additional fluorine hyperfine splitting was detected in the high temperature spectra.

In alkane cations the SOMO is more delocalised than in the cases discussed above. Furthermore, it is mostly confined into σ-bonds in the framework of the carbon-carbon bonds which prevents the formation of a complex from sterical reasons. The physical surrounding or the matrix has a more indirect influence on the cation structure. For example: i) the propane cation has been found to take a $^2B_1$ ground state in SF_6 but $^2B_2$ in CCl_2FCClF_2 [64], ii) the SOMO in the isobutane cation was found to change irreversibly from an a' orbital in C_s symmetry to an a_1 orbital in C_{3v} by increasing the temperature from 4.2 to 77 K [64], iii) the cation of *cis*1,3-dimethylcyclohexane was recently found to take different ground states depending on matrix [52b]. The matrix dependence on alkane cation conformations is discussed in other chapters of this book by Shiotani/Lund, and by Toriyama. The origin of these effects are difficult to understand in detail, more theoretical as well as systematic experimental studies seem necessary.

4.3.1. *Matrix Interactions in the ESR Signal of the Cyclohexane Cation.*

The information deduced from the ESR hf patterns of alkane radical cations are limited owing to the relatively broad line width, typically 10-30 G for rigid cations. It is therefore generally difficult to experimentally obtain the detailed SOMO as well as interactions with the surrounding matrix. A sophisticated method to measure small hyperfine interactions in alkane cations was introduced recently by using the ESEM (electron spin echo modulation) technique [66]. For details on an experimental setup and theoretical considerations we refer to the chapter by Kevan and to a recent report [67]. The amplitude of the modulations which are formed in ESEM patterns becomes stronger when the anisotropic hf interaction is large compared with the isotropic one. This situation arises when radicals are trapped in solids having low spin density on the solvating molecules where the interactions with surrounding magnetic nuclei becomes small and almost purely anisotropic. Thus, the ESEM method is particularly useful to study the small anisotropic interactions between matrix stabilised alkane cations and the surrounding matrix nuclei, since such interactions normally are confined into the ESR linewidth. In this context the cyclohexane cation at 4.2 K was studied using different matrices [66]. CCl_2FCClF_2 and $CClF_2CClF_2$ gave clear ESE modulations from anisotropic hf interactions due to matrix fluorine. By comparing the fourier transforms of experimental ESEMs with simulated ones, it was possible to obtain information of the distance from the cation to the matrix nuclei. The simulations were made by calculating the dipolar couplings [68] of the SOMO of the cation to hypothetical arrangments of the surrounding nuclei. It was found that the matrix fluorine has to be less than 3Å away from the center of the cyclohexane ring in order to explain the ESEM patterns

Using $CFCl_3$ or CCl_3CF_3 as matrix to stabilise the cyclohexane cation, ESEM patterns could not be detected. The reasons might have been that the cations in these cases were interacting too strongly or too weakly with the fluorine nuclei in the matrix. We believe that further systematic studies adopting the method can give additional important information of solvation structure of radical cations.

4.4. THE DECAY REACTIONS OF SUBSTITUTED CYCLOHEXANE CATIONS

It is well established that radical cations of alkanes stabilised in halocarbon matrices such as CCl_2FCClF_2, SF_6 or CF_3-cC_6F_{11} decay at elevated temperatures under the formation of neutral alkyl radicals [56,64,69-73]. The structure of the alkyl radicals can be considered as deprotonated forms of the corresponding alkane cations. The temperature generally depends on the solvent as it is related to a phase transition of the solid matrix. In early reports on structure and reactions of alkane cations in halocarbons, it was proposed that a deprotonation from the cation occurred at the C-H bond having the highest spin density [64,71-73]. It was stated that 1-alkyl radicals are formed by deprotonation of the *trans* hydrogen on one of the methyl groups in an extended conformer of *n*-alkane cations, and that 2-alkyl radicals are formed from the methylene hydrogens having high spin density in the corresponding *gauche* conformations. Basically the same reaction mechanism was proposed also in more recent reports [71], including studies of alkane cations in zeolites [74]. Experimental evidence which contradicted that the earlier proposed mechanism is universally valid has been given [69]. For example, *n*-butane/*n*-butane$^+$ in $CFCl_2CF_2Cl$ exclusively yielded 2-butyl radicals [69], although the *gauche* conformation of the cation has high spin density on the hydrogen on the methyl group coaxial with the C2-C3 bond [55,75].

It should also be pointed out that apart from a study on the decay of the ethane cation in SF_6 [72], attempts to quantitatively relate the alkyl radical formation to a reaction involving the alkane cation, have not been published. Furthermore, the reactions referred to above has only been detected in a few matrices. Using solvents such as CF_3CCl_3 or $CFCl_3$ the corresponding alkyl radical formation has not been reported, however, photodecomposition to alkene cations can occur in certain cases [70,73,76]. Thus, it seems desirable to investigate this subject more in detail. Here we make a brief comment on our recent result [77] of the decay of the methyl-substituted cyclohexane cations in CF_3-c-C_6F_{11}.

4.4.1. Cis/Trans Effects among Methyl-Substituted Cyclohexanes. The radical structures which are formed during the decay of selected cations of methyl-cyclohexanes at 175K in the CF_3-c-C_6F_{11} matrix, are summarised in Fig. 4, right column [77]. The electronic structure and associated hf splittings of the cations are presented to the left, and it is easy to see that the site of the alkyl radicals cannot be explained by a deprotonation from the C-H bonds having high spin density (large hf splittings). The objection was also valid in experiments of other cations, such as isopropylcyclohexane and the *cis* and *trans* isomers of 1-methyl,4-isopropylcyclohexane.

By examining the radical structures with respect to the conformation of the neutral molecules it is seen that all solutes of dimethylcyclohexane having at least one axial methyl group in the chair conformation of lowest energy give tertiary alkyl radicals. The other cases give secondary alkyl radicals with the unpaired electron located on the ring structure. In other words, there is a *cis-trans* isomer effect in the formation of alkyl radicals in halocarbons containing the associated radical cations. This site selectivity cannot be explained by the earlier model [64,71-73] in which an alkane cation deprotonates at a carbon-hydrogen bond of high spin density. Until further studies have been made it is not possible to make a new model for alkyl radical formation from alkane cations in halocarbons.

5. Radical Cations of Bicyclic Systems

5.1. BICYCLOALKANES

A considerable number of bicycloalkane cation radicals has been the subject for halocarbon matrix ESR studies, spiro[2.5]octane [7], bicyclo[2.2.1]heptane, *norbornane* [78], bicyclo[2.2.2]octane [79], bicyclo[2.2.0]hexane [80], bicyclo[2.1.0]pentane [79,81], bicyclo[1.1.0]butane [82], spiro[2.2]pentane [81], methylenecyclobutane [81] and *cis*-decalin [83]. Here we will discuss the cations of bicyclo[2.1.0]pentane and bicyclo[2.2.0]-hexane, which participate in interesting intramolecular hydrogen transfer reactions.

5.1.1. *Bicyclo[2.2.0]hexane.* Radiolytic oxidation of bicyclo[2.2.0]hexane (**VII**) in

$CFCl_2CF_2Cl$ and CF_3CCl_3 gave rise to an seven-line ESR hf pattern (a(6H)=12.0G, g= 2.0026) [80]. The spectrum was identified as the cation radical of cyclohexane-1,4-diyl (see below) which is a delocalised species with one electron shared between the 2p orbitals at C1 and C4 rather than a *distonic* cation radical consisting of non-interacting alkyl radical and carbonium ion centers. The six hydrogens were attributed to two equivalent α-hydrogens at C1 and C4 and four equivalent β-hydrogens at C_i (i=2,3,5,6). The latter hydrogens taking the axial position in a chair structure, see **VII-d$_2$$^+$** below.

Information about the stereochemistry of the formation of **VII$^+$** was obtained using a specificially ^{2}D-labelled compound, *exo-cis*-2,3-dideuterio-bicyclo[2.2.0]hexane (**VII-d$_2$**). The six line spectrum of this system was attributed to a chair structure (**VII-d$_2$$^+$**), in which the two deuteriums occupy equatorial and axial positions, respectively, see below. Thus, it is concluded that **VII** is oxidised to give the *cis* enantiomers of the chair structure although cleavage necessarily proceeds through an incipient boat like geometry.

VII-d$_2$ **VII-d$_2$$^+$**

An ESR spectrum identical to that of **VII$^+$** was observed by radiolytic oxidation 1,5-hexadiene **VIII** in a halocarbon matrix [79,84]. Furthermore, by photoillumination with λ > 620 nm or annealing the sample above 90K in CF_3CCl_3, the species **VII$^+$** was found to convert into the cyclohexene cation (**IX$^+$**) by a intramolecular hydrogen transfer. To summarise, in the radiolytic oxidations of **VII** and **VIII**, the chair form of **VII$^+$** represents a common intermediate along the pathways to **IX$^+$**, see below.

5.1.2. *Bicyclo[2.2.0]pentane.* The five membered analogue, bicyclo[2.2.0]pentane (**X**), gives the cyclopentane-1,3-diyl cation radical (**X$^+$**) as the initial product of radiolytic oxidation [79]. The ESR hf values are: 44.9 G (1H), 33.5 G (2H) and 11.7 (2H) in CF_3CCl_3 at 90 K. The authors argued that the formation of **X$^+$** is accompanied by retention of C_s symmetry with a planar radical center at the bridgehead positions. **X$^+$** undergoes subsequent isomerisation to the cyclopentene cation **XI+**. In contrast to the systems **VII** and **VIII**, the attempted cyclisation of the penta-1,4-diene cation failed [79]. Ushida et. al. observed in independent experiments that **X+** isomerises to **XI+** in $CFCl_3$ even at 4K [81].

X	**X+**	**XI+**

5.2. BICYCLIC SYSTEMS WITH HETEROATOMS

The bicyclic systems which have been studied are: 2,3-diazabicyclo[2.2.2]oct-2-ene [65,84], cyclopentaneoxide [17], cyclohexaneoxide [17], 9,10-octalinoxide [19] and *syn*-sesquinobornene [19]. Only 2,3-diazabicyclo[2.2.2]oct-2-ene will be considered here, the others were discussed in relevance to the oxirane cation in sections 2.2.1 and 2.2.2.

5.2.1. *2,3-diazabicyclo[2.2.2]oct-2-ene.* 2,3-diazabicyclo[2.2.2]oct-2-ene (**XII**) oxidised in $CF_2ClCFCl_2$ and CF_2ClCF_2Cl has an ESR hf structure corresponding to a quintet of quintets. The splittings have the following values: a(2N)=2a(4H)=31.0G at *ca* 90K, and the hydrogen splittings were assigned to the four equivalent *anti* positions (*anti* with respect to the nitrogens). The interpretation was verified by using a specifically deuterated compound, *cis-anti*-5,6-dideuterio-**XII**. The radical, **XII+**, has a σ-structure with a b_2 SOMO as depicted below.

XII	**XII+**	**IX+**

Photoconversion (λ>415nm) to the cyclohexene radical cation (**IX+**) was observed. It was assumed to proceed *via* the intermediate **VII+**, even though it was not detected. The ESR signal of **XII+** reappeared on warming the $CF_2ClCFCl_2$ matrix from 110 to 115K, indicating that the cyclohexene cation, **IX+**, is a more powerful oxidant than **XII**.

6. References

1. For reviews, see: a) M. C. R. Symons: *Chem. Soc. Rev.* (1984) 393, b) M. Shiotani: *Mag. Res. Rev.*, **12**, 33 (1987), c) A. Lund, M. Lindgren, S. Lunell, J. Maruani: *Molecules in Physics, Chemistry and Biology, Vol III*, p. 259 Ed. J. Maruani. Acad. Publ. (Dordrecht 1988) , c) T. Shida, E. Haselbach, T. Bally: *Acc. Chem. Res.*, **17**, 180 (1984).

2. K. Toriyama, K. Nunome, M. Iwasaki, T. Shida, K. Ushida: *Chem. Phys. Lett.,* **122(1),** 118 (1985).
3. M. Iwasaki, K. Toriyama, K. Nunome: *J. Chem. Soc. Chem. Commun.,* 202 (1983).
4. K. Ohta, H. Nakatsuji, H. Kubodera, T. Shida: *Chem. Phys.,* **76,** 271 (1983).
5. S. Lunell, L. Yin, M.-B. Huang: *Chem. Phys.,* **139,** 293 (1989).
6. T. Shida, Y. Takemura: *Radiat. Phys. Chem.,* **21,** 157 (1983).
7. X.-Z. Qin, Ff. Williams: *Tetrahedron,* **42,** 6301 (1986).
8. X. -Z. Qin, Ff. Williams: *Chem. Phys. Lett.,* **112(1),** 79 (1984).
9. X. -Z. Qin, L. D. Snow, Ff. Williams: *J. Am. Chem. Soc.,* **106,** 7640 (1984).
10. M. C. R. Symons: *Chem. Phys. Lett.,* **117(4),** 381 (1985).
11. X. -Z. Qin, L. D. Snow, Ff. Williams: *Chem. Phys. Lett.,* **117(4),** 383 (1985).
12. W. R. Down: *NASA Technical Note,* D-4327 (1968).
13. L. D. Snow, J. T. Wang, Ff. Williams: *Chem. Phys. Lett.,* **100(2),** 193 (1983).
14. M. C. R. Symons, B. W. Wren: *Tetrahedron Letters,* **1983,** 2315.
15. X. -Z. Qin, L. D. Snow, Ff. Williams: *J. Phys. Chem.,* **89,** 3602 (1985).
16. X. -Z. Qin, L. D. Snow, Ff. Williams: *J. Am. Chem. Soc.,* **107,** 3366 (1985).
17. J. Rideout, M. C. R. Symons, B. W. Wren: *J. Chem. Soc., Faraday Trans. I,* **82,** 167 (1986).
18. C. J. Rhodes, M. C. R. Symons: *Chem. Phys. Lett.,* **140(6),** 611 (1987).
19. Ff. Williams, S. Dai, L. D. Snow, X. -Z. Qin: *J. Am. Chem. Soc.,* **109,** 7526 (1987).
20. L. D. Snow, Ff. Williams: *Chem. Phys. Lett.,* **143(6),** 521 (1988).
21. T. Bally, S. Nitsche, E. Haselbach: *Helv. Chim. Acta,* **67,** 86 (1984).
22. X. -Z. Qin, Ff. Williams: *J. Phys. Chem.,* **90,** 2292 (1986).
23. X. -Z. Qin, Q. Meng, Ff. Williams: *J. Am. Chem. Soc.,* **109,** 6778 (1987).
24. Ff. Williams, X. -Z. Qin: *Radiat. Phys. Chem.,* **32(2),** 299 (1988).
25. X. -Z. Qin, Q. -X. Guo, J. T. Wang, Ff. Williams: *J. Chem. Soc., Chem. Commun.,* **1987,** 1553.
26. K. Ushida, T. Shida, M. Iwasaki, K. Toriyama, K. Nunome: *J. Am. Chem. Soc.,* **105,** 5496 (1983).
27. D. N. R. Rao, M. C. R. Symons, B. W. Wren: *J. Chem. Soc., Perkin Trans. 2,* **1984,** 1681.
28. X. -Z. Qin, Ff. Williams: *J. Chem. Soc., Chem. Commun.,* **1987,** 257.
29. X. -Z. Qin, Ff. Williams: *J. Am. Chem. Soc.,* **109,** 595 (1987).
30. D. O. Harris, G. G. Engerholm, C. A. Tolman, A. C. Luntz, R. A. Keller, H. Kim, W. D. Gwinn: *J. Chem. Phys.,* **50(6),** 2438 (1969). E. Diez, A. L. Esteban, F. J. Bermejo, M. Rico: *J. Phys. Chem.,* **84,** 3191 (1980). A. C. Legon: *Chem. Rev.,* **80,** 231 (1980).
31. J. E. Kilpatrick, K. S. Pitzer, R. Spitzer: *J. Am. Chem. Soc.,* **69,** 2483 (1947). W. J. Adams, H. J. Geise, L. S. Bartell: *J. Am. Chem. Soc.,* **92(17),** 5013 (1970). L. A. Carrera, G. J. Jiang, W. B. Persson, J. N. Willis, Jr.: *J. Chem. Phys.,* **56(4),** 1440 (1972). L. E. Bauman, J. Laane: *J. Phys. Chem.,* **92,** 1040 (1988).
32. K. Kimura, S. Katsumata, Y. Achiba, T. Yamasaki, S. Iwata,:*Handbook of HeI Photoelectron Spectra of Fundamental Organic Molecules,* Japan Scientific Society Press (Tokyo 1981).
33. K. Ohta, H. Nakatsuji, H. Kubodera, T. Shida: *Chem. Phys.,* **76,** 271 (1983).
34. M.-B. Huang, S. Lunell, A. Lund: *Chem. Phys. Lett.,* **99,** 201 (1983).
35. S. Lunell, L. Sjöqvist, J. Maruani: Private Communication
36. L. Sjöqvist, A. Lund, J. Maruani: *Chem. Phys.,* **125,** 293 (1988).
37. M. Shiotani, Y. Nagata, M. Tazaki, J. Sohma: Japanese ESR Symposium, Abstract p.45 (1982).
38. H. Kubodera, T. Shida, K. Shimokoshi: *J. Phys. Chem.,* **85(18),** 2583 (1981).
39. X. -Z. Qin, Ff. Williams: *J. Chem. Soc., Chem. Commun.,* **1987,** 450.
40. M. Shiotani, L. Sjöqvist, A. Lund, S. Lunell, L. Eriksson, M. -B. Huang: *J. Phys. Chem.,* Accepted.
41. K. Toriyama, K. Nunome, M. Iwasaki: *J. Chem. Soc. Chem. Commun.,* **1984,** 143.
42. M. Iwasaki, K. Toriyama, K. Nunome: *Faraday Discuss. Chem. Soc.,* **78,** 19 (1984).
43. S. Lunell, M. B. Huang, O. Claesson, A. Lund: *J. Chem. Phys.,* **82,** 5121 (1985).

44. S. Lunell, M. B. Huang, A. Lund: *Faraday Discuss. Chem. Soc.*, **78**, 35 (1984).
45. M. Shiotani, M. Lindgren, T. Ichikawa: *J. Am. Chem. Soc.*, **112**, 967 (1990).
46. M. Lindgren, M. Shiotani, N. Ohta, T. Ichikawa, L. Sjöqvist: *Chem. Phys. Lett.*, **161**, 127 (1989).
47. M. Shiotani, M. Lindgren, N. Ohta, T. Ichikawa: To be published.
48. K.S. Pitzer: *Chem. Rev.* **27**, 39 (1940).
49. W.S. Johnson, V.J. Bauer, J. L. Margrave, M. A. Frisch, Ll. H. Dreger, W. N. Hubbard: *J. Am. Chem. Soc.* **83**, 606 (1961). J.L. Margrave, M.A. Frisch, R.G. Bantista, R.L. Clarke, W.S. Johnsson: *J. Am. Chem. Soc.* **85**, 546 (1963).
50. F.A.L. Anet, M. Squillacote: *J. Am. Chem. Soc.*, **97**, 3244 (1975).
51. D.K. Dalling, D.M. Grant: *J. Am. Chem. Soc.*, **94**, 5318 (1972).
52. E. L. Eliel, N. L. Allinger, S. J. Angyal, G. A. Morrison: *Conformational Analysis*, Interscience Publishers Inc. (New York 1965).
53. D.K. Dalling, D.M. Grant: *J. Am. Chem. Soc.*, **89**, 6612 (1967).
54. a) M. Lindgren, M. Shiotani: *Japanese Symposium on Molecular Structure*, Abstract 1P42, Sept. (1989), b) M. Lindgren, M. Shiotani, To be submitted.
55. M. Lindgren, *Electron Spin Resonance Studies of Some Radical Ions*. Thesis ISBN 91-7870-381-6 (University of Linköping, Sweden, 1988).
56. M. Shiotani, N. Ohta, T. Ichikawa: *Chem. Phys. Lett.*, **149**, 185 (1988).
57. L. Sjöqvist, M. Lindgren, M. Shiotani, A. Lund: *J. Chem. Soc. Faraday Trans.*, Accepted May (1990).
58. D. Becker, K. Plante, M. D. Sevilla: *J. Phys. Chem.*, **87**, 1648 (1983).
59. J. Rideout, M. C. R. Symons: *J. Chem. Res. (S)*, **1984**, 268.
60. X. Qin, B. W. Walther, Ff. Williams: *J. Chem. Soc. Chem. Commun.*, **1984**, 1667.
61. L. D. Snow, Ff. Williams: *Chem. Phys. Lett.*, **100**, 198 (1983).
62. J. Fujisawa, S. Sato, K. Shimokoshi, T. Shida: *Bull. Chem. Soc. JPN*, **58**, 1267 (1985).
63. L. Sjöqvist, M. Shiotani, A. Lund: *Chem. Phys.*, **141**, 417 (1990).
64. K. Toriyama, K. Nunome, M. Iwasaki: *J. Chem. Phys.*, **77**, 5891 (1982).
65. S. C. Blackstock, J. K. Kochi: *J. Am. Chem. Soc.*, **109**, 2484 (1987).
66. J. Westerling, A. Lund: *Chem. Phys.*, **140**, 421 (1990).
67. J. Westerling: *ESR, ENDOR and ESEM Studies of Free Radicals and Construction of an ESE-spectrometer*. Thesis ISBN 91-7870-560-6 (University of Linköping, Sweden, 1990).
68. O. Edlund, A. Lund, M. Shiotani, J. Sohma, K.-Å. Thoumas: *Mol. Phys.*, **32**, 49 (1976).
69. M. Lindgren, A. Lund, G. Dolivo: *Chem. Phys.*, **99**, 103 (1985).
70. G. Dolivo, A. Lund: *Z. Naturforsch.*, **40a**, 52 (1984).
71. K. Toriyama, K. Nunome, M. Iwasaki: *J. Phys. Chem.*, **90**, 6836 (1986).
72. M. Iwasaki, K. Toriyama, K. Nunome: *Radiat. Phys. Chem.*, **21**, 147 (1983).
73. K. Toriyama, K. Nunome, M. Iwasaki: *Chem. Phys. Lett.*, **105**, 414 (1984).
74. K. Toriyama, K. Nunome, M. Iwasaki: *J. Am. Chem. Soc.*, **109**, 4496 (1987).
75. M. Lindgren, A. Lund: *J. Chem. Soc. Faraday Trans I*, **83** 1815 (1987)
76. M. Tabata, A. Lund: *Radiat. Phys. Chem.*, **23**, 545 (1984).
77. M. Shiotani, M. Lindgren, F. Takahashi, T. Ichikawa: *Chem. Phys. Lett.*, In Press (1990).
78. K. Toriyama, K. Nunome, M. Iwasaki: *J. Chem. Soc., Chem. Commun.*, **1983**, 1346.
79. Ff. Williams, Q. -X. Gou, T. M. Kolb, S. F. Nelsen: *J. Chem. Soc., Chem. Commun.*, **1989**, 1835.
80. Ff. Williams, Q. -X. Gou, D. C. Bebout, B. K. Carpenter: *J. Am. Chem. Soc.*, **111**, 4133 (1989).
81. K. Ushida, T. Shida, J. C. Walton: *J. Am. Chem. Soc.*, **108**, 2065 (1986).
82. F. Gerson, X. -Z. Qin, C. Ess, K. -S. Else: *J. Am. Chem. Soc.*, **111**, 6456 (1989).
83. V. I. Melekhov, O. A. Anisimov, A. V. Veselov, Y. N. Molin: *Chem. Phys. Lett.*, **127**, 97 (1986).
84. Ff. Williams, Q. -X. Gou, P. A. Petillo, S. F. Nelsen: *J. Am. Chem. Soc.*, **110**, 7887 (1989).
85. J. T. Wang, Ff. Williams: *Chem. Phys. Lett.*, **82**, 177 (1980).

DEUTERIUM LABELLING STUDIES OF CATION RADICALS

MASARU SHIOTANI
Department of Applied Physics and Chemistry, Faculty of Engineering
Hiroshima University, 0724 Higashi-Hiroshima, Japan

ANDERS LUND
Department of Physics and Measurement Technology, University of Linköping
S-581 83 Linköping, Sweden

1. Introduction

The cations formed by one electron oxidation of neutral molecules are radicals and hence amenable to ESR study. However, their highly reactive nature and difficulties with their generation have been obstacles to the study of representative organic and inorganic cation radicals This changed with the development of halocarbon and noble gas matrix isolation techniques combined with ionizing radiations as applied to ESR spectroscopy by Shida [1] and Knight [2] about one decade ago. Since then, a large number of novel cation radicals have been investigated by this method. Among them, the studies on hydrocarbon cations are prominent. Most of the isotope labelled hydrocarbon cations studied are deuterium (^{2}D) labelled ones, little having been reported so far on carbon-13 (^{13}C) labelled cations. In this chapter we are concerned with the ^{2}D-labelling studies of hydrocarbon cations. The reader is referred to previous review articles [3,4,5,6,7.8] and other chapters in this book for a more complete coverage of other aspects of cation radical spectroscopy.

An unequivocal assignment of an ESR spectrum is often possible only by using selectively ^{2}D-labelled compounds. ^{2}D-labelling reduces the corresponding ^{1}H hyperfine (hf) coupling constant by a factor of 6.5, the ratio of the magnetic moment of ^{1}H to that of ^{2}D. This generally causes significant changes in the ESR hf pattern, which enable us to determine precise hf constants and assign the spectrum correctly. Because of the mass difference, the ^{2}D labelling sometimes gives rise to important effects on electronic and geometrical structures of cation radicals in solid matrices at low temperate. Such isotope effects are occasionally more pronounced in alkane cations since these have several electronic levels in a narrow energy range [9]. In addition a ^{2}D-isotope effect can sometimes be seen in thermal and photoinduced rearrangement and/or fragmentation reactions of cations. The synthesis of ^{2}D-labelled compounds is often difficult and troublesome, time consuming and expensive. These are the main reasons why only a limited number of molecules have been subjected to ^{2}D and/or other isotope-labelling studies.

A. Lund and M. Shiotani (eds.), Radical Ionic Systems, 151–176.

This chapter will deal mainly with cation radicals of [2]D-labelled hydrocarbon, *i.e.*, alkane, alkene and alkynes, some important cations with oxygen atoms being included. Mentioned are ESR studies on the electronic and geometrical structure, dynamics, and reactions of the cation radicals together with molecular orbital (MO) calculations by *ab-initio* and/or *semi-empirical* methods. Some results for ^{13}C-labelled alkene cations are presented since the ^{13}C-labelling studies are especially helpful to discuss the electronic structure of hydrocarbon cations.

Knight and his collaborators have extensively studied rather simple and fundamentally important inorganic cation radicals. They cover species with various isotopes possessing magnetic moment such as [15]N (nuclear spin $I=1/2$), [17]O ($I=5/2$), [29]Si ($I=1/2$) together with [2]D ($I=1$) and [13]C ($I=1/2$). The details have been presented in a previous chapter.

2. Structure and Reaction of Normal and Branched Alkane Cations

It is known that alkane cations may have different conformations depending on their chain length and on the matrix [10]. In some instances several conformations are present in halocarbon matrices. The use of selectively [2]D-labelled *n*-alkanes, together with MO calculations makes it possible to probe the detailed nature of these conformers. Thermal and photoinduced fragmentation reactions of cations have been investigated. Here, isotope effects on the reactions are sometimes significant. The cations were usually produced by ionizing or vacuum UV irradiation in noble gas as well as halocarbon matrices at low temperature. Experimental ESR data are summarized in Table 1 and 2. Details are presented below.

2.1. STRUCTURE OF NORMAL AND BRANCHED ALKANE CATIONS

2.1.1. *Methane.* The methane cation cannot be produced by irradiation in halocarbon matrices due to its high ionization potential. The partially [2]D-labelled methane cation, $CD_2H_2^+$, was produced by neon discharge photoionization at 17 eV in a neon matrix at 4K. The ESR spectrum with $a_H = 121.7$ G (2H) and $a_D = (-)2.22$ G (2D) is consistent with a C_{2v} distorted structure. This conclusion was obtained by the combined experimental and theoretical work of Knight et al. [11]. The equilibrium geometry deduced from *ab-initio* calculations gives the hf constants, $a_H = 137$ G (2H) and $a_D = -2.58$ G (2D) [12a], which are in excellent agreement with the experimental values.

According to the theoretical analysis [12a,12b], dynamic averaging between different structures does not occur in $CD_2H_2^+$ in contrast to the *protiated* methane cation, CH_4^+. The preferential conformation is with the deuterium atoms at the short bonds (structure I) , confirming the suggestion by Knight et al.

2.1.2. *Ethane.* The structure of [2]D-labelled ethane cations has been investigated in a sulphur hexafluoride *(SF₆)* matrix [10,13,14]. As mentioned in a previous chapter (Toriyama), radiolytically produced ethane cation exhibits a C_{2h} distortion giving ^2A_g state whose ESR spectrum shows a large hyperfine structure from two hydrogens, one on each carbon atom with the C-H bonds in the same plane. The methyl groups tilt inward such that the C-C-H bond angle for the *in-plane* hydrogens becomes smaller than the tetrahedral angle. A *normal* C-C distance was implied, contradicting theoretical predictions of a long C-C bond of approximately 1.9 Å [15].

Table 1. ^{2}D-labelled *n*-alkane cations. The experimental ^{1}H hf couplings are compared with the calculated ones. The data were taken from refs [10,13,14,16,17,23-26].

Cation (matrix)	Temp. (K)	Hyperfine Couplings (G) [a]		Geometry (State)
		Exp.	Calc.	
$CH_2D_2^+$ (*neon*)	4.2	121.7 (2H) (-)2.22 (2D)	137 -2.58	C_{2v} ($^2B_1$) r(C-D) = 1.16 Å $\angle$D-C-D = 60°
$CH_3CD_3^+$ (*SF$_6$*)	4.2	140.5 (1H) 21.9 (1D) 9.0 (2H)	131.3	C_{2h} (^2A_g) r(C-C) = 1.58 Å
$CD_3CD_3^+$ (*SF$_6$*)	4.2	23.0 (2D)		
$CHD_2CHD_2^+$ (*SF$_6$*)	4.2	151.0 (2H)		
$CH_3CD_2CH_3^+$ (*SF$_6$*)	4.2	98.0 (2H)	88.6	C_{2v} ($^2B_1$) r(C-C) = 1.58 Å
$CD_3CH_2CD_3^+$ (*CF$_2$ClCFCl$_2$*)	4.2	105.5 (2H) 8.0 (4D)		C_{2v} ($^2B_2$) r(C-C) = 1.47 Å
$(CH_3CD_2)_2^+$ (*CF$_3$CCl$_3$*)	77	60.0 (2H) 10.4 (2H) 5.0 (2H)	39.6 5.9 3.7	*gauche* r(C$_2$-C$_3$) = 2.03 Å
$(CD_3CH_2)_2^+$ (*CF$_3$CCl$_3$*)	77	8.9 (2D) 7.4 (2H)	6.1 0.7	
$(CH_3CD_2)_2^+$ (*CF$_2$ClCFCl$_2$*)	93	60.5 (2H) 7.6 (4H)	40.0 4.7	*extended* r(C$_2$-C$_3$) = 2.02 Å
$(CH_3CD_2)_2CH_2^+$ (*CF$_2$ClCFCl$_2$*)	77	57.0 (2H)	48.9	*extended* r(C-C) = 1.55 Å
$(CD_3CH_2)_2CH_2^+$ (*CF$_2$ClCFCl$_2$*)	77	96.0 (1H)	74.5	*gauche*
$(CH_3CH_2)_2CD_2^+$ (*CF$_3$CCl$_3$*)	138	85.0 (2H) 13.0 (6H)	88.2 25.0, 11.0, 1.2	*gauche-gauche*

[a] 10 G = 1 mT

Recent calculations suggest a C_{2h} structure with a *normal* C-C bond length of 1.579 Å [16]. The calculated isotropic coupling is 131.3 G (2H) in good agreement with the experimental value. In the CHD_2CHD_2 cation [14] the C-H bonds preferentially occupy the *in-plane* positions (structure II). A similar observation has been made for partially ^{2}D-labelled *n*-butane cations (see below).

$^2B_1(C_{2v})$ $^2A_g(C_{2h})$ $^2B_1(C_{2v})$ $^2B_2(C_{2v})$

I   II   III   IV

($\longleftrightarrow$ indicates long, $\blacktriangleright\!\!\blacktriangleleft$ short bonds)

2.1.3. Propane. The structure of the propane cation has been investigated by Toriyama et al. [10] using the selectively ^{2}D-labelled compounds, $CH_3CD_2CH_3$, and $CD_3CH_2CD_3$. In SF_6, $CH_3CD_2CH_3^+$ gave exactly the same triplet spectrum as that of $C_3H_8^+$, while $CD_3CH_2CD_3^+$ gave a quintet spectrum due to two equivalent deuterons with $a_D = 14.8$ G (equivalent to $a_H = 96.2$ G). The results show conclusively that the hf coupling is due to the in-plane protons, one on each methyl group in a cation of C_{2v} symmetry (structure III). The unpaired electron occupies a singly occupied molecular orbital (SOMO) of b_1 symmetry. The possibility of a lower C_s structure with one long and one short C-C bond with a rapid averaging between two equivalent long-bond structures has been investigated [15]. According to a recent theoretical analysis [17], however, the proton hf coupling obtained by such an averaging would be much too small, approximately 20 G, and the possibility of a lower symmetry must be ruled out.

The $CD_3CH_2CD_3$ cation in halocarbon matrices gave ESR spectra differing from those obtained in SF_6 [10]. It was shown that the SOMO was such as to give spin density on the two methylene protons ($a_H = 105.5$ G) and the four out of plane methyl deuterons ($a_D = 8.0$ G corresponding to $a_H = 52.5$ G), see structure IV. INDO calculations showed that the unpaired electron is delocalized in a SOMO of b_2 symmetry composed of *out-of-plane* orbitals, approximately expressed by $(1/2)(\phi_1 + \phi_3) - (1/\sqrt{2})\phi_2$ were ϕ_i is the pseudo $\pi(CH_2)$ orbital belonging to the *i-th* carbon atom. A thorough theoretical investigation has recently been made which essentially confirms the original assignment [17].

2.1.4. n-Butane The *protiated* n-butane cation (nB-h^+) has been prepared by radiolysis in halocarbon matrices and its electronic and geometric structure has been investigated by ESR [10,18,19]. The data have been analyzed in terms of a C_{2h} extended *trans* structure to show that the SOMO is of σ-type and delocalized over a planar carbon chain. The observed hf coupling of 61 G (2H) was attributed to the two *in-plane* H atoms on the end methyl groups. The model could be somewhat refined based on experiments with specifically ^{2}D-labelled compounds [21]. By comparing experimental and simulated ESR spectra from $CH_3CD_2CD_2CH_3^+$ (nB-2,3-d_4^+) and $CD_3CH_2CH_2CD_3^+$ (nB-1,4-d_6^+) in CF_3CCl_3 (Fig. 1), it was concluded that the hf couplings of the methyl protons were 60.5, 10.4 and 5.0 G, respectively (structure V). The average coupling of the two latter *out-of-plane* protons, 7.4 G, agrees almost exactly with that derived from the ESR dynamics of a methyl rotation in $CFCl_3$ between 77 and 149K [20]. The non-equivalence of the hf couplings in the former case could be explained by an *out-of-plane* position of the axial hydrogens of about 4° [8].

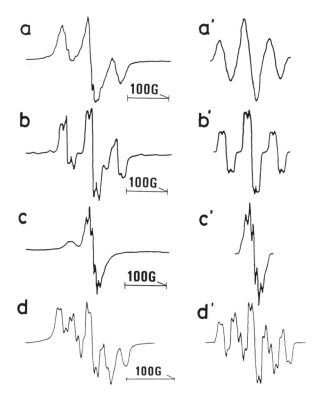

Fig. 1. ESR spectra of\
a) $CH_3CH_2CH_2CH_3^+$, b) $CH_3CD_2CD_2CH_3^+$, c) $CD_3CH_2CH_2CD_3^{+,}$ and d) $CHD_2CH_2CH_2CHD_2^+$ in CF_3CCl_3 at 77K. a')-d') are simulated spectra. The data were taken from [21].

Another possibility considered by the same authors [8] was that the C_2-C_3 bond is long and that the carbon chain becomes non-planar by a rotation about the weakened central C-C bond. The experimental data could be analyzed with two coupling constants for the methylene protons, one with 7.4 G (2H) and the other unresolved, which would support a non-planar carbon chain. However, the analysis of the ESR dynamics is in accord with a planar structure with equal splittings of the methylene protons of about 5 G. Both types of analysis are consistent with an elongated C_2-C_3 bond.

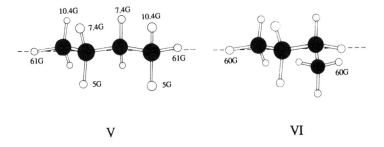

V   VI

The geometric structure of nB^+ has been further discussed by Lindgren [21] from an experimental, and by Huang and Lunell [22a] from a theoretical point of view. In both cases a comparatively small barrier for *trans-gauche* isomerization (*ca.* 0.5 kcal/mol), was noted, explaining the occurrence of the *gauche* form in matrix-isolation ESR experiments with *n*-alkane cations of, for instance, *n*-pentane and *n*-hexane (see below). The assignment of a *gauche* form of nB^+ depends critically on the assumption that the spectra are not affected by dynamic effects at 77K, and that the hf couplings are isotropic. These assumptions have been questioned on experimental and theoretical grounds [20,22].

Cations of $CH_2D(CH_2)_2CH_2D$ (*nB*-1,4-d_2) and $CHD_2(CH_2)_2CHD_2$ (*nB*-1,4-d_4) may be trapped as rotational isomers and an isotope effect similar to that for ethane cations has been observed [21,23,32]. The *nB*-1,4-d_4 cation was observed in the *gauche* conformation (structure VI) in CF_3CCl_3. The possible rotational isomers are VII, VIII, and IX which are indicated in the insert below. The hf couplings observed are attributed to the *in-plane* hydrogens (or deuteriums) H_5 and H_6 of the terminal methyl groups, the plane being defined by $(H_5-)C_1-C_2-C_3$ and $C_2-C_3-C_4(-H_6)$ which are twisted about the C_2-C_3 bond. The experimental abundance ratio of the three isomers, VII, VIII, and IX, is 3:5:2 based on the ESR spectral simulation, the ratio differing from the statistically expected one, i.e., 1:4:4.

The *ab-initio* MO calculations (UHF/3-21G level) for the *gauche* conformation of nB-h^+ resulted in the longer C-H bond of 1.098 Å for the *in-plane* hydrogens (H_5 and H_6) in the terminal methyl groups and the shorter bonds of 1.084 Å and 1.085 Å for the *out-of-plane* hydrogens [22a, 22b]. This result implies that the deuterium atoms will preferentially occupy the *out-of-plane* positions, since these have the higher force constants which cause a larger decrease in vibrational energy upon deuteration. Very recently Lunell et al. evaluated a theoretical abundance ratio of 2:5:2, which is very close to the experimental one, based on the *zero-point* vibrational energy for these isomers calculated by quantum chemical ab-initio methods under the assumption of a Bolzmann distribution [22b].

2.1.5. C_5-C_8 n-alkanes.

Selectively ^{2}D-labelled *n*-pentane (*nC5*) cations, *n*-$CH_3CH_2CD_2CH_2CH_3^+$ (*nC5*-3-d_2^+), *n*-$(CH_3CD_2)_2CH_2^+$ (*nC5*-2,4-d_4^+), and *n*-$CD_3(CH_2)_3CD_3^+$ (*nC5*-1,5-d_6^+) produced in various halocarbon matrices have been studied by ESR at 77K and up to the temperature where they decay [24,25]. A striking observation was that the geometry of $nC5^+$ depends critically on the ^{2}D-labelling. In the $CF_2ClCFCl_2$ matrix about 90 % of *nC5*-1,5-d_6^+ is in a *gauche* conformation obtained by a rotation of 120° about the C_3-C_4 bond of the *trans* geometry [26]. The *nC5*-2,4-d_4^+ has an extended *trans* geometry in the same matrix. The spectra and the proposed geometries are displayed in Fig. 2. The central part of the spectrum in Fig. 2b and the triplet in Fig. 2a are attributed to the *trans* geometry of $nC5^+$ [21,24]. Matrix induced distortions might also be responsible for the observations on *nC5*-3-d_2^+ in CF_3CCl_3 (Fig. 2c). Cations of specifically ^{2}D-labelled *n*-hexane, *n*-heptane and *n*-octane have also been produced by radiolysis in the above-mentioned matrices [25]. Also in these cases, the geometry depended on the ^{2}D-labelling and the matrix. The 120° *gauche* conformation of cations depended on the ^{2}D-labelling and the matrix. The 120° *gauche* conformation of cations with CD$_3$ end groups were often stabilized compared to the *extended* geometry, similar to the case of *n*-pentane cations. This behavior was related to the smaller *van der Waals* radius of CD$_3$ compared to CH$_3$.

In conclusion, experiments with ^{2}D-labelled *n*-alkanes have helped to obtain the electronic and geometric structure of the cations trapped in matrices. Deviations from the extended *trans* geometric structure occur, possibly to some extent in *n*-butane and definitely in longer *n*-alkane ions. The actual geometries depend on the matrix and on the position of the ^{2}D-labelling. Thus, the relative yield of different conformers varies and might not be the same as in the pure liquid after pulse radiolysis, see other chapters (Trifunac and Mehnert). As seen from Table 1, long carbon-carbon bonds are sometimes predicted in the *free* cations with Hartree-Fock type calculations. Whether they appear in matrices and in more accurate calculations is still an unresolved issue. A model to predict the geometric and electronic structure of *n*-alkane cations based on simple MO theory has been proposed [21].

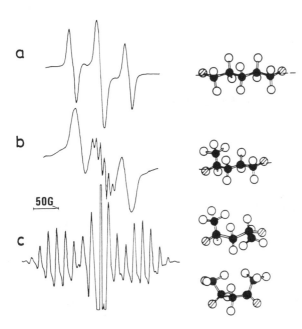

Fig. 2. ^{2}D-isotope and matrix dependences on the ESR spectra of *n*-pentane cations. a) *n*-(CH$_3$CD$_2$)$_2$CH$_2$$^+$ (*n*C5-2,4-*d$_4$*$^+$) in CF$_2$ClCFCl$_2$ at 77K (first derivative): extended *trans*. b) *n*-CD$_3$(CH$_2$)$_3$CD$_3$$^+$ (*n*C5-1,5-*d$_6$*$^+$) in CF$_2$ClCFCl$_2$ at 77K (first derivative): *gauche*. c) *n*-CH$_3$CH$_2$CD$_2$CH$_2$CH$_3$$^+$ (*n*C$_5$-3-*d$_2$*$^+$) in CF$_3$CCl$_3$ at 138K (second derivative): *gauche-gauche-trans* (upper) and *gauche-gauche-cis* (lower). The data were taken from [21,24].

50G

2.2. STRUCTURE OF METHYL-SUBSTITUTED ALKANE CATIONS

Cation radicals of methyl-branched alkanes have been extensively studied by several groups, as described in a previous chapter (Toriyama). The most important conclusion obtained from *protiated* alkanes is that the unpaired electron is rather confined to one of the C-C bonds such as to maximize the *hyperconjugative* effect. Consequently, ^{1}H hf couplings in the range of ca. 29 G to *ca.* 63 G have been observed for the C-H protons

158

trans with respect to the C-C bond in which the unpaired electron is mainly confined [10,27]. This conclusion was slightly modified by ^{2}D-labelling studies. In this section we are concerned with ^{2}D-labelled 2-methylbutane$^+$ (2MB$^+$), 3-methylpentane$^+$ (3MP$^+$) and 3-methylhexane$^+$ (3MH$^+$).

The cation of 2MB-4-d_3 gave a triplet with a ^{1}H hf coupling of 42.9 G in CF$_2$ClCFCl$_2$ at 77K [30,31]. This observation supports the [G.g] conformation (Fig. 3), while the [T.g] structure, which had been previously proposed based on only spectra of 2MB-h^+ [27], can be ruled out.

The spectrum of protiated 3MP cation (3MP-h^+) in CF$_2$ClCFCl$_2$ at 77K consists of a quintet with an averaged coupling of 47.3 G [28,29]. Iwasaki and Toriyama [28] suggested that 3MP-h^+ has a *gauche-trans* [GT] skeletal conformation, more exactly either [GT.gg] or [GT.tg] (see Fig. 3), and the unpaired electron is mainly localized at the C$_3$-C$_4$ bond, yielding magnetically equivalent hf splittings of 47.3 G for the three trans C-H protons bonded to the methylene (C$_2$), branched (C$_3$'), and terminal (C$_5$) atoms. In subsequent experiments using selectively ^{2}D-labelled 3-methylpentanes, Ohta and Ichikawa [29] precisely determined the hf couplings of 3MP$^+$. The experimental results are summarized in Table 2

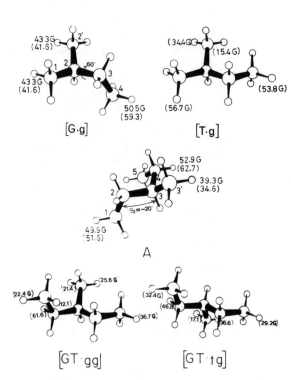

Fig. 3. The conformations of 2MB$^+$ (upper) and 3MP$^+$ (lowers). Experimental ^{1}H hf couplings are compared with theoretical ones (in parentheses). 3BP$^+$: conformation [G.g] was proposed instead of the previously reported structure, [GT]; 3MP$^+$: conformation *A* was proposed instead of the previously reported structures, [GT.gg] and [GT.tg]. Taken from [30].

for 3MP-h^+ suggests a mechanism other than hyperconjugation. Based on the ESR results together with INDO MO calculations an alternative conformation was proposed for 3MP$^+$: structure A in Fig. 3 in which the dihedral angles of C_2-C_3 and C_3-C_3' with respect to C_4-C_5 are ca. 40° and ca. 80°, respectively, and C_1-C_2 is *gauche* with respective to C_3-C_4 [30].

Furthermore it was pointed out that the inequivalency in the couplings originated from considerable participation of the p_z orbitals of C_2, C_3', and C_5 to the SOMO [30].

In addition to 3MP$^+$, a ^{2}D-labelling study was carried out for the 3MH cations using the four selectively ^{2}D-labelled compounds given in Table 2 [29]. It was revealed that 3MH$^+$ in CF$_2$ClCFCl$_2$ has non-equivalent ^{1}H hf splittings of 57.1 G, 39.1 G, and 61.9 G on the C_2, C_3', and C_5 atoms, respectively, although the *protiated* cation gives four hf lines with an averaged coupling of 52.7 G. The result means a negligible hf interaction of the unpaired electron with the two terminal methyl protons, indicating that the local conformation of 3MH$^+$ is essentially the same as that of 3MP$^+$.

We conclude that the hyperconjugative effect proposed by Iwasaki and Toriyama [10,27] qualitatively explains the origin of proton hf couplings in branched alkane cations and is very helpful to assign the observed ESR features. However, the effect is smaller than expected for some branched alkane cations, and delocalization of the spin density must be involved to explain the large and inequivalent proton hf couplings.

Table 2. Experimental ^{1}H hf couplings of selectively deuterated branched alkane cations in CF$_2$ClCFCl$_2$ matrix at 77 K. The data were taken from refs [29,30].

Cation	Hyperfine Couplings (G)[a] (Assignment)	
CH$_3$CH(CH$_3$)CH$_2$CH$_3$$^+$	45 (3H) 43 (2H: C$_1$,C$_2'$)	51 (1H: C$_4$)[b]
CH$_3$CH(CH$_3$)CH$_2$CD$_3$$^+$	43.0 (3H)	
(CH$_3$CH$_2$)$_2$CHCH$_3$$^+$	47.3 (3H) 39.3 (1H: C$_3'$)	52.9 (1H: C$_2$)[c]
(CH$_3$CH$_2$)$_2$CDCH$_3$$^+$	47.3 (3H)	49.9 (1H: C$_1$)
(CH$_3$CH$_2$)$_2$CHCD$_3$$^+$	51.3 (2H)	
(CH$_3$CD$_2$)$_2$CHCH$_3$$^+$	44.5 (2H)	
(CD$_3$CH$_2$)$_2$CHCH$_3$$^+$	46.0 (2H)	
(CH$_3$CH$_2$)(CH$_3$CH$_2$CH$_2$)CHCH$_3$$^+$	52.7 (3H) 39.1 (1H: C$_3'$)	57.1 (1H: C$_2$)[d]
(CH$_3$CH$_2$)(CH$_3$CH$_2$CH$_2$)CHCD$_3$$^+$	59.2 (2H)	61.9 (1H: C$_5$)
(CH$_3$CD$_2$)(CH$_3$CH$_2$CD$_2$)CDCH$_3$$^+$	50.5 (2H)	
(CH$_3$CD$_2$)(CH$_3$CD$_2$CD$_2$)CDCH$_3$$^+$	46.0 (3H)	

a) 10 G = 1 mT; b) C$_1$-C$_2$(C$_3'$)-C$_3$-C$_4$; c) C$_1$-C$_2$-C$_3$(C$_3'$)-C$_4$-C$_5$; d) C$_1$-C$_2$-C$_3$(C$_3'$)-C$_4$-C$_5$-C$_6$

2.3. REACTIONS OF N-ALKANE CATIONS

Thermal reactions of *n*-alkane cations to form neutral alkyl radicals and olefinic cations have been studied by ESR using partially ^{2}D-labelled alkanes in SF_6 and halocarbon matrices [10]. Conversion to alkyl radicals occurs in SF_6 and $CFCl_2CF_2Cl$ at 100-110 K.

In the case of $CH_3CD_3^+$ conversion to $\cdot CH_2CD_3$ and $CH_3CD_2\cdot$ occurred in the ratio 1.3:1, indicating a small isotope effect. Primary and secondary alkyl radicals were formed from partially ^{2}D-labelled propane cations in SF_6 and $CFCl_2CF_2Cl$, respectively, showing that deprotonation might be mainly governed by the unpaired electron distribution and not by isotope effects. Conversion of propane$^+$ to propene$^+$ occurs in $CFCl_3$ at 140 K by elimination of H_2 (or two hydrogen atoms). In the halocarbon matrices, the spin density is predominantly in the central C-H bonds in SF_6 on the terminal *in-plane* C-H bonds. It may therefore be concluded that the alkane cations undergo breaking of the bond in which the unpaired electron is highly localized. In the cases of *n*-butane and *n*-pentane, 2-alkyl radicals were observed after thermal decomposition of the cation, as confirmed by measurements on the partially ^{2}D-labelled *n*-alkane cations in $CF_2ClCFCl_2$ at 100 K [32].

The mechanism of selective deprotonation has been further studied in *n*-alkane cations trapped in SF_6 and in zeolites, in which case 1-alkyl radicals were observed [33,34]. In these media the parent cations have an extended planar carbon chain. The 2-alkyl radicals observed in halocarbon matrices may be formed from cations with a *gauche* geometry.

The yield of alkyl radicals from ^{2}D-labelled C6 - C8 *n*-alkane cations was lower than with the *protiated* ones [32], indicating a strong isotope effect. The photoinduced formation of alkene cations from *n*-alkane cations observed previously with *protiated* alkanes could be reproduced using partially ^{2}D-labelled *n*-butane in CF_3CCl_3 [32]. Thus, $CH_3CD=CDCH_3^+$ was formed from $CH_3CD_2CD_2CH_3^+$ indicating that D_2 is eliminated.

Photoinduced fragmentation could not be observed with the ^{2}D-labelled C6-C8 *n*-alkane cations in contrast to the *protiated* ones [26]. The origin of this isotope effect is not known.

3. Structure, Dynamics and Reactions of Some Cyclocalkane cations

Cyclic alkanes usually have a degenerate HOMO (highest occupied molecular orbital). Due to the instability of degenerate electronic states, cyclic alkane cation radicals will therefore undergo a distortion, an effect which was first recognized by Jahn and Teller (*J-T*) and is usually named after them. ESR spectroscopy is particularly helpful to provide experimental evidence on such distortions as outlined in other chapters of this volume (Toriyama and Lindgren/Shiotani) and in previous reviews [7,8]. Only a few ^{2}D labelling studies have been reported for the cyclic alkane cations so far. The ESR data are summarized in Table 3. Details follow below.

3.1. STRUCTURE OF CYCLOPROPANE

The radical cation of *protiated* cyclopropane ($cC3$-h^+) is known to distort from the D_{3h} symmetry of the neutral molecule to an obtuse triangle of C_{2v} symmetry [35,36], see also a previous chapter (Toriyama) and reviews [7,8]. In this case the SOMO is a $6a_1$ and most of unpaired electron is located on the two basal carbon atoms (structure X).

Table 3. Experimental ^{1}H hf couplings in ^{2}D-labelled cycloalkane cations together with the corresponding geometry and/or electronic state. The data were taken from [35,37,41,42,45,47].

Cation[a)]	Matrix	Temp (K)	^{1}H Couplings (G)[b)]	Geometry (State)
cC3-h^+	$CFCl_3$	4.2	24 (2H) (-)11 (4H)	C_{2v} ($^2A_1$)
cC3-1-d_2^+ (X)[c)]	CF_3CCl_3	4.2	(-)12 (4H)	C_{2v} ($^2A_1$)
" (XI)[c)]	CF_3CCl_3	4.2	24 (2H) (-)12 (2H)	C_{2v} ($^2A_1$)
Me-cC6-h^+	CF_3-cC_6F_{11}	4.2	$ca.$74 (2H)[d)] 34 (1H)	(^2A_g)[e)]
Me-cC6-h^+	CF_3-cC_6F_{11}	140	48.2 (2H) 42.2 (2H) 20.1 (2H)	
Me-d_3-cC6$^+$	CF_3-cC_6F_{11}	4.2	ca.74 (2H)	(^2A_g)[e)]
Me-d_3-cC6$^+$	CF_3-cC_6F_{11}	140	48.2 (2H) 42.2 (2H)	
Me-cC6-2,6-d_4^+	CF_3-cC_6F_{11}	4.2	$ca.$74 (1H) 34 (1H)	(^2A_g)[e)]
Me-cSiC5-h^+ (Me-d_3-cSiC5$^+$)[f)]	CF_3-cC_6F_{11}	4.2	93 (1H) 25 (2H)	C_s[g)] ($^2A'$)
Me-cSiC5-2,6-d_4^+	CF_3-cC_6F_{11}	4.2	93 (1H)	C_s[e)] ($^2A'$)

a) cC3: cyclopropane; Me-cC6: methylcyclohexane; Me-cSiC5: Methylsilacyclohexane; b) 10 G = 1 mT; c) See structures X and XI; d) The triplet coupling was recently refined as a_1 = 78 G (1H: C_5) and a_2 = 70 G (1H: C_2) based on the simulation of the dynamic process [45]; e) The state refers to that of the unsubstituted cyclohexane ring with C_{2h} symmetry; f) The same hf couplings as Me-cSiC5-h^+ were observed, but with narrower line width; g) The symmetry refers to that of the unsubstituted silacyclohexane ring.

Recently Matsuura *et al.* reported two positional isomers of the cation of cyclopropane-1-d_2 (cC6-1-d_2^+) both an a_1 SOMO [37]. One is an obtuse triangle with the CD_2 group at the top (structure X) and the other with the CD_2 group at one of the basal positions (structure XI). The ESR spectrum of structure X shows a hf splitting of $ca.$ 12 G due to four equivalent hydrogens, whereas structure XI has two pairs of equivalent hydrogens of ca. 12 G and ca. 24 G. After annealing the sample above 20K, only structure X was observed. This indicates that the preferred position of the CD_2 group in cC3-1-d_2^+ is at the top of the obtuse triangle. Furthermore, the authors suggest that cC3-1-d_2^+ has a tendency to more distorted C-H than C-D bonds, as in the case of selectively ^{2}D-labelled methane, ethane and *n*-butane cations [10-14,21-23,32] (see also previous section)

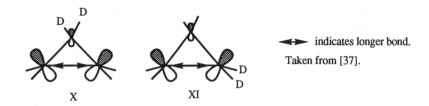

indicates longer bond.

Taken from [37].

X XI

The cations of ²D-labelled C4, C5 and C6 cycloalkanes have never been reported.

3.2. METHYLCYCLOHEXANE

Neutral cyclohexane (cC6) has a chair conformation of D_{3h} symmetry and a degenerate HOMO, e_g. As already described in other chapters (Toriyama and Lindgren/Shiotani), ESR studies on cC6 cation have attracted much attention [36,38-40] since this cation can be expected to provide important experimental information on the static and dynamic J-T effect as a consequence of the the instability of degenerate electronic states. The introduction of substituents in certain positions is another way of removing the orbital degeneracy. Here we are concerned only with the radical cations of ²D-labelled methylcyclohexane (Me-cC6), the simplest mono-substituted alkyl-cyclohexane, in relation to the structure, dynamics and reactions.

3.2.1. *Structure.* The cation radicals of three ²D-labelled methylcyclohexanes have been prepared by ionizing radiation in a CF_3-cC$_6$F$_{11}$ matrix at 77K [41-43]. The experimental hf couplings are given in Table 3. The assignment of the hf couplings is shown in Fig. 4, together with the proposed SOMO. The proposed electronic structure for Me-cC6⁺ is based on the ESR data and quantum chemical calculations for a ²A$_g$ state of cC6⁺, one of the possible states caused by the J-T distortion of a D_{3d} cyclohexane ring [40,44]. In this structure the methyl group is equatorially attached to one of the elongated C-C bonds in a distorted A$_g$ cyclohexane structure. The larger triplet of *ca.* 74 G is attributed to two equatorial hydrogens at C_2 and C_5, respectively, whereas the doublet of 34 G is assigned to one of the methyl hydrogens which is located at the trans position with respect to the elongated C_1-C_6 bond. The mechanism by which ¹H hf splittings arise is similar to the *trans* effect or *hyperconjugative* effect which is generally accepted to be the origin of ¹H hf splitting of extended and branched alkane cations [18,27,30,33,34].

Semiempirical MNDO/INDO MO calculations support the structure proposed for Me-cC6⁺ [43]. The calculated ¹H hf splittings for the ²A$_g$ like structure of Me-cC6⁺ reproduce rather well the experimental ones. The C_1-C_2 bond length is calculated to be as long as 1.942 Å. The bonds adjacent to the elongated C_1-C_2 bond are slightly shorter than those in the neutral molecules, 1.54 Å , the other bond lengths being nearly the same as the neutral ones. Contrary to the ²A$_g$ structure of cC6⁺ having two elongated C-C bonds [44], the calculations predict that in Me-cC6⁺ elongation occurs predominantly in one of the bonds, *i.e.*, the bond containing the carbon attached to the methyl group. The geometrical distortion as calculated by the MNDO method seems, however, rather severe and more sophisticated theoretical studies are required to discuss the geometry more in detail.

An electronic structure resembling the ²A$_g$ state of cC6⁺ has also been observed for radical cations of 1,1-dimethylcyclohexane (1,1-Me$_2$-cC6⁺) and a series of mono-substituted cyclohexane cations such as Et-cC6$^+$, n-Pr-cC6$^+$ and i-Bu-cC6$^+$ [43], as presented in another chapter (Lindgren/Shiotani).

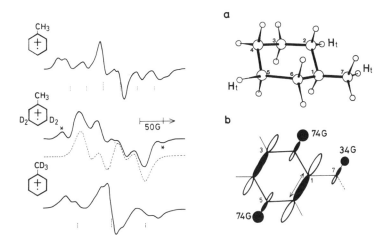

Fig. 4. Left column: ESR spectra of Me-cC6$^+$ (*upper*), Me-cC6-3,5-$d_4$$^+$ (*middle*) and Me-d_3-cC6$^+$ (*lower*) in CF$_3$-cC$_6$F$_{11}$ at 4.2K. The hf parameters in Table 3 are used for the simulation (*dashed curve*) and the stick plots. Right column: (a) The geometry of Me-cC6$^+$. (b) A schematic representation of the a_g like SOMO. $\leftrightarrow$ indicates longer bond. Taken from [42].

3.2.2. Dynamics.

Temperature-dependent ESR spectra have been observed for the three isotopically labelled methylcyclohexane cations, which further support the *distorted* ^2A_g structure proposed [41-45]. The 4K spectrum of Me-d_3-cC6$^+$ with a hf coupling of *ca.* 74 G (2H) changes successively with raising temperature into a triplet of triplets with a_1= 48.8 G (2H) and a_2 = 42.7 G (2H) at 140K. The spectrum of Me-cC6-h^+ has an additional triplet with a_3 = 20.1 G (2H) at the same temperature. This result indicates that two hydrogens of the methyl group become magnetically equivalent so as to change the doublet (34 G) at 4K into a triplet of 20.1 G. Surprisingly, no methyl group rotation was observed even at such a high temperature as 140K.

These temperature-dependent ESR spectra can be reasonably explained in terms of an exchange between two different but energetically equivalent structures, see the insert below.

Based on this two site exchange model, the simulated spectra of Me-d_3-cC6$^+$ have been successfully calculated [45]. It has been suggested that the dynamics take place even at 4K, probably through a thermally activated process with a very shallow energy barrier. The activation energy evaluated for the dynamical process is as small as 0.2 kcal/mol which is very close to the values reported for the inequivalent site jumping of cC6$^+$ (E_A=0,17 and 0.24 kcal/mol) [36,39]. Consistent with the results of Me-d_3-cC6$^+$, the ESR spectra of Me-cC6-h^+ in the temperature range between 41K and 155K are analyzed using the same model. However, the spectrum below 41K can not be successfully analyzed, suggesting unknown isotope effects.

indicates longer bond.
Taken from [42].

Similar temperature dependent ESR line shapes have been observed for $1,1\text{-Me}_2\text{-}cC6^+$ and successfully analyzed by the model proposed for the Me-cC6 cation [45].

3.2.3. *Reactions.* Alkyl radical formation in halocarbon matrices from the parent alkane cations has been studied for various alkyl-substituted cyclohexanes [41,46], as outlined in the chapter by Lindgren and Shiotani. Upon annealing samples above the phase transition of $CF_3\text{-}cC_6F_{11}$ (*ca.* 175K) and $CF_2ClCFCl_2$ (*ca.*110K), respectively, deprotonation of alkane cations generally occurs. By using selectively ^{2}D-labelled methylcyclohexanes, three different neutral alkyl radicals (XIV, XV and XVI) were found to form with almost equal relative yields. The yields of the other candidate radicals, XVII and XVIII, are negligibly small.

This observation suggests that, contrary to the reaction scheme suggested earlier for *n*-alkane cations [10,33,34], no direct relationship seems to exist between the location of high hydrogen spin density in the cation and site of deprotonation.

3.3. OTHERS

Electronic structures of other cycloalkane cations, methylsilacyclohexane [47] and bicyclo[1.1.0]butane [48] cations, are presented here. The structure of Me-cC6$^+$ is known to be a distorted A_g as described above. It is of interest to investigate changes in the structure caused by a replacement of one ring carbon for one silicon in the methylcyclohexane cation. The spectrum of protiated methylsilacyclohexane cation, Me-cSiC5-h^+, in $CF_3\text{-}cC_6F_{11}$ at 4K consists of a doublet of triplets with isotropic hf couplings of 93 G (1H) and 25 G (2H) [47]. The same spectrum has been observed for Me-d_3-cSiC5$^+$, but with narrower line width.

This result suggests that the methyl protons do not contribute to the spectrum of Me-cSiC5-h^+, whereas, Me-cSiC5-2,6-d_4^+ gives rise to a doublet of 93 G. Essentially the same experimental spectra have been observed for the methylsilacyclohexane cations in various matrices such as CF_3Cl, $CF_2ClCFCl_2$ and CCl_3CF_3. Based on these experimental results and on the *chair* form of neutral methylsilacyclohexane [49-51], the triplet (25 G) and doublet (93 G) splittings can be attributed to the two equatorial hydrogens at the C_2 and C_6 positions and to the equatorial hydrogen at the C_4 position, respectively. A $^2A'$ state in C_s symmetry is proposed for Me-cSiC5$^+$ whose SOMO is schematically depicted above. This assignment has been further supported by observing the spectrum of the silacyclohexane cation, cSiC5-h^+ which consists of a doublet of triplets with 100 G (1H) and 29 G (2H) in CF_3-cC_6F_{11}.

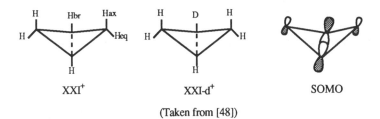

Taken from [47].

XIX   XX

The ESR spectrum of bicyclo[1.1.0]butane cation, XXI$^+$, has been observed in a $CFCl_3$ matrix at 160 K [48]. The spectrum consists of a well-defined triplet of quintets with 77.1 G (2H) and 11.4 G (4H) splittings, respectively. An unequivocal assignment of the hf splittings to sets of equivalent protons has been provided by observing the spectrum of 1-deuteriobicyclo[1.1.0]butane cation, XXI-d^+.

In the spectrum of XXI-d^+, the quintet is replaced by a quartet with 11.4 G. This finding leads to the conclusion that the 11.4 G splitting is attributed to the two bridge head protons (H_{br}) as well as to the two equatorial methylene protons (H_{eq}), while the 77.1 G coupling is assigned to their two axial counterparts (H_{ax}). The strongly differing values of $a(H_{eq})$ and $a(H_{ax})$ indicate a puckered geometry of XXI$^+$. The SOMO is located primarily in the transannular C_1-C_2 bond as depicted in the figure.

XXI$^+$ XXI-d^+ SOMO

(Taken from [48])

Radiolytic oxidation of the five and six membered analogues, bicyclo[2.2.0]pentane and bycyclo[2.2.0]hexane, gives the cyclopentane-1,3-diyl and cyclohexane-1,4-diyl cation radicals, respectively, whose SOMOs are very close to that of the bicyclo[1.1.0]butane cation, as presented in a previous chapter (Lindgren/Shiotani).

4. Structure and Reactions of Alkene and Alkyne Cations

4.1. STRUCTURE OF ALKENE CATIONS

A limited number of ^{2}D and ^{13}C labelled alkene cations have been studied. They are listed in Table 4. The labelling has been used to determine precise hf couplings and assign them correctly. The most important question on the structure of alkene cations is the *twisting* around the C=C double bond. ESR data of some *protiated* alkene cations have already been summarized in a previous review paper [7].

4.1.1. *Ethylene*.
The ethylene cation radical, $C_2H_4^+$, has attracted much attention and a number of theoretical and experimental studies have predicted a non-planar twisted structure [8,52-59]. For example, *ab-initio* SCF-MO-CI [52] and MNDO [53] calculations have predicted *ca.* 25° twist structure, whereas calculations by electronic force theory [54] and MINDO/3 [55] have resulted in 45° twist. Experimentally the twist angle of 25° has been reported by vacuum ultraviolet studies [58] and photoelectron spectrum [59] in the gas phase. However, no definitive ESR studies have been reported on $C_2H_4^+$ stabilized in the solid state.

One of the present authors reported the first ESR study on the irradiated $CH_2=CH_2$ in $CF_2ClCFCl_2$ [60]. The spectrum observed at 77 K was characterized by a quintet with an isotropic ^{1}H hf splitting of 23.3 G (4H) and was assigned to the radical cation of $CH_2=CH_2$. However, the spectral assignment was questioned by Fujisawa et al [61], who suggested an alternative assignment, *i.e.*, the quintet being attributed to a propagating radical, $-(CH_2-CH_2)_n-CH_2-CH_2\cdot$ formed by a reaction of neutral ethylene to 1-butene cation, the latter cation being initially formed through an ion-molecular reaction. We are reinvestigating the ethylene cation using X-rays irradiation with the neon matrix ESR technique [63a]. Our preliminary results show hf couplings of ca. 3.0 G for the four equivalent protons of $CH_2=CH_2^+$. Consistent with this result the perdeuteriated ethylene, $CD_2=CD_2^+$, gives a singlet. However, the ^{13}C coupling is missing even in the ^{13}C enriched ethylene.

Consistent with some previous calculations [52,53], very recent *ab-initio* MO calculations by Lunell and Huang [62] have predicted a twist around the C-C bond. The calculated isotropic ^{1}H and ^{13}C hf couplings for 25°are (-)4.5 and 3.0 G, respectively. Full agreement between theoretical and experimental ^{1}H couplings is obtained for a twist angle of 28°.

4.1.2. *Propylene*.
ESR studies of the propylene radical cation ($CH_3CH=CH_2^+$) have been reported by two groups [10,64,65]. Toriyama *et al.* first reported the ESR spectrum of protiated propylene cation, $CH_3CH=CH_2^+$, stabilized in CFCl$_3$.[10]. They analyzed the observed spectrum under the assumption that rotation of the methyl group was hindered [10]. However, using partially deuteriated propylene, $CH_3CH=CD_2$, the spectral analysis was altered by observing free rotation of the methyl group at 77K in CFCl$_3$ and SF$_6$ matrices [64]. The ^{1}H hf couplings (Table 4) are in agreement with the calculations for a planar π–type structure [64]. The spectrum of propylene cation depends considerably on both matrix and temperature, suggesting non-planar structure in the rigid limit in a CFCl$_3$ matrix [64,65].

Table 4. ^{2}D- and ^{13}C-labelled alkene cations. The experimental hf couplings are compared with theoretical ones.

Cation (matrix)	Temp. (K)	Hyperfine (G)	Ref	Twist Angle at Opt[b] (deg)	Hyperfine at Opt[b] (G)	Hyperfine at 0° (G)	Ref
$CH_2=CH_2^+$ (*neon*)	4	3 (4H)	[63a]	25	(-)4.5 (4H) (-)3.0 (*C)		[62][c]
$CH_3CH=CH_2^+$ {$CH_3CH=CD_2^+$} (*SF_6*)	77	24.0 (CH_3) 16.0, 9.0 (CH_2) 9.0 (CH)	[64]	40	18.7 (CH_3) 1.0, 0.0 (CH_2) 11.6 (CH)	27.4 (-)13.3, (-)13.4 (-)7.1	[57][d]
$CH_3CH=CH_2^+$ {$CH_3CH=CD_2^+$} (*CFCl_3*)	130	24.0 (CH_3) 16.0, 14.0 (CH_2) 9.0 (CH)	[64]				
cis-$CH_3CH=CHCH_3^+$ (*CFCl_3*)	104	23.5 (CH_3) 9.5 (CH) 6.6 (*CH_3)	[66]	24	24.8 (CH_3) (-)4.0 (CH) (-)7.5 (*CH_3)	28.7 (-)9.6	[57][d]
$(CH_3)_2C=C(*CH_3)(CH_3)^+$ (*CFCl_3*)	77	17.1 (CH_3) 5.6 (*CH_3)	[66]	0			[57][d]
$(CH_3)_2C=*C(*CH_3)(CH_3)^+$ (*CF_3CCl_3*)	138	17.5 (CH_3) 5.5 (*CH_3) ca.9.4 (*C=)[f]	[66]			17.8 (CH_3) (-)6.0 (*CH_3) 10.5 (*C=)	[66][e]
[1,2-dimethylcyclobutene cation structure: CH_3 CH_3 / + \ ==, CH_2 CH_2 \C/ H_2] (*CF_2ClCFCl_2*)	4	43.3 (-CH_2 *ax*) 29.1 (-CH_2 *eq*) 15.2 (CH_3)	[66]			42.3 (-CH_2 *ax*) 25.1 (-CH_2 *eq*) 16.7 (CH_3) 5.0 (*CH_3)	[66][e]
[1,2-dimethylcyclobutene cation structure: CH_3 CH_3 / + \ ==, CH_2 CH_2 \C/ H_2] (*CF_3CCl_3*)	77	42.0 (-CH_2 *ax*) 26.4 (-CH_2 *eq*) 16.7 (CH_3)	[66]				
[1,2-dimethylcyclobutene cation structure: CH_3 *CH_3 / + \ ==, CH_2 CH_2 \C/ H_2] (*CF_3CCl_3*)	157	34.2 (-CH_2)[g] 16.7 (CH_3) 5.9 (*CH_3)	[66]				

a) *: ^{13}C labelling, 10 G = 1 mT; b) Opt: optimization; c) *ab-initio* (CI level using the program package MELD); d) semiempirical AM1/UHF; e) *semiempirical* INDO; f) Anisotropic ^{13}C splitting was observed: (//) 28.4 G, ($\perp$) *ca*. 0 G; g) The axial and equatorial β-^{1}H couplings become magnetically equivalent due to a rapid ring inversion.

Clark and Nelsen [57] have recently reported AM1 *semiempirical* calculations on propylene cation which confirmed that the experimental couplings listed in Table 4 are reasonably explained by the planar structure, although a twist angle of 40° is calculated for the optimized geometry.

4.1.3. *^{13}C-labelled alkene cations.*

Since ^{13}C hf splittings are known to be sensitive to the geometrical structure of cations, the couplings have been observed for certain ^{13}C labelled alkene cations, *i.e., cis*-2-butene$^+$ (*c*-2B$^+$), tetramethylethylene$^+$ (TME$^+$), and 1,2-dimethylcyclopentene$^+$ (DMCP$^+$). The experimental data are summarized in Table 4 together with the results calculated by *ab-initio* [62] and *semiempirical* AM1/UHF [57] and INDO MO [66] methods. The optimized geometry of *c*-2B$^+$ by AM1 resulted in a 24° twistwd structure, but the experimental hf data are in better agreement with the planar structure. The INDO calculations for the planar structure also reproduce well the experimental ^{13}C and ^{1}H couplings.

The AM1 calculations of TME$^+$ resulted in a planar structure. The ^{13}C and ^{1}H couplings by INDO for the planar structure are in good agreement with the experimental ones. The couplings for DMCP$^+$ are also explained in terms of a non-twisted structure around the C=C bond.

Therefore, the alkene cations with a well-documented twisted structure are the ethylene cation itself and the trimethylsilylethylene [67] and cis-3-hexene [63b] cations

In conclusion, in contrast to earlier MNDO results [53], recent AM1 *semiempirical* calculations [57] predict that increasing alkyl substitution leads to a decresae in the *twist* angle around the central double bond and that the barrier to rotation increases concomitantly. These results are in accord with experimental observations.

4.2. REACTIONS OF ALKYNE CATIONS

The methylacetylene (MA) cation is stabilized in halocarbon matrices not as an isolated monomer cation, but as a complex cation with a chlorine atom of the matrix, whereas the isolated monomer cation has been observed for dimethylacetylene (DMA). Upon warming the samples containing MA and DMA in $CF_2ClCFCl_2$, the dimer cations of MA_2^+ and DMA_2^+ are formed, probably due to a cycloaddition of the parent molecule to the corresponding cation. A large D-isotope effect has been observed in the proton elimination reaction from the DMA dimer cation.

4.2.1. *Acetylene.*

The acetylene radical cation, the simplest Jahn-Teller active alkyne cation, can not be generated by irradiation in halocarbon matrices due to its high ionization potential (Ip$_1$: 11.4 eV [68]). The cation was not even stabilized in SF$_6$ [69] in spite of the higher ionization potential of the latter matrix (Ip$_1$ of SF$_6$: 15.7 eV [70]). Instead, a well resolved isotropic ESR spectrum of neutral fluorovinyl radical, CFH=CH·, was observed and the detailed electronic structure was discussed using the experimental data derived from the ^{2}D- and ^{13}C-labelled radicals, CFD=CD· and ^{13}CHF=^{13}CH· [69]. It may be worthwhile to try stabilizing the isolated acetylene cation in the neon matrix combined with ionizing radiation.

4.2.2. *Methylacetylene.*

Upon irradiating methylacetylene (MA) in halocarbon matrices, one does not obtain the isolated monomer cation, but instead a complex with a chlorine atom of the matrix [71]. Experimental spectra of three isotopic methylacetylenes,

$CH_3C\equiv CD$ (MA-d_1), $CD_3C\equiv CH$ (MA-d_3) and $CH_3C\equiv CH$ (MA) in $CFCl_3$ recorded at 4K were successfully analyzed using the ESR parameters listed in Table 5. These suggest that the unpaired electron is delocalized in the orbitals of MA and a p-orbital of one chlorine (spin density $ca.$ 0.37) of the matrix $CFCl_3$ to form a complex cation radical between the solute and solvent molecules, $i.e.$, $[(CH_3C\equiv CH)\cdots Cl\text{-}CCl_2F]^+$. Complex cations have been formed in other halocarbon matrices such as $CF_2ClCFCl_2$ and CF_2ClCF_2Cl, but with less resolution. Irradiation at 4K gave essentially the same ESR spectra at 77K, indicating that the MA cation prefers the complex form over an isolated monomer, probably due to the small difference in the ionization potentials between solute and matrix molecules (Ip_1s of MA, $CFCl_3$, $CF_2ClCFCl_2$ and CF_2ClCF_2Cl being 10.4 [72], 11.8 [1], 12.0 [7,73] and 12.7 eV [7,73], respectively). Such interaction of a solute cation with a matrix chlorine atom has been observed for σ-type solute cations possessing the SOMO localized mainly on one atom (oxygen, sulfur or halogen), see [4,7,8]. The present results demonstrate that such complex cations can be formed even for solute cations with delocalized π-type SOMO.

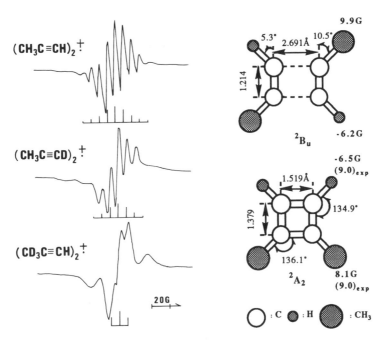

Fig. 5. Left column: ESR spectra of the dimer cations of three isotopic methylacetylenes at ca. 105K in CF_2ClCF_2Cl. Right column: Optimized geometries by the *ab-initio* calculations for two possible structures, $^2A_2$ and ^2B_u, of the dimer cation, MA_2^+. The calculated hf splittings are compared with the experimental ones. Taken from [71,74].

Upon warming the $[MA\cdots Cl\text{-}CCl_2F]^+$ complex cation is irreversibly converted to the cation radical of the MA dimer, MA_2^+: the six equivalent (2CH$_3$) and two equivalent (2CH=) protons give accidentally the same hf coupling (9.0 G). *Ab-initio* (3-21G) MO calculations predict that either ^2B_u or $^2A_2$ is the electronic ground state of the dimer cation calculations predict that either ^2B_u or $^2A_2$ is the electronic ground state of the dimer cation [74]. The former (^2B_u) is an unstable charge-resonance (*CR*) π-complex type state in which the unpaired electron occupies an *in-plane* $(\pi\text{-}\pi)^*$ orbital between two MA units with the two methyl groups in opposite positions, i.e., a*head-to-tail* form. The latter ($^2A_2$) corresponds to a cyclobutadiene cation with the two methyl groups in adjacent positions. Although the $^2A_2$ state is energetically favored, a definitive conclusion for the cyclization can not be given until more decisive experimental evidence is obtained. Upon further warming the dimer cation is converted into a propagyl radical, $H_2\bar{C}\text{-}C\dot{=}CH$.

Table 5. ^{2}D-labelled alkyne cations. For the dimer cations of methylacetylene and dimethylacetylene the experimental hf couplings are compared with calculated ones. The data were taken from [71,74,75]

Cation (*matrix*)	Temp (K)	g-value	Hyperfine Couplings (G) Exp		Cal	Geometry (State)
		$(//)$ $(\perp)$	$(//)$	$(\perp)$		
$[(CD_3C\equiv CH)\cdots Cl\text{-}CCl_2F]^+$ (*CFCl₃*)	4.2	2.0015, 2.0065	9.8 65.9 54.8	6.0 (^{1}H) 10.0 (^{35}Cl) 8.3 (^{37}Cl)		
$[(CH_3C\equiv CD)\cdots Cl\text{-}CCl_2F]^+$ (*CFCl₃*)	4.2	2.0015; 2.0065	17.9 65.9 54.8	16.5 (3H) 10.0 (^{35}Cl) 8.3 (^{37}Cl)		
$(CD_3C\equiv CH)_2^+$ (*CF₂ClCF₂Cl*)	105	2.0030		9.0 (2H)		
$(CH_3C\equiv CD)_2^+$ (*CF₂ClCF₂Cl*)	105	2.0030		9.0 (6H)		
$(CH_3C\equiv CH)_2^+$ (*CF₂ClCF₂Cl, CF₂ClCFCl₂*)	105	2.0030		9.0 (8H)	9.9 (6H) (-)6.6 (2H) ----- 8.1 (6H) (-)6.5 (2H)	C_{2h}[a) (^2B_u) ----- C_{2v}[b) ($^2A_2$)
$CH_3C\equiv CCH_3^+$ (*CF₂ClCFCl₂*)	77	2.0031 2.0071	22.0	19.7 (6H)		
$CH_3C\equiv CCD_3^+$ (*CF₂ClCFCl₂*)	77	2.0031 2.0071	22.0	19.7 (3H)		
$(CH_3C\equiv CCH_3)_2^+$ (*CF₂ClCFCl₂*)	105	2.0026		9.0 (12H)	7.9 (12H) ----- 8.8 (12H)	D_{2h}[c) ($^2B_{2u}$) ----- D_{2h}[d) ($^2B_{2g}$)
$CH_3CH_2C\equiv CH^+$ (*CF₂ClCF₂Cl*)	100	2.0025			15.9 (3H) 23.8 (2H) 9.5 (1H)	

[a)] Charge-Resonance (*CR*) type complex cation with the *head-to-tail* form; [b)] Dimethylcyclobutadiene cation with the *head-to-head* form; [c)] *CR*-type complex cation; [d)] Tetramethylcyclobutadiene cation.

4.2.3 *Dimethylacetylene.* Direct ESR evidence for cycloaddition has been observed for the dimethylacetylene (DMA) cation [75,76]. An irradiated solid solution of DMA in $CF_2ClCFCl_2$ gives a monomer cation, DMA^+, whose spectrum at 77K is characterized by the ESR parameters given in Table 5. The 4K spectrum of DMA^+ shows less resolved complex hf pattern, indicating a distortion due to the Jahn-Teller effect. On annealing the sample to about 100K in the same matrix, DMA^+ was converted into the dimer radical cation, DMA_2^+, with isotropic hf coupling of 9.0 G (12H). The formed DMA_2^+ is attributed to tetramethylcyclobutadiene radical cation, $cC_4(CH_3)_4^+$, which was formed by a $DMA^+ + DMA$ [1+2] cycloaddition reaction of the DMA^+ to the mother DMA molecule. This conclusion was derived from the good correlation of the 9 G ^1H coupling with the experimental couplings of $cC_4(CH_3)_4^+$ (8.7 G) in solution [77,78]. The *ab-initio* MO calculations support the cycloaddition ($^2B_{2g}$ state), although the *CR* type cation ($^2B_{2u}$ state) can not completely be ruled out [74,75].

A similar, but intermolecular cycloaddition has been reported for irradiated deca-2,8-diene in $CFCl_3$ at 77K [79].

By further annealing the sample to about 125K in $CF_2ClCFCl_2$, $cC_4(CH_3)_4^+$ decomposes to give rise to the methylpropagyl radical, $H_2\bar{C}\text{-}\bar{C}\text{=}\bar{C}\text{-}CH_3$. A large ^2D-isotope effect has been found in this proton elimination. The partially ^2D-labelled dimethylacetylene solute, $CH_3C\equiv CCD_3$, gave the radicals, $H_2\bar{C}\text{-}\bar{C}\text{=}\bar{C}\text{-}CD_3$ and $D_2\bar{C}\text{-}\bar{C}\text{=}\bar{C}\text{-}CH_3$, in a ratio of approximately 9:1. This large ^2D-isotope effect cannot be explained classically, but may be understood in terms of quantum-mechanical tunneling mechanism [80,81].

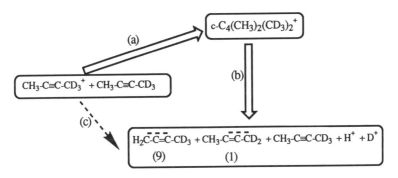

(Reactions *a* and *b* are predominant pathways to give the methylpropagyl radicals. Taken from [76])

The ethylacetylene monomer cation is stabilized in CF_2ClCF_2Cl matrix at 77K [71]. Upon warming, the cation is directly converted into ethylpropagyl radical, $H\text{-}\bar{C}\text{=}\bar{C}\text{-}\bar{C}HCH_3$, without forming a dimer cation.

Another example related to the reaction of acetylenic cations is a Cope rearrangement of 1,5-hexadiyne cation through the cyclic six-membered transition structure to 1,2,4,5-hexatetraene (bis(allene)) cation in halocarbon matrix at 77K studied by Dai *et al.* [82]. The partially deuteriated 1,5-hexadiyne, 1,6-dideuterio-1,5-hexadiyne, helped very much to confirm this reaction.

5. Reaction of Oxygen-Containing Cations

A number of [2]D-and ^{13}C-labelled oxygen-containing cation radicals have been subjected to halocarbon matrix ESR studies. They are cations of aldehydes [83-85], esters [86-94], ethers [95-100], and ketons [101]. Some of them have already been outlined in previous review papers [4,7,8]. In this section we are concerned only with methyl and ethyl formate ester cations.

5.1. METHYL FORMATE

A 4K matrix ESR study by Iwasaki *et al.* revealed that the primary radical cation of methyl formate is the oxygen centred non-bonding σ-type (n_σ) cation (XXIII) which is too reactive to convert to the oxygen-centred π-type (n_π) cation (XXV) in the CFCl$_3$ matrix [89]. The 77K irradiation gives rise to the complex cation (XXIV) which is formed by the thermal reaction of the n_σ cation with a chlorine atom in the matrix [86,89]. Employing selectively [2]D-labelled methyl formates, CH$_3$OCDO, CD$_3$OCHO, and CD$_3$OCDO, unequivocal evidence for the matrix and solute cation interaction has been experimentally confirmed [86,91]. The observed large chlorine hf splitting, $A_{//} = ca.$ 83 G (^{35}Cl) is characteristic of such interaction. The unpaired electron is mainly localized in an anti-bonding σ^*-orbital consisting of atomic orbitals of a chlorine atom in the matrix, CFCl$_3$, and an oxygen atom in the solute molecule. Similar complex cations with the matrix chlorine have been observed for a number of aldehyde, ketone, and ketene radical cation [4,7,8]. Upon further annealing above 77K or photobleaching the complex and n_π cation dissociated to a carbon-centred radical cation (XXVI) formed by an intra-molecular hydrogen atom transfer to the carbonyl oxygen atom [89,91]. This *intra-molecular* hydrogen shift is the first direct experimental evidence for the rearrangement of ester cation postulated by McLafferty in mass-spectroscopic fragmentation processes [102]. The proposed reaction schemes are summarized below.

Deuterium isotope effects have been reported for dissociation of the σ^*-complexes of methyl formates depending on the state of deuteriation of the methyl group, but not the formyl site [91]. The complex thermally dissociates at ca. 100K with the *protiated* methyl compounds in a CFCl$_3$ matrix, whereas with the methyl deuteriated compounds,

CD_3OCHO and CD_3OCDO, the complex are stable to near the melting point of the matrix, *ca.* 150K. The spectra found after photobleaching (or warming) of the methyl-deuteriated complexes, in each case a singlet at $g = 2.0017$, are characteristic of the methoxyoxomethyl radical, CD_3OCO. The authors proposed an *inter-molecular* hydrogen atom transfer for the reaction.

5.2. ETHYL FORMATE

Similar to the methyl formate cation, the protiated ethyl formate cation is unstable in a *CFCl₃* matrix at 77K and is easily followed by the intra-molecular hydrogen atom transfer to the carbonyl oxygen to form the carbon-centered radical, $\cdot CH_2CH_2OC(OH^+)H$ [89,91]. Employing selectively ^2D-labelled ethyl formates, CH_3CD_2OCDO and CD_3CH_2OCDO, thermal reactions of the carbon centered radical were studied [91,92]. The reaction process was found little affected by a ^2D-labelling at the formyl proton, however, a profound isotope effect on the reaction was found by a ^2D-labelling at different sites of the ethyl group. Irradiated CH_3CD_2OCDO in CFCl₃ at 77K gave a 22.5 G triplet due to the CH_2 group of $\cdot CH_2CD_2OCD(OH^+)$, whereas CD_3CH_2OCDO gave a large 40 G doublet due to one of the CH_2 group of $\cdot CD_2CH_2OD(OD^+)$ (XXVIII). Upon warming the latter sample to 140K, the 40 G doublet irreversibly changed to a 22.5 G triplet. The corresponding spectral change was not observed for the CH_3CD_2OCDO system. Based on these results, the authors proposed that the $\cdot CD_2CH_2OD(OD^+)$ radical underwent an intra-molecular alkyl radical attack on the carbonyl oxygen with concomitant C-O bond cleavage to yield $\cdot CH_2CD_2OCD(OD^+)$ (XXIX).

XXVII   XXVIII   XXIX

In conclusion, ester cations generally undergo an *intra-molecular* hydrogen shift in the solid state at low temperature. The only isolated ester cations known to be stable at 77 K are those of the neopentyl ester of formic, acetic, and propionic acids in which the spin density is localized mainly on the neopentyl alkyl group [88,92].

References

1. T. Shida, Y. Nosaka, and T. Kato: *J. Phys. Chem.* **82**, 695 (1978).
2. L.B. Knight, Jr. and J. Steadman: *J. Chem. Phys.* **77**, 1750 (1982).
3. T. Shida, E. Haselbach, and T. Bally: *Acc. Chem. Res.* **17**, 180 (1984)
4. M.C.R. Symons: *Chem. Soc. Rev.* 393 (1984).
5. T.K. Kemp: *A Specialist Periodic Report Electron Spin Resonance* 10A (senior Reporter: M.C.R. Symons). Chap.2. The Royal Society of Chemistry, Burlington House, London (1986).
6. L. B. Knight: *Acc. Chem. Res.* **19**, 313 (1986).
7. M. Shiotani: *Mag. Res. Rev.* **12**, 333 (1987).
8. A. Lund, M. Lindgren, S. Lunell, and J. Maruani: *Molecules in Physics, Chemistry and Biology* (Ed. J. Maruani), **III**, 259 (1989).
9. K. Kimura, S. Katsumata, Y. Yamazaki, and S. Iwata: *Handbook of HeI Photoelectron*

174

10. K. Toriyama, K. Nunome, and M. Iwasaki: *J. Chem. Phys.* **77**, 5891 (1982).
11. L.B. Knight, J. Steadman, D. Feller, and E.R. Davidson: *J. Amer. Chem. Soc.* **106**, 3700 (1984).
12. (a) M. N. Paddon Row, D-J. Fox, J.A. Pople, K.N. Houk, and D.W. Pratt: *J. Amer. Chem. Soc.* **107**, 7896 (1985); (b) R.F. Frey and E.R. Davidson: *Chem. Phys.* **88**, 1775 (1988).
13. M. Iwasaki, K. Toriyama, and K. Nunome: *J. Amer. Chem. Soc.* **103**, 3591 (1981).
14. M. Iwasaki, K. Toriyama, and K. Nunome: *Chem. Phys. Lett.* **111**, 309 (1984).
15. D.J. Bellville and N.L. Bauld: *J. Amer. Chem. Soc.* **104**, 5700 (1982).
16. S. Lunell and M.-B. Huang: *J. Chem. Soc. Chem. Commun.* **1989**, 1031 (1989).
17. S. Lunell, D. Feller and E.R. Davidson: *Theor. Chim. Acta.* **77**, 111 (1990).
18. J.T. Wang and Ff. Williams: *J. Phys. Chem..* **84**, 3156 (1980).
19. M. Tabata and A. Lund: *Rad. Phys. Chem.* **23**, 545 (1984).
20. K. Matsuura, K. Nunome, K. Toriyama, and M. Iwasaki: *J. Phys. Chem.* **93**, 149 (1989).
21. M. Lindgren: *Electron Spin Resonance Studies of Some Radical Ions, Electronic Structure and Reactions.* Linköping Studies in Science and Technology. *Dissertation,* No.195 (1988).
22. (a) M.-B. Huang and S. Lunell: *J. Molec.* Structure *(Theochem),* **205**, 317 (1990); (b) S. Lunell, L. Worstbrock, and L.A. Eriksson: submitted to *J. Amer. Chem. Soc.* (1990).
23. M. Lindgren and A. Lund: *J. Chem. Soc. Faraday Trans. I*, **83**, 1815 (1987).
24. G. Dolivo and A. Lund: *J. Phys. Chem.* **89**, 3977 (1985).
25. G. Dolivo and A. Lund: *Z. Naturforsch.* **40a**, 52 (1985).
26. A. Lund, M. Lindgren, G. Dolivo and M. Tabata: *Rad. Phys. Chem.* **26**, 491 (1985).
27. K. Nunome, K. Toriyama, and M. Iwasaki: *J. Chem. Phys.* **79**, 2499 (1983).
28. K. Toriyama, K. Nunome, and M. Iwasaki: *Chem. Phys. Lett.* **132**, 456 (1986).
29. N. Ohta and T. Ichikawa: *J. Phys. Chem.* **91**, 373 (1987).
30. M. Shiotani, A. Yano, N. Ohta and T. Ichikawa: *Chem. Phys. Lett.* **147**, 38 (1988).
31. N. Ohta, T. Ichikawa and M. Shiotani: *31th Japanese Symposium on Radiation Chemistry,* Abstract p.95 (1988).
32. M. Lindgren, A. Lund and G. Dolivo: *Chem. Phys.* **99**, 103 (1985).
33. K. Toriyama, K. Nunome and M. Iwasaki: *J. Phys. Chem.* **90**, 6836 (1986).
34. K. Toriyama, K. Nunome and M. Iwasaki: *J. Am. Chem. Soc.* **109**, 4496 (1987).
35. M. Iwasaki, K. Toriyama, and K. Nunome: *J. Chem. Soc. Chem. Comun.* **1983**, 717 (1983).
36. M. Iwasaki, K. Toriyama, and K. Nunome: *Faraday. Discuss. Chem. Soc.* **78**, 19 (1984).
37. K. Matsuura, K. Nunome, M. Okazaki, K. Toriyama, and M. Iwasaki: *J. Phys. Chem.* **93**, 6643 (1989).
38. M. Tabata and A. Lund: *Chem. Phys.* **75**, 379 (1983).
39. K. Toriyama, K. Nunome, and M. Iwasaki: *J. Chem. Soc. Chem. Comun.* **1984**, 143 (1984).
40. S. Lunell, M. G. Huang, O. Claesson, and A. Lund: *J. Chem. Phys.* **82**, 5121 (1985).
41. M. Shiotani, N. Ohta, and T. Ichikawa: *Chem. Phys. Lett.* **149**, 195 (1988).
42. M. Lindgren, M. Shiotani, N. Ohta, T. Ichikawa, and L. Sjöqvist: *Chem. Phys. Lett.* **161**, 127 (1989).
43. M. Shiotani, M. Lindgren, N. Ohta, and T. Ichikawa: to be submitted.
44. M. Shiotani, M. Lindgren, and T. Ichikawa: *J. Amer. Chem. Soc.* **112**, 967 (1990).
45. L. Sjöqvist, M. Lindgren, M. Shiotani, and A. Lund: *2nd Japan-China Bilateral ESR Symposium,* Abstract p.80. Dec. (1989).; *J. Chem. Soc. Faraday Trans. I,* in press (1990).
46. M. Shiotani, M. Lindgren, F. Takahashi and T. Ichikawa: *28th Japanese Symposium on ESR,* Abstract p.148 (1989) and *32th Japanese Symposium on Radiation Chemistry,* Abstract p.51 (1989); *Chem. Phys. Lett.* in press (1990).
47. M. Shiotani, L. Sjöqvist, J. Ohshita, and M. Ishikawa: to be submitted.
48. G. Fabian, Q. Xue-Zhi, E. Casper, and E. Kloster-Jensen: *J. Amer. Chem. Soc.* **111**, 6456 (1989).
49. E. Carler, I. Van den Enden, H.J. Geise, and F.C. Mijhoff: *J. Mol. Str.* **50**, 345 (1978).
50. Q. Shen, R.L. Hilderbrandt, and V.S. Mastryukov: *J. Mol. Str.* **54**, 121 (1979).
51. R. Carleer and M. J. O. Anteunis: *Org. Mag. Res.* **12**, 673 (1979).
52. A. J. Lorquet and J. C. Lorquet: *J. Chem. Phys.* **49**, 4955 (1968).
53. D.J. Bellville and N. L. Bauld: *J. Amer. Chem. Soc.* **104**, 294 (1982).
54. H. Nakatsuji: *J. Amer. Chem. Soc.* **95**, 2084 (1973).

55. M.J.S. Dewar and H.S. Rzepa: *J. Amer. Chem. Soc.* **99**, 7432 (1977).
56. T. Clark: *A Handbook of Computational Chemistry* , Chap. 4, Wiley, New York (1985).
57. T. Clark and S. Nelsen: *J. Amer. Chem. Soc.* **110**, 868 (1988).
58. A.J. Merer and L. Schoonveld: *J. Chem. Phys.* **48**, 522 (1968).
59. H. Koppel, W. Domcke, L.S. Cederbaum, and W. Von Niessen: *J. Chem. Phys.* **69**, 4252 (1978).
60. M. Shiotani, Y. Nagata, and J. Sohma: *J. Amer. Chem. Soc.* **106**, 4640 (1984).
61. J. Fujisawa, S. Sato, and K. Shimokoshi: *Chem. Phys. Lett.* **124**, 391 (1986).
62. S. Lunell and M.-B. Huang: *Chem. Phys. Lettt.* **168**, 63 (1990).
63. (a) L. Sjöqvist, M. Shiotani, and A. Lund: to be published; (b) L. Sjöqvist, M. Shiotani, and A. Lund: *Chem. Phys.* **141**, 417 (1990).
64. M. Shiotani, Y. Nagata, and J. Sohma: *J. Phys. Chem.* **88**, 4078 (1984).
65. K. Toriyama, K. Nunome, and M. Iwasaki: *Chem. Phys. Lett.* **107**, 86 (1984).
66. M. Shiotani, K. Ohta, and A. Berndt: unpublished data.
67. M. Kira, H. Nakazawa, and H. Sakurai: *J. Amer. Chem. Soc.* **105**, 6983 (1983).
68. (a) C. Baker and D.W. Turner: *Proc. R. Soc. London, Ser. A*, **326**, 165 (1972); (b) C.Baker and D.W. Turner: *Chem. Commun.* 797 (1967); (c) M.I. Al-Joboury and D.W. Turner: *J. Chem. Soc.* **1964**, 4434 (1964).
69. M. Shiotani, Y. Nagata, and J. Sohma: *J. Phys. Chem.* **86**, 413 (1982).
70. A.W. Potts, H.J. Lempka, D.G. Streets, and W.C. Price: *Philos. Trans. R. Soc. London, Ser A*. **268**, 59 (1970).
71. (a) K. Ohta, M. Shiotani and J. Sohma: *25th Japanese Symposium on ESR*, Abstract p.31 (1986); (b) K. Ohta: *Dissertation, Hokkaido University* (1986).
72. W. Ensslin, H. Bock, and G. Becker: *J. Amer. Chem. Soc.* **96**, 2757 (1984).
73. S. Katumata and M. Shiotani: Unpublished Data.
74. H. Tachikawa and M. Shiotani: *Symposium on Molecular Structure (Japan)*, Abstract p.222 (1988), to be submitted.
75. M. Shiotani, K. Ohta, Y. Nagata, and J. Sohma: *J. Amer. Chem. Soc.* **107**, 2562 (1985).
76. K. Ohta, M. Shiotani, J. Sohma, A. Hasegawa, and M.C.R. Symons: *Chem. Phys. Lett.* **136**, 465 (1987).
77. Q.B. Broxterman, H. Hogeveen, and D.M. Kok: *Tetra. Lett.* **22**, 173 (1981).
78. Q.B. Broxterman and H. Hogeveen: *Tetra. Lett.* **24**, 639 (1983).
79. J.L. Courtneige, A.G. Davies, S.M. Tollerfield, J. Rideout, and M.C.R. Symons: *J. Chem. Soc. Commun.* **1985**,1092 (1985).
80. E.F. Caldin: *Chem. Rev.* **69**, 135 (1964).
81. R. Hudson, M. Shiotani, and Ff. Williams: *Chem. Phys. Lett.* **48**, 193 (1977).
82. S. Dai, R.S. Pappas, G.-F. Chen, Q.-X. Gou, J. T. Wang, and Ff. Williams: *J. Amer. Chem. Soc.* **111**, 8757 (1989).
83. L.D. Snow and Ff. Williams: *Chem. Phys. Lett.* **100**, 198 (1983).
84. P.J. Boon, M.C.R. Symons, K. Ushida, and T. Shida: *J. Chem. Soc. Perkin. Trans. II*, **1984**,1213 (1984).
85. L.B. Knight Jr, B.W. Gregory, S.T. Cobranchi, Ff. Williams, and X.-Z. Qin: *J. Amer. Chem. Soc.* **110**, 327 (1988).
86. D. Becker, K. Plante, and M.D. Sevilla: *J. Phys. Chem.* **87**, 1648 (1983).
87. M.D. Sevilla, D. Becker, C.L. Sevilla, and S. Swarts: *J. Phys. Chem.* **88**, 1701 (1984).
88. M.D. Sevilla, D. Becker, C.L. Sevilla, K. Plante, and S. Swarts: *Farad. Diss. Chem. Soc.* **78**, 71 (1984).
89. M. Iwasaki, H. Muto, K. Toriyama, and K. Nunome: *Chem. Phys. Lett,* **105**, 586 (1984).
90. H. Muto, K. Toriyama, K. Nunome, and M. Iwasaki: *Chem. Phys. Lett,.***105**, 592 (1984).
91. M.D. Sevilla, D. Becker, C.L. Sevilla, and S. Swarts: *J. Phys. Chem.* **89**, 633 (1985).
92. D. Becker, S. Swarts, and M.D. Sevilla: *J. Phys. Chem.* **89**, 2638 (1985).
93. J. Rideout, M.C.R. Symons, S. Swarts, B. Besler, and M.D. Sevilla: *J. Phys. Chem.* **89**, 5251 (1985).
94. J. Rideout and M.C.R. Symons: *J. Chem. Soc. Perkin Trans II*, **1985**, 652 (1985).
95. W.K. Musker, T.L. Wolford, and P.B. Roush: *J. Amer. Chem. Soc.* **100**, 6416 (1978).
96. X.-Z. Qin, L.D. Snow, and Ff. Williams: *J. Amer. Chem. Soc.* **107**, 3366 (1985).

176

97. X.-Z. Qin, L.D. Snow, and Ff. Williams: *J. Phys. Chem.* **89**, 3602 (1985).
98. C.J. Rhodes and M.C.R. Symons: *Chem. Phys. Lett.* **140**, 611 (1987).
99. M.C.R. Symons and J.L. Wyatt: *Chem. Phys. Lett.* **146**, 473 (1988).
100. L.D. Snow and Ff. Williams: *Chem. Phys. Lett.* **143**, 521 (1988).
101. P.T. Boon, L. Harris, M.T. Olm, J.L. Wyatt, and M.C.R. Symons: *Chem. Phys. Lett.* **106**, 408 (1984).
102. F.W. McLafferty: *Interpretation of Mass Spectra. An Introduction*, W.A. Benjamin, New York (1967).

RADICAL CATIONS OF ALIPHATIC ETHERS

A Case Study by ESR and Theoretical Analyses

Tadamasa SHIDA, Takamasa MOMOSE, and Michio MATSUSHITA
Department of Chemistry, Faculty of Science, Kyoto University, Kyoto 606, Japan

1. Introduction

We will focus our subject to aliphatic ethers for the reasons; 1) they are one of the fundamental types of organic molecules and 2) despite their relatively simple composition, radical cations thereof expose many facets interesting enough to basic scientists and educative enough to students who want to deepen their understanding of vividness and subtlety of molecules. To be more specific, the radical cation of dimethylether provides a nice case for studying the internal rotation of the methyl group while the cations of acetals and dioxanes are a suitable system for analyzing the orbital interaction, for example. In the succeeding sections we will deal with several ethers dividing them into sections according to the topic to be emphasized.

2. Dimethylether -*"What is the E-line and what can we learn from it ?"*

The radical cation of the prototype ether is an ideal systems to study the internal rotation of the methyl group and the hyperfine interaction of methyl protons. The radical cation is observed for γ-irradiated dimethylether in $CFCl_3$ at 100K [1]. The ESR spectrum shows a septet signal with the binomial intensity ratio and $a(H_\beta) \approx 43G$, which indicates that the methyl groups are subject to a rotation rapid enough in the ESR time scale.

Upon cooling, the irradiated sample shows a reversible spectral change as demonstrated representatively in Fig. 1 [2]. The change will be discussed in terms of the appearance of the E-line, the concept of which is thoroughly established [3], but experimental data which demonstrate its appearance as unambiguously as in the present system are not so abundant in the literature. Furthermore, an introductory explanation of the E-line is uncommon in textbooks. Therefore, it may be meaningful to analyze the observed spectra from a somewhat instructive point of view. It will be attempted to get information on the internal rotation of the *two* methyl groups attached to a single oxygen atom.

This section is divided into two parts, the first being a general explanation of the E-line and the second a specific discussion on the dimethylether system. Those who are familiar with the theory of E-line may skip the first.

A. Lund and M. Shiotani (eds.), Radical Ionic Systems, 177–194.

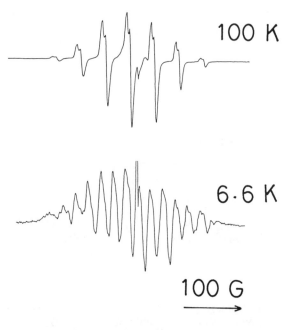

100 K

6·6 K

100 G

Fig. 1. ESR spectra of the radical cation of dimethylether at 100K and at 6.6K.

2.1. THE "E"-LINE

The explanation below is similar to the benchmark work by Freed [3a]. Consider a radical in a rigid environment where the overall rotation is prohibited. If a methyl group is attached to an adjacent atom bearing an unpaired electron in a $2p$-orbital, the wavefunction of the system is approximated by $\Phi = \phi_E \phi_V \phi_R \phi_S \phi_I$ where the five components describe the electronic wavefunction, the vibrational wavefunction, the rotational wavefunction of the methyl group, the electron spin and the nuclear spin wavefunctions, respectively. These components can be analyzed in terms of the permutation group, and the subgroup consisting of cyclic permutations of the protons will be used. Since this subgroup is isomorphous with the point group C_3, there are the irreducible representation A and separably degenerate representations Ea and Eb. Since the cyclic permutation (123) of the three protons corresponds to two successive permutations of the Fermions, the total wavefunction must belong to species A according to nuclear statistics. Therefore, the product $\phi_R \phi_I$ must be of A symmetry because at low temperatures we can assume that the electronic and the vibrational states are in the ground state of A symmetry in the complete nuclear permutation group. Hereafter, we consider the product $\phi_R \phi_I$ along with the pertinent Hamiltonian, $\mathcal{H}$, given as the sum of the internal rotation of the methyl group $\mathcal{H}_R$ and the spin Hamiltonian $\mathcal{H}_S$, $\mathcal{H} = \mathcal{H}_R + \mathcal{H}_S$. Under the high field approximation $\mathcal{H}_S$ is written as

$$\mathcal{H}_S = g_{ZZ}\beta S_Z H + \sum_{i=1}^{3} S_Z T_{ZZ;i} I_{Z;i} \, ,$$

where the summation is performed over the three protons. The second term of $\mathcal{H}_S$ denoted as $\mathcal{H}_{hf}$ represents the hyperfine interaction and can be written in the symmetry-adapted form as

$$\mathcal{H}_{hf} = \frac{1}{3} S_Z (\, T_{ZZ}^A I_Z^A + T_{ZZ}^{Ea} I_Z^{Eb} + T_{ZZ}^{Eb} I_Z^{Ea} \,)$$

using the relations,

$$O^A = O_1 + O_2 + O_3, \quad O^{Ea} = O_1 + \varepsilon \, O_2 + \varepsilon^* O_3 \, , \text{ and } O^{Eb} = O_1 + \varepsilon^* \, O_2 + \varepsilon \, O_3.$$

Here, O stands for T_{ZZ} and I_Z, and $\varepsilon = \exp(2\pi i/3)$. With the cyclic nuclear permutation of the protons, T_{ZZ} and I_Z transform as, $O^A \to O^A$, $O^{Ea} \to \varepsilon \, O^{Ea}$, and $O^{Eb} \to \varepsilon^* \, O^{Eb}$. As for the nuclear spin functions we use the following bases which are also symmetry-adapted and are the eigenfunctions of I_Z.

$$|M_I=3/2, A\rangle = |\alpha\alpha\alpha\rangle$$

$$|M_I=1/2; A\rangle = (|\beta\alpha\alpha\rangle + |\alpha\beta\alpha\rangle + |\alpha\alpha\beta\rangle \,)/ \sqrt{3}$$

$$|M_I=1/2; Ea\rangle = (|\beta\alpha\alpha\rangle + \varepsilon |\alpha\beta\alpha\rangle + \varepsilon^* |\alpha\alpha\beta\rangle \,)/ \sqrt{3}$$

$$|M_I=1/2; Eb\rangle = (|\beta\alpha\alpha\rangle + \varepsilon^* |\alpha\beta\alpha\rangle + \varepsilon |\alpha\alpha\beta\rangle \,)/ \sqrt{3} \, .$$

The functions for $M_I = -3/2$ and $-1/2$ are obtained by exchanging α and β above. As for the rotational states we take into account only the three lowest states, $i.e.$, the ground state of A symmetry and the doubly degenerate first excited states of E symmetry, because the system under study is at low enough temperatures. Since the product $\phi_R \phi_I$ must be of A symmetry as mentioned, the eight nuclear spin functions of $M_I = \pm 3/2$ and $\pm 1/2$ given above must be paired with rotational fuctions of proper symmetries to make the product correspond to A symmetry.

With all these eight products, the matrix elements of the Hamiltonian, $\mathcal{H} = \mathcal{H}_R + \mathcal{H}_S$ will be evaluated. In the evaluation of the matrix elements of $\mathcal{H}_{hf}$ in $\mathcal{H}_S$ we note the relations of $A \otimes$

$Ea = Ea$, $Ea \otimes Eb = A$, and $Ea \otimes Ea = Eb$ by which we have non-zero diagonal matrix elements such as $<A|A|A>$ and $<Ea|A|Ea>$, and non-zero off-diagonal elements such as $<A|Ea|Eb>$ and $<Eb|Ea|Ea>$. (Note that the symmetry of $<Eb|$, the c.c. of $|Eb>$, is Ea). On account of the nuclear spin operator I_Z, the 8 x 8 matrix is divided into four blocks, i.e., two 1 x 1 for $M_I = \pm 3/2$ and two 3 x 3 for $M_I = \pm 1/2$. With the above relations the non-zero diagonal elements of I_Z are $<M_I;X|I_Z^A|M_I;X> = M_I$ with X being any symmetry, and the non-zero off-diagonal elements are given as,

$$\langle \pm 1/2;A |I_Z^{Ea}| \pm 1/2;Eb \rangle = \langle \pm 1/2;Eb |I_Z^{Eb}| \pm 1/2;A \rangle^* = \mp 1$$

$$\langle \pm 1/2;A |I_Z^{Eb}| \pm 1/2;Ea \rangle = \langle \pm 1/2;Ea |I_Z^{Ea}| \pm 1/2;A \rangle^* = \mp 1$$

$$\langle \pm 1/2;Ea |I_Z^{Eb}| \pm 1/2;Eb \rangle = \langle \pm 1/2;Eb |I_Z^{Ea}| \pm 1/2;Ea \rangle^* = \mp 1$$

With these results for the operator I_Z the matrix elements of $\mathcal{H}_{hf}$ are given as, for example,

$$\langle ROT,A |\langle M_I=1/2;A |H_{hf}| M_I=1/2;Eb \rangle| ROT, Ea \rangle = - \langle ROT,A |T_{ZZ}^{Eb}| ROT,Ea \rangle.$$

In this way we obtain the 3 x 3 block matrix of $M_I = \pm 1/2$ for the sum, $\mathcal{H}_R + \mathcal{H}_{hf}$, i.e., the Hamiltonian $\mathcal{H}$ above without the Zeeman term.

$$
\begin{array}{l}
|M_S\rangle |M_I=\pm 1/2;A\rangle |ROT,A\rangle \\
|M_S\rangle |M_I=\pm 1/2;E_b\rangle |ROT,E_a\rangle \\
|M_S\rangle |M_I=\pm 1/2;E_a\rangle |ROT,E_b\rangle
\end{array}
\begin{pmatrix}
E_R(A) \pm M_S\, p & \pm M_S\, r^* & \pm M_S\, r \\
\pm M_S\, r & E_R(E) \pm M_S\, q & \pm M_S\, s^* \\
\pm M_S\, r^* & \pm M_S\, s & E_R(E) \pm M_S\, q
\end{pmatrix}
$$

where

$$p = \tfrac{1}{6} \langle ROT,A |T_{ZZ}^A| ROT,A \rangle, \quad q = \tfrac{1}{6} \langle ROT,Ea |T_{ZZ}^A| ROT,Ea \rangle,$$

$$r = -\tfrac{1}{3} \langle ROT,Ea |T_{ZZ}^{Ea}| ROT,A \rangle, \text{ and } s = -\tfrac{1}{3} \langle ROT,Eb |T_{ZZ}^{Ea}| ROT,Ea \rangle,$$

with the M_S in the matrix representing the expectation value of the operator S_Z.

Evaluation of the matrix elements and diagonalization give the split levels for the triply degenerate nuclear spin states. To evaluate the elements we need an explicit dependence of $T_{ZZ;1}$, $T_{ZZ;2}$, and $T_{ZZ;3}$ on the rotational angle θ of the methyl group. Ignoring the minor contribution of the angular-independent spin polarization effect, the hyperfine tensor can be approximated by the so-called $\cos^2\theta$ relation to obtain the results,

$$T_{ZZ;1} = 2B\cos^2\theta, \quad T_{ZZ;2} = 2B\cos^2(\theta + 2\pi/3), \quad \text{and} \quad T_{ZZ;3} = 2B\cos^2(\theta + 4\pi/3),$$

from which we have,

$$T_{ZZ}^A = 3B, \quad T_{ZZ}^{Ea} = \frac{2}{3}B\,\exp(2i\theta), \quad \text{a nd} \quad T_{ZZ}^{Eb} = \frac{2}{3}B\exp(-2i\theta)$$

and accordingly,

$$p = q = B/2, \quad r = -\frac{1}{2}B\langle ROT, Ea|\exp(2i\theta)|ROT, A\rangle,$$

$$\text{and} \quad s = -\frac{1}{2}B\langle ROT, Eb|\exp(2i\theta)|ROT, Ea\rangle.$$

At this stage we need explicit rotational functions. If the potential height for the internal rotation is small, the free rotor wavefunctions, i.e., $1/\sqrt{2\pi}\exp(im\theta)$ ($m = 0, \pm1, \pm2, ...$), may be used as the base function. The ground state of A symmetry corresponds to $m = 0$ and the first excited states of Ea and Eb to $m = -1$ and $m = +1$, respectively.

Using the relation, $\langle ROT, n|\exp(\pm2i\theta)|ROT, m\rangle = \delta_{n;m\pm2}$, we have,

$$
\begin{array}{l}
|M_S\rangle\,|M_I{=}\pm1/2;A\rangle\,|ROT,0\rangle \\[4pt]
|M_S\rangle\,|M_I{=}\pm1/2;E\rangle\,|ROT,-1\rangle \\[4pt]
|M_S\rangle\,|M_I{=}\pm1/2;E\rangle\,|ROT,+1\rangle
\end{array}
\left(
\begin{array}{ccc}
E_R(A)\pm\frac{1}{2}BM_S & 0 & 0 \\[6pt]
0 & E_R(E)\pm\frac{1}{2}BM_S & \mp\frac{1}{2}BM_S \\[6pt]
0 & \mp\frac{1}{2}BM_S & E_R(E)\pm\frac{1}{2}BM_S
\end{array}
\right)
$$

Diagonalization of the 2 x 2 E block yields the eigenvalues of $E_+(\pm1/2) = E_R(E)$ and $E_-(\pm1/2) = E_E \pm BM_S$. This is a typical E-line splitting of the two degenerate rotational states of E symmetry via the hyperfine interaction, and the ESR pattern changes from 1:3:3:1 to 1:1:1:2:1:1:1.

Another limiting case is that the potential barrier is so high that the three protons are subject to oscillation in the vicinity of the potential minimum. If the potential is assumed nearly parabolic, the harmonic oscillator wavefunction can be used as the base. At low temperatures the three protons are in the vibrational ground state described by the zero-*th* order Hermite polynomial. The symmetry-adapted rotational wavefunctions of the methyl group can, then, be represented as,

$$|ROT,A> = (H_1 + H_2 + H_3)/\sqrt{3}$$

$$|ROT,E_a> = (H_1 + \varepsilon H_2 + \varepsilon^* H_3)/\sqrt{3}$$

$$|ROT,E_b> = (H_1 + \varepsilon^* H_2 + \varepsilon H_3)/\sqrt{3}$$

where H_i denotes the Hermite polynomial. These are degenerate to the zero-th , but the penetration of the wavefunction to the neighboring potential wells lifts the degeneracy, and the lowest state of A symmetry and the first excited states of E symmetry are split by an energy of 3Δ which is related to the "tunneling frequency" of $1/\Delta$. Diagonalization of the previous matrix of $\mathcal{H}_R + \mathcal{H}_{hf}$ gives again the E-line splitting.

2.2. THE CASE OF TWO METHYL GROUPS

Since the ESR spectrum of the radical cation of dimethylether exhibits a large hyperfine splitting of 43 G per each proton and since only the six protons are magnetic, the system is suited for the study of the effect of internal rotation of the methyl groups on the hyperfine interaction. The primary interest here is to get any information on the relation between the two groups. The appropriate rotational Hamiltonian is now given as ,

$$H_R = - \frac{\hbar^2}{2I_{CH_3}} \left(\frac{\partial^2}{\partial\theta_1^2} + \frac{\partial^2}{\partial\theta_2^2} \right) + V\left(\theta_1,\theta_2\right)$$

with

$$V(\theta_1,\theta_2) = \sum_{i=1}^{2} V_3(1 - \cos 3\theta_i) + \frac{V_{33}}{2} \cos3\theta_1\cos3\theta_2 + \frac{V'_{33}}{2} \sin3\theta_1\sin3\theta_2 + \text{higher terms.}$$

Accordingly, the argument in the preceding subsection can be carried on by doubling the basis set. The three rotational wavefunctions considered are now expanded to nine, and as before, we consider the nine A symmetry products of the nuclear spin and rotational parts such as
$$|M_{I;1}M_{;2};AA\rangle|ROT,AA\rangle,|M_{I;1}M_{I;2};AEb\rangle|ROT,AEa\rangle,.......|M_{I;1}M_{I;2};EaEb\rangle|ROT,EbEa\rangle.$$

The 9 x 9 matrix of the Hamiltonian, $\mathcal{H}_R + \mathcal{H}_{hf}$, is blocked out into 1 x 1 of AA, 4 x 4 of AEa, AEb, EaA, EbA, and 4 x 4 of $EaEa$, $EbEb$, $EaEb$, $EbEa$ where the symbols such as AEa signifies that the symmetries of the rotational part of the first and the second methyl groups are A and Ea, respectively. Instead of the two kinds of the off-diagonal elements, r and s, for the case of a single methyl group in the preceding subsection we now have ten kinds of elements corresponding to the all possible combinations such as AA-AEa, AA-EaA, ...$EaEa$-$EaEb$, $EaEa$-$EbEa$. Among the off-diagonal elements those connecting two different rotational states may be regarded as vanishingly small because the energy difference between rotationally different states is typically of the order of GHz which is much greater than the energy of the hyperfine interaction of the order of 10-100 MHz. However, the off-diagonal elements connecting AEa with AEb, and those connecting EaA with EbA could cause the E-line splitting because they connect two rotationally degenerate states. As for the combinations of $EaEa$-$EaEb$, two possibilities arise; if two rotational states of $EaEa$ and $EaEb$ types are degenerate or nearly so, the off-diagonal elements between them are effective in inducing another E-line splitting. If, on the other hand, the two states are different in the rotational energy region, there should occur no E-line splitting for the reason mentioned above.

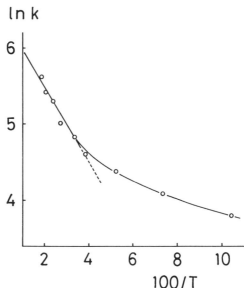

Fig. 2

Arrhenius plot of the rotational rate constant [see Ref. 2].

We have shown that the first case of $EaEa$ and $EaEb$ being degenerate is equivalent to the condition of $V'_{33} = 0$ (see Appendix of Ref [2]). Since the second and the third terms of $V(\theta_1,\theta_2)$ above can be rewritten as,

$$\tfrac{1}{4}(V_{33} - V'_{33}) \cos 3(\theta_1 + \theta_2) + \tfrac{1}{4}(V_{33} + V'_{33}) \cos 3(\theta_1 - \theta_2),$$

the condition of $V'_{33} = 0$ leads to the conclusion that both the "conrotation" and the "disrotation" in a classical picture suffer the same potential [2]. Since the states of $EaEa$ and $EaEb$ correspond to the con and dis rotations, respectively, the equivalence of the potential for the two rotations is self-consistent with the assumed degeneracy of the two states. It is found that the ESR spectra observed at 6-140K can be most reasonably interpreted by assuming this case of $V'_{33} = 0$ [2].

By analyzing the temperature dependent spectra the dependence of the rotational rate constant k on temperature was also studied to obtain the result shown in Fig. 2, which demonstrates a kink at about 25K and an activation energy of about 100 cal/mol for the thermally induced internal rotation. At temperatures below 25K the slope decreased to a quarter of that at the higher temperatures. From the above result we can conclude that the two methyl groups rotate, in a classical picture, in such a manner as the two different modes of con and dis rotations are indistinguishable and that the activation energy for the thermally induced rotation is only 100 cal/mol which is remarkably small compared with the corresponding energy of 2.6 Kcal/mol experimentally determined for the neutral molecule [4].

3. Cyclic Ethers -*"How much are they individualistic ?"*

In this section we should like to demonstrate that a series of three-, four-, five-, and six-membered cyclic ethers exhibit conspicuous differences as revealed by ESR spectra. We start with a story in our laboratory in the beginning: when we discovered that $CFCl_3$ gave resolved ESR spectra for unsaturated aliphatic hydrocarbons [5] and were astonished by unexpectedly large hyperfine coupling constants of β-protons, $a(H_\beta)$, we were enticed to get a "world-record" value of $a(H_\beta)$ for the radical cation of the hydrocarbons by studying cyclopropene. After synthesizing cyclopropene, we measured the ESR spectrum of γ-irradiated cyclopropene in $CFCl_3$ at 77K to get an expected triplet-triplet pattern with $a(H_\beta)$=89 and 40 G which are really unprecedentedly large. However, we were soon disappointed by finding that the signal was due to tetrahydrofuran (THF) used as the solvent in synthesis of cyclopropene. The negative result led us to explore the radical cation of ethers, however. The spectrum observed for THF is straightforwardly analyzed invoking the ring puckering whose activation energy is determined as 1.65 Kcal/mol [1a].

The three-membered ether is characterized by its ease of ring opening. Therefore, it is natural that the radical cation thereof also suffers a facile ring opening as has been studied by several workers [6]. The ring opening is demonstrated by experiments on the radical cations of some methylated derivatives [6c,7].

For example, the ESR and optical spectra observed for 2,3-dimethylated derivative shown in Fig. 3 can be understood in terms of *cis-trans* isomerization of the ring-opened radical cation. Both spectra change reversibly upon bleaching with photons of selected wavelengths as explained in the caption. Such a photoreversible process of radical ions in low temperature matrices was noticed many years ago for the radical anion of dimethyl disulfide [8]. Comprehensive reviews are given by Andrews and Bally in other chapters.

The four- [6a,9] and five-membered ethers [1a,9b] yield the cyclic radical cations in which the odd electron is mainly distributed over the oxygen atom and the adjacent methylene groups as is apparent from the ESR spectra shown in Fig. 4. In both the overall hyperfine splitting amounts to about 260 G as in the case of dimethylether in Fig.1 [1].

By a simple analogy one might expect that the overall hyperfine splitting of the radical cation of 1,4-dioxane be about the same as the above value (note that the spin density on each oxygen atom is halved but the splitting due to the β-protons will be doubled because of the equivalence of the protons at C2 and C6 with those at C3 and C5 whether the radical cation is in a chair or a boat form). However, as shown at the bottom of Fig. 4 the experiment reveals a much smaller splitting than expected. To search for the cause of this failure in our analogy, we will dwell a little upon the system in the following.

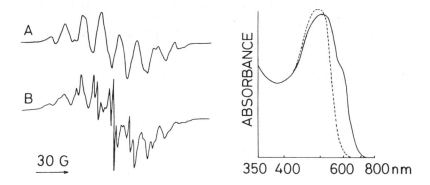

Fig. 3. ESR and optical spectra of the radical cation of 2,3-butyleneoxide in CFCl₃ at 77K. The upper ESR spectrum and the optical spectrum in dashed curve were observed by photobleaching the γ-irradiated sample with λ > 550 nm. The lower ESR spectrum and the optical spectrum in solid curve were observed by subsequently photobleaching the above sample with λ > 520 nm.

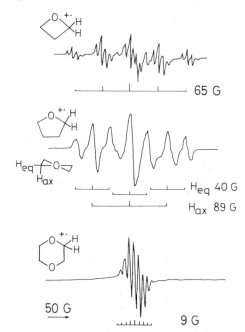

Fig. 4
ESR spectra of the radical cations of some cyclic ethers in CFCl₃ at 77K.

186

The small value of $a(H_\beta)$ has been discussed before but the argument was speculative based on an assumption that the radical cation was in a boat form [6b]. Thus, it is necessary to know the geometry of the radical cation before investigating the spin density distribution. Currently we are benefitted by availability of numerous programs for theoretical calculations of the optimum geometry. Considering the lack of experimentally determined geometries for most of radical ions, the benefit becomes even more valuable.

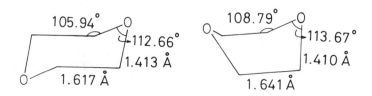

Fig. 5. Optimized geometries of the radical cation of 1,4-dioxane calculated by the UHF-STO3G approximation. The total energies for the chair and the boat forms are -301.726018 and -301.696588 hartree, respectively.

Fig. 5 shows the optimized geometries of the radical cation of 1,4-dioxane calculated under the condition given in the caption. To get a feeling of reliability of the calculation, we have optimized the geometry of the parent neutral molecule also with the same condition of computation to find that the result agrees with that obtained by electron diffraction within a few percent [11]. Both the calculation and the experiment indicate definitely that the chair form is favored in the neutral molecule. The total energy calculation for the radical cation also indicates that the chair form is much more stable as given in the caption for Fig. 5. Thus, we must conclude that the previous argument based on the assumed boat form [6b] loses its standing. Incidentally, the present conclusion is diametrically contrasting with the result of our study on the sulfur analogue of 1,4-dioxane which favors a boat form [12]. As for the electronic structure of the parent molecule, relevant information is provided by photoelectron spectroscopy, which indicates that the first two bands split by more than 1 eV [13]. A strong "through-bond" interaction via the central C-C bonds permissible for a chair form was invoked to interpret the large split [13], which was also confirmed by a standard MO analysis [14].

Fig. 6 shows the distribution of the odd electron in the ground state radical cations which gives an approximate image of the highest occupied orbital of the parent molecules. It is clearly seen that the out-of-plane lone pair orbital tilts significantly to favor the "through-bond" interaction in 1,4-dioxane whereas in oxetane the lone pair orbital is nearly orthogonal to the C-O-C plane. As is well known, the interaction of the odd electron with the β-proton is approximately proportional to $\cos^2\theta$ where θ is the angle between the axes of the lone pair orbital and the C-H bond. With the calculated geometries the values of $\cos^2\theta$ for the axial β-proton of the radical cations of oxetane, THF, and 1,4-dioxane turn out to be @0.7, 0.8, and 0.3, respectively. Thus, it is quite plausible that one reason for the small value of $a(H_\beta)$ in 1,4-dioxane is due to the geometry of the radical cation. Indeed, our recent calculation of the spin density by the method described in section 5 revealed a conspicuously small value of $a(H_\beta)$ for the dioxane in a chair form [15].

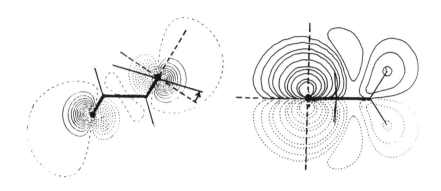

Fig. 6. The contour map of the odd electron density of the radical cations of 1,4-dioxane and oxetane. The black dots indicate the center of the oxygen atom.

4. Acetals -*"A cooperative effect of two oxygen atoms intensifies the hyperconjugation"*

4.1. HYPERCONJUGATION

1,3-Dioxaalkanes are called acetals. As in the case of 1,4-dioxane, the lone pair orbitals on the oxygen atoms interact as revealed in the photoelectron spectrum[13].

Fig.7 shows ESR spectra of the radical cations of two acetals [6b,10]. As for the conformation of the radical cation of the cyclic acetal, there are two possibilities, *i.e.*, twisted (C_2) and bent (C_2) forms. However, the spectrum in Fig. 7 indicates that the latter is realized and further that the singly-occupied molecular orbital (SOMO) is characterized by the in-phase combination, $n_1 + n_2$, of the out-of-plane lone pair orbital n of the two oxygen atoms: if the conformation were of C_2 symmetry, the ESR pattern would have to be composed of triple triplets due to two equivalent protons at C2 and two pairs of the axial and the equatorial protons, Hax and Heq, at C4 and C5 in the case of SOMO being of $n_1 + n_2$ nature, whereas in the case of $n_1 - n_2$ SOMO, double triplets due to the two Hax and Heq at C4 and C5.

The observed spectrum is immediately analyzed in terms of one proton with $a = 166$ G and another with $a = 134$ G plus four protons with $a = 11$G as indicated by the stick spectrum, which can only be compatible with a Cs conformation and the SOMO being $n_1 + n_2$. The result is supported by a geometry optimization with the UHF-STO3G approximation

188

which predicts that the dihedral angle between the planes of O-C-C-O and O-C-O is 20 degrees as shown in the insets of Fig. 7[16].

Two points of importance emerge from the above results, *i.e.*, the "natural" order of $n_1 - n_2$ and $n_1 + n_2$ is reversed and the hyperfine splitting due to the methylene protons at carbon site 2 amount to 300 G, implying about 60 % of residence of the odd electron on the methylene protons. In sections 3 and 4 we have seen that the hyperfine splitting of the methylene protons adjacent to the oxygen atom in the monoethers sum to a common value of about 260 G, i.e., 43 G x 6 = 258 G for dimethylether, 65 G x 4 = 260 G for oxetane, and (89 + 40) G x 2 = 258 G for THF. Thus, the result of the cyclic acetal above implies that the two flanking oxygen atoms cooperatively intensify the localization of the odd electron at the sandwiched methylene group. It may also be stated that the hyperconjugation between O and CH_2 are cooperatively intensified in acetals.

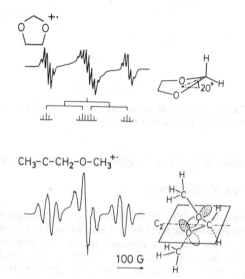

Fig. 7

ESR spectra of the radical cations of 1,3-dioxacyclopentane and acetal in CFCl3 at 77K. The inset for the former shows the optimized geometry calculated by the UHF-STO3G approximation. The lower inset shows the geometry of the parent molecule of acetal determined by electron diffraction[17]. The geometrical parameters are as follows. $r(O-CH_2)=1.382$Å, $r(O-CH_3)=1.432$Å, $<(O-CH_2-O)=114.3°$, $<(CH_2-O-CH_3)=114.6°$, $<(O-C-H$ in $CH_3)=110.3°$, and the dihedral angle between the planes of $O-CH_2-O$ and CH_2-O-CH_3 equal to 63.3°.

As for the reversal of the order of orbital energy, we will briefly review a standard method for the analysis of orbital interaction later in this section. Here, we continue the investigation of the radical cation of 1,3-dioxapentane whose ESR spectrum is in Fig. 7. The conformation of this prototype acetal has been determined by electron diffraction as shown in Fig. 7 [17]. If the conformation of the radical cation is also of a C_2 symmetry, the two methylene protons become equivalent in the interaction with the oxygen atoms. The ESR spectrum, though a bit broadened compared with the cyclic acetal, is featured by a symmetric triplet with $a \approx 135$ G. The symmetry implies that the conformation is really of C_2 symmetry. The large hyperfine coupling constant ascribed to the methylene protons is in accordance with the above argument of the cooperatively intensified hyperconjugation. The minor triplet with $a \approx 34$ G superimposed on the major must, then, be interpreted as that the methyl groups are hindered to rotate freely and one proton of the methyl groups is in near eclipse with the lone pair orbitals of the oxygen atoms. This means that the unit of O-CH$_2$-O dictates the conformation of the terminal methyl groups also. The triplet-triplet pattern in Fig. 7 was observed at temperatures at 4 to ~110K. Thus, the radical cation may be considered as quite a rigid molecule dictated by an extraordinary effect of the hyperconjugation. The rigidity contrasts with the facile rotation of the methyl groups in the radical cation of dimethylether in section 2.

4.2. ANALYSIS OF ORBITAL INTERACTION

From the ESR information in the preceding subsection it is concluded that the "natural" order of the highest occupied molecular orbitals of acetals is reversed. In this subsection we will discuss the orbital interaction to understand the reversal of the order. An analysis of orbital interaction reported in a previous paper [16] will be recapitulated briefly. Consider that we know the geometry of dioxacyclopentane .Then, we obtain canonical MO's ϕ by diagonalyzing the Fock matrix. The canonical MO's are, then, unitary transformed to a set of localized orbitals ϕ^L. One can choose several criteria to obtain localized MO's [16]. In any case, the Fock matrix $\mathbf{F}$ is diagonal for the canonical orbitals but becomes a full matrix $\mathbf{F}^L$ for the localized MO's. If we nullify all the off-diagonal matrix elements of $\mathbf{F}^L$, the remaining diagonal elements correspond to the energies of the localized orbitals without any interaction among them. Now, if we restore any nonzero matrix element between the m-th and the n-th localized orbitals, L_{mn} and its conjugate, leaving all the other off-diagonals as zero, diagonalization of this new matrix yields the orbital energies for the m-th and the n-th localized orbitals split by a certain amount. If the two orbitals are taken as the lone pair orbitals n_1 and n_2 in our previous notation, these are degenerate in the absence of any interaction but now they are split by $\pm \Delta$. This sort of direct interaction between two orbitals are conventionally called "through space interaction" [18].

Now, if we consider a third localized orbital, say ϕ_p^L in addition to the two orbitals, $n_1 = \phi_m^L$ and $n_2 = \phi_n^L$ and restore the elements L_{mp}, L_{pm}, L_{np}, and L_{pn}. Then, diagonalization of the matrix shifts the energies of the two previous split orbitals according to the magnitude of the newly introduced matrix elements. The shifts caused by this procedure may be regarded as an indirect effect on the interaction between n_1 and n_2 due to the third localized orbital ϕ_p^L. Continuing this stepwise procedure we can see the effect of the added matrix elements on the interaction of the initially chosen orbitals of n_1 and n_2. Of course, the diagonalization of the full matrix gives the canonical orbitals and their energies (see the illustration at the bottom of Fig. 8).

With this procedure we found, for the molecule of 1,3-dioxacyclopentane in a C_S conformation, that the indirect effect caused by the orbital localized on the axial methylene C-H bond at carbon 2 is exceedingly large to reverse the initial order of n_1- n_2 and $n_1 + n_2$. In contrast, the effect of all the other localized orbitals is too small to cause the reversal [16]. A part of the computational results is represented illustratively in Fig. 8 (see the lower energy diagram). In this way it is concluded that the reversal of the order of the two orbitals

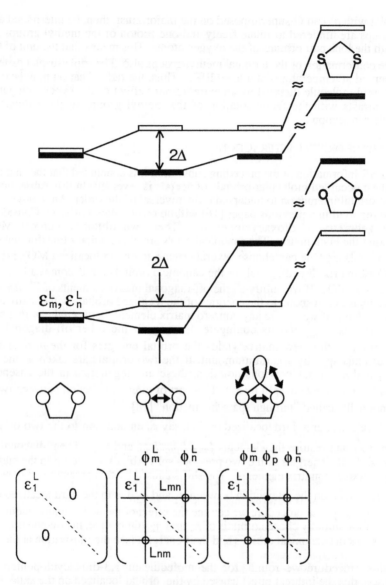

Fig. 8. Orbital energy diagrams for the first two highest occupied MO of 1,3-dithiacyclopentane and 1,3-dioxacyclopentane. The solid and the open horizontal bars represent the orbitals $n_1 + n_2$ and $n_1 - n_2$ in the text, respectively. The matrices illustrate the relevant matrix elements discussed in the text.

in 1,3-dioxacyclopentane is caused essentially by the extraordinarily large effect of the axial C-H bond at carbon 2. The conclusion requires the correction of a previous assignment of the first two photoelectron bands [13].

Supporting evidence for the above conclusion is obtained for the radical cations of 1,3-dioxacyclopentane derivatives whose central methylene proton was replaced with either a methyl or an ethyl group. The ESR spectrum observed for the radical cations of these alkylated acetals showed exclusively the signal of the methyl or the ethyl radical, which can be understood by considering that the SOMO of these radical cations is related to the bonding orbital characterized by $n_1 + n_2$ [19].

4.3. COMPARISON WITH THIO ANALOGUES

Somewhat digressive but interesting discussions can be made by comparing the result of 1,3-dioxa system with its sulfur analogue. By the same procedure described above the orbital interaction in 1,3-dithiacyclopentane is also analyzed to obtain the result shown in Fig. 8(the upper energy diagram). The split due to the direct interaction (the second pattern from the left) is much larger than the oxygen analogue. This is understood in terms of a larger overlap of the sulfur $3p$ orbitals. Owing to this large split the effect of the introduction of the C-H localized orbital is short of jacking up the $n_1 + n_2$ orbital above the level of the $n_1 - n_2$ orbital (the third pattern from the left). This orbital order is maintained up to the final canonical orbitals shown at the right. In consistence with this result, the ESR spectrum of the radical cation of 1,3-dithiacyclopentane indicates definitely that the SOMO is of the nature of $n_1 - n_2$ bisecting through the methylene group at carbon 2.

5. Spin Density Calculation - *"Semiempirical methods are incompetent while brute force ab initio approaches are not necessarily rewarding"*

Familiar semiempirical MO methods such as INDO are based on the parametrization of calculated results to best fit experimental data. For the isotropic hyperfine coupling constant of proton, Pople *et al.* collected various experimental data for fitting [20]. However, comparison of the experimental data with the calculated results reveals that the discrepancy is significantly large for several heteroatom containing radicals [see Table 4.23 of Ref.20]. There is no guarantee that the parameters which reproduce the experimental results of many hydrocarbons are equally transferable to heteroatom containing radicals.

In this section we will describe the present situation of MO calculations for the spin density of radicals of the size of chemists' interest such as the radical cation of dimethyether. A number of attempts to calculate the spin density has been made since 1960's. Although perturbation theoretic treatments [21] were tried in early days, progress in computer techniques prompted the CI approach which included Multi-Reference SDCI and SDTQ CI [22], pseudo-orbital and SAC-CI [23], finite field coupled cluster methods [24], field theoretic treatments [25]. All in all, these approaches gave satisfactory results for the α-proton and magnetic nuclei near the radical center.

However, as for the β-proton and nuclei at a distance from the radical center, calculations are not quite successful. Thus, in their paper entitled, "Difficulties in *ab initio* CI calculations of the hyperfine structure of small radicals" Davidson *et al.* demonstrated that even after inclusion of vibrational averaging effects, MCSCF orbitals, and large CI reference spaces, the result of calculations on β-protons, $e.g.$, that of the radical cation of formaldehyde deviates from the experimental data significantly [22]. Since the spin density is a local property to be described by an operator of δ-function type, an accurate estimation of the expectation value is naturally difficult in comparison with calculations for other properties such as the geometry and the energy. In general, large basis sets and an appropriate treatment of electron correlation are required. In order to obtain expectation values of any operator of a δ-type function the wavefunction has to satisfy the cusp condition at the nucleus [26, 23c]. Unfortunately, however, the Gaussian type orbital (GTO), suited for computer calculi, never satisfies the cusp condition. In an attempt to avoid this obstacle we proposed replacing the δ-function type operator with a spatially distributed one [27]. So far this approach has been explored to the extent that most of the fundamental problems have been solved, but not to the extent that an easy application to realistic molecules is feasible.

Among the existing approaches, however, the symmetry-adapted cluster expansion (SAC)-CI method developed by Nakatsuji and Hirao is capable for giving fair results in spite of the use of the conventional δ-function type operator [28]. By this approach we have shown that the experimental hyperfine coupling constant of the β-protons of the radical cations of dimethylether and the cyclic ethers discussed in section 3 can be reproduced to about 90 % [15,28]. For these systems both semiempirical INDO and ab $initio$ UHF calculations failed fatally to reproduce the experimental results [15]. The reasons for the success are 1) the SAC-CI method takes into account the electron and the spin correlations adequately, and 2) it includes configurations which contribute significantly to the hyperfine interaction even if they are not so important for energy and other properties. Owing to the rational setup of the theory the dimension of matrices to be diagonalized is reduced drastically [29], so that the SAC-CI method does not require any arbitrary configuration selection. As for the basis set, it is shown that Dunning's double zeta set is the best choice for the use in combination with the method to describe properties at and near nuclei such as the spin density. Since properties at and near nuclei are, in general, transferable from molecule to molecule, the adequacy of the present method is favorable for predicting or confirming the spin density of newly found radicals.

6. Concluding Remark

We have confined ourselves to the study of a specific class of aliphatic ethers rather than attempting to cover a wide range. Nevertheless, we hope that we have been able to present several topics for a deeper understanding of molecular systems. It is to be noted that even commonplace aliphatic ethers shed light into details of the nature of molecules only once they are ionized.

In concluding, I (T.S.) retrospect those good old days in 1960's when Bill Hamill was pioneering the new field of ionic species produced by high energy radiation. His request to me to detect trapped electrons in molecular crystals, not in well studied amorphous solids, could not be answered positively. Instead, however, exploration of molecular cations by using chloride matrices was initiated [30]. The extension of the matrix from the initial carbon tetrachloride [30] to a Freon [31] was attempted by Tatsuhisa Kato, which resulted in discovering a rich harvesting ground for ESR information of many new radical cations including those of aliphatic ethers. George Simpson, who was carrying out his PhD work on the same floor of my laboratory at Notre Dame, was inspired by the chloride matrix and later introduced a useful Freon matrix [32] for the optical study of radical cations. This beneffited me to compile optical spectra of radical ions [33]. To these people, among many others, I should like to thank cordially.

References

1. a) H. Kubodera, T. Shida, and K. Shimokoshi: *J. Phys. Chem.* **85**, 2583 (1981). b) J.T. Wang and Ff. Williams: *J. Amer. Chem. Soc.* **103**, 6994 (1981).
2. M. Matsushita, T. Momose, and T. Shida: *J. Chem. Phys.* **92**, 4749 (1990).
3. a) J.H. Freed: *J. Chem. Phys.* **43**, 1710 (1965). b) S. Clough and F. Poldy: *J. Chem. Phys.* **51**, 2076 (1969). c) R.B. Davidson and I. Miyagawa: *J. Chem. Phys.* **52**, 1727 (1970). d) S. Kubota, M. Iwaizumi, Y. Ikegami, and K. Shimokoshi: *J. Chem. Phys.* **71**, 477 (1979). e) M. Geoffroy, L.D. Kispert, and J.S. Wang: *J. Chem. Phys.* **70**, 4238 (1979).
4. F.J. Lovas, H. Lutz, and H. Dreizler: *J. Phys. Chem. Ref. Data* **8**, 1051 (1979).
5. T. Shida, Y. Egawa, H. Kubodera, and T. Kato: *J. Chem. Phys.* **73**, 5963 (1980).
6. a) L.D. Snow, J.T. Wang, and Ff. Williams: *Chem. Phys. Lett.* **100**,193(1983); *J. Amer. Chem. Soc.* **104**, 2062 (1984). b) M.C.R. Symons and B.W. Wren: *J. Chem. Soc. Chem. Commun.* **1982** 817 (1982); *Tetrahedron Lett.* **24** 2315 (1983). c) T.Bally, S. Nitsche, and E. Haselbach: *Helv. Chim. Acta* **67** 86 (1984).
7. K. Ushida, T. Shida, and K. Shimokoshi: *J. Phys. Chem.* **93** 5388 (1989).
8. T. Shida: *J. Phys. Chem.* **72** 2597 (1968).
9. a) M.C.R. Symons and B.W. Wren: *Tetrahedron Lett.* **24**, 2315 (1983). b) *idem.* : *J. Chem. Soc. Perkin Trans.* **2**, 511 (1984).
10. L.D. Snow, J.T. Wang, and Ff. Williams: *J. Amer. Chem. Soc.* **104**, 2062 (1982).
11. M. Davis and O. Hassel: *Acta Chem. Scand.* **17** 1181 (1963).
12. a) T. Momose, T. Suzuki, and T. Shida: *Chem. Phys. Lett.*. **107**,568 (1984); b) T. Shida and T. Momose: *J. Mol. Struct.* **126**, 159 (1985).
13. D.A. Schweigart and D.W. Turner: *J. Amer. Chem. Soc.* **94**, 5592 (1972).
14. D. Gonbeau, M. Loubt, and G. Pfister-Guillouzo: *Tetrahedron* **36**, 381 (1980).
15. T. Momose, J. Takahashi, M. Yamaguchi, and T. Shida: to be published.
16. T. Momose, R. Tanimura, K. Ushida, and T. Shida: *J. Phys. Chem*. **91**, 5582 (1987).
17. E.E. Astrup: *Acta Chem. Scand.* **27**, 3271 (1973).
18. a) R. Hoffmann, A. Imamura, and W.J. Hehre: *J. Amer. Chem. Soc.* **90**, 1499 (1968). b) R. Hoffmann: *Acc. Chem. Res.* **4**, 1 (1971).
19. K. Ushida and T. Shida: *J. Amer. Chem. Soc.* **104**, 7332 (1982).
20. J.A. Pople and D.L. Beveridge: *"Approximate Molecular Orbital Theory"*, McGraw-Hill, New York (1970).
21. a) J.-P. Malrieu: *J. Chem. Phys.* **46**, 1654 (1966). b) A.L.H. Chung: *J. Chem. Phys.* **46**, 3144 (1966). c) S.Y. Chang, E.R. Davidson, and G. Vincow: *J. Chem. Phys.* **49**, 529 (1968). d) G. Vincow: *J. Phys. Chem.* **75**, 3400 (1970). e) A. Denis and J.-P. Malrieu: *Mol. Phys.* **23**, 581 (1972). f) P. Millie, B. Levy, and G. Berthier:*Int. J. Quantum Chem.* **4**, 155 (1972).
22. D. Feller and E.R. Davidson: *Theor. Chim. Acta* **68**, 57 (1985).

23. a) K. Ohta, H. Nakatsuji, K. Hirao, and T. Yonezawa: *J. Phys. Chem.* **73,** 1770 (1980). b)   H. Nakatsuji, K. Ohta, and T. Yonezawa: *J. Chem. Phys.* **87,** 3068 (1980). c) H. Nakatsuji and M. Izawa: *J. Chem. Phys.* **91,** 6205 (1989).
24. H. Sekino and R.J. Bartlett: *J. Chem. Phys.* **82,** 4225 (1985).
25. S. Aono, K. Nishikawa, and K. Deguchi: *Bull. Chem. Soc. Japan* **53,** 1238 (1980).
26. T. Kato: *Commun. Pure and Appl. Math.* **10,** 151 (1957).
27. T. Momose and T. Shida: *J. Chem. Phys.* **87,** 2832 (1987); *J. Chem. Phys.* **88,** 7258 (1988).
28. T. Momose, H. Nakatsuji, and T. Shida: *J. Chem. Phys.* **89,** 4185 (1988).
29. a) H. Nakatsuji and K. Hirao:*Chem. Phys. Lett.* **47,** 569 (1977). b)   H. Nakatsuji and K. Hirao: *J. Chem. Phys.* **68,** 2053 (1978). c) K. Hirao and H. Nakatsuji: *Chem. Phys. Lett.* **79,** 292 (1981). d) H. Nakatsuji: *Chem. Phys. Lett.* **59,** 362 (1978); *Chem. Phys. Lett.* **67,** 329 (1979); *Chem. Phys. Lett.* **67,** 334 (1979).
30. T. Shida and W.H. Hamill: *J Chem. Phys.* **44** 2369 (1966); *J. Chem. Phys.* **44** 2375 (1966).
31. T. Shida and T. Kato: *Chem. Phys. Lett.* **68,** 106 (1979).
32. A. Grimison and G.A. Simpson: *J. Phys. Chem.* **72,** 1776 (1968).
33 T. Shida: *"Electronic Absorption Spectra of Radical Ions"* , Physical Sciences Data 34, pp.446, Elsevier Science Publishers, Amsterdam (1988).

STUDY OF RADICAL CATIONS BY TIME-RESOLVED MAGNETIC RESONANCE

A. D. TRIFUNAC, D. W. WERST
Chemistry Division, Argonne National Laboratory, 9700 S. Cass Avenue
Argonne, IL 60439 USA

1. Introduction

This review focuses on the study of hydrocarbon radical cations by time-resolved magnetic resonance methods. These studies attempt to elucidate the early chemical events induced by ionizing or photoionizing radiation. The development of specialized time-resolved magnetic resonance tools has allowed us to dispense with a key element that has been vital to a wide range of radical cation studies using static EPR, i.e., a stabilizing (rigid, chemically inert, electron capturing) matrix (e.g., frozen halocarbons, inert gases). The time-resolution and sensitivity achieved with the methods described in this chapter allow observations of very reactive radical cations in essentially neat systems (pure hydrocarbons) and in fluids up to room temperature. These methods open the way for the study of chemical intermediates such as hexamethyl(Dewar benzene)‡ where implications and insights provided by theory are in need of further experimental data on radical cation structure and behavior under conditions less influenced by solvent/matrix interactions.

1.1 RADICAL CATIONS IN RADIOLYSIS AND PHOTOLYSIS

Radical ions have been recognized as important intermediates in many areas of chemistry. Their widespread occurrence is becoming increasingly appreciated with the advent of faster and better tools, allowing detection of very short-lived species generated by pulses of energy. The electron-loss species, radical cations, are ubiquitous in chemistry induced by ionizing and photoionizing radiation.

Energetic radiation induces charge separation as it interacts with matter, and the subsequent fate of the charge pairs represents a first and often decisive step in the chemistry that follows. Such "high-energy chemistry" represents an important subset of photochemistry where radical ions are known to be important intermediates induced by photodriven electron transfer. It is thus not surprising that considerable effort (much of it reviewed in this book) has been expended in order to study the chemistry, structure and dynamics of radical cations. This considerable research effort has been directed for the most part towards the study of "stabilized" radical cations in low temperature matrices, where details of structure could be obtained by EPR and optical studies [1-3]. Many ideas of various chemical transformations, activated thermally or photolytically, were advanced in studies of such "stabilized" radical cations.

On the other hand, the observations of radical cations in reacting/dynamic systems have been few. A limited number of optical absorption and emission studies have been carried out [4-8], all of which have suffered from the complications which are common in optical

A. Lund and M. Shiotani (eds.), Radical Ionic Systems, 195–229.

absorption/emission studies in the condensed phase. These difficulties include poor spectral resolution and overlapping signals from different species. A few conductivity studies have been carried out [9,10], and some very intriguing processes like fast hole mobility were ascribed to alkane radical cations. The sum of the optical and conductivity work is a confusing pattern of observations testifying to the diversity of radical cation chemistry even in simple saturated hydrocarbon systems. Many processes have been enumerated, but a coherent mechanism of, e.g., alkane radiolysis, has been elusive.

Several years ago two groups, one in Novosibirsk[11,12] and our group at Argonne [13,14] proceeded to develop and apply an optically detected magnetic resonance method for the study of radical ion pairs in liquids. Time-resolved Fluorescence Detected Magnetic Resonance (FDMR) was developed at Argonne in order to study radical cation chemistry in pulse radiolysis and, more recently, in laser photoionization [15]. Furthermore, at Argonne a comprehensive set of studies has been undertaken with state-of-the-art time-domain tools, utilizing the shortest available electron beam and UV laser pulses, to examine, in a coordinated fashion, the earliest events in pulse radiolysis and laser photoionization in hydrocarbons. We will outline here the main conclusions of the time-resolved magnetic resonance studies of radical cations and some very recent picosecond optical work. The optical work confirms the findings of the magnetic resonance studies that two processes dominate the chemistry of radical cations in radiolysis/photoionization: ion-recombination and ion-molecule reactions.

In this chapter we intend to provide a review of the observations of radical cations in the framework of these dominant processes occurring after energy deposition by ionizing radiation. These dominant processes can be regarded as "clocks" which define the observation time window and the time scale on which other processes must compete. For example, in the studies of radical ions by FDMR one only observes geminate ion-pair recombination, and the time frame of geminate ion-recombination is the dominant overall clock. To the extent that we can slow down recombination, we are able to widen the observation time window as well as extend the time available for other processes (e.g., radical cation transformation, ion-molecule reactions) to occur. We will describe various studies which depend on the conversion of the highly mobile electrons into less mobile species (e.g., scavenger radical anions, solvated electrons) in order to extend the time frame of ion recombination from picoseconds to nanoseconds.

In all condensed media, polar (e.g., water (eqn 1)) and nonpolar (e.g., hydrocarbons (eqn 2)), the initial event following energy deposition by ionizing radiation is the production of a charge pair, with the average pair separation, pair recombination lifetime and escape probability depending on the solvent polarity as well as the degree of anisotropy of the solvent molecules and method of ion pair generation.

$$H_2O + e^- \text{ beam} \longrightarrow H_2O^{+} \cdot + e^- \tag{1}$$

$$RH + e^- \text{ beam} \longrightarrow RH^{+} \cdot + e^- \tag{2}$$

In polar systems fast ion-molecule reactions (e.g., proton transfer) convert the primary holes into neutral radicals and diamagnetic positive ions [16].

$$H_2O^{+} \cdot + H_2O \longrightarrow HO \cdot + H_3O^+ \tag{3}$$

$$ROH^{+} \cdot + ROH \longrightarrow RO \cdot + ROH_2^+ \tag{4}$$

The process in water (3) or alcohols (4) is so fast that even the fastest real-time tools have not allowed direct observation of the primary cations. The proton transfer reaction of the primary cations in water and in alcohols occurs quickly because there are many proton acceptor sites in the vicinity of the radical cation. We will attempt to assess the relative importance of ion-molecule reactions in the overall radiation chemistry of nonpolar systems.

1.2 TIME-RESOLVED MAGNETIC RESONANCE

The challenge of species identification is always equal to, and often transcends, that of the detection of very short-lived species in the condensed phase. In this regard, magnetic resonance excels. Magnetic resonance methods (e.g., electron paramagnetic resonance) provide unique fingerprints (hyperfine structure) of paramagnetic species which can be deciphered to yield precise details of electronic structure and molecular geometry. Compared to magnetic resonance, the spectral resolution of optical methods in the condensed phase is very inferior. The ability of optical absorption and emission spectroscopy to provide structural information for species in the condensed phase is often defeated by the sheer numbers of spectral lines (transitions) which coalesce into broad (halfwidths of hundreds of cm^{-1}), featureless bands that merely map the general contours of the underlying vibrational and rotational structure. The spectral extent of the vibronic bands makes it very difficult to resolve absorptions of different species which are close in energy.

The shortcomings of magnetic resonance methods are limited sensitivity and time resolution. The magnetic moment of the electron and thermal population differences between electron spin levels require observations of changes of about one part in one thousand. This, coupled with the fact that species with unpaired electron spin are usually quite reactive, forces one to try to somehow stabilize these reactive radicals and radical ions or create sufficient steady state concentrations to allow detection by conventional static EPR, or to use pulsed generation of transients combined with time-resolved EPR detection.

Many chapters of this volume attest to the numerous possibilities where some host matrix is used to stabilize radical ions at low temperatures. Our goal is to study short-lived paramagnetic species in real time and in "real" chemical systems since in many instances the intrinsic reactivity of radical ions is masked by strong interactions with the host matrix. Strong medium effects dilute the value of matrix isolation studies for predicting the behavior of radical ions in organic materials. While static methods have been utilized to study "stabilized" radical ions and some radicals in fluid systems, for the most part, very reactive radicals and radical ions in fluids can only be studied using time-resolved magnetic resonance or special variants like optically detected magnetic resonance.

How do we realize magnetic resonance observations in real time? First, one must have a reliable method for generation of transient paramagnetic species. Electron accelerators and pulsed lasers can provide sufficient pulse-to-pulse stability and repetition rates. There are advantages and disadvantages of radiolytic and photolytic generation of paramagnetic intermediates. In radiolysis the bulk solvent must absorb energy, and energy transfer and charge transfer from the solvent to the solute allow generation of many species (solvent ions, solute ions, excited states, neutral radicals). In photolysis one can directly excite the desired solute, so that the solvent is not so intimately involved in the energy deposition/transfer process. The solvent does play a very significant role, as it determines the nature of the charge dynamics, mobility, solvation, etc., and thus, in either case represents a very significant factor in the study of radical ions or radicals in condensed phases.

The time-resolved detection of transient paramagnetic species has been realized in at least two ways, so far. Direct methods for performing the time-resolved EPR experiment use either cw [17] or pulsed microwaves [18] and determine the change in the reflected microwave power from the cavity or observe a transient response of magnetization such as Free Induction Decay (FID) or an echo. The need for time-resolved observation requires broad-band amplification with the resultant degradation in the spectrometer sensitivity using a conventional 100 kHz modulated EPR spectrometer as a benchmark. Thus, any means of enhancing the signal intensity from transient species by observing Chemically Induced Dynamic Electron Polarization (CIDEP) [19] is welcome even though the complexity of analysis is increased.

Many examples illustrate that transient radicals in liquids are easily studied by time-resolved EPR [17,18,20]. Pulsed methods allow easy and convenient kinetic and relaxation studies of radicals [21] However, while time resolution in the range of 10^{-8} s is achievable, sensitivity remains a major shortcoming in many time-resolved EPR studies.

The very reactive nature of radical ions, especially cations, means that some live for only nanoseconds or less and require methods of study which can be considered to be "indirect" magnetic resonance experiments. In such experiments, magnetic fields and microwaves are utilized to affect the spin dynamics of paramagnetic species, and observation is made of some perturbable property of the system, like light emission. One advantage of such experiments is considerable improvement in sensitivity since visible photons are detected rather than microwave photons. Also, the process giving rise to the light emission is usually highly specific; thus, the indirect methods are a more selective means of detection than the direct EPR techniques.

What kind of mechanism will allow such sensitive and selective means of detecting radical ions? We will discuss fluorescence detected magnetic resonance of the time-resolved variety, but our discussion of the concepts involved would be applicable to other "indirect" magnetic resonance methods.

Fluorescence Detected Magnetic Resonance (FDMR) uses fluorescence emission from excited states of aromatic scintillators to detect magnetic resonance [13-15]. The process which produces fluorescence (excited states) must involve paramagnetic species and be perturbable by microwaves at the magnetic field of resonance of such paramagnetic species. This appears to be a rather tall order, but it turns out that in many instances nature is obliging, and there are many processes which involve reactions of radical ions which yield excited molecules. For example, the reaction of an aromatic radical cation $A^{\ddot{+}}$ and radical anion $A^{\ddot{-}}$ will give the excited molecule A* and the ground state molecule (5).

$$A^{\ddot{+}} + A^{\ddot{-}} \longrightarrow A^* + A \tag{5}$$

This is the result of the fact that charge neutralization produces enough excess energy to excite aromatic and aliphatic molecules. If A is an aromatic like anthracene, we could observe the photons emitted from the excited anthracene singlet (fluorescence) or triplet state (phosphorescence), S_1 or T_1 (6).

$$A^* \longrightarrow A + h\nu \tag{6}$$

Charge neutralization processes are common where radical ions are generated and can react with one another, e.g., radiolysis, photoionization, electrochemical generation of ion pairs, etc. However, generation of radical ion pairs which occur by random encounters of radical ions is not sufficient for the optical detection of magnetic resonance. For FDMR and related experiments, we must have pairwise generation of radical ions with a well defined spin phasing (singlet or triplet) at the instant of creation. Such radical ion pair systems do not have all their spin levels populated equally, so that transitions induced by microwave perturbation at resonant magnetic field produce a change in the singlet/triplet population of the radical ion pair ensemble. In other words, the microwave/magnetic field perturbation changes the relative number of singlet versus triplet radical ion pairs. This is reflected in an increase or decrease of emission from the excited state resulting from the radical ion pair annihilation (5). If one is monitoring fluorescence, then the change in signal intensity reflects the change in the yield of the excited singlet state, $^1A*$:

$$\text{excess singlet pairs at t=0} \xrightarrow[\text{magnetic field}]{\text{microwaves}} \text{decrease of } {}^1A*$$

$$\text{excess triplet pairs at t = 0} \xrightarrow[\text{magnetic field}]{\text{microwaves}} \text{increase of } {}^1A*$$

Thus, by noting whether the intensity of fluorescence increases or decreases at resonance, it can be determined whether the radical ion pairs were created initially with a triplet or singlet spin pairing.

Another important point is that charge/spin transfer occurs very quickly without changing the spin multiplicity. That means that once a radical ion pair is created, its descendents, which can occur by charge transfer, retain the spin multiplicity of the original pair and, as we shall see, the memory of microwave/magnetic field perturbation as long as there are no other factors affecting the spin coherence. The radical ion electron spin relaxation occurs on the microsecond timescale, and most radical ion reactions occur on a submicrosecond timescale. Except in some very special circumstances where Heisenberg spin exchange can occur [14], we will not have to worry about electron spin relaxation in radical ion pairs in a time frame of tens to hundreds of nanoseconds.

1.2.1. *FDMR in Pulse Radiolysis.* Ionizing and photoionizing radiation invariably produce spin-correlated radical ion pairs. In the pulse radiolysis of nonpolar solutions like alkanes the first detectable transients are singlet-phased radical ion pairs composed of solvent holes and electrons (2). At room temperature in a liquid like cyclohexane recombination occurs in a few picoseconds (7) [22]. Fluorescence from RH^*

$$RH^+\cdot + e^- \longrightarrow RH^* \tag{7}$$

is weak but measurable [23]; however, the lifetime of such pairs is so short that no magnetic field or microwave/magnetic field effect can be observed.

In the presence of small amounts of aromatic scintillators, charge scavenging reactions compete with ion recombination. Since the electrons are much more mobile than solvent holes (radical cations), electrons are scavenged first (8) to give radical ion pairs consisting of $RH^+\cdot$ and $A^-\cdot$, retaining the spin multiplicity of the original pair.

$$e^- + A \longrightarrow A^-\cdot \tag{8}$$

Radical cations are scavenged somewhat less efficiently (9) to give the $A^+\cdot + A^-\cdot$ descendent pair,

$$RH^+\cdot + A \longrightarrow RH + A^+\cdot \tag{9}$$

which also has the identical spin-pairing as existed in the original solvent hole/electron ion pair.

The conversion of the electron into the less mobile radical anion A^- greatly extends the time frame of geminate recombination for that (small) fraction of ion pairs scavenged. It is now possible, on the time scale of tens to hundreds of nanoseconds, to affect the spin evolution of geminate ion pairs by applying magnetic fields alone, or by microwave/magnetic field perturbation - FDMR. Furthermore, extending the radical ion pair lifetime allows more time for competing processes to occur. For example, one can observe transformations of radical cations like the loss of H_2 (10).

$$RH^{+} \longrightarrow Ol^{+} + H_2 \tag{10}$$

The olefin radical cation (Ol^{+}) would still be a partner of a spin-correlated radical ion pair, yielding the excited state of the scintillator upon recombination (11).

$$Ol^{+} + A^- \longrightarrow Ol + A^* \tag{11}$$

The spin multiplicity of A^* produced in reactions such as (5) and (11) depends on the relative spin orientation of the radical ion pair at the moment of recombination. Recombination of singlet-phased pairs gives $^1A^*$ (fluorescence); recombination of triplet-phased pairs gives $^3A^*$ (no fluorescence). In pulse radiolysis the FDMR signal is detected as a decrease in the fluorescence intensity since the application of a microwave pulse at resonant magnetic field reduces the number of singlet recombining pairs. Microwave-induced EPR transitions between the doublet levels of the separated ions accelerates the mixing between the initially populated singlet pair states and triplet pair states. Thus, the FDMR spectrum, fluorescence intensity as a function of magnetic field, contains the superimposed EPR spectra of radical ions which were present during the microwave pulse (20 to 150 ns) and which recombined (or whose "descendent" recombined) with their geminate partner to give A^*.

1.2.2. *FDMR in Laser Photoionization.* FDMR experiments using laser photoionization differ from FDMR in pulse radiolysis in that the solvent is not directly involved in the process of energy deposition. Instead, aromatic solutes are the absorbing species. Absorption of one, two, or more UV photons leads to photoionization to give the solute radical cation plus an electron (12-16).

$$A \xrightarrow{2h\nu} A^{+} + e^- \tag{12}$$

$$A \xrightarrow{h\nu} {}^1A^* \tag{13}$$

$$^1A^* \xrightarrow{h\nu} A^{+} + e^- \tag{14}$$

$$^1A^* \xrightarrow{ISC} {}^3A^* \tag{15}$$

$$^3A^* \xrightarrow{h\nu} A^{+} + e^- \tag{16}$$

Charge transfer reactions analogous to those in pulse radiolysis are important, and, in polar systems, solvation (17) is important. Fluorescence resulting from recombination of geminate radical ion pairs is again the observable quantity in the experiment.

$$e^- + (S)_n \longrightarrow e_s^- \qquad (S = solvent) \qquad (17)$$

In pulse radiolysis, in nonpolar systems, most of the fluorescence comes from radical ion recombination. In photoionization, the fluorescence from ionic processes is a small part of the fluorescence which comes primarily from the direct photoexcitation of A into the $^1A*$ state. Due to the multiphotonic nature of the process resulting in delayed fluorescence from radical ion recombination, measures are taken to minimize fluctuations in the ion yield resulting from shot-to-shot fluctuations in the laser intensity. Low solute concentrations and high laser powers are used so that the number of photons is far in excess of that needed to saturate the scintillator $S_1 \leftarrow S_0$ transition. Under these conditions the ion yield will vary linearly with the laser power, rather than as the square as expected for a consecutive two-photon process. The use of solutes with short emission lifetimes allows better separation of the prompt and delayed fluorescence signals.

Another, very significant difference between photoionization and radiolysis is the charge-pair separation distance distribution. The major difference is that in radiolysis one finds a significant number of geminate ion pairs with large separation distances. In photoioniza-tion, the use of monoenergetic radiation produces a narrower radial distribution of cation-electron separation distances [24]. Furthermore, in photoionization, one cannot deposit as much energy in the ejected electron, so that one does not see geminate ion-pairs with substantial separation distances.

As we will discuss in some detail later, we cannot observe FDMR from photoionization in nonpolar solutions where the bulk of the radiolysis FDMR work is carried out. The FDMR studies in photoionization are possible only when the electron is quickly scavenged or solvated as is the case with polar solvents like alcohols [15]. An additional limitation of FDMR/photoionization studies is the consequence of the fact that only dilute (typically 5×10^{-5} to 10^{-4} M) solutions of the absorbing solute/scintillator can be used for reasons of laser pulse fluctuations mentioned above. This, in turn, does not allow prompt electron scavenging, which becomes more critical when cation-electron separation distances are smaller. Additionally, one must judiciously select solutes which absorb at the wavelength of the exciting light and scavengers which do not. This becomes more difficult as one uses light of shorter and shorter wavelength.

In spite of these complications, the ability to study radical ions in photoionization offers many possibilities for the study of chemistry of highly excited species with the advantages of the information content inherent in magnetic resonance spectroscopy. Carried out in conjunction with pulse radiolysis studies, the photoionization studies allow us to compare the processes induced by UV photons versus high energy electrons. By manipulating other parameters in the FDMR experiments (e.g., solvent viscosity and polarity, solute concen-trations), we can enhance production of various ions and obtain resonant signatures of the radical ion pairs by applying the microwave/magnetic field perturbation at the appropriate time when a particular ion pair is present. Thus, as we will discuss in considerable detail in the various parts of this chapter, we can obtain EPR spectra of many radical ions and learn a great deal about radical ion structure and reactivity by following the changes in spectra or kinetics in the time frame before the radical ions disappear by reaction with their geminate partners. Such optically detected magnetic resonance experiments are the most sensitive means yet conceived for making magnetic resonance measurements of transients in the condensed phase.

1.2.3. *Experimental.* Time-resolved FDMR experiments are carried out with the pulsed X-band EPR spectrometer originally developed in our laboratory for time-resolved EPR studies [18a]. Provision for the detection of fluorescence emitted from the sample which is irradiated in situ in the EPR cavity is essentially all that was required to adapt the spectrometer for FDMR experiments. The components of the pulsed microwave bridge used for FDMR are shown schematically in Figure 1. They consist of a source of variable frequency

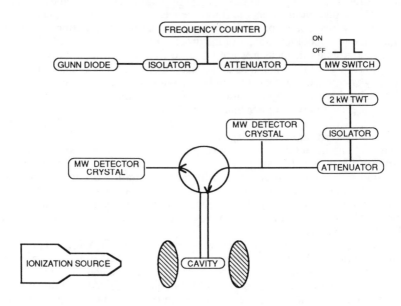

Figure 1. Schematic diagram of the pulsed spectrometer used for time-resolved FDMR experiments.

microwave pulses coupled to the resonant microwave cavity. A fast switch (on-off-on time ~10 ns) gates the microwaves from the Gunn diode source. The variable-width pulses are amplified by a pulsed traveling wave tube amplifier capable of generating microwave fields H_1 at the sample as high as tens of Gauss. The usual phase-sensitive microwave-detection components have been omitted from Figure 1 as they are not relevant for FDMR experiments. The ionization source is a 3 MeV pulsed Van de Graaff electron accelerator (minimum pulse duration = 5 ns) or a pulsed excimer laser (pulse duration 10-20 ns).

Fluorescence collection is accomplished either by means of lenses as shown in Figure 2 or, alternatively, via a suprasil quartz rod light guide. In the former case a hole provided in the side wall of the cavity allows light transmission. The hole is masked by a short length of rectangular waveguide to minimize energy losses from the cavity. Use of a 3 mm cylindrical light guide inserted axially into the sample (e.g., from the top of the cavity) is equally efficacious as a means of light collection and reserves the horizontal axis, for example, for the entrance of a laser beam. A colored glass filter or a monochromator may be placed before the photomultiplier in order to discriminate between two emitting species and/or block scattered laser light.

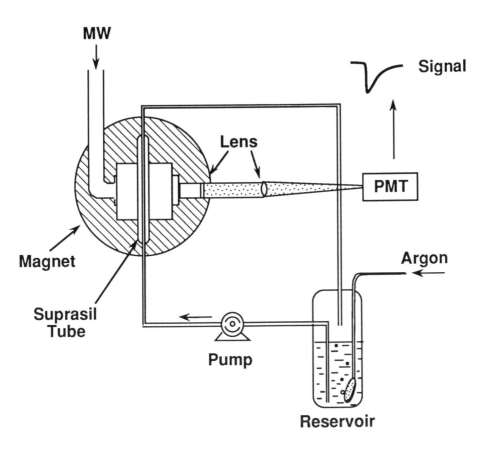

Figure 2. Detail drawing of the EPR cavity illustrating sample handling and light collection.

In liquid phase FDMR experiments the sample is continuously circulated through the cavity (see Figure 2), a measure necessitated by the heating effects and sample degradation caused by repetitive pulsing by the ionization source. In solid phase experiments these same effects can be partially obviated by vertical translation of the solid sample. The stepper motor-driven translation mount shown in Figure 3 has been used in conjunction with a liquid helium transfer cryostat to carry out FDMR experiments in solids at temperatures down to 10K. A detailed description of the sample translation device has been given elsewhere [25].

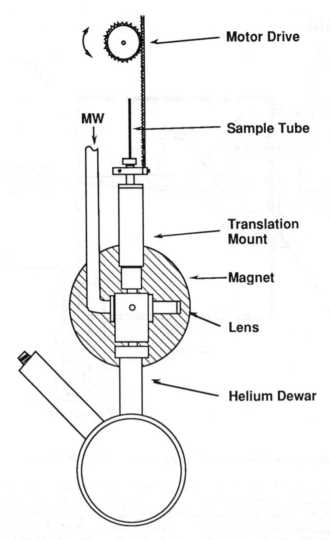

Figure 3. Detail drawing of the solid sample handling assembly which allows temperature control and sample translation.

The microwave pulse (20 ns or longer) is applied at some time after the arrival of the ionization pulse, and the fluorescence signal is integrated by a boxcar detector (gate width typically 100 ns) with the boxcar gate applied immediately following the microwave pulse. This timing scheme is depicted in Figure 4.

Historically, FDMR was first used in the study of geminate radical ions in pulse radiolysis of nonpolar media in recognition of the sensitivity advantage in solvents such as hydrocarbons where more than 95% of the radical ion pairs undergo geminate recombination. Initial efforts characterized the signals from the aromatic radical ions arising from the scintillator used. Section 4 describes some results of these studies including the tendency of some aromatic radical cations to react with the neutral parent molecules to form dimer-multimer radical cations and leading to excimer formation.

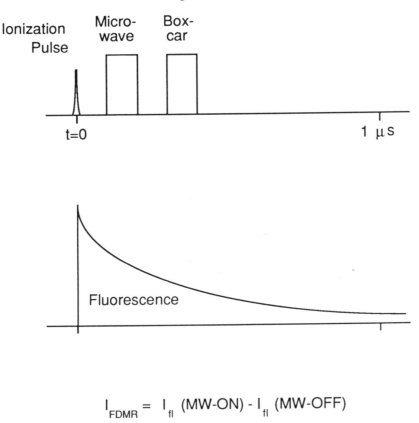

$$I_{FDMR} = I_{fl}(MW\text{-}ON) - I_{fl}(MW\text{-}OFF)$$

Figure 4. Experimental timing diagram for time-resolved FDMR.

The main emphasis of subsequent work has been to make observations of the solvent holes themselves, to better understand the primary events involved in condensed phase radiolysis. Solvent-derived radical cation signals were observed in electron-irradiated neat alkane liquids, but they proved not to be the solvent radical cations. Instead, these signals could be assigned to parent-minus-H_2 olefin radical cations. These findings are discussed in section 3 along with the conclusions of concentration and temperature dependence studies of olefin radical cation signals in alkane mixtures demonstrating the origin of the olefin radical cations to be dissociation of the parent radical cations.

Progress in observing saturated hydrocarbon radical cations in electron-irradiated alkanes has come from two approaches: (1) formation of $RH^{+\cdot}$ by charge transfer to a guest molecule (with lower ionization potential) present at low concentration in a host alkane, and (2) studies in very low temperature alkane solids. These approaches have shed much light on the fate of alkane radical cations in a hydrocarbon environment and are described in section 2.

Recent forays into alternative systems, i.e., more polar solvents, have proved that FDMR is not limited to nonpolar media. In section 5, FDMR observations of the radiolysis of alcohols and ethers is discussed along with parallel experiments using photoionization. Section 6 describes one example illustrating how structural information about radical cations obtained from FDMR studies can circumvent complexities which may arise in matrix EPR studies due to strong cation-matrix interactions, making the FDMR method a very useful means for the study of radical cation structure and reactivity.

A recurring theme in the FDMR investigation of radical cations is the importance of ion-molecule reactions which provide reaction pathways for radical cations besides neutralization. The relative longevity of radical cations may be principally determined by their propensity to undergo ion-molecule reactions with surrounding solvent molecules. A surprising degree of diversity exists in the reactivity of different alkane radical cations as judged from their ease of observation by FDMR. The importance of ion chemistry to the overall radiolysis mechanism of hydrocarbons is discussed in section 7, along with recent results of picosecond optical studies of alkane solvent cation transformations.

2. Alkane Radical Cations

Alkane radical cations result from the removal of one electron from a σ molecular orbital. The spatial extent of the singly-occupied molecular orbital (SOMO) varies with the structure of the molecule. For example, the SOMO may be largely localized at a single C-C bond as in radical cations of highly branched alkanes [26], but it can also extend over the entire molecular frame (e.g., cyclohexane$^+$ [27] and n-alkane radical cations [28]). This information is directly obtained from the hyperfine couplings observed in the EPR spectrum of the radical cation which are a measure of the unpaired electron spin density at different nuclei (protons) in the molecule. The EPR spectrum is thus a highly specific structural signature of an individual radical cation, far superior for identification purposes than, for example, optical absorption spectra in liquids or solids which consist of broad unstructured bands appearing in the UV, visible or near IR, depending again on the spatial extent of the SOMO [29].

Since its advent, EPR spectroscopy has been used to study many radical ions. Initially, relatively stable radical ions (those with conjugated π-systems, aromatics), sufficiently large concentrations of which could be generated in the liquid phase by redox reactions, electrolysis or photolysis, were studied [30]. The investigation of the radical cations of alkanes and other smaller organic molecules lagged due, in part, to their shorter lifetimes and higher ionization potentials. This situation changed a decade ago with the development of improved matrix isolation techniques. Shida and coworkers first showed that radical cations of alkanes could be stabilized in frozen solutions of halogenated solvents containing the alkanes [31]. In this method an ionizing source (^{60}Co γ-rays, x-rays) is used to produce the solvent positive ion and an electron. The ejected electrons are prevented from rapidly returning to the positive ion by dissociative electron capture reactions with matrix molecules, and the positive charge migrates through the matrix via electron transfer until it becomes trapped on a solute molecule. The ionization potential of the halogenated solvents used (freons, SF_6, etc.) is higher than that of most organic molecules (I.P. $\cong 10\,eV$ for alkanes in the gas phase). In a related experiment, Knight and coworkers prepared matrix samples of cations by gas phase ionization of small molecules diluted in neon followed by deposition at 4 K onto a surface [3]. The method of Shida has found wider application than that of Knight in investigations of alkane radical cations largely because of its greater simplicity and the greater temperature range possible with halocarbon matrices. Recent studies in this group have shown that radical cations in xenon matrices with electron scavengers can be observed by static EPR [32] and that observations of a variety of radical cations generated in zeolite Na-Y are possible even at room temperature in some instances [33]. Room temperature studies of several aromatic radical cations have been successfully carried out in synthetic superacid membranes [34].

The last ten years have seen much progress in the electronic and structural characterization, as well as the enumeration of many principal reactions, of alkane radical cations owing to matrix EPR studies. A long-time goal has been to gain a detailed understanding of the role of alkane radical cations in hydrocarbon radiolysis as a prototype for radiation-induced chemistry in organic materials. Unfortunately, the extrapolation from matrix EPR studies is not trivial. Experiments carried out under static conditions can mask the details of early events and gloss over subtle, but crucial, differences in reactivity. In most real systems (i.e., organic materials) multiple reaction pathways are presented to the radical cation. The natural branching between these various channels is subverted in the artificial environment of the matrix. Interactions between the matrix and solute radical cations are imprinted on the observed chemistry leading, in many cases, to markedly different behavior in one matrix compared to another.

Measurement of difference spectra upon photobleaching has allowed Ichikawa and Ohta to observe several alkane radical cations in γ-irradiated alkane mixtures at 77K using static EPR spectroscopy [35]. The development of time-resolved FDMR made possible for the first time the direct EPR observation of radical cations in hydrocarbon systems on the nanosecond time scale. In section 2.1 we describe the observations of alkane radical cations in electron-irradiated, liquid and solid hydrocarbon solutions. The emphasis is on the experimental conditions, temperature and dilution, which allow the observation of alkane radical cations on the time scale of ten's to hundred's of nanoseconds. Illustrative examples are given in place of an exhaustive treatment of the compounds that have been studied. In section 2.2 we discuss the implications of these observations for the decay mechanisms of alkane radical cations, i.e., ion-molecule reactions.

2.1. OBSERVATIONS IN LIQUIDS AND SOLIDS

Alkane radical cations are not observed in FDMR experiments carried out in neat alkane liquids over the temperature range 140K to room temperature for reasons that shall be discussed below [36,37]. However, alkane radical cations are observed under dilute conditions [37,38]. Figure 5a shows the FDMR spectrum obtained at 185K in n-pentane containing 10^{-2} M cis-decalin and 10^{-4} M anthracene-d_{10}. The microwave pulse was applied from $t = 0$ to $t = 100$ ns and the fluorescence intensity was integrated during the boxcar detector window, $t = 100$ ns to $t = 200$ ns. The spectrum consists of the characteristic intense central peak due to the unresolved EPR lines of the scintillator radical ions superposed on the 5-line spectrum of the cis-decalin radical cation. This assignment is verified by comparison to the static EPR spectrum of cis-decalin^{+} obtained in a $CFCl_3$ matrix [39].

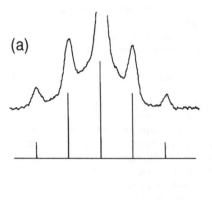

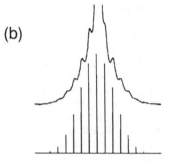

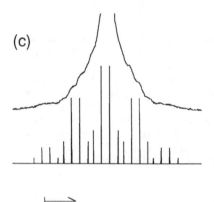

$\overset{\longrightarrow}{\underset{50\ G}{\vdash\quad\quad}}$

Figure 5: FDMR spectra observed at 185K in n-pentane containing 10^{-4} M anthracene-d_{10} and 10^{-2} M (a) cis-decalin, (b) 2,2,3,3-tetramethylbutane and (c) methylcyclohexane. The hyperfine parameters used for the simulated stick spectra are given in reference [38].

The cis-decalin radical cation was first observed in dilute solution by workers in Novosibirsk by static optically detected EPR [40]. The observation was repeated in studies using time-resolved FDMR which showed that cis-decalin^{+} could be observed in a variety of host alkane solvents over a wide temperature range (140K to room temperature) [37]. Consistent with earlier negative results in neat cis-decalin, the intensity of the cis-decalin^{+} FDMR signal was found to decrease and eventually disappear when the concentration of cis-decalin was gradually increased beyond 0.1 M.

A clear requirement for FDMR detection of a radical cation in a mixed system is that the solute molecule have a lower ionization potential than the solvent. In addition, the detection sensitivity is affected by the number of lines and degree of resolution in the radical cation EPR spectrum. An EPR signal of only a few widely spaced lines is easier to detect than one where the EPR intensity is shared between many lines which are poorly resolved. This is well illustrated by the three examples in Figure 5.

A number of other alkane radical cations have now been observed by time-resolved FDMR in dilute solutions, the most comprehensive treatment being found in reference [38]. While the concentration dependence has not been systematically studied for each case, the gradual disappearance of the radical cation FDMR signal at solute concentrations exceeding 0.1 M seems to be a general trend, as illustrated in Figure 6 for the case of the bicyclopentyl radical cation.

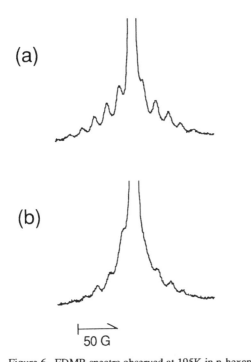

Figure 6. FDMR spectra observed at 195K in n-hexane containing 10^{-4} M anthracene-d_{10} and (a) 0.1 M or (b) 0.5 M bicyclopentyl. The spectra in (a) and (b) have been normalized with respect to the central peak height.

50 G

The FDMR response in different solvent mixtures can vary dramatically. Some alkane radical cations are detected with greater facility than others even after differences, such as the number of EPR lines or ionization potential, are taken into account. As spin relaxation processes in the radical ion pairs are not significant factors in the loss of spin coherence in the submicrosecond time regime, chemical reactions and/or transformations of the radical cations (which can compete with geminate recombination) must account for the observed differences in FDMR results. Therefore, a systematic comparison of the FDMR response of closely related systems like alkanes can shed light on the relative reactivity of the radical cations under study.

Two cations which would be expected (based on known gas phase ionization potentials) to be observable in dilute n-pentane solutions, for example, but are not observed, are trans-decalin^{+} and cyclohexane^{+}. Addition of dilute quantities (10^{-2} M) of trans-decalin or cyclohexane to n-pentane gave rise to no new spectral features attributable to the respective solute radical cations, but instead caused a quenching of the overall fluorescence intensity and a decrease in the intensity of the central FDMR peak, suggesting the occurrence of other reactions [38].

FDMR detection of alkane radical cations becomes possible in the neat alkane (i.e., single alkanes containing scintillators - e.g., 10^{-3} M anthracene-d_{10}) at sufficiently low temperatures (150K or lower) in the solid. This was first demonstrated for decalins [40,41] and subsequently extended to a large number of structurally diverse alkanes [38]. Three examples are shown in Figure 7. The most remarkable feature of the solid phase studies, besides the ability to detect alkane radical cations even in neat alkane samples, is the variation in the observed FDMR intensity in different alkanes.

In a survey of more than 30 different alkanes the observed FDMR intensity (height of the central FDMR peak) in low temperature (35K) solids varied by at least two orders of magnitude depending on the alkane [38]. The variation of the FDMR response from alkane solids is consistent with liquid phase FDMR experiments. That is, alkane radical cations which give weak (or negligible) FDMR signals in dilute liquid solutions also give weak signals in frozen solids.

210

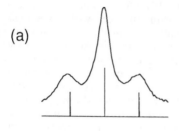

(a)

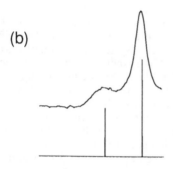

(b)

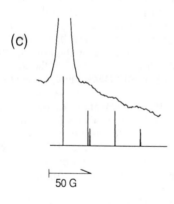

(c)

⊢————⟶
50 G

Figure 7. FDMR spectra observed at 35K in (a) n-hexane, (b) 2-methylhexane and (c) cis-1,2-dimethylcyclohexane. The anthracene-d_{10} scintillator concentration was 10^{-3} M. Only the low field and high field portions of the spectra are displayed in (b) and (c), respectively. The hyperfine parameters used for the simulated stick spectra are given in reference [38].

As an example, consider cis- and trans-decalin. In liquid phase experiments the cis-decalin radical cation is observed (under dilute conditions) and the trans-decalin radical cation is not. At 35K the FDMR intensity observed in neat cis-decalin is an order of magnitude greater than that in trans-decalin. The intensity of the trans-decalin$^+$ spectrum is somewhat enhanced (in the solid) by diluting trans-decalin with a second alkane. The contrasting behavior is particularly striking for these two compounds because of their seemingly great similarity in other respects.

2.2. ION-MOLECULE REACTIONS

Unification of the sundry observations by FDMR of alkane radical cations in electron-irradiated hydrocarbons is possible in a framework of ion-molecule chemistry. The conditions under which alkane radical cations can be observed in the FDMR detection window (20 - 100ns) include isolation of RH$^+$ from RH (low concentrations) and/or low temperatures. It is clear that bimolecular reactions, in competition with geminate recombination, provide an alternate decay pathway for alkane radical cations. Possible reactions between RH$^+$ and RH are resonant charge transfer (18) and ion-molecule reactions yielding neutral radicals (19 or 20).

$$RH^+\cdot + RH \longrightarrow RH + RH^+\cdot \quad (18)$$

$$RH^+\cdot + RH \longrightarrow R\cdot + RH_2^+ \quad (19)$$

$$RH^+\cdot + RH \longrightarrow RH_2^+ + R\cdot \quad (20)$$

Fast resonant charge transfer (18) has been advanced to explain observations indicating fast positive charge transport in several hydrocarbons, with cyclohexane and trans-decalin being the most notable examples [42]. Fast charge transfer would tend to delocalize the electron spin and, in the limit of fast electron hopping, narrow the radical cation EPR spectrum into a single, narrow line. This mechanism, however, is ruled out by a variety of experimental observations and is not consistent with the FDMR results. Here we just mention the relevant FDMR results, most significant of which is the failure to observe the FDMR spectrum of certain alkane radical cations (e.g., trans-decalin⁺) in a solution of n-pentane where the solute is the species with the lowest ionization potential. If only resonant charge transfer prevented the observation of radical cations, then the radical cation of any alkane diluted in a solvent which has a higher ionization potential should be observed. Furthermore, there is no evidence for the onset of spectral narrowing (e.g., loss of resolution or broadening of individual lines) in liquid phase FDMR spectra with increasing solute concentration. On the contrary, as in the case of bicyclopentyl⁺ shown in Figure 6, the solute radical cation signal only decreases in intensity. The persistence of resolved hyperfine structure is indication that the radical cation remains localized. Finally, fast charge transfer cannot explain the temperature dependence of the FDMR results in neat alkanes. To explain the results in a solvent such as cis-decalin, one would have to argue that resonant charge transfer slows down with decreasing temperature, which is very improbable [38].

There is considerable experimental evidence that the conversion of alkane radical cations into neutral alkyl radicals occurs via reactions such as (19), proton transfer, and/or (20), H atom abstraction. Perhaps the most relevant work illustrating the importance of ion-molecule reactions of alkane radical cations in the condensed phase is the low temperature studies of Iwasaki and Toriyama who have investigated the static EPR signals induced by ionizing radiation in solutions of alkanes in halogenated matrices and zeolites and in neat n-alkane crystals [28,43-46]. These studies establish that most of the alkyl radicals are formed from alkane radical cations and that the process involves a bimolecular reaction between the cations and alkane molecules. Optical pulse radiolysis studies of alkanes have observed alkyl radicals on the picosecond time scale indicating that radicals are produced by fast processes [7]. In another study using time-resolved EPR it was shown that the cyclohexyl radical yield is not affected by cation or electron scavengers, also consistent with the idea that conversion of radical cations into radicals is fast [47].

The failure to observe alkane radical cations in neat alkane liquids by FDMR is, therefore, explained by the occurrence of fast ion-molecule reactions between the radical cations and neutral solvent molecules (19 and/or 20). Dilution or lower temperatures tend to decrease the rate of such reactions. The majority of solute radical cations react with their neutral parent molecules but not with solvent molecules. However, there are some exceptions (e.g., trans-decalin⁺, cyclohexane⁺) where solute radical cations cannot be observed in liquids. This observation plus the great variation of the FDMR response at very low temperature in different alkane solids strongly suggests that the propensity for undergoing ion-molecule reactions varies greatly for different alkane radical cations. The kinetic and thermodynamic factors controlling alkane radical cation reactivity are not yet well understood.

3. Alkene Radical Cations

3.1. Observation and Origin

Olefin radical cations are also important intermediates in alkane radiolysis [5]. Olefin radical cations are observed on the picosecond time scale in optical pulse radiolysis studies of alkanes [6], and signals due to "parent minus H_2" olefin radical cations are observed by FDMR experiments in neat alkane liquids and alkane mixtures [36,37]. Two examples are shown in Figure 8. The sharp multiplets observed in the FDMR spectra obtained in neat 2,3-dimethylbutane (Figure 8a) and cyclopentane (Figure 8b) belong to the EPR spectra of tetramethylethylene$^+$ and cyclopentene$^+$, respectively. Olefin radical cation formation in neat alkane liquids has been observed over the entire liquid range up to room temperature [37].

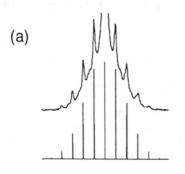

(a)

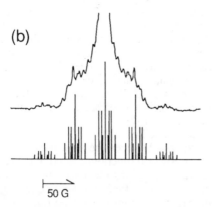

(b)

50 G

Figure 8. FDMR spectra observed at 180K in (a) 2,3-dimethylbutane; and (b) cyclopentane. The anthracene-d$_{10}$ scintillator concentration was 10^{-4} M. The hyperfine parameters used for the simulated stick spectra are given in reference [37].

What is the origin of olefin radical cations observed in electron-irradiated alkanes? Trivial mechanisms involving hole transfer to unsaturated species present in the alkanes as impurities or as the result of the accumulation of radiolysis products were carefully ruled out [36]. Therefore, a mechanism is required that explains the prompt (occurring during the geminate radical cation lifetime) formation of olefin radical cations. The simplest mechanism involving direct excitation by the electron beam and prompt formation of olefin radical cations is dissociation of parent alkane radical cations (21).

$$RH^+ \cdot \longrightarrow Ol^+ \cdot + H_2 \qquad (21)$$

This mechanism can explain the temperature and concentration dependence of FDMR observations in (liquid) alkane mixtures, e.g., cis-decalin/n-pentane, 2,3-dimethylbutane/n-pentane, cyclohexane/n-pentane. For example, it was found that at 180K olefin radical cations derived from the solute alkane only begin to be observed at solute alkane concentrations >10%. In contrast, solute-derived olefin radical cations are observed at solute alkane concentrations one to two orders of magnitude lower than this at room temperature. This is opposite the trend of the FDMR detection sensitivity with temperature and can only mean that the conversion of the solute alkane radical cations into olefin radical cations is temperature dependent.

The temperature dependence of olefin radical cation formation can be understood if we assume that fragmentation of alkane radical cations (21) depends on the energy content of the cation. Alkane radical cations formed via direct excitation by the electron beam pulse may initially possess significant amounts of excess (electronic, vibrational) energy, sufficient to cause dissociation. Dissociation of "hot" cations will not depend on the sample temperature. However, in alkane mixtures solute radical cations will be formed via hole transfer from solvent radical cations, in which case they will possess, at most, small amounts of excess energy commensurate with the difference in ionization potential (0.1 - 1.0 eV). Therefore, the observation of solute alkane-derived olefin radical cations at lower solute concentrations at room temperature than at low temperature is explained by the tendency of alkane radical cations formed by hole transfer to dissociate at room temperature but not at low temperature. That is, the fate of these cations depends on their thermal energy. At solute concentrations greater than 10% the probability is greater for direct ionization of solute alkane molecules, and these can be expected to dissociate at any temperature.

A second possible "prompt" mechanism for the formation of olefin radical cations is charge transfer from alkane radical cations to olefins produced in close proximity (in the radiolysis "spur") to the charge pair (22).

$$RH^* \longrightarrow Ol + H_2 \tag{22a}$$

$$RH^+ \cdot + Ol \longrightarrow RH + Ol^+ \cdot \tag{22b}$$

This mechanism, however, offers a less satisfactory explanation of the FDMR data because it implies that dissociation of alkane excited states, produced by either direct excitation (23) or charge neutralization (24), is temperature dependent.

$$RH + e^- \text{ beam} \longrightarrow RH^* \tag{23}$$

$$RH + e^- \text{ beam} \longrightarrow RH^+ \cdot + e^- \longrightarrow RH^* \tag{24}$$

It is certain that olefin radical cation formation occurs in the primary stages of alkane radiolysis, most likely as a result of the transformation of the initial alkane radical cations. Unfortunately, relative yield estimates cannot be obtained from the FDMR experiment, and therefore the relative importance of this reaction pathway to the overall radiolysis mechanism still needs to be determined.

3.2. ION-MOLECULE REACTIONS

FDMR [37] and optical studies [4-6] indicate that olefin radical cations formed in liquid alkanes are longer lived than saturated radical cations. Olefin radical cations are apparently quite inert with respect to reaction with alkane solvent molecules, and their predominant fate will be recombination.

Olefin radical cations will undergo well-established ion-molecule reactions with neutral olefin molecules to form aggregate radical cations (25 and 26) [5,48-52].

$$Ol^+ \cdot + Ol \longrightarrow Ol_2^+ \cdot \tag{25}$$

$$Ol_n^+ \cdot + Ol \longrightarrow Ol_{n+1}^+ \cdot \qquad (n=2,3,4,...) \tag{26}$$

At 10^{-2} M olefin concentration the conversion of monomer radical cation into dimer radical cation is observed on the time scale of hundreds of nanoseconds by FDMR [48]. This was illustrated, for example, for the case of tetramethlyethylene (TME) whose monomer and dimer radical cation EPR spectra can be easily resolved. Shown in Figure 9a is the FDMR spectrum observed at 185K in methylcyclohexane containing 10^{-3} M TME and 10^{-4} M anthracene-d_{10}. The microwave pulse delay (relative to the electron beam pulse) was zero. Under these conditions the monomer radical cation TME$^+ \cdot$ ($a_{12H} = 17.1$ G) is clearly observed. When the concentration is increased to 10^{-2} M TME (Figure 9b, microwave pulse delay = 0 ns) a new FDMR signal appears with a coupling constant of 8.2 G belonging to the dimer radical cation $(TME)_2^{\ddagger}$. Utilizing the time-resolved capability of FDMR, we can see how this system evolves in time. In Figure 9c the spectrum obtained in the same sample as Figure 9b is shown to be comprised of predominantly the dimer cation signal at a microwave delay of 800 ns.

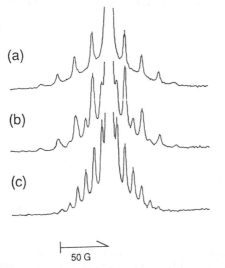

(a)

(b)

(c)

50 G

Figure 9. FDMR spectra observed at 185K in methylcyclohexane containing 10^{-4} M anthracene-d_{10} and tetramethylethylene (TME): (a) 10^{-3} M TME, microwave delay = 0 ns; (b) 10^{-2} M TME, microwave delay = 0 ns; (c) 10^{-2} M TME, microwave delay = 800 ns

Reduction of the coupling constant in the dimer radical cation compared to the monomer radical cation is the result of greater delocalization of the unpaired electron as it is now shared between the two molecules. The trimer radical cation $(TME)_2^{\ddagger}$ can be identified at TME concentrations of 0.1 M, and even higher order aggregate cations can form at higher concentrations (0.1 to 0.5 M) as evidenced by the gradual total collapse of the EPR spectrum toward the center. A very unexpected result was obtained at the extreme concentration of 1.0 M TME where again a resolved FDMR spectrum (coupling constant = 14.8 G) is observed [48]. A possible interpretation of this signal is to assign it to a higher order aggregate cation, $n \geq 4$, which is asymmetrical, with the unpaired electron shared less equally between the constituent molecules, remaining more or less localized on a single TME molecule.

EPR studies of TME radical cations in $CF_2ClCFCl_2$ reveal that dimer radical cations undergo unimolecular dissociation to form the 1,1,2-trimethylallyl radical, casting $(TME)_2^{\ddagger}$ in the role of an intermediate in ion-molecule reactions similar to those implicated in studies of alkane radical cations [53]. From the FDMR results, however, it must be concluded that if such ion-molecule reactions involving olefin radical cations occur in the hydrocarbon systems, then they are much slower than the analogous reactions of alkane radical cations.

4. Aromatic Radical Cations

4.1. OBSERVATIONS

The earliest observations of FDMR signals were of aromatic radical ions in alkane solutions [11-14, 54], and, as the previous sections have illustrated, subsequent observations of alkane solvent and alkene radical cations depended upon reactions of these species with aromatic radical ions to generate sufficient numbers of emitted photons and achieve acceptable signal-to-noise levels. The overlapping of the EPR lines of the aromatic radical ions with those of solvent radical cations is usually only a minor problem since the aromatic radical ion hyperfine couplings are quite small and the EPR spectra consequently narrow. While this is an advantage for FDMR in general, the small magnitude of the coupling constants of aromatic radical ions makes them difficult to measure by time-resolved FDMR whose frequency resolution is limited due to the pulsed nature of the excitation and high microwave powers typically used. Resolution of the hyperfine structure of aromatic radical cations has been achieved in time-resolved FDMR experiments, but at the sacrifice of sensitivity and time-resolution, by using low microwave fields (<0.5 G) and very wide pulses (>500 ns) [12].

The EPR spectra of many aromatic radical cations have been characterized by (high resolution) cw EPR spectroscopy [30]. Owing to their relative stability, radical cations of aromatic compounds can be generated in fairly high concentrations, even in liquids, by electrochemical means or by photolysis. The relative stability of aromatic radical ions made them ideal probes for use in early diagnostic studies of FDMR. For example, the time dependence of the FDMR intensity measured in aromatic solutions corroborated the expectation that the FDMR effect is due to a resonant change in the yield of singlets produced by geminate recombination (and not homogeneous ion recombination) [12]. Other mechanistic details were also demonstrated, for example, concerning the spin dynamics of the ion pair spin states under the influence of the hyperfine induced S-T_0 mixing in the radical ion pairs [55].

In section 5 we will examine the use of aromatics to study photoionization processes in liquids using time-resolved FDMR. Aromatic compounds are well suited for photoionization studies because of their high extinction coefficients (e.g., in the near UV) and low ionization potentials. In the remainder of this section we discuss aromatic radical cation aggregate formation, the study of which has shed light on the role of dimer-multimer cations in the process of excimer formation.

4.2. ION-MOLECULE REACTIONS

Aromatic radical cations (A^{+}) in alkane solutions may undergo ion-molecule reactions analogous to reactions (25) and (26) with their neutral parent molecules [54].

$$A^{+}_{\cdot} + A \longrightarrow A^{+}_{2\cdot} \tag{27}$$

$$A^{+}_{n\cdot} + A \longrightarrow A^{+}_{n+1\cdot} \quad (n = 2,3,4...) \tag{28}$$

The dimer-multimer cations will also participate in ion recombination events and give rise to EPR features in the FDMR spectrum. The spectral narrowing caused by spin delocalization serves to distinguish the EPR spectra of dimer-multimer radical cations from that of the monomer radical cation.

A dramatic example of EPR-linewidth narrowing was observed for pyrene in cyclohexane solvent [54]. The FDMR linewidth (fwhm) for pyrene in cyclohexane at 20°C changed from 14 to 5 G over the pyrene concentration range 10^{-4} to 8×10^{-2} M. The linewidth measured at the highest pyrene concentration suggests aggregates with possibly as many as eight pyrene molecules. However, as the FDMR spectrum consists of the superposition of monomer and aggregate radical cation signals ($\Delta g \cong 0$), the observed linewidth reflects the relative contributions of all species (present at the time of application of the microwave pulse) to the overall FDMR signal. (The EPR spectra of the aromatic radical anions may or may not contribute to the FDMR linewidth depending on the width of the anion spectrum relative to that of the cation.)

Further evidence for the formation of aggregate aromatic radical cations was obtained from studies using the time and wavelength-resolved capabilities of FDMR. Kinetic studies revealed that the FDMR signal in pyrene/cyclohexane solutions persists for longer times at higher pyrene concentrations, consistent with the longer diffusion times required for the recombination of aggregate radical cations [54]. As expected, the kinetics of the FDMR signal for biphenyl, which does not form aggregate radical cations [56], were independent of the biphenyl concentration.

Different ion recombination kinetics are observed when the time dependence of the pyrene FDMR signal is measured while monitoring the excimer emission wavelength (485 nm) instead of the monomer emission wavelength (385 nm). The FDMR decay kinetics are slower for 485 nm detection, consistent with dimer-multimer cation participation. Excimer (A_2^* or A_n^*) formation can result from a variety of pathways (29-31)

$$A_2^{+} + A^{-} \longrightarrow A_2^{*} + A \qquad\qquad (29)$$

$$A_n^{+} + A^{-} \longrightarrow A_n^{*} + A \qquad n = (2,3,4...) \qquad (30)$$

$$A^{*} + A \longrightarrow A_2^{*} \qquad\qquad (31)$$

including the usual one (31) which occurs even in the absence of ion recombination. The FDMR kinetics results clearly show that in the electron-irradiated system recombination of aggregate radical cations is a significant source of excimer emission.

5. Polar Solutions

For some time it was debated whether radical cations existed as primary species in the radiolysis of polar fluids like alcohols. One approach was to try to detect excited states (with or without scintillators) formed by recombination of radical ion pairs. An early study of alcohols and ethers subjected to pulse radiolysis concluded that, with the possible exception of tetrahydrofuran, no excited states were formed by ion recombination [57]. FDMR provides a much more direct way of answering this question since an FDMR signal is only detected in the event of excited state formation due to geminate recombination reactions. In section 5.1 we describe FDMR results in several alcohols and ethers which unequivocally show that geminate radical ion recombination can be observed in polar liquids and that solvent radical cations can be scavenged by solutes.

In section 5.2 we further examine geminate ion recombination processes observed in polar liquids with a discussion of FDMR experiments carried out in alcohols using laser photoionization of aromatic solutes [15]. In this case the solvent itself is not ionized but plays an important role in the recombination dynamics.

5.1. RADIOLYSIS

There are various factors which might hinder the observation of an FDMR signal in radiolysis of polar solvents such as alcohols. In particular, as stated in section 1, fast ion-molecule reactions (eqn 4) can lead to the early demise of the solvent radical cation, preventing the formation of the scintillator radical cation and recombination of the solvent radical cation with the scintillator radical anion. A second factor is the effect of dielectric constant on the escape probability of geminate ion pairs. Because FDMR only detects events involving geminate ion recombination, its sensitivity is directly affected by an increase in the escape probability given by

$$P = \exp(-r_c/r_0) \tag{32}$$

where r_c is the Onsager radius (geminate ion separation distance at which the Coulomb energy equals the thermal energy, i.e., $e^2/4\pi\varepsilon_0\varepsilon_r r_c = k_B T$) and r_0 is the initial separation between the positive and negative ions [42]. In hydrocarbons, where $\varepsilon_r \cong 2$, the Onsager radius is on the order of 300 Å and the fraction of ions which escape is exceedingly small (<5%). In alcohols, where ε_r may range as high as 35 or greater, the Onsager radius is about 20 Å (methanol $\cong$ 16 Å; ethanol $\cong$ 19 Å; 2-propanol $\cong$ 30 Å), and the fraction of escaped ions can easily exceed 50%.

The Onsager radius also influences the geminate ion lifetime, γ, given by

$$\gamma = 0.43 r_c^2/\mu \tag{33}$$

where μ is the sum of positive and negative ion mobilities [42]. The value of γ decreases with the shrinking of the Onsager radius because of the truncation of the ion pair separation distance distribution for geminate ions. The numerator of the expression for γ is smaller by two orders of magnitude or more for polar solvents compared to hydrocarbons. For positive ion/electron geminate pairs, however, this effect is more than compensated in most polar systems by a decrease in μ attributable to solvation of the electron. For example, in comparing room temperature ethanol ($\varepsilon_r = 25$, $\mu \approx \mu_e = 0.25 \times 10^{-3}$ cm^2 V^{-1} s^{-1}) [58] and n-hexane ($\varepsilon_r = 2$, $\mu \approx \mu_e = 0.08$ cm^2 V^{-1} s^{-1}) [42] we have γ(ethanol)/γ(n-hexane) ≈ 2. In the case of solvent radical cation/scintillator radical anion pairs (or scintillator radical cation/scintillator radical anion pairs) where the value of μ does not vary as dramatically with solvent polarity, the geminate lifetime is predicted to be significantly shorter in polar solvents.

FDMR signals are observed in polar systems notwithstanding the effects of solvent polarity on charge recombination [59]. The expected trend with increasing ε_r is observed. The FDMR signal intensity roughly decreases with increasing ε_r. (The solvent viscosity must also be taken into account.) This effect is shown for several ethers and alcohols in Table 1. Higher scintillator concentrations are generally required than in hydrocarbons to generate appreciable signals. And the observed enhancement of the FDMR intensity upon lowering the temperature is much more dramatic in polar liquids than in hydrocarbons.

Figure 10 shows FDMR spectra obtained at different temperatures in ethanol containing 10^{-2} M PPO (2,5-diphenyl-1,3-oxazole). As in FDMR experiments in neat alkane liquids (section 2), no resolvable EPR features attributable to the solvent radical cations are observed. Undoubtedly, this is because of the same reason advanced for alkanes, i.e., fast ion-molecule reactions between solvent holes and neutral solvent molecules which compete with geminate recombination. Nevertheless, the FDMR results are positive proof that at sufficient scintillator concentrations solvent radical cations can be scavenged and that some (perhaps a small fraction) of the scavenged pairs undergo geminate recombination.

TABLE 1. The effect of dielectric constant on the FDMR signal intensity observed at 20°C in solvents containing 10^{-2} M PPO.

Compound	ε_r	r_c (Å)	Viscosity (@ 20°C, cP)	FDMR intensity (rel. units)
1,4-Dioxane	2.2	264	0.80	>4000
THF	7.6	76	0.51	1500
1-Butanol	17.8	33	2.95	900
2-Propanol	18.3	31	2.60	1100
1-Propanol	20.1	29	2.25	1100
Ethanol	25.0	23	1.20	520
Methanol	35.0	17	0.61	a

[a]Less than the noise level.

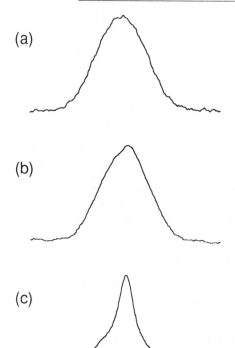

(a)

(b)

(c)

10 G

Figure 10. FDMR spectra observed at (a) 290K, (b) 243K and (c) 203K in ethanol containing 10^{-2} M PPO. The signals have been approximately normalized.

From the data shown in Figure 10 it is possible to assign the FDMR signal observed in ethanol to two processes. A broad component (fwhm = 12.5 G) is due to recombination of the PPO positive and negative ions (34). A narrow component (fwhm = 2.5 G), which grows in relative intensity with decreasing temperature, is due to recombinations between PPO radical cations and solvated electrons (35).

$$PPO^{+\bullet} + PPO^{-\bullet} \longrightarrow PPO^* + PPO \qquad (34)$$

$$PPO^{+\bullet} + e_s^- \longrightarrow PPO^* \qquad (35)$$

The EPR signature of the solvated electron is readily recognized by its characteristic width and g factor (g = 2.0020) [60]. The relative contributions of reactions (34) and (35) to the FDMR signal also depend on the scintillator concentration and the microwave delay time (relative to the electron beam pulse).

Following the discussion at the beginning of this section, it is remarkable that it is possible to observe processes (34) and (35) at all in the polar liquids. Because of the lack of absolute yield information from the FDMR data, it is not possible to estimate anything like the efficiency of scavenging of the solvent radical cations or the fraction of ions which are detected. However, it may not be an overestimation of the sensitivity of FDMR to postulate that the formation of PPO$^{\dot{+}}$ is mainly limited to the "static" scavenging of solvent radical cations (homogeneous scavenging being too slow to compete with ion-molecule reaction (eqn 4)), and that the observed FDMR signal arises from the (minuscule) fraction of longer-lived ion pairs whose initial separation distances exceed the Onsager radius and which nonetheless recombine with preservation of the well-defined spin correlation which existed at the time of cation/electron pair creation.

5.2. PHOTOIONIZATION

Figure 11 shows the FDMR spectrum obtained at room temperature in 2-propanol containing 5x10^{-5} M TMPD (N,N,N',N'-tetramethyl-p-phenylenediamine) which was irradiated with pulses of 308 nm photons from an excimer laser [15]. The spectrum consists of two separate features: a single narrow peak and a weaker broad spectrum. The broad spectrum can be simulated using the reported coupling constants for the TMPD radical cation [61], and the narrow central peak is readily assigned to the solvated electron. Thus, the FDMR signal clearly arises from recombination reactions between TMPD radical cations and solvated electrons. No other technique used for studying geminate ion recombination provides such unequivocal identification of the radical ions generated by photoionization.

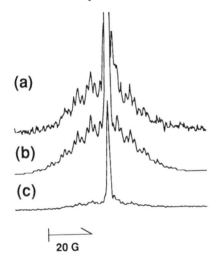

(a)

(b)

(c)

|—————⟩
20 G

Figure 11. FDMR spectrum observed at room temperature in 2-propanol containing 5 x 10^{-5} M TMPD: (a) experimental spectrum; (b) simulated spectrum using EPR parameters given in reference 15; (c) experimental spectrum of part (a) times 1/40.

FDMR signals have been studied for TMPD in various alcohols irradiated with either 308 nm or 248 nm photons, and a reasonably strong correlation is found between the signal intensity and reported values for the lifetime of the solvated electrons in the respective solvents [15]. In neat alkane solvents (cyclohexane, n-pentane) no FDMR signal is observed. However, addition of small quantities of alcohol (e.g., 1% ethanol in cyclohexane) gave rise to an easily observable FDMR signal due to recombination of the scintillator radical cations and solvated electrons. These observations indicate that stabilization of geminate ion pairs by electron solvation is a key requirement for the observation of an FDMR signal by photoionizing aromatic solutes.

The contrast between the FDMR results obtained by photoionization versus pulse radiolysis gives promise that further work may provide insights into the inherent differences in energy deposition by ionizing and photoionizing radiation and into the mechanism of photoionization. For example, how strongly is the initial distribution of cation/electron separation distances coupled to the mode of excitation (and amount of excess energy), and to what extent is the geminate ion lifetime affected? Is photoionization monophotonic or biphotonic; and, if biphotonic, does it occur by simultaneous or sequential absorption of two photons? A biphotonic process involving ionization from the triplet state to give a triplet radical ion pair would give an inverse FDMR response (increase in fluorescence intensity) and tend to cancel with the FDMR response from singlet radical ion pairs. Finally, the possibility of chemical transformations of highly excited aromatics or aromatic radical cations resulting from multiphoton excitation should not be ruled out.

6. Structural Studies of Radical Cations by FDMR

During most of our discussion in this chapter of FDMR observations of radical cations, the emphasis has been placed on the identity of the observed species, the conditions which affect radical cation longevity, transformations of radical cations, etc., with a view toward ascertaining mechanistic details of the physical and chemical events induced by ionizing or photoionizing radiation. Consideration of the spectroscopic information as it pertains to radical cation structure has been kept to a minimum.

In this section we illustrate the application of time-resolved FDMR as a spectroscopic tool and its value as a source of information about radical cation structure not obtainable by other means. The principal advantage of FDMR, as stated above, is its ability to observe short-lived radical cation species in nonpolar, fluid media where interactions between the radical cation and its environment are quite weak. A wide variety of problems exist in the literature where conflicting conclusions from theoretical and experimental studies of radical cations call for additional experimental data on radical cation structure under conditions where the cation-matrix interaction cannot be a significant factor. Here we use a single example, the hexamethyl(Dewar benzene) radical cation (HMDB⁺·), to illustrate the complexity which has surrounded certain radical cation species and the clarification possible from FDMR observations.

The likelihood that HMDB⁺· serves the role of intermediate in the photosensitized conversion of HMDB to hexamethylbenzene [62-66] has attracted considerable interest to HMDB⁺· for experimental and theoretical study. Recently, the electronic structure of HMDB⁺· has been studied extensively, but the assignment of its ground state is still in doubt. Electron removal must occur from one of the two highest occupied molecular orbitals of HMDB, the b_2 and a_1 orbitals [66,67], to give either the $^2B_2$ state or the $^2A_1$ state

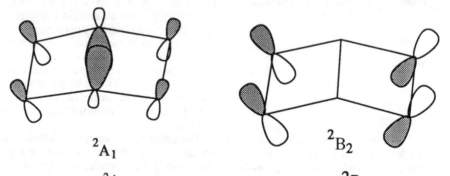

$$^2A_1 \qquad\qquad ^2B_2$$

of HMDB⁺·. In the $^2B_2$ state spin density is confined to the bonds between the two pairs of olefinic carbons, while in the $^2A_1$ state spin density is localized in the bond between the two transannular carbons. Ab initio calculations by Roth et al. support the $^2B_2$ assignment of the ground state, predicting a separation of 8 kcal/mol between the $^2B_2$ state and the $^2A_1$ state [66]. However, a MINDO/3 calculation by Bews and Glidewell predicts that $^2A_1$ has the minimum energy geometry [68].

This confusion was not resolved by initial experimental studies of $HMDB^+$. In a CIDNP study of photoreactions of HMDB with excited triplet electron acceptors, Roth et al. presented indirect evidence for the formation of both the $^2B_2$ state and the $^2A_1$ state of $HMDB^+$ by competing pathways, reaching no conclusion as to the ground state assignment [66]. EPR studies of $HMDB^+$ in freon matrices have been carried out by Rhodes [69]. He observed $^2B_2$ $HMDB^+$ in $CFCl_3$ and $^2A_1$ $HMDB^+$ in $CF_2ClCFCl_2$. The proposal by Rhodes was that the $^2A_1$ state observed in the less rigid $CF_2ClCFCl_2$ matrix be assigned to the ground state, and that the $^2B_2$ state is observed in $CFCl_3$ because the matrix is more tightly packed and prevents $HMDB^+$ from relaxing via elongation of the transannular bond to give the $^2A_1$ state. Recently, however, this assignment of $^2A_1$ $HMDB^+$ was shown to be in error [70]. Instances have been reported for other radical cations where the choice of the matrix may dictate which electronic state is observed [28b, 71].

$HMDB^+$ has been studied by FDMR in n-pentane and cyclopentane solvent.[72] Figure 12 shows the FDMR spectrum obtained at 205K in cyclopentane containing 10^{-2} M hexamethyl(Dewar benzene) and 10^{-4} M anthracene-d_{10}. There was no change in the spectral parameters over the temperature range 205K to 245K. Eleven lines out of a 13-line pattern with binomial intensity distribution (simulated stick spectrum) can be observed in the FDMR spectrum. The coupling constant is 9.2 G. This spectrum is consistent with $HMDB^+$ in the $^2B_2$ state and is inconsistent with $HMDB^+$ in the $^2A_1$ state, which should give rise to an EPR spectrum with the major hyperfine coupling to only six protons (i.e., seven lines). Assignment of the FDMR spectrum in Figure 12 to the hexamethylbenzene radical cation, to which HMDB may rearrange upon removal of an electron, can also be excluded. FDMR observations between 205K and 298K in cyclopentane containing 10^{-3} M hexamethylbenzene show a spectrum with a coupling constant of 6.7 G [73] in agreement with the reported value for hexamethylbenzene$^+$ [74].

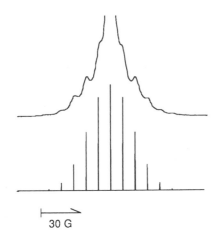

30 G

Figure 12. FDMR spectrum observed at 205K in cyclopentane containing 10^{-2} M hexamethyl(Dewar benzene) and 10^{-4} M anthracene-d_{10}. The hyperfine parameters used for the simulated stick spectrum are given in reference [72].

In order to probe the longevity of $HMDB^+$, a study was carried out by delaying application of the microwave pulse with respect to the electron beam pulse [72]. The intensity of the FDMR spectrum decreased with increasing time delay from 0 to 1000 ns. But even after 1000 ns the multiplet due to $HMDB^+$ could be clearly recognized, indicating that $^2B_2$ $HMDB^+$ can persist for a microsecond or longer (before recombining with A^-) under these experimental conditions. In the FDMR time frame, no features that can be assigned to $HMDB^+$ in the $^2A_1$ state were observed, and no evidence of the hexamethylbenzene radical cation spectrum could be seen. These FDMR results show that the lifetime of $HMDB^+$ in alkane solvents is significantly longer than the reported value of less than 15 ns estimated in a study of the isomerization of HMDB to hexamethylbenzene in polar solvents [64].

The relative longevity of the HMDB radical cation obtained in the FDMR study leads to the conclusion that the observed cation is the ground state species, which is the $^2B_2$ state. No evidence was found of $^2A_1$ HMDB‡ in the FDMR study even though one expects charge transfer from the solvent radical cation to populate both states of HMDB‡ since the energy difference between them is expected to be small. If both states are produced and the $^2A_1$ state quickly rearranges to hexamethylbenzene‡, then one would expect to see the hexamethylbenzene‡ FDMR spectrum, which is not the case. It is concluded, therefore, that either there is no involvement of the $^2A_1$ state or there is a fast conversion of the $^2A_1$ state into the $^2B_2$ state and the reverse is not thermally feasible. Such an interconversion is state-symmetry forbidden [75] but is not without precedent [76].

7. Early Events in Radiolysis: Radical Cation Transformations

Our perceptions of the primary events in the radiolysis of nonpolar condensed phase systems need to be updated. In particular, a keener appreciation is needed of the chemistry involving radical cations following the charge separation event caused by energetic radiation. The simple picture of how radiolysis of a hydrocarbon like cyclohexane proceeds must be modified in view of recent work which points out that radical cation transformations play a crucial role, the understanding of which is necessary to interpret studies of radical cation dynamics, especially ones carried out on the time scale of nanoseconds or slower.

This conclusion is the consequence of two different studies at Argonne. We have already discussed the FDMR studies of alkane radical cations which have established that all alkane radical cations studied to date do not persist in neat hydrocarbon solutions for more than a few nanoseconds. We have discussed the ion-molecule reactions which can account for the demise of alkane radical cations (i.e., proton transfer or H atom abstraction). While the reasons for the remarkable range of reactivities remains to be understood, FDMR studies have established that radical cations react very fast by ion-molecule reactions.

In order to obtain independent confirmation of the fast transformation of radical cations, a picosecond fluorescence study was undertaken in our group which has measured the rates of transformation of the solvent radical cations in cyclohexane and n-hexane. Using pulses from the electron LINAC (5 to 6 ps fwhm) and a streak camera (2 ps resolution), the time dependence of fluorescence resulting from the production of excited states in the ion recombination process was measured. An aromatic scintillator is dissolved in the hydrocarbon to probe the recombination event. As illustrated above, typical scintillators capture electrons and positive charge, and upon recombination the fluorescent state of the scintillator is produced. The time at which the excited state is formed depends on the time at which the electron and/or positive ion reacts with the scintillator and on the initial separation distance between the electron and the positive ion. Comparison of the results of Monte Carlo model calculations with experiment allow conclusions to be drawn about the distribution of separation distances and the chemistry of the solvent radical cations [77].

The results indicate that the chemistry of the initially formed radical cation, prior to its geminate recombination or reaction with the scintillator, strongly affects the dynamics of the production of the fluorescent state of the scintillator. Figure 13 shows the cumulative excited state concentration (calculated from the experimental data using the fluorescence lifetime) versus time for several scintillators at 10^{-3} M concentration in the two solvents. The results for the different scintillators are normalized to the same level at 4 to 5 ns. There is no dependence on the scintillator used. Calculated curves, based on the reactions and mobilities of the ions and a given distance distribution (an exponential distribution was used here), are shown in Figure 14 and are compared with the experimental results of Figure 13, represented by the broad, gray lines. (The ordinate scales correspond to the calculated curves and are in units of fluorescent scintillator states, ^1A, formed per ion pair.) The Monte Carlo model calculation includes the excitation transfer from excited solvent molecules to the scintillator. It is assumed that all the excited solvent molecules arise from ion recombination (prior to scavenging) since the direct excitation is shown to be less than 10% in cyclohexane [78,79].

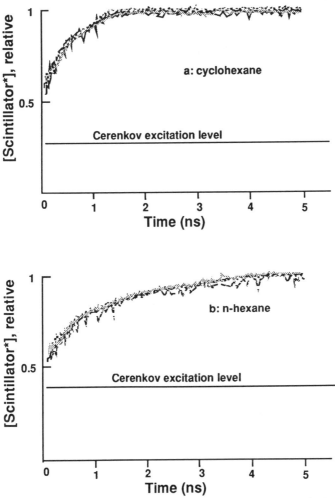

Figure 13. Concentration (relative units, nomalized at 4-5 ns) of scintillator excited state vs. time for 10^{-3} M solutions of 2,5-diphenyloxazole, p-terphenyl, anthracene and diphenylanthracene: (a) cyclohexane solvent; (b) n-hexane solvent. The horizontal lines in (a) and (b) represent the contribution from excitation of the scintillator by Cerenkov light generated in the sample and are based on the results from 10^{-3} M solutions in ethanol.

Figure 14, part a, shows the results for n-hexane. The dashed line was calculated assuming that the n-hexane cation undergoes no chemistry other than reaction with negative ions or charge transfer to the scintillator. The experimental data rises much more quickly than the calculation, implying that the recombination reaction that gives the fluorescent species occurs much more quickly or is terminated more quickly than the normal geminate reaction. The solid curve was calculated assuming that the n-hexane radical cation undergoes a transformation with a rate constant of 2×10^8 s^{-1} to a form which does not lead to the fluorescent state of the scintillator. A good fit with experiment is obtained.

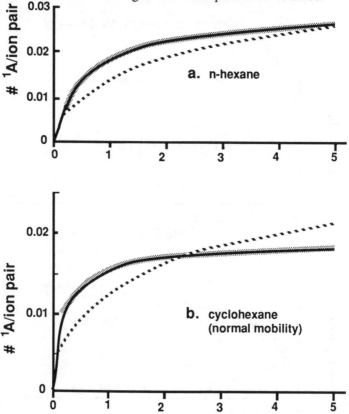

Figure 14. Comparison of the experimental results from Figure 13 with the results calculated by a Monte Carlo model. The broad, gray curves represent averages of the experimental results for the four scintillators and are normalized to the solid curves. The dashed curves and the solid curves are described in the text: (a) n-hexane solvent (the ordinate values for the dashed curve have been multiplied by 0.7); (b) cyclohexane solvent, (the ordinate values for the dashed curve have been multiplied by 0.64).

For cyclohexane a similar situation exists. Figure 14, part b, shows the results for cyclohexane assuming that the initially formed radical cation of cyclohexane has "normal" mobility. Clearly, the shape of the calculated curve for the situation where the radical cation undergoes no transformation (dashed curve) does not agree with experiment. The same shape as experiment can be obtained using a transformation rate constant of 3×10^9 s^{-1}. (It can be noted that the calculated result assuming that the initially formed radical cation of cyclohexane has high mobility, as suggested by other studies [9,10], also does not agree with experiment unless one includes a transformation reaction with a rate constant of 1×10^9 s^{-1}.)

The conclusion from the comparison of the experimental and simulated results at 10^{-3} M scintillator is that the solvent radical cation undergoes a transformation which prohibits its participation in formation of the scintillator excited state at later times. (A concentration of

10^{-3} M is ideal for observations of processes involving the solvent radical cation because positive charge transfer to the scintillator is relatively unimportant in the first 5 ns at this concentration.) One possibility for such a transformation is ion-molecule reaction with the solvent, i.e., equation (19) or (20). In order to explain the experimental results, the ion RH_2^+ cannot lead to production of the scintillator excited state. The first order transformation rate constants postulated to bring the simulated curves into agreement with the experimental results would correspond to the quantity $k_2[RH]$, where k_2 is the second order rate constant for reaction of RH^+ with RH. The values of 3×10^9 s^{-1} (cyclohexane) and 2×10^8 s^{-1} (n-hexane) mean that in neat room temperature cyclohexane and n-hexane the parent radical cation lifetime is ~0.3 and ~5 ns, respectively, very much in line with the FDMR conclusions. An alternative transformation, dissociation of the parent radical cation resulting in the loss of H_2, cannot explain the fluorescence results because FDMR studies have shown clearly that olefin radical cations can lead to the formation of scintillator excited states [36,37,48].

With the addition of these picosecond fluorescence results to the growing body of experimental data, one can begin to envision a comprehensive framework which accommodates the range of events occurring in the initial phases of radiolysis of scintillator solutions in hydrocarbons (i.e., such as occur in an FDMR experiment). In Figure 15 we have depicted such a scheme for n-hexane containing 10^{-4} M scintillator.

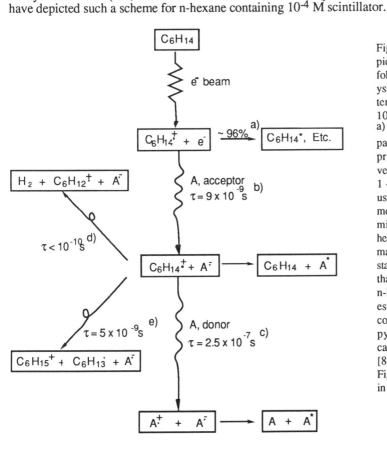

Figure 15. Scheme depicting ionic processes following electron radiolysis of n-hexane (at room temperature) containing 10^{-4} M scintillator (A). a) Fraction of geminate pairs which recombine prior to electron scavenging approximated by $1 - (\alpha C)^{1/2}/[1+(\alpha C)^{1/2}]$, using the empirical parameter $\alpha = 13$ M^{-1} determined for CH_3Br in n-hexane (ref. [42]); b) estimated from the rate constant for reaction of naphthalene with electrons in n-hexane (ref. [80]); c) estimated from the rate constant for reaction of pyrene with solvent cations in n-hexane (ref. [81]); d) ref. 6; e) see Figure 14 and discussion in section 7.

The scheme will vary for other hydrocarbon systems for which individual reaction rates will differ. Figure 15 clearly illustrates the competition among various transformation pathways of the parent radical cations and charge transfer/recombination reactions. As we have tried to show, this competition determines the observability of alkane radical cations in the FDMR experiment. All studies of radical cations in hydrocarbons must take into account fast transformation reactions of alkane radical cations. Indeed, the import of the FDMR and the picosecond fluorescence studies discussed above is that experiments in room temperature hydrocarbons by nanosecond techniques, whether optical or conductivity, cannot observe alkane radical cations after several nanoseconds.

8. Summary

This chapter has illustrated that it is feasible to observe a considerable variety of transient organic radical cations by magnetic resonance methods. The Optically Detected Magnetic Resonance method or, more appropriately, Time-Resolved Fluorescence Detected Magnetic Resonance, allows observation of a large variety of radical cations provided they are created in a pairwise, geminate manner by ionizing or photoionizing radiation. By manipulating the geminate ion recombination event, e.g., by slowing down the electron by converting it to an aromatic scintillator anion, the time frame of the geminate recombination event can be extended. Temperature and viscosity variation are also feasible as we have shown that this experiment can be realized even near liquid helium temperatures.

The wide-ranging possibilities of this experimental method allow the study of virtually all classes of radical cations. We have discussed alkanes [37,38], alkenes [36,37,48] and aromatics [11-14,54]; others that were studied but were not included in this chapter are dienes [48,82] and aliphatic amines [83]. Polar and nonpolar liquids and solids are accessible. One can ionize the bulk solvent, as with ionizing radiation, or a specific solute by laser photoionization.

A considerable effort has been expended towards examining various aspects of ion-molecule reactions of alkane radical cations in condensed phases. The fate of alkane radical cations in hydrocarbons is primarily determined by the competition between geminate recombination and the ion-molecule reaction, most likely proton transfer, with the parent neutral. These conclusions are buttressed by the recent picosecond optical studies of the same systems. What is revealed is the surprising diversity of reaction rates exhibited by alkane radical cations which are otherwise very similar. Cis- and trans-decalin and cyclo- and n-hexane are the prototypes of this diversity of exhibited rates of cation reactions.

These findings necessitate a substantial revision of our notions regarding the mechanism of hydrocarbon radiolysis. Most importantly, these studies show that any conclusions that were derived from the nanosecond conductivity and optical observations which assumed that no alkane cation transformation occurred must be re-evaluated.

On a broader plain, we have illustrated how time-resolved magnetic resonance methods can be used to provide additional insights regarding structure and stability of a variety of radical cations, some of which are the paradigms of our theoretical and experimental concepts of chemical reactivity of this important class of reactive intermediates.

ACKNOWLEDGMENTS. First, we acknowledge the contributions of Drs. Joseph P. Smith, Stephen M. Lefkowitz and Marc F. Desrosiers in the development and early applications of time-resolved FDMR technique at Argonne. Second, we acknowledge the current work of Drs. Martin G. Bakker and X.-Z. Qin in photodriven FDMR and FDMR applications, respectively. Third, we wish to thank our colleagues, Drs. Myran C. Sauer, Jr. and Charles D. Jonah for many discussions and critical comments and especially for sharing with us the results of their picosecond work on alkane radical cations. Drs. Sauer, Jonah, Bakker and Qin are all thanked for reading and commenting on this manuscript. Last, we acknowledge and are indebted to Robert H. Lowers for providing us with the reliable Van de Graaff electron beam and other essential support. Work performed under the auspices of the Office of Basic Energy Sciences, Division of Chemical Science, US-DOE under contract number W-31-109-ENG-38.

References

1. Shiotani, M. , *Mag. Reson. Rev.* **12**, 333-381 (1987).
2. Shida, T., Haselbach, E., and T. Bally, *Acc. Chem. Res.* **17**, 180-186, (1984).
3. Knight, Jr., L. B., *Acc. Chem. Res.* **19**, 313-321 (1986).
4. Klassen, N. V. and Teather, G. G., *J. Phys. Chem.* **83**, 326-329 (1979); Teather, G. G. and Klassen, N. V., *J. Phys. Chem.* **85**, 3044-3046 (1981); Klassen, N. V. and Teather, G. G., *J. Phys. Chem.* **89**, 2048-2053 (1985).
5. Mehnert, R., Brede, O. and Naumann, W., *Ber. Bunsenges. Physik. Chem.* **88**, 71-80 (1984); Mehnert, R., Brede, O. and Cserep, G., *Radiat. Phys. Chem.* **26**, 353-363 (1985).
6. LeMotais, B. C. and Jonah, C. D., *Radiat. Phys. Chem.* **33**, 505-517, (1989.
7. Tagawa, S., Hayashi, N., Yoshida, Y., Washio, M. and Tabata, Y., *Radiat. Phys. Chem.* **34**, 503-511 (1989).
8. Sauer, Jr., M. C., Romero, C. and Schmidt, K. H., *Radiat. Phys. Chem.* **29**, 261-273 (1987).
9. de Haas, M. P., Warman, J. M., Infelta, P. P. and Hummel, A., *Chem. Phys. Letters* **31**, 382-386 (1975); de Haas, M. P., Hummel, A., Infelta, P. and Warman, J. M., *J. Chem. Phys.* **65**, 5019-5020 (1976).
10. Sauer, Jr., M. C., Schmidt, K. H., and Liu, A., *J. Phys. Chem.* **91**, 4836-4839 (1987); Sauer, Jr., M. C. and Schmidt, K. H., *Radiat. Phys. Chem.* **32**, 281-285 (1988).
11. Anisimov, O. A., Grigoryants, V. M., Molchanov, V. K. and Molin, Yu. N., *Chem. Phys. Lett.* **66**, 265-268 (1979).
12. Molin, O. A., Anisimov, Yu. N., Grigoryants, V. M., Molchanov, V. K. and Salikhov, K. M., *J. Phys. Chem.* **84**, 1853-1856 (1980).
13. Trifunac, A. D. and Smith, J .P., *Chem. Phys. Lett.* **73**, 94-97 (1980).
14. Smith, J. P. and Trifunac, A. D. *J. Phys. Chem.* **85**, 1645-1653 (1981).
15. Percy, L. T., Bakker, M. G. and Trifunac, A. D., *J. Phys. Chem.* **93**, 4393-4396 (1989).
16. Klassen, N. V. in "Radiation Chemistry", Farhataziz and M. A. J. Rodgers (eds.), VCH Publishers, New York, pp. 29-64 (1987).
17. Fessenden, R. W. , *J. Chem. Phys.* **58**, 2489-2500 (1973); McLauchlan, K. A. and Stevens, D. G., *Molecular Phys.* **57**, 223-239 (1986).
18. Trifunac, A. D., Norris, J. R. and Lawler, R. G., *J. Chem. Phys.* **71**, 4380-4390 (1979); b) Mims, W. B., in "Electron Paramagnetic Resonance" G. Geshwind (ed.), Plenum, New York, pp. 263-351 (1972).
19. Freed, J. H. and Pedersen, J. B., *Adv. Magn. Reson.* **8**, 1-84 (1976).
20. Trifunac, A. D., Lawler, R. G., Bartels, D. M. and Thurnauer, M. C., *Prog. Reaction Kinetics* **14**, 43-156 (1986).

228

21. Bartels, D. M. and Lawler, R. G., *J. Chem. Phys.* **86**, 4843-4855 (1987); Bartels, D. M., Lawler, R. G. and Trifunac, A. D., *J. Chem. Phys.* **83**, 2686-2707 (1985); Han, P. and Bartels, D. M., *Chem. Phys. Letters* **159**, 538-542 (1989).
22. Warman, J. M., Infelta, P. P., de Haas, M. P. and Hummel, A., *Chem. Phys. Letters* **43**, 321-325 (1976).
23. Rothman, W., Hirayama, F. and Lipsky, S., *J. Chem. Phys.* **58**, 1300-1317 (1973).
24. Schmidt, K. H., Sauer, Jr., M. C., Lu, Y. and Liu, A., *J. Phys. Chem.*, **94**, 244-251 (1990).
25. Werst, D. W., Percy, L. T. and Trifunac, A. D., *J. Magn. Reson.* **82**, 588-591 (1989).
26. Wang. J. T. and Williams, F., *Chem. Phys. Letters* **82**, 177-181 (1981); Nunome, K., Toriyama, K. and Iwasaki, M., *J. Chem. Phys.* **79**, 2499-2503 (1983).
27. Tabata, M. and Lund, A., *Chem. Phys.* **75**, 379-388 (1983); Iwasaki, M., Toriyama, K., Nunome, K., *Faraday Discuss. Chem. Soc.* **78**, 1-15 (1984).
28. a) Toriyama, K., Nunome, K. and Iwasaki, M., *J. Phys. Chem.* **85**, 2149-2152 (1981); b) Toriyama, K., Nunome, K., and Iwasaki, M., *J. Chem. Phys.* **77**, 5891-5912 (1982).
29. Shida, T. and Takemura, Y., *Radiat. Phys. Chem.* **21**, 157-166 (1983).
30. "Radical Ions", E. T. Kaiser and L. Kevan (eds.), Interscience, New York (1968). "Magnetic Properties of Free Radicals", Landolt-Bornstein, Vol. 9, H. Fischer and K.-H. Hellwege (eds.), Springer-Verlag, Berlin, part d (1985).
31. Shida, T., Nosaka, Y. and Kato, T., *J. Phys. Chem.* **82**, 695-698 (1978).
32. Qin, X.-Z. and Trifunac, A. D., *J. Phys. Chem*, in press.
33. Qin, X.-Z. and Trifunac, A. D., *J. Phys. Chem*, in press.
34. Jacob, S. L., Craw, M. T., Depew, M. C. and Wan, J. K. S., *Research on Chemical Intermediates* **11**, 271-279 (1989).
35. Ichikawa, T. and Ohta, N., *J. Phys. Chem.* **91**, 3244-3248 (1987); Ichikawa, T., Shiotani, M., Ohta, N. and Katsumata, S., *J. Phys. Chem.* **93**, 3826-3831 (1989).
36. Werst, D. W., Desrosiers, M. F. and Trifunac, A. D., *Chem. Phys. Letters* **133**, 201-206 (1987).
37. Werst, D. W. and Trifunac, A. D., *J. Phys. Chem.* **92**, 1093-1103 (1988).
38. Werst, D. W., Bakker, M. G. and Trifunac, A. D., *J. Am. Chem. Soc.*, **113**, 40-50 (1990).
39. Werst, D. W., Desrosiers, M. F. and Trifunac, A. D. unpublished results.
40. Melekhov, V. I., Anisimov, O. A., Veselov, A. V. and Molin, Yu. N., *Chem. Phys. Letters* **127**, 97-100 (1986).
41. Werst, D. W., Percy, L. T. and Trifunac, A. D., *Chem. Phys. Letters* **153**, 45-51 (1988).
42. Warman, J. M. in "The Study of Fast Processes and Transient Species by Electron Pulse Radiolysis" J. H. Baxendale and F. Busi (eds.), Reidel, Boston, pp. 433-534 (1981).
43. Iwasaki, M., Toriyama, K., Fukaya, M., Muto, H. and Nunome, K., *J. Phys. Chem.* **89**, 5278-5284 (1985).
44. Nunome, K., Toriyama, K. and Iwasaki, M., *Tetrahedron* **42**, 6315-6323 (1986).
45. Toriyama, K., Nunome, K. and Iwasaki, M., *J. Am. Chem. Soc.* **109**, 4496-4500 (1987).
46. Toriyama, K., Nunome, K. and Iwasaki, M., *J. Phys. Chem.* **90**, 6836-6842 (1986).
47. Werst, D. W. and Trifunac, A. D., *Chem. Phys. Letters* **137**, 475-481 (1987).
48. Desrosiers, M. F. and Trifunac, A. D., *J. Phys. Chem.* **90**, 1560-1564 (1986).
49. Badger, B. and Brocklehurst, B., *Trans. Faraday Soc.* **65**, 2576-2581 (1969).

50. Ichikawa, T., Ohta, N. and Kajioka, H., *J. Phys. Chem.* **83**, 284-294 (1979).
51. Mehnert, R., Brede, O. and Cserep, G., *Radiochem. Radioanal. Lett.* **47**, 173-188 (1981).
52. Saik, V., Anisimov, O., Lozovoy, V. and Molin, Y., *Z. Naturforsch A* **40**, 239-245 (1985).
53. Williams, F. and Qin, X.-Z., *Radiat. Phys. Chem.* **32**, 299-308 (1988).
54. Desrosiers, M. F. and Trifunac, A. D., *Chem. Phys. Letters* **121**, 382-385 (1985).
55. Smith, J. P. and Trifunac, A. D., *Chem. Phys. Letters* **83**, 195-198 (1981).
56. Kira, A., Arai, S. and Imamura, M., *J. Phys. Chem.* **76**, 1119-1124 (1972).
57. Baxendale, J. H., Beaumond, D. and Rodgers, M. A. J., *Chem. Phys. Letters* **4**, 3-4 (1969).
58. Fowles, P., *Trans. Faraday Soc.* **67**, 428-439 (1971).
59. Percy, L. T., Werst, D. W.and Trifunac, A. D., *Radiat. Phys. Chem.* **32**, 209-213 (1988).
60. Shirashi, H., Ishigure, K. and Morokuma, K., *J. Chem. Phys.* **88**, 4637-4649 (1988).
61. Knolle, W. R. PhD. Thesis, University of Minnesota, (1970). As reported in: Wertz, J. E., Bolton, J. R. (1986), "Electron Spin Resonance", Chapman and Hall, New York, pp. 466.
62. Taylor, G. N., *Z. Phys. Chem. N. F.* **101**, 237-254 (1976).
63. Evans, T. R., Wake, R. W. and Sifain, M. M., *Tetrahedron Letters*, 701-704 (1973)
64. Peacock, N. J. and Schuster, G. B. , *J. Am. Chem. Soc.* **105**, 3632-3638 (1983).
65. Al-Ekabi, H. and de Mayo, P. J., *J. Phys. Chem.* **90**, 4075-4080 (1986).
66. Roth, H. D., Schilling, M. L. M. and Raghavachari, K., *J. Am. Chem. Soc.* **106**, 253-255 (1984).
67. Bieri, G., Heilbronner, E., Kobayashi, T., Schmelzer, A., Goldstein, M. J., Leight, R. S. and Lipton, M. S., *Helv. Chim. Acta* **59**, 2657-2673 (1976).
68. Bews, J. R. and Glidewell, C., *J. Mol. Struct. (THEOCHEM.)* **86**, 197-204 (1982).
69. Rhodes, C. J., *J. Am. Chem. Soc.* **110**, 446-4447 (1988). Rhodes, C. J., *J. Am. Chem. Soc.* **110**, 8567-8568 (1988).
70. Williams, F., Guo, Q.-X. and Nelson, S. F., *J. Am. Chem. Soc.*, in press.
71. Shiotani, M., Yano, A., Ohta, N. and Ichikawa, M., *Chem. Phys. Letters* **147**, 38-42 (1988).
72. Qin, X.-Z., Werst, D. W. and Trifunac, A. D., *J. Am. Chem. Soc.*, in press.
73. Qin, X.-Z., Werst, D. W. and Trifunac, A. D., unpublished results.
74. Hulme, R. and Symons, M. C., *J. Chem. Soc.*, 1120-1126 (1965).
75. Goldstein, M. J. and Leight, R. S., *J. Am. Chem. Soc.* **99**, 8112-8114 (1977).
76. Haselbach, E., Bally, T., Lanyiova, Z. and Baertschi, P., *Helv. Chim. Acta* **62**, 583-592 (1979); Raghavachari, K., Haddon, R. C. and Roth, H. D., *J. Am. Chem. Soc.* **105**, 3110-3114 (1983); Gebicki, J.L., Gebicki, J. and Mayer, J., *Radiat. Phys. Chem.* **30**, 165-167 (1987).
77. Sauer, M. C. Jr., Jonah, C. D. and Naleway, C. A. to be submitted.
78. Choi, H. T., Haglund, J. A.; Lipsky, S. *J. Phys. Chem.* **87**, 1583 (1985)
79. Sauer, M. C., Jr.; Jonah, C. D.; Le Motais, B. D.; Chernovitz, A. C., *J. Phys. Chem.* **92**, 4099 (1988).
80. Baxendale, J. H. and Rasburn, E. J., *J. Chem. Soc. Faraday Trans. I* **70**, 705-717 (1974).
81. Zador, E., Warman, J. M. and Hummel, A., *Chem. Phys. Lett.* **23**, 363-366 (1973).
82. Desrosiers, M. F. and Trifunac, A. D., *Chem. Phys. Lett.* **118**, 441-443 (1985).
83. Lefkowitz, S. M. and Trifunac, A. D., *J. Phys. Chem.* **88**, 77-81 (1984).

RADICAL CATIONS IN PULSE RADIOLYSIS

R. MEHNERT

Central Institute of Isotope and Radiation Research of the Academy of Sciences of the G.D.R.,
Permoserstr. 15,
DDR-7050 Leipzig

1. Introduction

If an electron is removed from a closed-shell molecule a radical cation is generated. Due to their unpaired electron and their charge, radical cations are usually very reactive. In condensed matter radical cations interact with their neighboring molecules forming solvation shells. This is also important for cations in non-polar liquids. Solvation of the cations occurs probably within picoseconds. For liquids such as alkanes or carbon tetrachloride which exhibit a dielectric constant of about 2, cation solvation energies of typically 1 to 2 eV are obtained.

In liquids the ionization potential of a molecule is decreased in comparison to the gas phase by the solvation energy. Moreover, in liquids excess energy of cations can be easily distributed to the surrounding molecules. This is one reason why in liquids parent radical cations show much less fragmentation than observed in the gas phase. Thus, radical cations can be prepared and studied in non-polar liquids.

This chapter is dealing with spectral and kinetic characteristics of radical cations which were generated by pulse radiolysis in liquids at room temperature. As it will be shown in the following section, the lifetime of radical cations is typically in a range of 10^{-12} to 10^{-6} s. Intense picosecond or nanosecond electron pulses must be used to generate a detectable concentration of radical cations. The transients can be followed by different detection methods with suitable time resolution. In use are optical absorption spectroscopy [1], D.C. conductivity [2,3], microwave conductivity [4] and optically detected magnetic resonance (ODMR) methods [5,6]. One of the latter methods, the optically detected ESR is extremely sensitive. As described in detail in the next chapter even 20 radical pairs within a sample can be detected. This implies that stationary irradiation can be used to produce very small steady-state concentrations of radical cations.

2. Formation and Fate of Radical Cations in Non-Polar Liquids

2.1 ION-PAIRS

When non-polar liquids are irradiated with low-LET radiation, such as fast electrons, ion-pairs are generated. Most of them are single pairs with separation distances of the partner ions well below the Onsager escape

A. Lund and M. Shiotani (eds.), Radical Ionic Systems, 231–284.
© 1991 *Kluwer Academic Publishers. Printed in the Netherlands.*

distance r_c. This is the length at which the coulomb attraction of the ions is equal to the thermal energy. These ion-pairs are often referred to as coulomb-correlated (geminate) pairs, and recombination of the ions occurs independently in each pair. A schematic representation of ion-pair formation and transformation as typically found in electron-irradiated alkanes is given in Fig. 1.

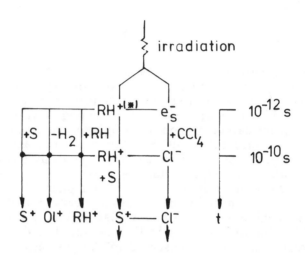

Fig. 1: Ion-pair recombination and transformation in liquid alkanes containing carbon tetrachloride as electron scavenger and the solute S as scavenger of the positive charge. Ol^+ = olefinic radical cation.

The electrons which are produced in the primary ionization event thermalize in liquid alkanes at times $<10^{-12}$ s, become solvated and remain paired with the (presumably vibrationally excited) parent positive ion $RH^{+(*)}$. The parent cation may transfer a proton to an alkane molecule RH, may decompose to form olefinic radical cations or may transfer its positive charge to the solute S. The initially existing ion-pair $RH^{+(*)} - e_s^-$ is finally transformed to different types of ion-pairs consisting of solute cations S^+, olefinic radical cations Ol^+ and protonated alkanes RH_2^+ on one side and electrons or anions on the other. Only a small fraction of ions escape from geminate recombination and combine homogeneously.

As can be seen from the scheme of Fig. 1 the kinetics of the geminate recombination affects such important reactions as charge transfer to solutes in a decisive manner. Therefore, different approaches to describe the geminate recombination have been developed which will be briefly discussed here.

The diffusive motion of two oppositely charged ions in each other's coulomb field in a non-polar liquid is described by the Debye-

Smoluchowski equation. For spherically symmetric interaction between the reactants the probability density $\rho(r,t)$ for finding at time t the partner ion at a distance r is given by

$$\partial\rho/\partial t = D/r^2 \, \partial/\partial r(r^2 \, \partial\rho/\partial r + r_c \, \rho) \tag{1}$$

where D is the sum of the diffusion coefficients of the reactants, $r_c = e^2/4\pi\varepsilon\varepsilon_0 kT$ the Onsager length, $\varepsilon\varepsilon_0$ being the dielectric constant of the medium, e the electron charge, k Boltzmann's constant and T the absolute temperature. Hong and Noolandi [7] presented an analytical solution of Eq. (1) which used the condition that the two ions are initially separated by a distance r_0

$$\rho(r,t=0/r_0) = 1/4\pi r_0^2 \, \delta(r-r_0) \tag{2}$$

and the boundary condition

$$\rho(r,t) = 0 \text{ for } r \le a \tag{3}$$

The ions recombine when they reach the distance a, the reaction radius. The (time-dependent) recombination rate is obtained by

$$R(t) = 4\pi a^2 \, D \, \partial\rho/\partial r/r=a \tag{4}$$

and the survival probability by

$$S(t) = 1 - \int_0^t R(t')dt' \tag{5}$$

At long times the survival probability approaches the value

$$S(t) = \exp(-r_c/r_0) \, (1+r_c / (\pi Dt)^{1/2}) \tag{6}$$

or if the yield G(t) is used

$$G(t) = G_{fi} \, (1 + r_c / (\pi Dt)^{1/2}) \tag{7}$$

G_{fi} is the yield of ions which escape their geminate partner and become distributed homogeneously.

If the Debye-Smoluchowski equation is solved numerically, arbitrary initial distributions of separation distances can be used. For liquid alkanes the experimental survival probabilities are better described if instead of the delta distribution (2) the exponential distribution

$$\rho(r,t=0) = (1/4\pi r^2 b)^{-1} \exp(-r/b) \tag{8}$$

is taken [8]. The parameter b has the meaning of a mean initial separation of the ions [9] and is assumed to be about 5.8 nm.

Using the conditions (3) and (8) the Debye-Smoluchowski equation was solved numerically and the survival probability S(t) was calculated.

Survival probabilities of various ion-pairs which are formed in electron irradiated n-heptane containing the electron scavenger carbon tetrachloride are shown in Fig. 2. The diffusion coefficient of the solvated electrons is known to be 1.2×10^{-3} cm^2/s [10]. That of the cations is estimated to be 0.65×10^{-5} cm^2/s as typical for "massive" ions [11]. Mobile positive ions ("holes") [12] could not be directly detected in n-heptane. Their existence in the subnanosecond time scale cannot be ruled out, however, Therefore, a diffusion coefficient of 0.5×10^{-3} cm^2/s close to that measured in cyclohexane [13] is assumed.

Fig. 2 contains the survival probability profiles of the ion pairs "hole" - e_s^-, $C_7H_{16}^+$ - e_s^-, and $C_7H_{16}^+$ - Cl^-. These types of ion-pairs are of general

importance in irradiated liquid alkanes.

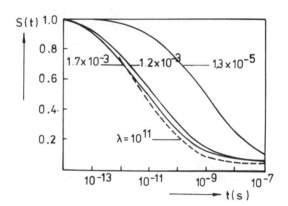

Fig. 2: Survival probabilities calculated from Eq. (1) for ion-pairs in n-heptane. The sum of the diffusion coefficients is indicated. r_c = 30 nm, b = 5.8 nm. Dashed line: survival probability S(t) = G(t)/G$_{gi}$ obtained from the empirical formula (10) using $\lambda = 10^{11}$ s^{-1}.

Computer simulations of the recombination of small groups of ions have been performed by Bartczak and Hummel [14]. They found in some cases significant differences between the single pair recombination kinetics and the simulated kinetics for the systems composed of two or three ion-pairs.

A second possible approach for calculating survival probabilities of geminate ion pairs has been proposed by Warman, Asmus and Schuler [15]. From the concentration dependence of ion scavenging in cyclohexane

$$G(c_p) = G_{fi} + G_{gi} (\alpha c_p)^{1/2} / (1 + (\alpha c_p)^{1/2}) \tag{9}$$

they derived an expression for the yield G of the surviving ions at time t

$$G(t) = G_{fi} + G_{gi} \exp(\lambda t) erfc(\lambda t)^{1/2} \tag{10}$$

In Eq. (9) α is a parameter characterizing the reactivity of a solute towards the cation or electron of the pair, c_p is the solute concentration and λ has the meaning of a characteristic ion-pair lifetime $\tau_c = 1/\lambda$. The survival probability of the electron-"hole" pair can be satisfactorily approximated by the expression $G(t)/G_{gi}$ taking $\lambda = 10^{11}$ s^{-1}. For small solute concentrations the relationship [16]

$$\lambda^{-1} = (r_c^2/D) (G_{fi}/G_{gi})^2 \tag{11}$$

can be used.

If transformation of the primary ion-pair takes place, the characteristic lifetime is increased by the ratio $r_D = D_p/D_s$, where D_p is the sum of the diffusion coefficients of the constituents of the primary ion-pair and D_s that of the secondary one. The characteristic lifetime is increased, e.g., after electron scavenging in n-heptane by a factor r_D of about 115 and with the corresponding λ-value of 8.6x10^9 s^{-1} a survival probability is calculated which approaches the right curve of Fig. 2.

2.2 KINETICS OF CHARGE SCAVENGING FROM ION-PAIRS. CONSEQUENCES OF THE PULSE RADIOLYSIS OF NON-POLAR LIQUIDS

In the pulse radiolysis of non-polar liquids like alkanes primary and secondary cations as well as electrons can be measured directly using spectroscopic and conductometric detection. From the time profiles obtained experimentally, information on the ion kinetics can be derived. To deduce rate constants of the reaction

$$RH^+ + S \rightarrow S^+ + RH \tag{12}$$

from measured time profiles of ionic species, which still are coulomb-correlated, the WAS equation must be modified in an appropriate manner [17].

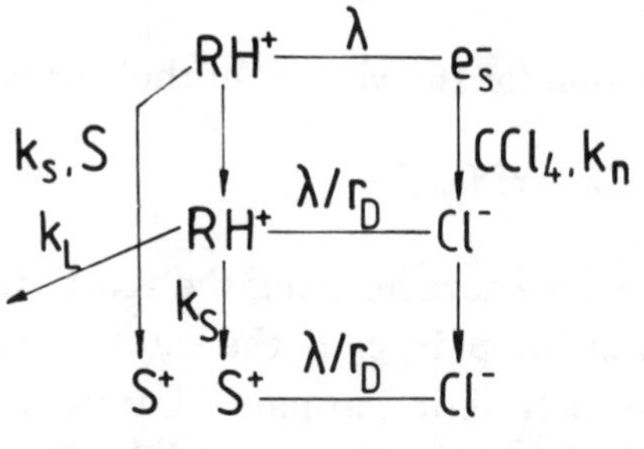

Fig. 3: Simplified scheme of the ion-pair transformation in liquid alkanes containing carbon tetrachloride as electron and the solute S as cation scavenger.

Most of the experiments to characterize radical cations have been done by using electron scavengers with concentrations high enough to form anions with reaction time constants of typically 10^{-10} s. As can be seen from the (simplified) reaction scheme of Fig. 3 at times longer than 10^{-9} s

the ion-pair RH^+ -- Cl^- can be taken as the primary one. The simple expression

$$G_{RH^+}(t) = \exp(-k_s c_s t) G_{gi} \exp(\lambda t) erfc(\lambda t)^{1/2} \tag{13}$$

is obtained for the time dependence of the G-value.

This simple approximation has to be modified if electron scavenging cannot be assumed to be fast. For the case represented by the reaction scheme of Fig. 3 the following expression can be derived:

$$G_{RH^+}(t) = \exp(-k_s c_s t) [G_{fi} + G_{gi} (\int_t^\infty f(\tau) d\tau$$

$$+ \int_{t/r_D}^t f(\tau) (1 - \exp(-k_n c_n (\tau/r_D - t/r_D - 1)) d\tau)), \tag{14}$$

where $f(\tau) = \lambda[(1/\pi\lambda\tau)^{1/2} - \exp(\lambda t) erfc(\lambda t)^{1/2}]$ is the ion-pair lifetime distribution function and k_n the rate constant for electron scavenging. Evaluation of the integrals leads to an expression [18], which can be used for a comparison with experimental time profiles of alkane radical cations. As an example, time profiles of the G-value of n-hexadecane radical cations are shown in Fig. 4. The corresponding radical cations were generated after electron-pulse irradiation in n-hexadecane solutions containing carbon tetrachloride and different amounts of pentylcyclo-hexane (PCH). The time profiles were calculated from Eq. (14)

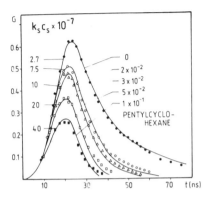

Fig. 4: G-values of n-hexadecane radical cations vs. time. Solid lines are calculated from Eq. (14) using the parameters given in the text. Points are experimental [19]. The pentylcyclohexane concentration is indicated.

using $\lambda = 8 \times 10^{10}$ s^{-1}, $r_D = 54$, $k_n = 3 \times 10^{11}$ dm^3 mol^{-1} s^{-1} [4b], $G_{fi} = 0.12$ and $G_{gi} = 3.8$ [19]. For comparison with the experimental time profiles the curves were convoluted with the Cerenkov pulse response, and the extinction coefficient $\varepsilon_{960} = 10.7 \times 10^3$ dm^3 mol^{-1} cm^{-1} was used as a scaling factor between calculated Gε-and experimental G-values. From the calculated $k_s c_s$-values and the known solute concentrations c_s the charge transfer rate constant k_s of the reaction

$$C_{16} H_{34}^+ + PCH \rightarrow PCH^+ + C_{16}H_{34} \qquad (15)$$

was deduced.

If it is assumed that the primary cations react with the solute to produce solute cations which have the same diffusion coefficient as the initial cations, then the characteristic lifetime of the ion-pair is unaffected. The growth and the decay of the solute cations will then be given by Eq. (14) with exp($-k_s c_s t$) replaced by (1-exp ($-k_s c_s t$)).

The transformation of both primary cations and electrons to slowly diffusing, more "massive" ions considerably complicates the description of the ion-pair kinetics. In Fig. 5 a scheme of such a mechanism is given, which can be used to describe the ion-pair recombination and transformation in, e.g., irradiated cyclohexane solution. Here, initially a mobile positive species ("hole") exists [12] which is transformed subsequently to massive radical cations.

The nature of the cationic species involved will be discussed in section 3.2.1.

From the standpoint of the ion kinetics only the changes in the diffusion coefficients of the ions are of importance. Infelta and Rzad [17] presented a solution of the problem outlined in Fig. 5 (formula B2 in [17]).

Competing pseudo-first order processes as charge transfer to solutes from different types of cations can be included into their phenomenological formalism.

In conclusion, one can say that the phenomenological model of ion-pair recombination and transformation is a very useful approach to describe the kinetics of ions as represented by the reaction schemes of Figs. 1, 3 and 5.

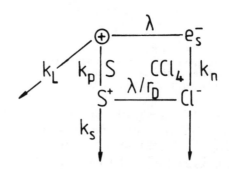

Fig. 5: Reaction scheme of ion-pair recombination and transformation if as primary ion a positive "hole" is generated.

2.3 HOMOGENEOUS ION RECOMBINATION

Ions are called free when they have escaped their mutual coulombic attraction and diffuse freely in the bulk medium. Warman [4b] has given a useful expression of the time required to escape with 90% probability: $\tau_{fi} = 16.7\, r_c^2/D$. Taking as an example cyclohexane, the escape times for the primary ions are estimated to be 23 ns and for secondary ions 6.8 µs.

In a second order reaction the ions disappear according to

$$c(t)^{-1} - c(0)^{-1} = k_r t, \tag{16}$$

where $c(0)$ is the initial ion concentration, $c(t)$ that at time t and k_r is the recombination rate constant. For non-polar liquids the recombination rate constant according to Debye [20] is given by

$$k_r = 4\pi\, r_c\, D. \tag{17}$$

Eq. (17) is equivalent to that for the rate constant of a diffusion-controlled reaction with the Onsager radius r replaced by the reaction radius.

3. Alkane Radical Cations

3.1 OPTICAL ABSORPTION SPECTRA OF ALKANE RADICAL CATIONS

Already in 1968 Louwrier and Hamill [21] were able to generate alkane radical cations in γ-irradiated glassy matrices of CCl_4 and 3-methylpentane containing alkanes. They studied the optical absorption spectra of some radical cations and made an attempt to estimate extinction coefficients. Nearly ten years later, by means of pulse radiolysis, similar optical absorption bands were observed in neat liquid alkanes (See Table 1) which were attributed to alkane radical cations [22, 23]. Reactions of alkane radical cations such as decomposition, proton transfer and electron transfer could now be studied in the liquid phase.

Table 1: Spectral characteristics of alkane radical cations as observed in pulse radiolysis

Alkane	λ(max) nm	Ext. coefficient $dm^3 \, mol^{-1} \, cm^{-1}$	
n-Hexane	480		[37]
n-Heptane	550	0.88×10^4	[35]
n-Decane	760	1.0×10^4	[35]
n-Dodecane	800	1.0×10^4	[35]
n-Tetradecane	840		[35]
n-Hexadecane	880	1.4×10^4	[35]
3-Methylheptane	560	0.8×10^4	[35]
3-Methyloctane	600	0.88×10^4	[39]
Cyclohexane	500		[35,43]
Methylcoclohexane	500		[35]
Squalane	1200	1.4×10^4	[39]

As shown in Fig. 2 the neutralization of ion-pairs composed of alkane radical cations and anions is nearly completed within a few nanoseconds. Time-resolved techniques in the picosecond and nanosecond domain, such as transient absorption and emission spectroscopy and conductivity methods, have to be applied to characterize spectroscopic and kinetic characteristics of alkane radical cations. Optical and especially conductivity techniques often suffer from the lack of selectivity. In many cases an unequivocal identification of the species observed makes it desirable to have more selective structural information. Magnetic resonance based techniques are able to provide the structural information lacking in the optical and conductometric studies. Particularly, the time-resolved fluorescence detected magnetic resonance (FDMR) method developed by Trifunac and coworkers [6] proved to be an extremely powerful tool to study structure and reactions of geminately recombining radical cations in the time interval from 30-150 ns. Radical cations can be observed by FDMR the geminated electrons of which have been converted to less mobile radical anions. The characteristic lifetime of the ion-pair is then extended to the time range accessible by the experiment.

Similar selective structural information and also some kinetic information derived from line width considerations can be obtained using stationary fluorescence (optically) detected EPR. This technique is called ODEPR (optically detected electron paramagnetic resonance) and was developed by Molin, Anisimov and coworkers [5,24]. In the past a large amount of information on geminate radical cations of alkanes, alkenes and aromatics as well as of anion radicals has been collected. These results are presented by Anisimov in the next chapter.

The use of the low-temperature matrix isolation technique and especially the employment of halocarbon and sulphur hexafluoride matrices initiated not only further investigations into optical properties of alkane radical cations [25,26,27] but enabled detailed EPR studies of solute radical cations. Unambiguous EPR spectra of a large number of alkane radical cations have been obtained since the first work of Symons [28], Williams [29] and Shida [30]. Much of our present knowledge on EPR spectra and electronic structure of alkane radical cations originates from their work and from studies done by Iwasaki´s group [31]. A comparison of the structural information obtained from the EPR spectra of normal and specifically deuterium-labelled alkane radical cations with molecular orbital calculations yielded valuable information on the geometry of alkane radical cations and matrix interaction effects [32].

Shida [27] first demonstrated that halocarbon matrices allow the parallel detection of optical and EPR spectra of alkane radical cations. From the thermal and photobleaching behavior of both kinds of spectra the structureless optical absorption bands observed in the visible and infrared range could be unequivocally assigned to alkane radical cations. This was in accordance with photodecomposition and photoelectron spectra of alkanes [33,34].

The optical transient absorption spectra which have been observed after electron-pulse irradiation of neat alkanes and alkane/alkane solutions are summarized in Figs. 6-8. The same structureless bands are obtained as found in the matrix studies. In all cases suitable electron scavengers were used to avoid superposition with the solvated electron absorption.

In all solutions absorption vs. time profiles were observed which consisted of a rapidly disappearing spike and a more slowly decaying background (see insets of Fig. 6). A spike lifetime of about 10^{-8} s was measured at room temperature. The background absorption disappeared within some 10^{-7} s.

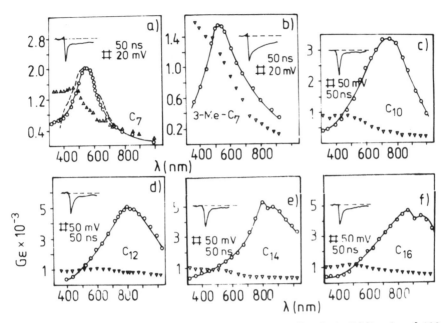

Fig. 6: Transient optical absorption spectra observed immediately (solid lines) and 100 ns (triangles) after nanosecond electron-pulse irradiation in a) n-heptane, b) 3-methylheptane, c) n-decane, d) n-dodecane, e) n-tetradecane and f) n-hexadecane containing carbon tetrachloride as electron scavenger [35]. In a) the ion-cyclotron resonance photodissociation spectrum of n-heptane radical cations is given as dashed line [33].

In Fig. 6 the spike lifetime is close to the electron pulse duration used. Picosecond pulse radiolysis studies by Jonah [36] revealed a similar behavior and proved that the spike measured in the nanosecond time domain is a "tail" originating from transients surviving up to nanoseconds.

The spectrum of the spike was different from that of the background. Nearly symmetrical absorption bands with maxima shifting gradually from 550 nm for n-heptane to 880 nm for n-hexadecane were obtained for the rapidly decaying component. The fast disappearing component of the transient optical absorptions shown in Fig. 6 is attributed to the corresponding alkane radical cations. This assignment is based on the following arguments:

- The absorption increases if electron scavengers are added.
- Band position and shape agree well with the optical absorption spectra found in glassy matrices and are attributed to alkane radical cations by parallel EPR measurements.

- The ICR photodissociation spectra available for the radical cations of n-heptane, n-decane and cyclohexane show a close resemblance to the absorption spectrum of the spike.
- A comparison of known liquid phase ionization potentials of alkanes with those estimated from the onset of charge transfer from the radical cation shows reasonable agreement.

In n-pentane and n-hexane we failed to observe a spike absorption. Using picosecond time resolution Lewis and Jonah [37] were able to detect n-hexane radical cations but they found no indication of n-pentane radical cations. This is in accordance with high-pressure mass spectroscopic studies of Meot-Ner et al. [38], who showed that the ionization entropy of n-alkanes changes from slightly positive values for carbon numbers six and less to negative values for higher carbon numbers. They interpret this result as due to an interaction between the end groups resulting in a more "cyclic" conformation which could prevent fragmentation of the alkane radical cations. The lifetime of the n-alkane radical cations observed in the neat liquid slightly increases with increasing carbon number. Their lifetime is also increased, if they are produced as solute cations by charge transfer from an alkane solvent exhibiting a higher ionization potential. An example of such an increase is given in Fig. 10. The lifetime of the n-hexadecane radical cation in n-heptane is much longer than that in n-hexadecane as solvent. A similar lifetime prolongation can be obtained if the temperature is decreased. This was studied for the case of 3-methylheptane. 3-Methylheptane is known to form transparent glasses at temperatures lower than 152 K. Therefore 3-methylheptane containing 2 vol% carbon tetrachloride was cooled down to 128 K and the absorption vs. time profile of the cationic transients was followed in liquid and solid state. It can be seen from Fig. 7 that the amplitude of the absorption is enhanced with decreasing temperature. The decay is slowed down but exhibits a spike also in the low-temperature liquid and even in the glass. The lifetime of all cationic species present is prolonged but the decay at 540 nm is faster than that of the solvated electrons the residual absorption of which can be followed at wavelengths larger than 900 nm. A similar result was reported by Klassen and Teather [39] who studied electron-pulse irradiated 3-methyloctane glasses. A possible explanation is that alkane radical cations can react with matrix molecules in an ion-molecule reaction. This is supported by recent FDMR results [40].

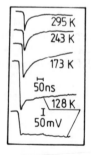

Fig. 7: Absorption vs. time profiles measured after electron-pulse irradiation of 3-methylheptane containing 2 vol% carbon tetrachloride. Temperatures are indicated. Upper four curves: λ = 540 nm, lowest curve: λ = 900 nm.

After electron-pulse irradiation of cyclohexane or methylcyclohexane containing carbon tetrachloride, time profiles of the optical absorption were measured (see Fig. 8) which also consisted of a small spike and a longer-lived background. The spike lifetime was very sensitive to impurities and olefins produced by the irradiation. To measure the spectra of Fig. 8 it was necessary to carefully purify the solvents [41] and to avoid pulse to pulse accumulation of radiolysis products. If the spectrum of the spike is displayed, bands are obtained for cyclohexane and methylcoclohexane radical cations which show maxima at about 500 nm.

All arguments given above to justify the assignment of alkane radical cations can also be applied to the transient spectra (open circles) shown in Fig. 8. Particularly, the known ICR photodissociation spectrum of cyclohexane radical cations [33] closely resembles that found as spike absorption in cyclohexane.

Picosecond pulse radiolysis experiments revealed that the spectrum of ionic transients in cyclohexane is superimposed by S_1 - S_n transitions from the first excited singlet state of cyclohexane [42].

But using cyclopropane as ion scavenger Le Motais and Jonah [43] were able to deduce a difference spectrum attributed to the cyclohexane radical cation which is in accordance with that given in Fig. 8.

The optical absorption spectra of 3-methyloctane, 2-methyldecane and squalane radical cations [44] are also given in Fig. 8. Squalane and 3-methyloctane radical cations were measured 100 ns after a 40 ns electron pulse in glassy squalane and 3-methyloctane at 6 K using N_2O as electron scavenger. The spectrum of the 2-methyldecane radical cation was detected 1 µs after the electron pulse in a 3-methylpentane matrix containing 2 vol% N_2O.

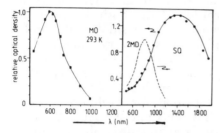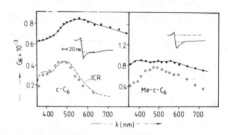

Fig. 8: Transient optical absorption spectra observed for radical cations of 3-methyloctane (MO), squalane (SQ), 2-methyldecane (MD), cyclohexane (open circles) and methylcyclohexane (open circles).

The optical absorptions of the alkane radical cations in the visible and near infrared wavelength region can be understood in molecular orbital terms. The photoelectron spectrum of the alkane molecules displays a closely packed set of molecular orbitals in the C_{2p} region extending from the highest occupied molecular orbital to valence molecular orbitals of about 7 eV higher binding energy [45]. In Fig. 9 this is illustrated for n-hexane.

In the corresponding radical cation the highest occupied molecular orbital is half-filled. Optical absorptions represent the energy needed to bring electrons from the valence orbitals to the half-filled one. Benz and Dunbar [33] have chosen MINDO/3 to calculate molecular orbitals for a selection of stable conformations of the n-hexane molecule. The conformations were equally weighted and were combined to give the density of states profile shown on the right side of Fig. 9. The optical absorption spectrum of the n-hexane radical cation reflects the density of states profile weighted by the optical transition probabilities.

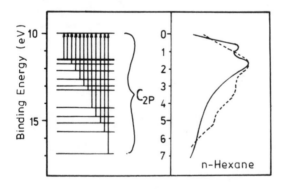

Fig. 9: Molecular orbitals of n-hexane. The arrows indicate possible electronic transitions responsible for the optical absorption spectrum of the radical cation.

The calculations reproduced the position of the n-alkane radical cation bands, their shape and the shift of the peak position to longer wavelengths with increasing carbon number.

3.2 REACTIONS OF ALKANE RADICAL CATIONS

3.2.1 *Electron Transfer to Alkane Radical Cations.* Alkane radical cations (RH^+) are able to react with solutes (S) by electron (charge) transfer:

$$RH^+ + S \rightarrow S^+ + RH \tag{18}.$$

The electron transfer reaction (18) is expected to proceed effectively, if the (gas phase) ionization potential difference $\Delta I = I(RH) - I(S)$ is positive. In various alkane mixtures electron transfer can be followed by the observation of both the solvent radical cation RH^+ and the solute cation S^+. The latter is generated synchronously with the disappearing solvent radical cation. As an example the electron transfer from n-hexadecane to the n-heptane radical cation at room temperature is shown in Fig. 10. The 560 nm band of the n-heptane radical cation is decreased by adding n-hexadecane and a new band grows in which in shape and position resembles that attributed to the n-hexadecane radical cation (cf. Fig. 6).

It can be seen from Fig. 10 a that the spike absorption at 560 nm was decreased and decayed faster on adding n-hexadecane. The longer-lived background remained unaffected. Time profiles of the absorption taken at 960 nm are given in Fig. 10 b. For the lowest n-hexadecane concentrations even an indication of a small growing part can be seen which is superimposed on a background absorption. A comparison of Fig. 10 b with Fig. 6 f shows that the lifetime of the n-hexadecane radical cation in n-heptane is considerably longer than that in neat n-hexadecane. Implications of this result will be discussed in connection with proton transfer from alkane radical cations in section 3.2.2.

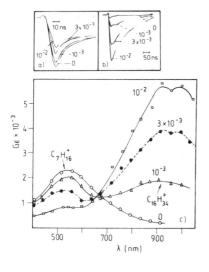

Fig. 10: Absorption vs. time profiles observed in the pulse radiolysis of n-heptane/5×10^{-3} mol dm^{-3} carbon tetrachloride after addition of 10^{-3}, 3×10^{-3} and 10^{-2} mol dm^{-3} n-hexadecane a) at 560 nm, b) at 960 nm, c) Spectral change observed after electron pulse irradiation in liquid n-heptane containing n-hexadecane and CCl$_4$. Gε-values given in (100 eV)$^{-1}$ dm^3 mol^{-1} cm^{-1}.

Solute cation formation was also studied in electron-pulse irradiated cyclohexane containing solutes as n-decane, n-dodecane, n-hexadecane, benzene and tetramethylethylene [41].

For the cyclohexane, n-alkane mixtures the electron transfer should be nearly thermoneutral (n-decane) or only slightly exothermic (n-dodecane, n-hexadecane), whereas for benzene as solute ΔI amounts to about -0.8 eV. For all solutes the well-known absorptions of the corresponding radical cations were observed (Fig. 11). In the case of benzene the absorption monitored around 950 nm belongs to the dimeric radical cation (Fig. 12).

The electron transfer rate constants are expected to increase from low values for thermoneutral and slightly exothermic reactions to a limiting value when the electron transfer step is more exothermic than -0.5 eV [46].

In the series of electron transfer spectra given in Fig. 11 and 12 this is reflected qualitatively. A concentration of 3×10^{-1} mol dm^{-3} of n-decane is needed to see some indication of the solute cation absorption. In the case of n-hexadecane more than one order in magnitude lower concentrations are sufficient to obtain a similar $G\varepsilon$-value. At 3×10^{-1} mol dm^{-3} n-hexadecane the G-value of the solute radical cation reaches 0.2 and its dependence on the solute concentration follows the WAS equation (9).

For benzene as solute concentrations of 10^{-3} mol dm^{-3} generate well observable benzene dimer radical cation absorptions. If one can assume that the extinction coefficients of the solute cations observed do not differ much (see Fig. 6), the conclusion must be drawn that the rate constants of the corresponding electron transfer reactions are varying over orders in magnitude (see Table 2).

Table 2: Rate constants for the reaction of the cyclohexane high mobility cation with different solutes.

Solute	Rate constant 10^{-11} dm^3 mol^{-1} s^{-1}	Solute	Rate constant 10^{-11} dm^3 mol^{-1} s^{-1}
n-Heptane	≤ 0.01[a]	Ethanol	0.47[a], 1.4[c]
Cyclooctane	≤ 0.01[a]	Methanol	1.65[c]
Isooctane	0.02[a]	Diethylether	1.3[c]
Methylcyclohexane	≤ 0.01[a]	Cyclopropane	1.0[c]
n-Decane	0.0002[b]	Ammonia	2.5[a], 2.6[c]
n-Dodecane	0.002[b]	Triethylamine	2.5[a], 1.3[c]
n-Hexadecane	0.006[b]	DMA	2.9[c]
t-Decalin	2.5[a]	Decalin (mixt.)	1.9[c]
Cyclohexene	1.3[a]	Benzene	1.9[c]
Ethylbromide	0.12[a]	n-Butylbromide	0.5[a]
Bicyclohexyl	2.2[a]	Biphenyl	1.3[c]

a) ref. 49, b) ref. 41, c) ref. 4b, DMA = dimethylaniline

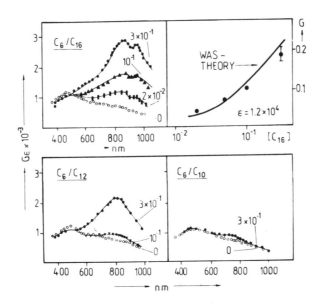

Fig. 11: Transient optical absorption spectra observed after electron pulse irradiation of a) cyclohexane containing 5×10^{-3} mol dm^{-3} CCl$_4$ and n-hexadecane, b) G-value of n-hexadecane radical cations as calculated from Eq. (9) (solid line). Points are experimental and are deduced from a) using $\varepsilon(max) = 1.4\times10^4$ dm^3 mol^{-1} cm^{-1}, c) and d) the same as a) but using n-decane and n-dodecane as solute.

Particularly, it is interesting that the limiting value of the rate constant of the charge transfer reaction (18) reaches a value of about 2×10^{11} dm^3 mol^{-1} s^{-1}. This result is confirmed by time-resolved microwave conductivity studies monitoring the solvent cation decay generated in electron-pulse irradiated cyclohexane [13], by stationary charge scavenging experiments [47], by the observation of the growing solute cation absorption in the pulse radiolysis of cyclohexane solutions [48] and recently by photoionization conductivity studies [49]. At present a large body of rate constants can be found in the literature which mainly were deduced from the decay of the positive charge carriers formed in cyclohexane. The type of reaction competing with charge recombination is not identified in all cases. Besides electron transfer, proton transfer to, e.g., ammonia, alcohols and amines, and H$_2$-transfer (cyclopropane) seem to play an important role.

The high rate constants measured point to the existence of a high mobility positive ion generated in irradiated cyclohexane. De Haas et al. [12] were able to make the first direct observation of a mobile positive ion in cyclohexane and methylcyclohexane and later in trans-decalin [50]. In other alkanes no mobile cations could be detected by microwave conductivity.

The nature of the mobile positive ion and the mechanism of its unusual fast migration are still under discussion. Warman [4b] has reviewed the experimental findings known until 1981 and came to the conclusion that the mobile positive charge carrier ("hole") is the cyclohexane radical cation $C_6H_{12}^+$. He presented a qualitative explanation of the fast charge transport which is based on the argument that the nuclear geometries of the cyclohexane radical cation and that of the cyclohexane molecule are not very different. Charge transfer from the cation to the neighboring molecule should not be hindered energetically, and a type of "resonance charge transfer" could take place. From the mobility of 10^{-2} cm^2 s^{-1} determined for the positive ion in cyclohexane [4b] a residence time of the positive charge at one distinct molecule of about 10 ps is estimated. This time of charge localization is long enough to induce only minor changes of the optical absorption spectrum of the positive hole in comparison to the "normal" positive ion. However, this short residence time should prevent the observation of a hyperfine structure in the ESR spectrum of a mobile cyclohexane radical cation.

Arguments in favor of Warman's assignment of the mobile hole as due to the cyclohexane radical cation can be summarized as follows:

- The onset of the electron transfer to n-alkanes, monitored by the appearance of the corresponding alkane radical cations, occurs for n-decane as solute (see Fig. 11). The liquid phase ionization potential of n-decane is only 0.1 eV lower than that of cyclohexane.[51].

  Species such as $C_6H_{11}^+$ or $C_6H_{13}^+$ are not able to transfer their positive charge to n-alkanes. The ionization potential of C_6H_{11} is known to be 7.7 eV. That of C_6H_{13} is expected to be similarly low.

- Addition of only 10^{-4} mol dm^{-3} of a good positive charge scavenger like benzene to cyclohexane or methylcyclohexane affects the decay of the spike absorption observed in the 500 nm region (see Fig. 12). Using ion-pair kinetics the estimation of the corresponding rate constants leads to values ranging between 1 and 2×10^{11} dm^3 mol^{-1} s^{-1}. The longer-lived background absorption remains unaffected using benzene or tetramethylethylene (TME) as solutes, but its decay is increased after addition of $> 5 \times 10^{-3}$ mol dm^{-3} triethylamine. The rate constant deduced from this decay is in the order of 10^{10} dm^3 mol^{-1} s^{-1}, indicating the reaction of a normally diffusing ionic species as expected for olefinic radical cations in cyclohexane.

On the other hand, Le Motais and Jonah [43] were able to measure in the 0 - 5 ns time domain in a broad optical absorption band around 500 nm and attributed it to the cyclohexane radical cation. This band was the part of the sum spectrum which disappeared after addition of cyclopropane.

Cyclopropane is known to react with cyclohexane radical cations by H_2-transfer [52] and does not quench the singlet excited state also absorbing in the 500 nm region. From the effect of cyclopropane and triethylamine on the decay of the 500 nm absorption they derived rate constants of about 10^{10} dm^3 mol^{-1} s^{-1} and suggested that the cyclohexane radical cation $C_6H_{12}^+$ exhibits a normal mobility.

The conclusion is drawn that the mobile hole should be the secondary cation $C_6H_{13}^+$ as also proposed by Trifunac et al. [53] on the basis of FDMR results. Sauer and Schmidt [49] studied the effect of various additives on the decay of the high mobility species in cyclohexane using the photoionization conductivity technique. The resulting quenching rate constants were correlated with thermochemical properties such as the ionization potential difference of the couples, the proton affinity (PA) difference $PA(C_6H_{11})$ - $PA(S)$ and $PA(C_6H_{12})$ - $PA(S)$. The latter difference should correlate with the proton transfer from $C_6H_{13}^+$, whereas the first one is an indication for electron transfer and the second one for proton transfer from the cyclohexane radical cation $C_6H_{12}^+$. The data show that charge and proton transfer are likely to occur from the cyclohexane radical cation in competition with ion recombination. But for two solutes (ethylbromide and n-butylbromide) only proton transfer from $C_6H_{13}^+$ is slightly exothermic. They concluded that because of the extrapolation from gas phase to liquid phase this correlation cannot be regarded as conclusive evidence concerning the nature of the high mobility species in cyclohexane. However, the secondary ion $C_6H_{13}^+$ should be considered as a possible candidate.

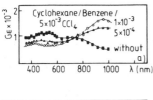

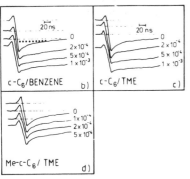

Fig. 12: a) Spectral change observed after electron-pulse irradiation in liquid cyclohexane containing benzene and carbon tetrachloride. Absorption vs. time profiles observed in the pulse radiolysis of cyclohexane containing the solutes benzene b) and tetramethylethylene (TME) c). d) methylcyclohexane /TME.

In order to make sure that electron transfer from solutes to alkane radical cations is really studied, both cations of the donor acceptor couple should be identified experimentally. Examples are given in Figs. 10-12. Here the decay of the solvent radical cation absorption is accompanied by the appearance of the corresponding solute cation absorption. Because for solutes such as amines the proton transfer reaction

$$RH^+ + S \rightarrow SH^+ + R \qquad (19)$$

is highly exothermic, the rate constants determined experimentally cannot be regarded as conclusive evidence for electron transfer. Using gas phase data proton transfer from alkane radical cations to alkanes or to alkenes should be endothermic or nearly thermoneutral. However, clear evidence exists [33,40,54] that in liquid neat alkanes and in alkane mixtures alkane radical cations disappear by proton transfer. This reaction is probably responsible for the faster disappearance of alkane radical cations if compared to that of solvated electrons (see Fig. 7). At room temperature reactions of alkane radical cations can only be studied in the time domain of 0 - 30 ns (see Fig. 6). Here the electron transfer is able to compete effectively with charge recombination, proton transfer (or the equivalent H-transfer) and H_2-transfer.

In a similar way, as shown in Fig. 10 for the couple n-heptane/n-hexadecane, a series of electron transfer reactions was studied in electron-irradiated n-hexadecane containing solutes such as cycloalkanes, alkenes and aromatics [18]. As in the case of n-heptane solutions [35] the charge transfer rate constant k_s was determined from the effect of the solutes on the time profile of the n-hexadecane radical cation absorption. The reaction scheme of Fig. 3 was adopted, and Eq. (14) was used to deduce k_s (for an example see Fig. 4).

To describe the electron transfer to alkane radical cations in a more quantitative way the following general reaction scheme is adopted [55]:

$$RH^+ + S \underset{k_{-d}}{\overset{k_d}{\rightleftharpoons}} (RH^+ \dots S) \underset{k_{-et}}{\overset{k_{et}}{\rightleftharpoons}} (RH \dots S^+) \underset{k_d}{\overset{k_{-d}}{\rightleftharpoons}} RH + S^+ \qquad (20)$$

In this equation k_d and k_{-d} are the rate constants for the formation and dissociation of the encounter pair ($RH \dots S^+$), k_{et} and k_{-et} represent the rate constants of the charge transfer step and its reversion. Using the reaction scheme (20), stationary conditions, and neglecting electron back transfer for products, the rate constant k_s of reaction (18) is obtained as

$$k_s = \frac{k_d}{1 + \dfrac{k_{-d}}{k_{et}}\left(1 + \dfrac{k_{-et}}{k_{-d}}\right)} \qquad (21)$$

If the relations

$$k_{-et}/k_{et} = \exp(\Delta G/RT) \tag{22}$$

$$k_{et} = k_{et}^0 \exp(-\Delta G^\#/RT) \tag{23}$$

are valid (ΔG is the free energy change of the forward electron transfer step, k_{et}^0 and $\Delta G^\#$ are the frequency factor and the free energy of activation of its rate constant, respectively, using the relations (22) and (23) formula (21) can be rewritten as

$$1/k_s = 1/k_d (1 + \exp(\Delta G/RT)) + k_{-d}/k_d \, k_{et}^0 (\exp(\Delta G^\#/RT)) \tag{24}$$

or

$$1/k_s = 1/k_d (1 + \exp(\Delta G/RT)) + 1/k_{et} \, \Delta V, \tag{25}$$

where k_d/k_{-d} represents the molar encounter volume ΔV.

To describe the dependence of the electron transfer rate constant k_s on the free energy change ΔG the empirical free energy relationship proposed by Marcus [56]

$$\Delta G^\# = \beta/4 (1 + \Delta G/\beta)^2 \tag{26}$$

and by Rehm and Weller [55]

$$\Delta G^\# = \Delta G/2 + ((\Delta G/2)^2 + (\beta/4)^2)^{1/2} \tag{27}$$

are usually applied. In Eqs. (26) and (27) β stands for the reorganization energy resulting from the solvent rearrangement and from the changes in the inner nuclear coordinates. The treatment of electron transfer processes in the framework of the absolute rate theory is believed to be a reasonable approach if only excitation of low energy modes ($\hbar\omega \ll kT$, e.g. solvent modes) is important for the electron transfer rate.

For non-polar media solvent reorganization is expected to play a minor role. Reorganization of high frequency intramolecular vibrational modes should determine the electron transfer rate.

To take into account the quantum nature of these high frequency vibrations the expression for k_{et} has to be modified. For this purpose we relate to a concept presented by Ulstrup and Jortner [57] and used to describe the observed k_s vs ΔG behavior observed for charge transfer in liquid n-butylchloride and n-alkanes [33,58]. Here the transfer rate k_{et} is calculated as the rate of a two state transition by first order perturbation theory. It is assumed that the ionization potential difference between the solvent and solute molecules affects the overlap between reactant and product states and in this way acts on the Franck-Condon part of k_{et}.

Taking as excess energy-accepting intramolecular modes the favored C-H modes of the solvent radical cation/solute couple only and treating the solvent part in a classical way as proposed by Levich and Dogonadze [59] we obtain the following formula for k_{et}:

$$k_{et}(\Delta G) = I_e^2 (\pi/\hbar^2 E_s kT)^{1/2} \exp(-2\mu) \sum_{n_s, n_d} \frac{\mu_s^n \mu_d^n}{n_s! \, n_d!}$$

$$* \exp -\frac{(\Delta G - \hbar\omega_s n_s - \hbar\omega_d n_d)^2}{4 E_s kT} \tag{28}$$

In this expression I_e stands for the transfer-inducing matrix element and ω_s and ω_d denote the mean frequencies of the C-H stretching and deformation modes. The parameter μ is related to the reorganization of the stretching modes which is equal to $\mu\hbar\omega_s$. E_s represents the reorganization energy of the solvent; n_s and n_d are vibrational quantum numbers.

Inserting k_{et} into Eq. (25) provides a rate constant value k_s which can be compared with the experimentally observed one. For such a comparison it is necessary to have a measure of the free energy of the forward electron transfer step ΔG, which can be obtained as follows: For an exothermic electron transfer the potential energy change (PE) from the system $(RH^+...S^+)$ plus environment is schematically plotted vs. a general nuclear coordinate in Fig. 13.

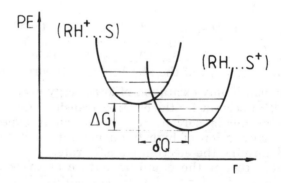

Fig. 13: Potential energy change for electron transfer

The potential energy profiles are approximated by harmonic oscillator potentials displaced by a distance ∂Q after the electron transfer. Equal vibrational frequencies are assumed for reactants and products. This means that entropy changes are assumed to be zero and ΔG as well as $\Delta G^{\#}$ contain energies rather than free energies. In this way ΔG is equivalent to the difference between the maxima of both potential curves. The free energy change can now be approximated by

$$- \Delta G \approx \Delta E = \left(I_{RH}^{g,a} + P_{RH}^{+} \right) - \left(I_{S}^{g,a} + P_{S}^{+} \right) \qquad (29)$$

In Eq. (29) $I_{RH}^{g,a}$ denotes the adiabatic gas phase ionization potential of the alkane RH, $I_{S}^{g,a}$ that of the solute S, P_{RH}^{+} the polarization energy of the medium induced by the ion RH$^+$ and P_{S}^{+} that of the solute. The adiabatic gas phase ionization potentials of n-C$_4$- to n-C$_{11}$ - alkanes are known with high precision [38]. The error limit of ΔE depends critically on the value of the polarization energy which can be calculated from the Born formula

$$P_i = e^2/2r_i \, (1 - 1/\varepsilon_{opt}), \qquad (30)$$

where r_i is an effective ionic radius, ε_{opt} the optical dielectric constant and e the elementary charge.

Another way to determine polarization energies is the measurement of photoionization thresholds I_{liq} in liquids. From $I_{liq} = I^{g,v} + V_0 + P$ polarization energies P can be obtained if the vertical gas phase ionization potential $I^{g,v}$ and the electron work function V_0 are known.

The electron transfer rate constants k_s can now be correlated with ΔE. The rate constants k_s increase from low values at thermoneutral reactions to a limiting value when the electron transfer is more exothermic than typically 0.5 eV. This limiting value of k_s can be compared with that estimated from the expression valid for a rate constant k of a bimolecular reaction in solution $k = 4\pi D r_0 P_{react}$, where D is the sum of the diffusion coefficients of the reactants, r_0 the reaction radius and P_{react} the reaction probability. For a diffusion-controlled reaction the reaction probability is unity. In Fig. 14 an attempt is shown to calculate for all electron transfer couples measured in cyclohexane, n-heptane and n-hexadecane the reaction probability according to $P_{react} = k_s /k^{lim}$ and to plot it as a function of ΔE.

The P_{react} vs. ΔE dependence obtained demonstrates that in the case of high exothermicities the rate constant of the actual electron transfer step $k_{et} \Delta V$ is higher than the diffusion-controlled rate constant k_d:

$$k_s = \frac{k_d \, k_{et} \, \Delta V}{k_d + k_{et} \, \Delta V}, \quad \frac{k_{et} \, \Delta V}{k_d + k_{et} \, \Delta V} = P_{react} \tag{31}.$$

Taking $E_s = 0.4$ and assuming that for all reactant pairs a mean encounter volume $\Delta V = 0.5 \, dm^3 \, mol^{-1}$ can be used, P_{react} was calculated for all electron transfer couples presented in Fig. 14. A reasonable fit is obtained for $2\pi \, I_e^2 \, h^2 ; \omega_d = (0.5 - 5) \times 10^{12} \, s^{-1}$ and $\mu = 2$ in Eq. (28). If the sum of the vibrational deformation energies is only 0.7 eV as in our case of $\mu=2$ at an excess energy $\Delta E \geq 0.4$ eV, the maximum efficiency of the electron transfer is obtained. As in the case of the Marcus theory a drop-off of k_{et} is predicted for highly exothermic reactions. This is masked by a "flattening" of the calculated profile resulting from the inequality $k_{et} \Delta V \gg k_d$.

In conclusion, it appears that the rate of all electron transfer reactions studied in liquid alkanes is not only determined by the electronic matrix element I_e^2 but also in an important way by the vibrational overlap (Franck-Condon factor) of reactant and product states. In contrast to electron transfer in polar media, in non-polar liquids such as alkanes the excitation of intramolecular vibrational modes is favored. The manner in which the excess energy is distributed over the intramolecular vibrational modes of the electron transfer couple determines the rate of transfer.

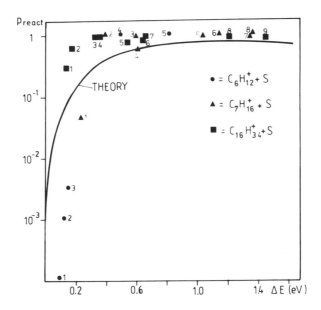

Fig. 14: Dependence of the reaction probability of the electron transfer reactions summarized below on ΔE.
Electron transfer from the solutes 1 n-decane, 2 n-dodecane, 3 n-hexadecane, 4 decalin (mixture) and 5 benzene to the cyclohexane radical cation; ▲ electron transfer from the solutes 1 cyclohexane, 2 n-hexadecane, 3 decalin (mixture) 4 heptene-1, 5 benzene, 6 cyclohexene, 7 toluene and 8 tetramethylethylene to the n-heptane radical cation; electron transfer from the solutes 1 pentylcyclohexane, 2 1,2,4-trimethylcyclohexane 3 heptene-1, 4 t-decalin, 5 cyclohexylbenzene, 6 benzene, 7 bicyclohexylbenzene, 8 biphenyl and 9 cyclohexylbiphenyl to the n-hexadecane radical cation.

3.2.2 *Proton Transfer from Alkane Radical Cations.*

The observation that alkane radical cations generated in neat liquid alkanes decay faster than the corresponding solvated electrons (see Fig. 7) implies that alkane radical cations disappear not only by recombination. An additional pseudo-frist order rate constant k_L (see Fig. 3) is needed to simulate the faster decay of alkane radical cations. Unimolecular reactions such as H_2 elimination or deprotonation cannot fully account for this additional reaction channel. H_2 elimination can be disregarded because the products of such an elimination would result in a delayed absorption around 280 nm. This is clearly not observed. The 280 nm peak is ascribed to alkene radical cations which are formed within the electron pulse duration (see section 4.2.2). No delayed absorption could be detected.

Although deprotonation was proposed to explain the disappearance of alkane radical cations formed in low-temperature matrices, this reactions seems to play a minor role in neat liquid alkanes. Usually the lifetime of alkane radical cations is prolonged if they are generated not in neat alkanes but in alkane mixtures. This stabilization effect cannot be understood if deprotonation would be important. It points to a possible bimolecular reaction channel such as proton transfer (or the equivalent H transfer)

$$RH^+ + RH \rightarrow RH_2^+ + R^\cdot \quad . \tag{32}$$

A similar conclusion was drawn by Trifunac et al., based on evidence obtained from FDMR experiments [40]. This is also in accordance with results of experiments in zeolite hosts containing alkanes which demonstrated the need of alkane molecules as proton acceptors [54].

From the simulation of the $G\varepsilon$ vs.time profiles of the n-alkane radical cations shown in Fig. 4 rates k_L of about 10^8 s^{-1} were estimated. In this series of n-alkanes the k_L values are not much different. At room temperature an upper limit of the rate constant of reaction (32) of about 10^7 dm^3 mol^{-1} s^{-1} is obtained. As shown in Fig. 7 this rate constant seems to decrease with decreasing temperature.

Alkyl radicals are generated as product of reaction (32). However, at the alkyl radical absorption at 240 nm (see Fig. 16) no measurable delayed absorption can be seen. This points to the importance of sources of alkyl radicals other than proton transfer. There is some evidence that alkyl radical formation via charge recombination [60-62] and fragmentation of vibrationally excited radical cations [63] is dominating.

Using gas phase data, the proton transfer reactions (32) are expected to be endothermic for C > 5 alkanes [64]. This is not very conclusive for the liquid phase. However, the low proton transfer rate constants of 10^7 dm^3 mol^{-1} s^{-1} observed in n-alkanes may imply that the proton transfer is nearly thermoneutral or even slightly endothermic.

4. Alkene Radical Cations

4.1 OPTICAL ABSORPTION SPECTRA OF ALKENE RADICAL CATIONS

Already in 1966 Shida and Hamill [65] showed that vinylene (-CH=CH-) and vinylidene ($H_2C=C<$) type olefins form radical cations in γ-irradiated alkyl chloride or alkane glasses. They observed broad optical absorption bands at about 600-800 nm ("red bands") which were attributed to the radical cations of the olefins studied. It was assumed that the positive charge generated by ionization migrates through the matrix (RX) and is trapped by the olefinic solute (S) forming radical cations (S$^+$):

$$RX^+ + S \rightarrow S^+ + RX \tag{33}$$

From similar experiments using cyclohexene as solute Badger and Brocklehurst [66] concluded that the red bands are due to dimer radical cations.

Typical red band spectra were also observed after electron-pulse irradiation of liquid alkanes containing olefins [67]. In these experiments monomer radical cations and cations of the vinyl type ($H_2C=CH-$) could also be detected.

A typical optical absorption spectrum of transients generated in electron-pulse irradiated cyclohexane solution containing heptene-1 and carbon tetrachloride is given in Fig. 15. Monomer and dimer heptene-1 radical cations exhibit optical absorption bands at 280, 320 and 650 nm. Monomer and dimer cations can be distinguished by their kinetic behavior: At alkene concentrations $\geq 10^{-3}$ mol dm^{-3} in cyclohexane solution alkene monomer radical cations (Ol$^+$) are formed within a 10 ns electron pulse (see insert of Fig. 15 and Fig. 16). Dimer cations (Ol$_2^+$ grow in slowly as expected from the bimolecular reaction

$$Ol^+ + Ol \rightarrow Ol_2^+ \ . \tag{34}$$

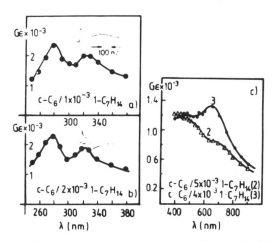

Fig. 15: Transient optical absorption spectra observed in solutions of heptene-1 in cyclohexane containing 10^{-2} mol dm^{-3} carbon tetrachloride. Insert: oscilloscope traces taken at 320 nm- Heptene-1 concentrations (in mol dm^{-3}) as indicated.

This behavior is demonstrated in Fig. 16b at 320 nm (growing part of the absorption).

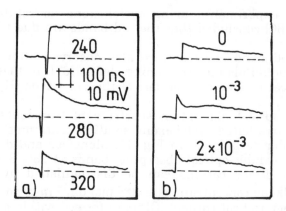

Fig. 16: a) Absorption vs. time profiles taken after electron pulse irradiation of n-heptane containing 5×10^{-3} mol dm^{-3} carbon tetrachloride at 240, 280 and 320 nm.
b) Absorption vs. time profiles taken at 320 nm after electron pulse irradiation of cyclohexane containing 10^{-2} mol dm^{-3} carbon tetrachloride and heptene-1. Heptene-1 concentrations (in mol dm^{-3}) as indicated.

Table 3: Spectroscopic and kinetic characteristics of alkene radical cations (m = monomer, d = dimer)

Solute	λ_{max} (nm) [67]	[65]	Rate conctant k_{34} (dm^3 mol^{-1} s^{-1}) [67]
Tetramethylethylene	900 d	866	5×10^9
Cyclohexene	270 m		1×10^{10}
	750 d	708	
1-Methylcyclohexene	870 d	820	7×10^9
3-Methylcyclohexene	790 d	745	1×10^{10}
4-Methylcyclohexene	770 d	715	5×10^9
4-Vinylcyclohexene	780 d	713	1×10^{10}
Methylenecyclohexane	630 d	720	4×10^9
Allylcyclohexane	650 d		5×10^9
2-Methylhexene-1	650 d		6×10^9
Heptene-1	280 m		6×10^9
	650 d		6×10^9
t-Hexene-2	280 m		8×10^9
	700 d		
Limonene	290 m		
	850 d		

In glassy solutions of 3-methylheptane and methylcyclohexane containing olefins, olefin monomer cations were found after irradiation, whereas the dimerization was too slow to result in a measurable amount of dimer cations [67].

Spectroscopic and kinetic properties of alkene radical cations are summarized in Table 3.

The assignment of the bands denoted by m in Table 3 to alkene monomer radical cations is based on the following arguments:

- These absorptions grew in synchronously with the decay and the decrease of the alkane radical cation absorption.
- Even in glassy solutions a remaining solute absorption part was observed which was formed during a 10 ns electron pulse.
- The appearance of an absorption maximum at about 280 ns is in agreement with quantum chemical predictions of a strong π - π^* transition within this wavelength range.
- The ion-cyclotron resonance photodissociation spectrum of the cis-2-pentene radical cation, which is believed to be characteristic for alkene radical cations, showed a peak position and structure similar to that observed for the 280 nm band (see Fig. 19a).

The assignment of the bands denoted by d in Table 3 to alkene dimer radical cations is supported by the following arguments:

- For all d bands a time delay between the disappearance of the alkane radical cation absorption and the growth of the d band absorption was observed.
- The rise time of the growing d band absorption decreased with increasing alkene concentration (see Fig. 21).
- The growth of the d band absorption and its lifetime were prolonged at decreasing temperature. The absorption disappeared in glassy matrices indicating that diffusive motion is essential for its formation.
- Quantum chemical calculations predict charge resonance bands with strong oscillator strengths in the region between 600 and 800 nm if a sandwich-like structure of the cation and the neutral molecule in the dimer is assumed [68,69].

4.2 FORMATION OF ALKENE RADICAL CATIONS

4.2.1 *Formation of Alkene Radical Cations via Charge Transfer.* Electron-pulse irradiation of alkane solutions generates absorption bands of alkane radical cations. The addition of the alkenes listed in Table 3 caused the alkane radical cation absorption to decrease (see Fig. 17). New optical absorption bands appeared in the wavelength region between 250 and 1000 nm (Fig. 16). This indicates that charge transfer from the alkane radical cation to the alkene solute effectively takes place:

$$RH^+ + Ol \rightarrow Ol^+ + RH \tag{35}$$

As discussed in section 3.2.1, the rate constants of reaction (35) are strongly dependent on the free energy change ΔG of the electron transfer step. For the alkane/alkene electron transfer couples mentioned in Fig. 14 the excess energy ΔE is larger than 0.4 eV and the limiting value of the electron transfer rate constant is already reached ($p_{react} = 1$). This means that charge transfer from alkane radical cations to alkenes proceeds with unit efficiency. The same holds for some halocarbon solvents [70].

Their low ionization potentials favor alkenes as scavengers of positive charges in a variety of solvents.

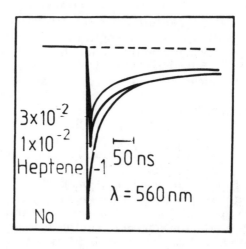

Fig. 17: Effect of heptene-1 on the n-heptane radical cation absorption. Oscilloscope traces taken at 560 nm. Heptene-1 concentrations given in mol dm^{-3}.

4.2.2 *Formation of Alkene Radical Cations in Neat Liquid Alkanes.* In Fig. 18 the optical absorption spectra of transients are shown which were generated in an electron-pulse irradiated n-heptane solution containing the electron scavenger carbon tetrachloride. Care was taken to avoid accumulation of stable radiolysis products as olefins. After each electron pulse the sample solution was replaced by a fresh one. At least three different transients can be identified from the spectrum displayed in Fig. 18:

- A very long-lived transient absorbing at 240 nm (and probably also at lower wavelengths). This component is assigned to alkyl radicals. Their optical absorption spectra are known from proton pulse radiolysis experiments [71].

- A short-lived absorption showing a pronounced peak at 280 nm. This absorption band corresponds to the well-known π-π*-bands of alkene radical cations. The background absorption of Fig. 6 showing a shoulder at 450 nm and some tailing up to the near infrared range exhibits the same decay time constant as observed at 280 nm.

Quantum chemical calculations predict several σ-π transitions in this wavelength range. Therefore, the background absorption (triangles in Fig. 18b and Fig. 6) is also attributed to alkene radical cations. The observation of alkene radical cations in neat liquid alkanes is consistent with the detection of parent minus H_2 alkene radical cations in electron-irradiated alkanes using FDMR [72,73]. The structural information obtained from these experiments is superior to that from optical absorption spectroscopy and confirms unambiguously that alkene radical cations are formed in electronpulse irradiated alkanes.

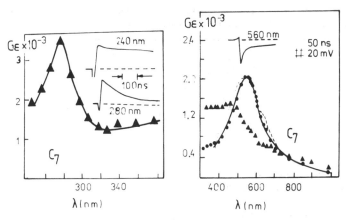

Fig. 18: Transient optical absorption spectra observed in electron-pulse irradiated n-heptane solution containing 10^{-2} mol dm^{-3} carbon tetrachloride. Insert: Oscilloscope traces taken at the wavelengths indicated.

Similar bands were observed in neat liquid n-hexane, n-hexadecane and less pronounced in cyclohexane (see Fig. 19).

As already discussed in section 3.1 the "spike" absorption (points of Fig. 18b) is ascribed to n-heptane radical cations. The question arises, how alkene radical cations are generated in irradiated neat alkanes and alkane mixtures. Does charge transfer occur to alkenes which are present in the alkanes either as impurity or as stable radiolysis product? Or can fragmentation of vibrationally excited or relaxed alkane radical cations contribute considerably to the alkene radical cation formation?

$$RH^+ + Ol \text{ (impurity, radiolysis product)} \rightarrow Ol^+ + RH \qquad (36)$$

$$RH^{+*} \rightarrow Ol^+ + (H_2, CH_4, ...) \tag{37a}$$

$$RH^+ \rightarrow Ol^+ + (H_2, CH_4, ...) \tag{37b}$$

In the pulse radiolysis experiments mentioned, reaction (36) can be excluded as main pathway of the alkene radical cation formation. The maximum residual olefin concentration in the alkane solvent was checked by UV spectroscopy and gas chromatography to be less than 5×10^{-5} mol dm^{-3}. The maximum end-of-pulse concentration of olefinic radiolysis products was 2×10^{-5} mol dm^{-3} [35].

The effect of small heptene-1 concentrations on the alkene radical cation band observed in cyclohexane is shown in Fig. 16. By addition of 10^{-3} and 2×10^{-3} mol dm^{-3} heptene-1 the 280 nm peak remained nearly unchanged but a dimeric band appeared at 320 nm. Addition of 10^{-4} mol dm^{-3} heptene-1 caused no detectable change in the spectrum. In neat cyclohexane the monomer band at 280 nm shows nearly the same $G\varepsilon$ value, but there is no indication of a dimer band (see Fig. 19).

The absence of a dimer radical cation peak in the neat alkane solution (Fig. 19) on one hand and the large monomer radical cation peak on the other hand implies the assumption that the alkene concentration is too low to allow effective charge transfer (reaction 36). Therefore, the results favor a possible fragmentation of alkane radical cations as source of alkene radical cations. However, decomposition of vibrationally relaxed alkane radical cations can be excluded as the main pathway, because alkane and alkene radical cations coexist in neat alkane solutions and the alkene radical cations are formed promptly even in the picosecond time domain [43]. It is also known from gas phase experiments [74] that only alkane radical cations decompose which carry some excess energy.

In conclusion, it appears that alkene radical cations are formed in irradiated neat liquid alkanes and alkane mixtures mainly via fragmentation of vibrationally excited alkane radical cations (reaction 37a).

There is evidence from FDMR experiments that in the liquid phase mainly parent minus H_2 olefin radical cations are generated [72], although the gas phase decomposition threshold for CH_4 elimination is lower [74]. Fragmentation of vibrationally excited n-alkane radical cations dominates over vibrational relaxation when the carbon number is lower than 5. In the nanosecond pulse radiolysis n-hexane radical cations could not be observed. Le Motais and Jonah [43] demonstrated that even in the picosecond time domain in neat n-pentane only pentene radical cations could be detected.

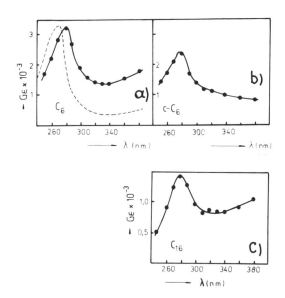

Fig. 19: Transient optical absorption spectra observed in electron pulse irradiated a) n-hexane, b) cyclohexane and c) n-hexadecane. The ion cyclotron resonance spectrum of the cis-2-pentene radical cation [33] is shown as dashed line.

4.3 FORMATION AND REACTIONS OF ALKENE DIMER RADICAL CATIONS

Using nanosecond pulse radiolysis, alkene dimer radical cations could be followed directly in the near infrared region [67,75] (Fig. 20). Rate constants could be determined by simulating the red band absorbance vs. time profiles at different alkene concentrations (Fig. 21, full lines). Homogeneous kinetics was applied and the rate equations were solved for the simplified mechanism

$$S^+ + S \rightarrow S_2^+, \qquad k_{38} \text{ calculated from fit} \tag{38}$$

$$S_2^+ + Cl^- \rightarrow \text{prod. 1}, \; k_{39} = 5 \times 10^{11} \text{ dm}^3 \text{ mol}^{-1} \text{ s}^{-1} \tag{39}$$

$$S_2^+ + S \rightarrow \text{prod. 2}, \; k_{40} \text{ deduced at high alkene concentrations} \tag{40}$$

Alkene monomer radical cations and chloride ions are assumed to be formed during the electron pulse. Alkene dimer radical cations disappear by recombination or a further reaction with alkenes. The dimerization rate constants k_{38} obtained by this procedure are also summarized in Table 3. Pronounced changes which correlate with the molecular structure of the alkenes were not observed. All rate constants determined are within or slightly below the diffusion-controlled range.

264

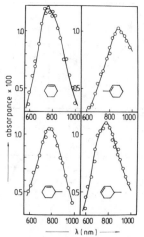

Fig. 20: Transient optical absorption spectra of alkene dimer radical cations observed in electron-pulse irradiated n-heptane solution containing carbon tetrachloride and cylohexene, 1-methylcyclohexene, 3-methylcyclohexene and 4-methylcyclohexene.

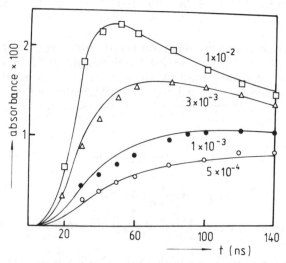

Fig. 21: Formation of cyclohexene dimer radical cations observed in electron-pulse irradiated n-heptane solution containing carbon tetrachloride and cyclohexene. Points are experimental, full lines are calculated using the mechanism (38 - 40). Concentrations in mol dm^{-3}.

The rate constants of reaction (40) were deduced from the decay of the red band absorptions at alkene concentrations $\geq 5 \times 10^{-2}$ mol dm^{-3} and are listed in Table 4. Some product absorptions appeared between 320 and 450 nm. The assignment of these bands is not clear. There is evidence from ODESR and FDMR [76,77] that higher aggregates, e.g., trimer radical cations of olefins, can be formed.

Table 4: Rate constants of the reaction $S_2^+ + S \rightarrow$ products (40)

Solute	Rate constant k_{40} $(dm^3\ mol^{-1}\ s^{-1})$	Product absorption at (nm)
Cyclohexene	6.0×10^6	320
1-Methylcyclohexene	6.5×10^7	380
3-Methylcyclohexene	1.8×10^8	350
4-Methylcyclohexene	1.8×10^7	380
4-Vinylcyclohexene	4.5×10^7	
Methylenecyclohexane	2.0×10^7	
Allylcyclohexane	2.0×10^8	
2-Methylhexene-1	2.0×10^7	
Limonene	1.0×10^9	450
t-Hexene-2	2.0×10^8	
Heptene-1	2.5×10^8	
Octene-1	2.0×10^8	

It is expected that alkene radical cations are able to donate a proton to alcohols. After addition of alcohols listed in Table 5 the red band absorption disappeared faster, indicating that there is a pseudo-first order reaction of the dimer radical cations with the alcohols. The observed correlation of the rate constants deduced from the cyclohexene dimer radical cation decay with the proton affinity of the alcohols is strong evidence that the proton transfer reaction

$$(c\text{-}C_6H_{10} \ldots c\text{-}C_6H_{10}^+) + ROH \rightarrow (c\text{-}C_6H_{10} \ldots c\text{-}C_6H_9^{\cdot}) + ROH_2^+ \tag{41}$$

takes place. The rate constants of the corresponding proton transfer reactions and the proton affinities are listed in Table 5. There is also experimental evidence that not only alkene dimer but also monomer radical cations undergo proton transfer reactions.

Table 5: Proton transfer from cyclohexene dimer radical cations to different alcohols (reaction 41)

Alcohol	Rate constant k_{41} (dm^3 mol^{-1} s^{-1})	Proton affinity (kJ mol^{-1})
Methanol	1.7×10^9	764
Ethanol	2.0×10^9	789
iso-Propanol	3.5×10^9	819
t-Butanol	1.0×10^{10}	865

5. Diene Radical Cations

5.1. FORMATION AND OPTICAL ABSORPTION SPECTRA

Optical absorption spectra of various diene radical cations have been determined in γ-irradiated sec-butylchloride glasses at 77 K [78]. These spectra showed pronounced maxima in the range between 420 and 460 nm and in the infrared between 1300 and 1400 nm. Using the Longuet-Higgins-Pople theory the visible band could be assigned to $\pi_2^2 \pi_2^1$ - $\pi_1^1 \pi_1^2$ and π_1^2 π_3^1 transitions, whereas the infrared bands were ascribed to intramolecular charge resonance transitions [79].

After electron-pulse irradiation of n-butylchloride solutions containing the dienes given in Table 6 similar optical absorption bands were observed [80]. The solvent n-butylchloride was used because its (gas phase) ionization potential of 10.67 eV is well above the diene ionization potentials. Furthermore, it is known from pulse radiolysis experiments done in neat liquid n-butylchloride that n-butylchloride radical cations show an absorption band around 520 nm. This absorption decays at room temperature with a time constant of about 100 ns. An underlying weaker absorption at 420 nm was attributed to the butene radical cation [81].

For comparison with the various diene cation transient absorptions, the spectra obtained in electron-pulse irradiated n-butylchloride are given in Fig. 22.

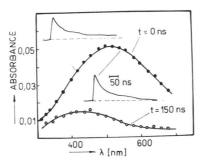

Fig. 22: Transient optical absorption spectrum observed after electron-pulse irradiation in neat liquid n-BuCl at room temperature

Concomitant with the BuCl⁺ decay at 600 nm, product absorptions were observed to grow in between 1000 and 1400 nm in solutions containing 1,3-pentadiene, isoprene and 1,4-cyclohexadiene. The growth could easily be measured, because practically no solvent cation absorption was observed in this wavelength range.

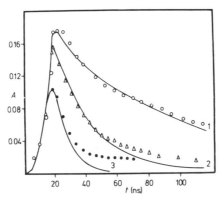

Fig. 23: Effect of 3×10^{-3} (2) and 1×10^{-2} mol dm^{-3} cis-1,3-pentadiene (3) on the BuCl⁺ decay at 600 nm, (1) pure BuCl.

As an example some oscilloscope traces taken at 1050 nm in irradiated n-butylchloride - isoprene solutions are given in Fig. 24.

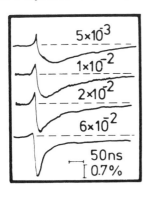

Fig. 24: Oscilloscope traces taken at 1050 nm after electron-pulse irradiation of n-BuCl solutions containing isoprene.
Concentrations given in mol dm^{-3}.

At diene concentrations equal to or larger than 10^{-2} mol dm^{-3} the solvent cation absorptions were strongly decreased, and product absorptions also appeared in the visible region. Pronounced absorption bands with typical maxima around 400 nm were obtained.

From the same kinetic behavior observed for the 400 nm and the infrared bands and from the simultaneous formation of both bands in electron-pulse irradiated glassy s-butylchloride solution (see Fig. 25c) it can be concluded that both bands belong to the same species and can be attributed to diene monomer radical cations. Their optical absorption spectra are given in Fig. 25 to 27. The inserts are oscilloscope traces taken at the wavelengths indicated.

Table 6: Spectral and kinetic characteristics of diene radical cations.

Solute	Peak wavelength (nm)		Formation rate constant (dm^3 mol^{-1} s^{-1})	Dimerization
cis-1,3-Pentadiene	m:440	1350	9×10^9	4×10^8
	d:380	780		
trans-1,3-Pentadiene	m:450	1350		
	d:380	780		
Isoprene	m:450	1320	5×10^9	5.5×10^8
	d:380	600		
1,3-Cyclohexadiene	m:400	480	1×10^{10}	
	d:400	800		
	CHD: 310			
1,4-Cyclohexadiene	m:400	480	1×10^{10}	4.5×10^9
	m:1100			
	d:400	800		
	CHD: 310			
1,5-Cyclooctadiene	m:680		1×10^{10}	
	d:not observed			

m = monomer, d = dimer, CHD = cyclohexadienyl radical

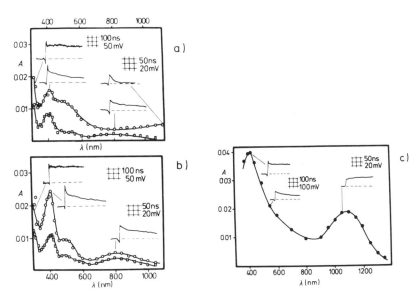

Fig. 25: Transient optical absorption spectra observed after electron-pulse irradiation of liquid (a and b, T = 293 K) and solid (T = 107 K) s-BuCl containing 1,4-cyclohexadiene. a) 10^{-2} mol dm^{-3}, b) 10^{-1} mol dm^{-3}, O immediately after the pulse, ☐ after 200 ns, c) 10^{-1} mol dm^{-3} ● after 300 ns. Inserts are oscilloscope traces.

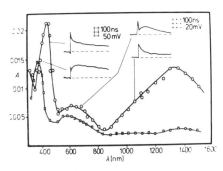

Fig. 26: Transient optical absorption spectra observed in electron-pulse irradiated n-BuCl containing 3×10^{-2} mol dm^{-3} isoprene. O immediately after the pulse, ☐ after 150 ns.

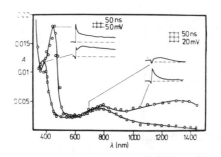

Fig. 27: Transient optical absorption spectrum observed in electron-pulse irradiated n-BuCl containing 5×10^{-2} mol dm^{-3} cis-1,3-pentadiene. ○ immediately after the pulse, □ after 200 ns.

5.2 REACTIONS OF DIENE RADICAL CATIONS

From Fig. 24 it can be seen that with increasing isoprene concentration the decay time constant of the isoprene monomer cation band decreases. On the other hand, growing parts of the transient absorption were observed at 800 and below 400 nm. These bands could not be observed in glassy s-BuCl solutions containing the corresponding dienes. There is strong evidence that diene radical cations dimerize. Rate constants of the dimerization reactions are included into Table 6. 1,4-Cyclohexadiene dimer radical cations (and probably higher multimers) were also detected by FDMR experiments [77].

For 1,3- and 1,4-cyclohexadiene solutions after irradiation an increase of the absorption in the vicinity of 320 nm was observed. Its growth became faster with increasing diene concentration as expected for a bimolecular reaction. ESR studies done in γ-irradiated frozen trichlorofluoromethane matrices containing high concentrations of cyclohexadiene proved unambiguosly that cyclohexadienyl radicals were formed [82].

Cyclohexadienyl radicals are known to absorb at 310 nm [83]. Therefore, it seems reasonable to assume that at least a part of the absorption measured at 320 nm can be ascribed to cyclohexadienyl radicals. Proton transfer from the cyclohexadiene radical cation to the corresponding solute is assumed as the source of these radicals. It is known from matrix isolation studies [84] that ring opening of cyclohexadiene radical cations occurs after photoillumination. Hexatriene cations were identified as products. These types of cations also exhibit pronounced optical absorption bands around 400 nm and can hardly be distinguished from the cyclohexadiene cations in pulse radiolysis experiments.

6. Radical Cations in Carbon Tetrachloride

6.1 TRANSIENTS OBSERVED IN NEAT LIQUID CCl₄

A typical transient absorption spectrum generated by nanosecond electron pulses in pure liquid CCl_4 is shown in Fig. 28. The dominant spectral characteristics: bands with peak wavelengths at 340, 370 and 475-500 nm and lifetimes at room temperature in the 100, > 300 and 20 ns range, respectively, are well established by results of different research groups [46,85-87].

In contrast to the other transients, the species absorbing at 370 nm is quenched by oxygen. This points to its radical nature. A comparison of the band with results obtained in acidic sodium chloride solution [88] shows that the 370 nm transient can be attributed to Cl_2^- produced by the reaction

$$Cl^{\cdot} + Cl^{-} \rightarrow Cl_2^{-} \ . \tag{42}$$

The assignment of the other transient bands is less unequivocal, however.

The main characteristics of the optical absorption bands and some critically reviewed assignments are summarized in Table 7.

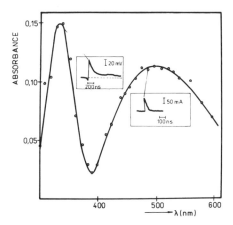

Fig. 28: Transient optical absorption spectrum generated in pure liquid carbon tetrachloride immediately after electron-pulse irradiation. Inserts: oscilloscope traces taken at 350 and 450 nm

6.2 THE PRIMARY CATION CCl_4^+

In 1966 Shida and Hamill [89] observed a weak absorption centered at 420 nm in γ-irradiated polycrystalline CCl_4 at 77 K and attributed it to a cation $(CCl_3 - Cl)^+$ with an increased nuclear separation. Shida and Takemura [27] showed by parallel optical and ESR experiments that a shoulder detected at 400 nm in γ-irradiated pure CCl_4 can be assigned to the primary cation CCl_4^+. The same band shifted to 380 nm in a CCl_3F matrix. In an irradiated 20 K argon matrix containing CCl_4 Andrews and Prochaska [90] found a band at 425 nm which was identified as CCl_4^+.

The photoelectron spectrum of CCl_4 indicates that possible one electron transitions should have energies of 1.8 and 5.0 eV, respectively. The photoelecton mass spectrum revealed no CCl_4^+, but CCl_3^+ appeared at 11.3 eV, i.e. 0.4 eV below the first vertical ionization energy. Using these findings and other spectral information Andrews et al. [91] suggested the potential curves for the ground and excited state of CCl_4^+ as given in Fig. 29. The $\widetilde{C} \rightarrow \widetilde{X}$ transition energy for a structure of the cation which is close to that of the CCl_4 molecule (tetrahedral structure) is given by the difference of the photoelectron bands of 5 eV. Structural relaxation in both $\widetilde{X}$ and $\widetilde{C}$ states results in minima shifted to larger C-Cl distances.

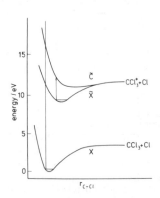

Fig. 29: Suggested potential curves for the ground $(\widetilde{X})$ and excited state $(\widetilde{C})$ of CCl_4^+ [91].

Table 7: Characteristics and assignments of the optical absorption bands observed in electron-pulse irradiated liquid CCl4.

Characteristics	UV-Band	Visible Band
λ(max)	335-350	475-500
decay	first order	first order
$\tau_{1/2}$ (20° C)	100 ns [87]	14-20 ns [85,87]
formation	during a ps pulse [92,93]	growth after a ps pulse [92,93]
reactions	reacts with alkanes	addition of alkanes causes a decrease of the band but has no effect on its decay
slow decay	no	small contribution

Assignments

Cooper and Thomas [85]	not observed	cationic species
Zürich group [86,94]	CCl_2	CCl_3^+, free or within an ion-pair
Leipzig group [87]	CCl_4^+	ion-pair (CCl_4^+ ...Cl^-)
Delft group [95]	-	product resulting from geminately recombining ions

Lengthening of the C-Cl bond is revealed experimentally by the infrared spectrum of CCl_4^+ in solid argon. However, lengthening of the C-Cl-bond shifts the absorption maximum of the cation to lower energies. Thus, the large deviation of the CCl_4^+ absorption maximum at different matrices can be understood qualitatively. The effect of the matrix on CCl_4^+ seems to be twofold: stabilization against decomposition to CCl_3^+ and structural relaxation of CCl_4^+. Similar effects are observed for $CHCl_3$ and CH_2Cl_2. Absorption maxima of the primary cations are found at 388 and 342 nm, respectively [96].

It cannot be excluded that the same structural relaxation is operative not only in frozen matrices but also in liquid carbon tetrachloride. The oxygen-insensitive, fast decaying absorption at 340 nm which is formed immediately within a ps electron pulse [92,93] could be attributed to the primary cation CCl_4^+. This assumption is further supported by the experimental finding, that the 340 nm transient reacts with a large number of solutes (S) such as alkylchlorides, alkanes, alkenes and aromatics as expected for the charge transfer reaction

$$CCl_4^+ + S \rightarrow S^+ + CCl_4 \qquad (43).$$

A pronounced dependence of the measured rate constant on the ionization potential difference $\Delta I = I(CCl_4) - I(S)$ was observed. The rate constant reached a peak value of about 3×10^9 dm^3 mol^{-1} s^{-1} at $\Delta I > 0.8$ eV. Well-known absorption bands of alkane and radical cations appeared at solute concentrations larger than 10^{-2} mol dm^{-3} [46].

The CCl_4 cation seems to be rather instable. In argon matrices it is easily bleached by 500 - 1000 nm photoillumination. It is very likely that a large portion of the CCl_4 primary cations initially possess excess energy. Excited cations either decompose, neutralize with its partner ion or become stabilized by their interaction with the matrix. A schematic representation of this process is given in the ion-pair scheme of Fig. 30. The possibility cannot be excluded that in irradiated carbon tetrachloride, as in alkanes, primary cations and decomposition products exist simultaneously.

Conductivity studies done by the Delft group [95] showed that if the absorption vs. time profile measured for the 340 nm band (100 ns decay) and that of the 475 nm band (20 ns decay plus underlying μs decay) are added, the conductivity time profile can be reproduced.

The arguments given above are in favor of the assumption that the primary cation CCl_4^+ is observable at 340 nm in the time range of 10^{-7} s. It is likely that decomposition of excited CCl_4 can occur in competition to energy relaxation and geminate recombination. A similar mechanism is operative in irradiated alkanes and in butylchloride.

For the 340 mm band an alternative assignment is proposed by Bühler et al [97]. They attribute the band to CCl_2 and conclude from kinetic calculations [94c] that the CCl_4^+ ion should have a lifetime of about 50 ps which excludes its observation at nanoseconds or longer.

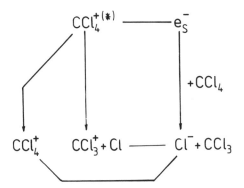

Fig. 30: Possible primary ion-pairs formed in liquid CCl4. Primary and secondary ion-pairs indicated by horizontal lines.

6.3 THE FRAGMENT CATION CCl_3^+

After irradiation of neat liquid carbon tetrachloride with nanosecond [85-87] or picosecond [92,93] electron pulses a transient absorbing between 475 and 500 nm was observed. The absorption showed a time-resolved buildup from which a formation rate of 1.4×10^{10} s^{-1} at 20 C° was deduced [93,94 b]. Its decay consisted of two components: a fast one disappearing by first order with a time constant of about 15 ns and a slower one decaying by second order within a few microseconds. The amplitude of the slow absorption part was about 5% of that of the fast one. If typical cation scavengers such as alkanes, alkenes, aromatics or amines were added to CCl4, two different effects could be observed after pulse irradiation: first, reduction of the amplitude of the 475 nm band at solute concentrations > 10^{-2} mol dm^{-3} but no effect on the decay time constant and second, reduction of the amplitude accompanied by an increase of the decay time constant. These experimental findings indicate that the 475 nm band has cationic nature. Alkanes such as n-hexadecane (gas phase ionization potential 10.1 eV) are able to compete with the 475 nm transient for a common precursor. As shown in Fig. 31 the 475 nm band is decreased on addition of n-hexadecane but its decay remains unchanged. This solute is unreactive towards the 475 nm transient. However, the well-known radical cation band of n-hexadecane is appearing around 900 nm. This shows that the charge transfer reaction

$$CCl_4^+ + \text{n-}C_{16}H_{34}^+ \rightarrow \text{n-}C_{16}H_{34}^+ + CCl_4 \tag{44}$$

takes place in competition to the formation of the 475 nm band.

If cation scavengers exhibiting lower ionization potentials than alkanes are used, an onset of charge transfer from the 475 nm transient is observed at solute ionization potentials of about 9 eV (e.g., for cyclohexene).

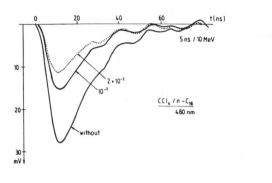 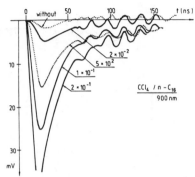

Fig. 31: a) Effect of n-hexadecane on the decay of the 480 nm band generated in electron-pulse irradiated liquid carbon tetrachloride. b) Formation of the n-hexadecane radical cation.

The experimental findings mentioned above support the suggestion of Bühler et al. [94] that the 475 nm band is related to a fragment cation of CCl_4. Its charge transfer onset at about 9 eV is compatible with the suggestion that CCl_3^+ is involved (ionization potential of CCl_3 = 8.9 eV). In the light of these experimental results some previous assignments proposed for the 475 nm band, CCl_4^+ [85] or (CCl_4^+ ...Cl^-) are unlikely.

However, the nature of the 475 nm transient is still under discussion. The question is raised: does this transition belong to an isolated carbonium ion CCl_3^+, to an ion-molecule complex CCl_3^+...CCl_4, to a contact ion-pair CCl_3^+...Cl^-, or is a solvent separated ion-pair $CCl_3^+//Cl^-$ involved?

Additional charge transfer bands which are affected by solvent and temperature are only expected for contact ion-pairs or ion-molecule complexes. An observed variation of the band maximum from 460 to 500 nm depending on solvent, CCl_4 concentration and temperature points to possible complexing [94d].

7. Cations in other Halomethanes

7.1 RADICAL CATIONS IN n-BUTYLCHLORIDE

The transient optical absorption spectrum observed in pure liquid n-butylchloride after electron pulse irradiation is shown in Fig. 22. From the decay kinetics (see inserts of Fig. 22) two intermediates can be distinguished. The faster decaying and the longer living components have peak wavelengths at 520 and 450 nm, respectively. These results agree well with matrix isolation studies [78] and low-temperature pulse radiolysis experiments [81]. Concluding from the effect of additives on both species and from the occurrence of charge transfer reactions Arai et al. [81] attributed the short-lived transient to n-butylchloride radical cations (BuCl$^+$) and the longer-lived to butene radical cations (Bu$^+$). The decay of the 520 nm transient became faster by addition of , e.g. n-heptane. The longer lived absorption was decreased but its decay remained unchanged. Rather similar effects were observed by adding cyclohexane, decalin, hexene-1 and benzene. If the other solutes mentioned in Table 8 were added, the lifetime of the 450 nm absorption was reduced in a similar way as observed for the 520 nm transient. In nearly all cases the well-known absorptions of monomer and dimer solute cations interfered with the underlying 450 nm absorption. As an example, the effect of benzene and biphenyl on the BuCl$^+$ decay is shown in Fig. 32.

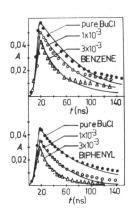

Fig. 32: Effect of benzene and biphenyl on the solvent cation decay in n-BuCl. Concentrations given in mol dm^{-3}.

The experimental results can be explained if the following reaction mechanism for some important ionic processes is adopted:

$$BuCl^+ + S \rightarrow S^+ + BuCl \tag{45}$$

$$Bu^+ + S \rightarrow S^+ + Bu \tag{46}$$

Using the ion-pair picture this mechanism is displayed in Fig. 33.

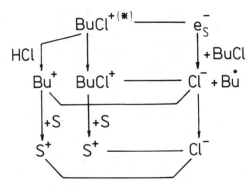

Fig. 33: Ion-pairs formed in irradiated n-BuCl solution containing the solute S.

Some rate constants of reaction (45) are summarized in Table 8.

Table 8: Rate constants of reaction (45) as a function of the ionization potential difference $\Delta I = I(BuCl) - I(S)$ [51]. $I(BuCl) = 10.67$ eV.

Solute	ΔI (eV) (eV)	k_{45} (dm^3 mol^{-1} s^{-1})
n-Heptane	0.32	1.8×10^7
c-Hexane	0.79	6.6×10^8
Decalin (mixt.)	1.07	1.8×10^9
Hexene-1	1.21	3.0×10^9
Benzene	1.43	2.9×10^9
Cyclohexene	1.77	5.9×10^9
Styrene	2.20	1.0×10^{10}
Biphenyl	2.43	8.8×10^9
Naphthalene	2.55	1.0×10^{10}
Triethylamine	3.17	1.0×10^{10}

The occurrence of charge transfer from cations to solutes offers a possibility to distinguish between different types of cations. The comparatively sharp "reaction onset" for ΔI from 0.5 to 1.0 eV points to a reaction of the BuCl parent radical cation BuCl$^+$. The reaction onset of the 450 nm transient seems to be obtained for the solute cyclohexene (I = 8.9 eV). This provides further evidence that butene radical cations are observed at 450 nm and reaction (46) becomes efficient. Already for biphenyl (I = 8.24 eV) diffusion-controlled rate constants are reached for reaction (46). A similar ΔI dependence as found for charge transfer from BuCl$^+$ probably also holds for that from butene radical cations.

On the product side, radical cations of the solutes hexene-1, benzene, cyclohexene etc. were identified in electron-irradiated n-butylchloride solution by comparison of their transient spectra with those obtained under similar conditions in liquid alkanes or from matrix isolation studies.

7.2 CATIONS IN METHYLENE CHLORIDE, FLUOROTRICHLORO-METHANE AND BROMOTRICHLOROMETHANE

As in CCl_4 in electron pulse-irradiated CH_2Cl_2, $CFCl_3$ and $CBrCl_3$, UV and visible transient absorption bands were observed which probably are of cationic nature. A comparison of their characteristics and assignments is given in Table 9.

Table 9: Characteristics and assignments of some transients observed after the pulse radiolysis of halomethanes. Formation and decay times measured at room temperature.

Substance	UV Band				Visible Band			
	λ(max)	Assign.	Form.	Decay	λ(max)	Assign.	Form.	Decay
CCl_4	340 nm	CCl^+ [87] CCl_2 [86]	<30ps [92,93]	100ns [87]	480nm	CCl_3^+ [86]	<200ps [92,92]	15ns [85]
CH_2Cl_2	350nm	$CH_2Cl^+//Cl^-$ [98]		8ns	560nm	$CH_2Cl^-//Cl^-$ [98]		8ns
$CFCl_3$	305nm	$CFCl_3^+$ [99]	<30ps	430nm	$CFCl_3^+//Cl^-$ [99]			<200ps
$CBrCl_3$	360nm	$CBrCl^+$ [99]		<30ps	520nm	$CBrCl_3^+//Cl^-$ [99]	<200ps	

The assignments of the visible bands of CFCl and CBrCl are questionable. The time-resolved buildup observed for these bands in ps pulse radiolysis experiments [99] indicates that fragmentation could be involved. The appearance of UV and visible bands after irradiation of halomethanes seems to be a general feature.

References

1.a J.P. Keene in: *Pulse Radiolysis*, ed. M. Ebert, Academic Press N.Y. (1965)
 b. G. Roffi in: *The Study of Fast Processes and Transient Species by Electron-Pulse Radiolysis*, ed. J.H. Baxendale and F. Busi, Reidel (1982)
2.a K.D. Asmus and E. Janata in: *The Study of Fast Processes and Transient Species by Electron-Pulse Radiolysis*, ed. J.H. Baxendale and F. Busi, Reidel (1982),
 b A. Hummel and W.F. Schmidt: *Radiat. Res. Rev.* **5**, 199 (1974)
 c K.H. Schmidt: *Int. J. Radiat. Phys. Chem.* **4**, 439 (1972)
3.a G. Beck: *Rev. Sci. Instrum.* **50**, 1147 (1979)
 b J.M. Warman and M.P. de Haas: *IRI Report* Nr 134-88-01, Delft (1988)
4.a P.P. Infelta, M.P. de Haas and J.M. Warman: *Radiat. Phys. Chem.* **10**, 353 (1977)
 b J.M. Warman in: *The Study of Fast Processes and Transient Species by Electron-Pulse Radiolysis,* ed. J.H. Baxendale and F. Busi, Reidel (1982)
5.a Yu. N. Molin, O.A. Anisimov, V.M. Grigoryants, V.K. Molchanov and K.M. Salikhov: *J. Phys. Chem.* **84**, 1853 (1980)
 b O.A. Anisimov, V.M. Grigoryants, V.K. Molchanov and Yu.N. Molin: *Chem. Phys. Letters* **66**, 265 (1979)
 c Yu.N. Molin and O.A. Anisimov: *Radiat. Phys. Chem.* **21**, 77 (1983)
6.a A.D. Trifunac, J.R. Norris and R.G. Lawler: *J. Chem. Phys.* **71**, 4380 (1979)
 b A.D. Trifunac and J.P. Smith: *Chem. Phys. Letters* **73**, 94 (1980)
 c A.D. Trifunac and J.P. Smith in: *The Study of Fast Processes and Transient Species by Electron-Pulse Radiolysis*, ed. J.H. Baxendale and F. Busi, Reidel (1982)
 d J.P. Smith, S.M. Lefkowitz and A.D. Trifunac: *J. Phys. Chem.* **86**, 4347 (1982)
 e A.D. Trifunac, R.G. Lawler, D.M. Bartels and M.C. Thurnauer: *Prog. React. Kinetics* **14**, 43 (1986)
7. K.H. Hong and J. Noolandi: *J. Chem. Phys.* **68**, 5163 (1978)
8. R.J. Friauf, J. Noolandi and K.M. Hong: *J. Chem. Phys.* **71**, 143 (1979)
9. A. Mozumder: *J. Chem. Phys.* **51**, 3020 and 3026 (1971)
10. W.F. Schmidt: *Can. J. Chem.* **55**, 2197 (1977)
11. J.M. Warman: *IRI-Report* 134-81-23 (1981)
12. M.P. de Haas, J.M. Warman, P.P. Infelta and A. Hummel: *Chem. Phys. Letters* **31**, 382 (1975)

13. J.M. Warman, P.P. Infelta, M.P. de Haas and A. Hummel: *Chem. Phys. Letters* **43**, 321 (1976)
14. W.M. Bartczak and A. Hummel: *Radiat. Phys. Chem.* **27**, 71 (1986)
15. J.M. Warman, K.-D. Asmus and R.H. Schuler: *J. Phys. Chem.* **73**, 631 (1969)
16. J.A. Crumb and J.K. Baird: *J. Phys. Chem.* **83**, 1130 (1979)
17. P.P. Infelta and S.J. Rzad: *J. Chem. Phys.* **58**, 3775 (1973)
18. R. Mehnert, O. Brede and W. Naumann: *Ber. Bunsenges. Phys. Chem.* **89**, 1031 (1985)
19. W.F. Schmidt and A.O. Allen: *J. Chem. Phys.* **52**, 2345 (1970)
20. P. Debye: *Trans. Electrochem. Soc.* **82**, 265 (1942)
21. P.W.F. Louwrier and W.H. Hamill: *J. Phys. Chem.* **72**, 3878 (1968)
22. H.A. Gillis, N.V. Klassen and R.J. Woods: *Can. J. Chem.* **55**, 2022 (1977)
23. R. Mehnert, O. Brede and J.Bös: *Z. Chem. (Leipzig)* **17**, 268 (1977)
24. O.A. Anisimov, V.M. Grigoryants, V.K. Molchanov, V.J. Melechov and Yu.N. Molin: *Dokl. Akad. Nauk SSSR* **260**, 1159 (1981)
25. M. Strobbe and J. Ceulemans: *Spectroscopy Letters* **19**, 207 (1986)
26. G. Wolput, M. Neyens, M. Strobbe and J. Ceulemans: *Radiat. Phys. Chem.* **23**, 413 (1984)
27. T. Shida and Y. Takemura: *Radiat. Phys. Chem.* **21**, 157 (1983)
28. M.C.R. Symons: *Chem. Phys. Letters* **69**, 198 (1980)
29. J.T. Wang and F. Williams: *J. Phys. Chem.* **84**, 3156 (1980)
30. T. Shida, H. Kubodera and Y. Egawa: *Chem. Phys. Letters* **79**, 179 (1981)
31.a M. Iwasaki, K. Toriyama and K. Nunone: *J. Am. Chem. Soc.* **103**, 3591 (1981)
   b K. Toriyama, K. Nunone and M. Iwasaki: *J. Phys. Chem.* **85**, 2149 (1981)
   c K. Toriyama, K. Nunone and M. Iwasaki: *J. Chem. Phys.* **77**, 5891 (1982)
   d M. Iwasaki, K. Toriyama and K. Nunone: *Chem. Phys. Letters* **111**, 309 (1984)
   e K. Nunone, K. Toriyama and M. Iwasaki: *Chem. Phys. Letters* **105**, 414 (1984)
   f K. Nunone, K. Toriyama and M. Iwasaki: *Tetrahedron* **42**, 6315 (1986)
   g K. Toriyama, K. Nunone and M. Iwasaki: *J. Phys. Chem.* **90**, 6836 (1986)
32.a M. Tabata and A. Lund: *Chem. Phys.* **75**, 379 (1983)
   b M.B. Huang, S. Lunell and A. Lund: *Chem. Phys. Letters* **99**, 201 (1983)
   c M. Tabata and A. Lund: *Radiat. Phys. Chem.* **23**, 545 (1984)
   d G. Dolivo and A. Lund: *Z. Naturforschung* **40a**, 52 (1985)
   e G. Dolivo and A. Lund: *J. Phys. Chem.* **89**, 3977 (1985)
   f A. Lund, M. Lindgren, G. Dolivo and M. Tabata: *Radiat. Phys. Chem.* **26**, 145 (1985)

282

g   A. Lund, M. Lindgren, M. Tabata, S. Lunell, G. Dolivo and
    T.Gäumann in: *Electronic Magnetic Resonance of the Solid State* ed.
    by J.A. Weil, Ottawa 1987
h   M. Lindgren: *Electron Spin Resonance Studies of Some Radical Ions.
    Electronic Structulres and Reactions* Dissertation, Linköping Studies
    in Science and Technology No. 195, Linköping (1988)
i   A. Lund, M. Lindgren, S. Lunell and J. Maruani in: *Molecules in
    Physics, Chemistry and Biology,* **III**, 259 (1989) Ed. J. Maruani
33. R.C. Benz and R.C. Dunbar: *J. Am. Chem. Soc.* **101**, 6363 (1979)
34. K. Kimura, S. Katsumata, Y. Achiba, T. Yamazaki and S. Iwata:
    *Handbook Of HeI Photoelectron Spectra of Fundamental Organic
    Compounds,* Japan Scientific Soc. Press, 1981
35. R. Mehnert, O. Brede and W. Naumann: *Ber. Bunsenges. Phys.
    Chem.* 88, 71 (1984)
36.a C.D. Jonah: *Radiat. Phys. Chem.* **21**, 53 (1983)
  b C.D. Jonah and M.A. Lewis: *Radiat. Phys. Chem.* **26**, 367 (1985)
37. M.A. Lewis and C.D. Jonah: *Radiat. Phys. Chem.* **33**, 1 (1989)
38. M. Meot-Ner (Mautner), L.W. Sieck and P. Ausloos: *J. Am. Chem.
    Soc.* **103**, 5342 (1981)
39. N.V. Klassen and G.G. Teather: *J. Phys. Chem.* **89**, 2048 (1985)
40.a D.W. Werst, L.T. Percy and A.D. Trifunac: *Chem. Phys. Letters* **153**,
    45 (1988)
  b A.D. Trifunac, D.W. Werst and L.T. Percy: *Radiat. Phys. Chem.* **34**,
    547 (1989)
  c D.W. Werst, M.G. Bakker and A.D. Trifunac: *J. Am. Chem. Soc.*
    112, 40-50 (1990)
41. R. Mehnert, O. Brede and W. Naumann: *Radiat. Phys. Chem.* **26**,
    499 (1985)
42. K. Kimura and J. Hormes: *J. Chem. Phys.* **79**, 2756 (1983)
43. B.C. Le Motais and C.D. Jonah: *Radiat. Phys. Chem.* **33**, 505 (1989)
44. J. Cygler, G.G. Teather and N.V. Klassen: *J. Phys. Chem.* **87**, 555
    (1983)
45.a E. Heilbronner: *Helv. Chim. Acta* **60**, 2234 (1977)
  b E. Heilbronner: *Helv. Chim. Acta* **60**, 2248 (1977)
  c J. Pireaux, S. Svenson, E. Basilier, P. Malmquist, U. Gelius, R.
    Caudano and K. Siegbahn: *Phys. Rev.* 14 A, 2133 (1976)
46. R. Mehnert, O. Brede, J. Bös and W. Naumann: *Ber. Bunsenges.
    Phys. Chem.* 83, 992 (1979)
47. E.L. Davids, J.M. Warman and A. Hummel: *JCS Faraday Trans. I*
    71, 1252 (1975)
48. E.Zador, J.M. Warman and A. Hummel: *Chem. Phys. Letters* **23**,
    363 (1973)
49. M.C. Sauer Jr. and K.H. Schmidt: *Radiat. Phys. Chem.* **32**, 281 (1988)
50. J.M. Warman, P.P. Infelta, M.P. de Haas and A. Hummel: *Can. J.
    Chem.* **55**, 2249 (1977)

51. V.J. Vedenev, L.W. Gurewich, V.H. Kondratev, V.B. Medvedev and E.L. Frankevich: *Energien Chemischer Bindungen, Ionisationspotentiale und Elektronenaffinitäten*, Leipzig 1981, trans. from Russian.
52. A.A. Scala, S.G. Lias and P. Ausloos: *J. Am. Chem. Soc.* **88**, 570 (1966)
53. A.D. Trifunac, M.C. Sauer and C.D. Jonah: *Chem. Phys. Letters* **113**, 316 (1985)
54. K. Toriyama, K. Nunone and M. Iwasaki: *J. Am. Chem. Soc.* **109**, 4496 (1987)
55. D. Rehm and A. Weller: *Ber. Bunsenges. Phys. Chem.* **73**, 834 (1969)
56. R.A. Marcus: *Ann. Rev. Phys. Chem.* **15**, 155 (1964)
57.a J. Ulstrup and J. Jortner: *J. Chem. Phys.* **63**, 4358 (1975)
   b J. Jortner: *J. Chem. Phys.* **64**, 4860 (1976)
58.a R. Mehnert, O. Brede and W. Naumann: *Ber. Bunsenges. Phys. Chem.* **86**, 525 (1982)
   b O. Brede, R. Mehnert and W. Naumann: *Chem. Phys.* **115**, 279 (1987)
59. V.G. Levich and R.R. Dogonadze: *Coll. Czech. Chem. Comm.* **26**, 193 (1961)
60. L.H. Luthjens, H.D.K. Codee, H.C. de Leng and A. Hummel: *Radiat. Phys. Chem.* **21**, 21 (1983)
61. H.T. Choi, K.C. Wu and S. Lipsky: *Radiat. Phys. Chem.* **21**, 95 (1983)
62. M.C. Sauer Jr., C.D. Jonah, B. Le Motais and A.C. Cheronvitz: *J. Phys. Chem.* **92**, 4099 (1988)
63. D.W. Werst and A.D. Trifunac: *Chem. Phys. Letters* **137**, 475 (1987)
64. P. Ausloos: *Radiat. Phys. Chem.* **20**, 87 (1982)
65. T. Shida and W.H. Hamill: *J. Am. Chem. Soc.* **88**, 5376 (1966)
66. B. Badger and B. Brocklehurst: *Trans. Faraday Soc.* **65**, 2576 (1969)
67. R. Mehnert, O. Brede and Gy. Cserep: *Radiat. Phys. Chem.* **26**, 353 (1985)
68. T. Shida, S. Iwata and M. Imamura: *J. Phys. Chem.* **78**, 741 (1974)
69. G. Hanschmann, W. Helmstreit and H.J. Köhler: *Z. Phys. Chem. (Leipzig)* **261**, 81 (1980)
70. R. Mehnert, O. Brede and W. Naumann: *Ber. Bunsenges. Phys. Chem.* **86**, 525 (1982)
71. T. Reitberger: *Thesis*, Royal Institute of Technology, Stockholm, (1975)
72. D.W. Werst, M.F. Desrosiers and A.D. Trifunac: *Chem. Phys. Letters* **133** 201 (1987)
73. D.W. Werst and A.D. Trifunac: *J. Phys. Chem.* **92**, 1093 (1988)
74. J.L. Holmes, P.C. Burger, M. Yu. A. Mullah and P. Wolkoff: *J. Am. Chem. Soc.* **104**, 2879 (1982)
75. Gy. Cserep, O. Brede, W. Helmstreit and R. Mehnert: *Radiochem. Radioanal. Letters* **38**, 15 (1978)
76. V. Saik, O. Anisimov, V. Lozovoy and Yu. N. Molin: *Z. Naturforschung* **A 40**, 239 (1985)
77. M.F. Desrosiers and A.D. Trifunac: *J. Phys. Chem.* **90**, 1560 (1986)
78. E.T. Kaiser and L. Kevan (Eds.): *Radical Ions*, Interscience, 1968

284

79. T. Shida, Y. Nosaka and T. Kato: *J. Phys. Chem.* **82**, 695 (1978)
80. R. Mehnert, O. Brede and Gy. Cserep: *Ber. Bunsenges. Phys. Chem.* **86**, 1123 (1982)
81. S. Arai, A. Kira and M. Imamura: *J. Phys. Chem.* **80**, 1968 (1976)
82. T. Shida, Y. Egawa and H. Kubodera: *J. Chem. Phys.* **73**, 5963 (1980)
83. O. Brede, W. Helmstreit and R. Mehnert: *J. Prakt. Chem. (Leipzig)* **316**, 402 (1974)
84. T. Bally, S. Nitsche, K. Roth and E. Haselbach: *J. Phys. Chem.* **89**, 2528 (1985)
85. R. Cooper and J.K. Thomas: *Adv. Chem. Ser.* **82**, 351 (1968)
86.a R.E. Bühler and B. Hurni: *Helv. Chim. Acta* **61**, 90 (1978)
   b H.U. Gremlich, Tac-Kyo Ha, G. Zumofen and R.E. Bühler: *J. Phys. Chem.* **85**, 1336 (1981)
87. O. Brede, J.Bös and R. Mehnert: *Ber. Bunsenges. Phys. Chem.* **84**, 63 (1980)
88. M. Anbar and J.K. Thomas: *J. Phys. Chem.* **68**, 3829 (1964)
89. T. Shida and W.H. Hamill: *J. Phys. Chem.* **70**, 1630 (1966)
90. L. Andrews and F.T. Prochaska: *J. Phys. Chem.* **83**, 368 (1979)
91. L. Andrews, B.K. Kelsall, J.H. Miller and B.W. Keelan: *JCS Faraday Trans. II* **79**, 1417 (1983)
92. T. Sumiyoshi, S. Sawamura, Y. Yoshikawa and M. Katayama: *Bull. Chem. Soc. Jpn.* **55**, 2341 (1989)
93. Y.Tabata, H. Kobayashi, T. Washio, S. Tagawa and Y. Yoshida: *Radiat. Phys. Chem.* **26**, 473 (1985)
94.a R.E. Bühler: *Radiat. Phys. Chem.* **21**, 139 (1983)
   b H.U. Gremlich and R.E. Bühler: *J. Phys. Chem.* **87**, 3267 (1983)
   c R.E. Bühler: *J. Phys. Chem.* **90**, 6293 (1986)
   d R.E. Bühler: *J. Radioanal. Nucl. Chem.* **101**, 288 (1986)
95.a C.A.M. van den Ende, L.H. Luthjens, J.M. Warman and A. Hummel: *Radiat. Phys. Chem.* **19**, 455 (1982)
   b M.P. de Haas, J.M. Warman and B. Vojnovic: *Radiat. Phys. Chem.* **23**, 61 (1984)
96. L. Andrews, F.T. Prochaska and B.S. Ault: *J. Am. Chem. Soc.* **101**, 9 (1979)
97. T.K. Ha, H.U. Gremlich and R.E. Bühler: *Chem. Phys. Letters* **65**, 16 (1979)
98. S.S. Emmi, G. Beggiato and G. Casalbore-Miceli: *Radiat. Phys. Chem.* **33**, 29 (1989)
99. T. Sumiyoshi, K. Tsugaru and M. Katayama: *Chem. Soc. Jpn. Chemistry Letters* 1431 (1982)

ION PAIRS IN LIQUIDS

O. A. ANISIMOV
Institute of Chemical Kinetics and Combustion,
Novosibirsk 630090, USSR

1. Introduction

In a homogeneous medium, radical ion pairs are formed by separation of opposite charges and, correspondingly, of spins of a neutral molecule. The charges are separated for different reasons, by chemical oxidation, ionizing irradiation, etc. The resulting radical ion pairs may have different distances between the opposite partners. If the distance is small, i.e. a few angstroms there is a strong electronic and magnetic interaction between the partners. This can lead to a substantial difference between the properties of ion pairs and isolated radical ions. In non-polar media these correlated pairs recombine very rapidly. In polar media they may be stabilized and exist for a long time due to solvation effects which weaken the Coulomb interaction. A large body of publications is devoted to the properties of such pairs (see, e.g. monograph [1].

Under ionizing radiation the distance between partners in most pairs is tens of angstroms. The interaction between the partners of these pairs is very weak. Therefore their properties are actually the same as those of isolated radical ions. At low radiation doses, however, an explicit spatial correlation between the partners takes place. Therefore the recombination rate of pairs (the geminate recombination rate) is much higher than the rate of radical ion recombination in the bulk. The geminate recombination kinetics differs markedly from the recombination kinetics in the bulk. A great body of work (see, e.g. [2]) on radiation chemistry is devoted to studying the peculiarities of geminate recombination.

In radical ion pairs, spin correlation is observable together with the spatial one. Having been coupled in the neutral molecule, the electron removed by ionization from the molecule and the unpaired electron of the radical cation will have anti-parallel orientations of spins. The interaction between electron spins and the medium is rather weak. The times of spin-lattice relaxation are e.g. for organic radical ions, 10^{-6} - 10^{-5} s. Therefore

A. Lund and M. Shiotani (eds.), Radical Ionic Systems, 285–309.
© 1991 *Kluwer Academic Publishers. Printed in the Netherlands.*

under certain conditions, the spin correlation is preserved until recombination.

Spin correlation and spin dynamics have been the subject of numerous studies of the processes associated with recombination of the pairs of neutral radicals. The results were generalized in a recent monograph [3]. Phenomena such as chemical polarization of nuclei and electrons as well as the magnetic field effects on the rate of radical reactions are based on spin effects.

Radical ion pairs in contrast to those of neutral radicals can form electron-excited products of recombination because of Coulomb interaction energy between the pair partners. The multiplicity of the excited molecule depends on the multiplicity of the pair prior to recombination. The singlet-excited molecule is formed from the recombination of a singlet pair with anti-parallel orientation of partner spins. Accordingly, the recombination of a triplet pair leads to the formation of the triplet excited molecule. The chemical and physical properties of the singlet- and triplet-excited molecules are quite different. Hence, the presence and loss of spin correlation in pairs may have a substantial effect on the processes involving radical ions.

This chapter is concerned with the behaviour of spin-correlated radical ion pairs in liquids, mostly nonpolar solutions under ionizing radiation.

Brocklehurst [4] was the first to consider spin effects in radical ion recombination in liquid solutions. He reported in detail [5] properties of spin effects such as the external magnetic field effect on radical ion recombination under radiolysis. In this review [5] he also discusses the interdependence between spin correlation in radical ion pairs and the fraction of geminate and cross recombinations in the radiation spur and the related ionization density as well as the rate of ion-molecular transfer involving one of the recombination partners and a number of other moments. Some of these issues will be discussed below. However, our attention will be focused on the coherent phenomena in pair recombination (quantum beats) and on the effect of resonant microwave fields on geminate recombination of radical ions (OD ESR method) as well as the novel data on the properties of radical ions in liquid solutions obtained using these methods.

2. Formation and Recombination of Radical Ion Pairs

Consider now the formation of radical ion pairs in solution under ionizing radiation. The main processes involved can be described by the following scheme:

$$S \xrightarrow{\gamma} S^+ + e^- \qquad (1)$$

$$e^- + A \rightarrow A^- \qquad (2)$$

$$S^+ + D \rightarrow D^+ + S \qquad (3)$$

$$S^+ + e^- \rightarrow S^* \qquad (4)$$

$$D^+ + e^- \rightarrow D^* \qquad (5)$$

$$S^+ + A^- \rightarrow S + A^* \qquad (6)$$

$$D^+ + A^- \rightarrow D^* + A \text{ or } D + A^* \qquad (7)$$

Here S is the solvent molecule; A and D are acceptors of electrons and holes, respectively. The first reaction corresponds to ionization of solvent molecules, the second and third describe the capture of electrons and holes by acceptors. As already mentioned, the primary spin state of the S^+/e^- pair is singlet. The spin correlation may be lost during formation of the secondary pairs. Most frequently this is due to the hyperfine interaction (*hfi*) between electrons and magnetic nuclei. The efficiency of this process depends on the external magnetic field strength. In the zero field the *hfi* may mix up the singlet pair level with three triplet sublevels. In the simplest case when all four sublevels are populated uniformly, only 25% of the pairs recombine in the singlet state (deviations from uniform population are discussed in detail in [5]). Contrary to the zero field, in the high magnetic field case only one triplet sublevel T_0 may be populated due to Zeeman splitting the T_+ and T_- sublevels remaining unpopulated. Thus, at high field 50% of the pairs recombine in the singlet spin state, i.e. twice as much as for the zero field.

Fig. 1 depicts the collective spin sublevels of a radical ion pair for different situations. The singlet sublevel S corresponds to the anti-parallel orientation of spin pairs; the parallel one corresponds to the triplet sublevels T_+, T_0, T_-.

As mentioned above, the recombination of singlet pairs forms singlet excited products and the triplet ones give triplet excited products. The singlet excited recombination products can unlike the triplet ones fluoresce. An increase in the recombination fluorescence solutions of aromatic acceptors has been observed [6-8] in accord with this scheme.

288

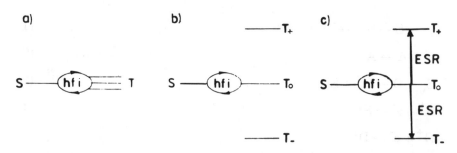

Fig. 1. Singlet-triplet transitions in a radical pair (a) in a low magnetic field, (b) in a high magnetic field plus microwave radiation; *hfi*-mixing due to hyperfine interactions [27].

2.1. SPIN DYNAMICS IN RECOMBINING PAIRS

Since the recombination fluorescence intensity is dependent on pair multiplicity at the moment of recombination, the character of singlet-triplet mixing in recombining pairs may be determined by the changes in the intensity on concrete conditions. This mixing is peculiar for each situation and depends on both the electron structure of the radical ions in a pair and their chemical reactions prior to recombination. Thus, the modulation of recombination fluorescence by spin interactions in the radical ion pair is a new way to get information about the structure and reactivity of radical ions.

A spin-lattice relaxation characterized by time T_1, is the most general reason for the loss of spin correlation in radical ion pairs. The singlet-triplet mixing resulting from spin-lattice relaxation has an irreversible character. Therefore various spin effects attenuate exponentially with a characteristic time T_1. As previously mentioned, for organic radical ions the times T_1 are usually rather large (about 10^{-6} - 10^{-5} s). In liquid solutions most pairs have enough time to recombine during shorter periods of time.

Within this time range of major importance is the singlet-triplet mixing with a dynamic character. The physical reasons leading to spin dynamics in radical ion pairs are different for different systems. The hyperfine interactions of unpaired electrons with magnetic nuclei often cause the partner spins to precess with different freuquencies which leads to periodic changes in the population of singlet and triplet sublevels of the pair. In high magnetic fields a difference in g-factors also causes a difference in the precession frequencies of the partners.

The spin dynamics in resonance conditions in cross magnetic and microwave fields is of a more complex form, which will be discussed in detail later.

If the microwave field is absent, the simplest spin dynamics form is observable in a high constant magnetic field. Spin dynamics is described in terms of a function $\rho_s(t)$ corresponding to the probability of the pair to be in the singlet state at moment t. The vector model for pair spins in high magnetic field is given in Fig. 2. The singlet-triplet transition is consistent with the change of the precession phase. If the pair at the initial moment of time was in a singlet state and the difference in the precession frequencies of the partners was $\Delta\omega$, then for $\rho_s(t)$, the following expression holds:

$$\rho_s(t) = (1/2)\,[1 + \cos(\Delta\omega t)] \tag{8}$$

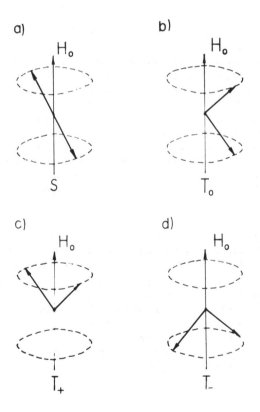

Fig. 2.   The vector model of electron spins for radical ion pair.

If the difference in the frequencies is caused only by the difference in the g_1 and g_2 values of g-factors, hence:

$$\rho_s (t) = (1/2) [1 + Cos \{(g_1 - g_2)\beta\hbar^{-1}Ht\}] \tag{9}$$

In the general case for a radical ion pair with a given configuration of nuclear spins the frequency of oscillations between the singlet and triplet states obeys the equation:

$$\omega_{ST_0} = |\Delta g\beta\hbar^{-1} H + \sum a_{1i} m_{1i} - \sum a_{2j} m_{2j} | \tag{10}$$

where $\Delta g = g_1 - g_2$ is the difference of radical g-factors; a_{1i} and a_{2j} are the *hfi* constants in radical 1 with nucleus i and in radical 2 with nucleus j; m_{1i} and m_{2j} are the projections of the corresponding nuclear spins onto the external magnetic field direction; β is the Bohr magneton.

The formula shows that the oscillation (quantum beats) frequencies are determined by the same parameters as the position of the lines in the ESR spectrum of the radical pair.

A real pattern of quantum beats in the radical pair with non-equivalent magnetic nuclei is the superposition of the set of oscillations which as a rule "blurs" the beats on the kinetic curve. The pattern of beats is simple in two cases. First, when the *hfi* of radicals with nuclei are weak and the g-factor difference is substantial. Consequently, the curve of beats is described by a simple cosine dependence the period of which depends on both the g-factor difference and the magnetic field.

Another simple case is when the g-factor difference is negligible, and the *hfi* between the spin and equivalent magnetic nuclei is dominating. In this case the singlet-triplet oscillations in different sub-ensembles occur with multiple frequencies. The curve of beats is represented by a succession of rather sharp peaks the period of which varies with the *hfi* constant.

Klein and Voltz [9] were the first to discuss the question of quantum oscillation in a radical recombination. In diluted solutions of scintillator 2b-PPD in cyclohexane they observed weak oscillations with a period of about 120 ns in the time dependence of the magnetic field effect in radiofluorescence. Unfortunately, since the system is complex, the experimental data [9] cannot be correlated to the *hfi* parameters.

The Novosibirsk group managed to observe quantum beats [10-14] in the kinetics of radical ion recombination and to demonstrate a good agreement between theory and experiment.

In the above papers the radiofluorescence kinetics was recorded by using the single photon counting technique.

3. Experimental Observation of Quantum Beats. Dependence of their Form and Amplitude on Radical Ion Reactions

3.1. Δg-MECHANISM-INDUCED BEATS

Fig. 3 depicts the radiofluorescence kinetics of diphenyl sulfide-d_{10} and 10^{-3} M *para*-terphenyl-d_{14} solutions in *trans* -decalin and squalane irradiated by fast electrons at different temperatures and concentrations of diphenyl sulfide in an external magnetic field of 3300 G [13]. This system was chosen because the *g*-factors of radical cations of sulphur-containing hydrocarbons substantially exceed those of radical ions of aromatic molecules [15] and the *hfi*

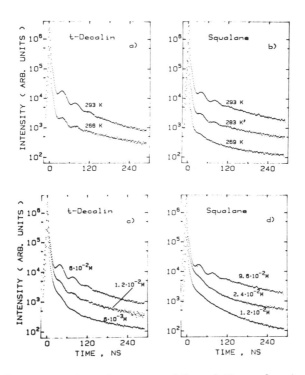

Fig. 3.   The fluorescence intensity curves of the solutions when irradiated by fast electrons in a magnetic field of 3300 G: (a) 10^{-3} M *para*-terphenyl-d_{14} and 6x10^{-3} M diphenyl sulfide-d_{10} in *trans* -decalin at different temperatures; (b) 10^{-3} M *para* - terphenyl-d_{14} and 9.6x10^{-2} M diphenyl sulfide-d_{10} in squalane at different temperatures; (c) 10^{-3} M *para* -terphenyl-d_{14} in *trans* -decalin at different concentrations of diphenyl sulfide-d_{10} at 293 K; (d) 10^{-3} M *para* -terphenyl-d_{14} in squalane at various concentrations of diphenyl sulfide-d_{10} at 293 K. For convenience, the curves are shifted relative to each other along the vertical axis [13].

constants are small for deuterated compounds. The oscillations are observed to appear and increase in amplitude with increasing concentration of diphenyl sulfide. Similar concentration and temperature changes are observed in other hydrocarbons solvents as well.

It is difficult to use the experimental curves of the radiofluorescence intensity $I(t)$ to analyze the beats because they are superimposed on the rapidly decaying function $F(t)$ which describes the lifetime-distribution of pairs prior to recombination. In addition, at short times, the function of radiofluorescence intensity $I(t)$ besides the component related to the recombination of radical ion pairs, includes the rapidly decaying luminescence caused by the sources of molecule excitation (electron-cation recombination, direct excitation, Cerenkov light, etc.)

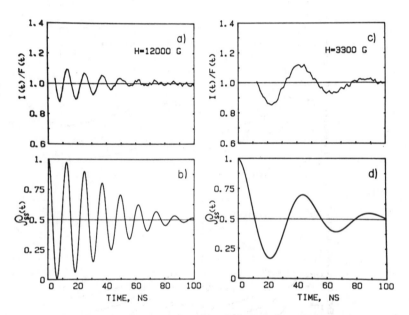

Fig. 4 Experimental curves (a, c) for quantum beats in the recombination of radical ion pairs in a solution of 6×10^{-2} M diphenyl sulfide and 10^{-3} M *para*-terphenyl-d$_{14}$ in *cis*-decalin; and calculated curves (b, d) for the time variation of the population $\rho_S(t)$ in the pair singlet state; (a, b) H = 12000 G, (c, d) H = 3300 G [11].

Taking into account that for the recombination part of the curve $I(t) \approx F(t)\rho_S(t)$, may be determined by division of $I(t)$ by $F(t)$. The resulting experimental data for the solution 6×10^{-3} M diphenyl sulfide and 10^{-3} M *para*-terphenyl-d$_{14}$ in *cis*-decalin for different values of a magnetic field are given in Fig. 4 [11]. The function $F(t)$ was chosen with taking into account its degree asymptotically [11].

From Fig. 4 it is evident that the period T of the experimentally observed oscillations is inversely proportional to the external magnetic field strength. This verifies the fact that the beats observed are defined by the Δg-mechanism.

Less amplitude of the observed oscillations in comparison with calculated ones is due to a big part of cross recombination in the spur at which there is no spin correlation in the recombining pairs.

3.2 RELATION BETWEEN THE SHAPE OF BEATS AND KINETIC PROCESSES

The theory predicts that the shape of beats induced by the pair D^+/A^- depends on the time of D^+ and A^- formation. If one of the ions is formed with some delay after ionization, this must lead to a shift in the maxima on the curve of beats towards long times and to a decrease of their amplitudes.

According to the classical scheme (1) - (7) the radical anion is formed practically on the spot due to a high electron mobility. The hole mobility is usually controlled by the diffusion. Consequently, in fairly viscous solvents, such as squalane or n-pentadecane, at the concentrations of hole acceptors of 10^{-3} M - 10^{-2} M the theoretically predicted shift of the extrema along the curve of beats approaches the maximum value equal to one fourth of the period of beats [13]. For the above (diphenyl sulfide-d$_{10}$)$^+$ / (para-terphenyl-d$_{14}$)$^-$ pair the predicted shift in the field of 3300 G is close to 10 ns. The experimentally observed shift does not, however, exceed 2 ns [13].

Within the frame of the scheme (1) - (7) such a small shift may be explained by assuming that the reaction radius of hole capture is very large (13-14 Å). In this case, however, the portion of the instantaneous capture will be very high [16], which is sure to lead to a weak dependence of the amplitude of beats on viscosity and temperature. At the same time, it was experimentally verified that the amplitude of beats drastically decreases with decreasing temperature (see Fig. 3).

Especially interesting are experimental results for trans-decalin solutions. As seen from Figs. 3 a,c, the behaviour of quantum beats in trans-decalin resembles that in squalane. At the same time within the frame of the model of hole capture an absolutely different behaviour should be expected in trans-decalin. According to experimental results for microwave abosorption [17] the hole mobility in cyclohexane and trans-decalin is not diffusion-controlled and is higher than the mobility of molecular ions (more than an order of magnitude). This mobility was shown [18] to be practically temperature-independent. This suggests that the curve of beats should also be temperature-independent and the concentration dependence should be weaker.

A large shift of the maxima on the curve of beats at small concentrations of hole acceptors is theoretically verified, if the holes do not participate in chemical reactions prior to their capture by an acceptor. However, when the hole is unstable and decays to form a product which fails to give a spin-correlated pair (*e.g.* during proton transfer) the predicted shift will not exceed the hole lifetime. Thus, a small experimentally observable shift may be assigned to the instability of the solvent radical cations which is valid for some solvents with short-lived radical cations (e.g. isooctane. However, for such hydrocarbons as squalane and pentadecane this shift is also very small. At the same time it is concluded that the holes in these cases are stable enough, e.g. in [19] they managed to record an OD ESR signal of the *n*-pentadecane hole. This means that its lifetime is equal to tens of nanoseconds. The quantum beats induced by the g-factor difference of holes and radical ions of *para*-terphenyl have been detected [12] by observing the recombination fluorescence in squalane and *n*-pentadecane solutions in strong magnetic fields. Such beats are detectable if the hole lifetime was no less than 10 ns. The data on optical absorption on radiolysis also testify to the stability of alkane radical cations with a long hydrocarbon chain.

The above analysis indicates that not all the data can be explained in terms of a simple model. An additional channel of $(DPS-d_{10})^+$ formation must exist. The portion of such a channel must be quite large, up to tens of percent. The diffusion-controlled amplitude and a small shift are typical for the experimentally observed quantum beats. Thus, the following conclusions were drawn. First, an additional channel must have a diffusion-controlled stage, and second, the time of $(DPS-d_{10})^+$ formation via this channel must be no more than 1-2 ns. This channel is likely to be due to the formation of excited radical cations or highly excited solvent molecules with less than 1 ns lifetime. The latter were believed to exist in liquids [20,21]. In this case the diphenyl sulfide molecules may be ionized by energy transfer from the above species to diphenyl sulfide molecules. It is evident that the information gained by the method of quantum beats in the present paper is insufficient to reliably identify the nature of the additional channel. This problem is still under investigation.

3.3 MIXTURE OF TWO HOLE ACCEPTORS

In [13] the method of quantum beats was employed to study the processes of charge transfer between two hole acceptors in the solution. The author studied the variations of beats curve in *trans*-decalin solutions containing 10^{-3} M *para*-terphenyl-d_{14} with different concentrations of the hole acceptor *tert*-butyl sulfide (TBS) and diphenyl sulfide-d_{10} (DPS-d_{10}), which give different periods of beats. The following conclusions were made [13] by analyzing and comparing experimental curves of the beats and theoretical calculations: (i) the ionization potential of TBS is a bit higher than that of diphenyl sulfide (7,8 eV); (ii) the radical cations of *tert*-butyl sulfide exist in non-polar solutions as dimers (*tert*-butyl sulfide)$_2^+$; (iii) when solutions containing TBS and DPS-d$_{10}$ molecules are irradiated, the reversible reaction of positive charge transfer occurs:

$$(TBS)_2^+ + (DPS\text{-}d_{10}) \underset{k_2}{\overset{k_1}{\rightleftharpoons}} (TBS)_2 + (DPS\text{-}d_{10})^+ \qquad (11)$$

The k_1/k_2 ratio is about 5 so that the equilibrium is shifted to the right. The experimental curves of beats observed in the mixture of two acceptors allow a direct record of the nonstationary stage of (11), which manifests itself in a successive increase of the period of beats with time.

3.4 THE *hfi*-MECHANISM-INDUCED BEATS

As has been mentioned, the systems with equivalent magnetic nuclei have favourable conditions to observe quantum beats. The best conditions appear in the pair where the *hfi* of one of the spins with equivalent magnetic nuclei dominate over the rest. This is true for the (tetramethylethylene)$^+$/(*para*-terphenyl-d$_{14}$)$^-$ pair formed under radiolysis of tetramethylethylene and *para*-terphenyl-d$_{14}$ solution in non-polar solvents [10]. The tetramethylethylene radical cation contains only equivalent hydrogen nuclei whereas in the radical anion of deuterated *para*-terphenyl the hyperfine interactions are small.

Fig. 5 depicts the curve of quantum beats for a *trans*-decalin solution containing 2.5×10^{-3} M tetramethylethylene (TME) and 3.7×10^{-4} M *para*-terphenyl-d$_{14}$.

If the number of equivalent protons is even, the period of beats, T, is determined from the relation:

$$T = 2\pi/\gamma a$$

where γ is the gyromagnetic ratio for the electron, a is the *hfi* constant for the equivalent cation protons. Substituting $a = 16.5$ G, known from [10], into this relation we obtain T = 21.6 ns. This value is close to that obtained from the curve in Fig. 5 which is equal to 22 ± 0.5 ns.

Experimental results show that in *trans*-decalin and squalane the observed shift of the maxima on the curve of beats is as small as in the case of Δg-mechanism. Thus, the above conclusion on the additional fast channel of the formation of acceptor radical cations in irradiated solutions is also valid for tetramethylethylene solutions.

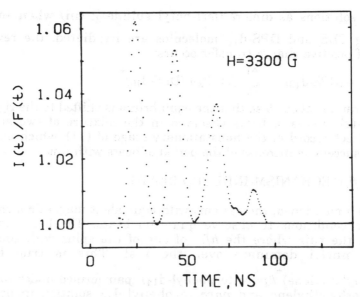

Fig 5. The experimental curve of quantum beats in a fast-electron-irradiated solution of 2.5×10^{-3} M tetramethylethylene and 3.7×10^{-4} M *para*-terphenyl-d_{14} in the field H = 3300 G, determined as the ratio of $I(t)$ to the smoothed function $F(t)$ which imitates the pair-lifetime distribution before recombination.

3.5 RF-FIELD-INDUCED BEATS

The beats appearing by Δg- or *hfi*-mechanisms may be observed only in rare specially chosen systems. To make the methods of beats suitable for a greater number of complex systems one may use the beats in the recombination fluorescence induced by resonance radiofrequency field.

Theoretical studies [3,22,23] show that radiofrequency (RF) or microwave (mw) fields under paramagnetic resonance must induce periodic changes in the population of a singlet state of the pair, and hence, in the recombination luminescence.

An applied radiofrequency field which is resonant to the spin of one of the pair partners causes the turning of the spin with a period:

$$T = 2\pi/\gamma H_1 \tag{12}$$

where H_1 is the amplitude of the magnetic component of the radiofrequency field. The RF transitions occur between T_0 and T_+, T_- sublevels. However, the *hfi* induces transitions between the S and T_0 sublevels. As a result, transitions occur between all sublevels of the pair.

In a complex system with a great number of non-equivalent nuclei the high-frequency beats induced by hyperfine interaction, are smoothed because they have different periods for different sub-ensembles. The period of RF-turning is, however, the same for all sub-ensembles and depends only on the amplitude H_1. Therefore, the modulation of recombination fluorescence by the RF-field is equally displayed in simple and complex systems.

Fig. 6 gives the curve of beats observed in the resonance RF-field for 10^{-3} M *para*-terphenyl-h_{14} solution in squalane irradiated by fast electrons [14]. The curve of beats is $\Delta I/I = [I(H_1, t) - I(0, t)]/I(0, t)$, where $I(H_1, t)$ and $I(0, t)$ are the intensities of the recombination fluorescence with and without RF-field, respectively.

The beats were recorded in a modified setup for recording radiofluorescence kinetics using a single photon counting technique with a radioactive source. A cell with the sample was placed inside the coil of a resonant LC-circuit tuned to a frequency of 94.6 MHz. The external magnetic field H = 34 G was produced by Helmholtz coils. The strength of the RF-field was 8.0 ± 0.5 G in the rotating coordinate system. The period of beats was determined to be 46 ns. This value is equal to that from formula (12). It was experimentally verified [14] that the change of the amplitude of the RF-field H_1 causes the changes in the period of beats according to (12).

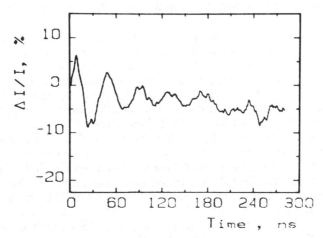

Fig. 6. Quantum beats induced by a resonant RF-field in squalane solution of 10^{-3} M *para*-terphenyl-h_{14} at room temperature, $H_1 = 8 \pm 1$ G [14].

4. OD ESR Method

From a theoretical viewpoint, transitions between the spin sublevels of pairs induced by resonant RF (or mw) field are most explicit in the phenomenon of quantum beats given earlier. However, for many practical applications it is more convenient to observe the integral decrease in the recombination fluorescence intensity during resonance. This decrease follows from the above discussion. Indeed, in non-resonance strong magnetic field only S and T_0 sublevels are populated. Since on resonance the number of populated sublevels increases, the population of the singlet sublevel decreases, which leads to the decrease in the recombination fluorescence intensity (see Fig. 1).

The idea of a possible microwave resonance field effect on the relation between the products of radical pair recombination was proposed in [24] for the first time and realized in the experiments involving photo-induced pairs in alkyl haloid crystals by Ruedin et al. [25] and in molecular crystals by Frankevich et al. [26]. In 1969 Brocklehurst [4] reported on the possibility of the mw resonance field effect on the products of the primary recombination of radical ion pairs under radiolysis of liquid solutions. However, no appropriate experiments were undertaken, particularly, because Brocklehurst under-estimated the role of hyperfine interactions in the mixing of singlet and triplet pair states. At the same time, he considered S-T mixing to depend on the spin-spin relaxation of the radicals of the pair. The experimental conditions for this case are difficult to realize. The development of the concepts on the role of hyperfine interactions in the singlet-triplet evolution of a radical pair [3] made the prediction more favourable. In 1979, the Novosibirsk group pioneered in observing the mw resonance field effect on the intensity of the recombination fluorescence on radiolysis of liquid napthalene solutions [27].

According to theoretical predictions, under ionizing radiation of scintillator solutions in an ESR spectrometer cavity with magnetic field sweeping, the intensity of the recombination fluorescence decreases when passing each ESR spectrum line of recombination partners. It is thus possible to optically detect electron paramagnetic resonance (OD ESR) of radical ion pair spectra. As in conventional ESR spectroscopy, the OD ESR spectra are often recorded as the first derivative of luminescence intensity. The OD ESR spectra are observable for the radical ions which are the partners of spin-correlated pairs, i.e. participate in geminate recombination. This is the main difference between the OD ESR and usual ESR spectra in which all paramagnetic particles of the sample are detected. The second distinctive feature of the OD ESR technique is its high sensitivity which is 10^9 times higher than that of the conventional ESR technique. The high sensitivity allows the OD ESR method to be used for recording ESR spectra of short-lived radical ions participating in geminate recombination in liquid solutions. The lifetime of such radical ions ranges from tens to hundreds of nanoseconds. However, this method can detect only those geminate pairs, the recombination products of which are represented by the excited molecules capable of fluorescing. In addition, it has insufficient resolution. The widths of individual lines in the OD ESR spectra are as a rule more than 0.5 G. Wide lines are due to the mw fields of relatively large amplitudes that must be used ($H_1 = 0.5 - 1$ G) to induce efficient transitions between the pair sublevels during its lifetime prior to recombination.

Different apparatus have been described to record OD ESR spectra [27-30]. The simplest variant for steady-state measurements is an attachment to a conventional ESR spectrometer [28]. A quartz cell with a sample is inserted into a cavity and is irradiated with an X-ray tube through the cavity front wall. The sample light passes via a vertical light guide which connects the cell bottom and the photomultiplier photocathode. Instead of using a crystal detector, the anode of the photomultiplier is connected with the input of the phase-sensitive detector. A time-resolved variant of the OD ESR technique is highly potent (another name is FDMR) [30]. Some results obtained using this variant are reported by the Argonne group [30-34].

The theory of OD ESR spectra has much in common with the theory of usual ESR spectra. At the same time there are some peculiarities which are under study [22, 35-39].

At present, the OD ESR technique has already been used to record various radical ions participating in the geminate recombination in liquid and glassy solutions. The primary radical cations (holes) of the solvent [19, 40, 41] and excess electrons [39,42] were detected resulting from ionization of solvent molecules. In [43, 44] the signals of solvated electrons were detected in polar solvents [43] as well as in non-polar hydrocarbons containing polar additives [44]. A few papers were devoted to the OD ESR study of organic radical cations and anions emerging from the capture of the primary holes and electrons by the acceptor molecules in solutions [27-34, 40, 45-53]. We are here going to consider only the results which may favour our understanding of the processes of the appearance and recombination of radical ion pairs in the liquid as well as those which are specific for this technique.

4.1 ELECTRON-CATION PAIRS

At room temperature the primary electron-cation pair recombination in hydrocarbons occurs within a picosecond time scale. With available mw field strengths, this time is insufficient to affect the mutual orientation of pair spins prior to recombination. However, since in most hydrocarbons the electrons have a temperature-activated mobility, decreasing the temperature we slow down the recombination to tens of nanoseconds. Under these conditions the OD ESR signal of an electron-cation pair is an intensive central narrow line which belongs to the electron, and the other lines refer to radical cations. These spectra were described theoretically in [39]. At low temperatures near the temperature of glassing as well as in glassy solutions satellite lines appear associated with the forbidden *flip-flop* transitions at simultaneous flips of electron and proton spins of the solvent [54]. The splitting between the central line and the satellite approaches the nuclear Zeeman splitting. Knowing the deviation from this value, the number of satellites in the spectrum and their relative intensities, one may calculate the mean number of protons in the nearest environment of the electron and the mean electron-proton distance. The number of protons in the nearest electron environment was determined to be close to 2 [54].

The electron mobility in liquid hydrocarbons decreases drastically not only with decreasing temperature but also with their solvation in the clusters consisting of the molecules of polar additives (water, alcohols). The OD ESR signal of solvated electron is a single line the width of which depends on the nature of molecules of the cluster [44]. For water clusters the line is especially narrow ($\Delta H \leq 0.5$ G). The analysis of the OD ESR signals [44] shows that in hydrocarbons with a small amount of water additive the electrons are solvated mostly in dimeric clusters. The lifetime of these clusters is long enough. The electron remains in the cluster up to the moment of recombination. The mobility of solvated electrons is close to that of molecular ions.

In polar solvents, e.g. in alcohols, the electrons are solvated directly by the solvent molecules. The OD ESR signal of the solvated electron was observed for this case in [43].

4.2 HOLE-ANION PAIRS

The properties of the primary radical cations of solvent (holes) S^+ may be substantially different from those of isolated radical cations of the same molecules. This difference is due to the possibility of fast ion-molecular charge transfer reactions between the holes and surrounding neutral molecules.

The OD ESR signals of the holes may be observed when the electron of the primary pair is captured by a luminescing molecule of the electron acceptor A. As a result the pair S^+/A^- is formed. The OD ESR signal of the hole may be detected by the fluorescence resulting from the recombination of this pair (see reaction (6)).

In liquid solutions, the OD ESR signals of the hole were recorded only in three solvents, namely benzene, para-xylene, n-pentadecane [19, 40]. In all the cases, the OD ESR signal of the hole is a narrow line with the width less than 1 G in benzene, and para-xylene, and less than 2.7 G in n-pentadecane. The lines are narrowed by the process of ion-molecular charge transfer. For para-xylene the charge transfer rate constant was estimated from the value of narrowing to be 3.2×10^9 M^{-1} s^{-1} which is close to the diffusion-limited one. For n-pentadecane the rate constant is much lower than the diffusion controlled one. Thus, the hole mobility observed using OD ESR is practically the same as that of molecular ions.

The OD ESR spectrum with a resolved hyperfine structure with characteristic splittings of 36 G and 8 G observed under irradiation with fast electrons of cyclohexane solution of 10^{-4} anthracene-d$_{10}$ was identified [33] as the spectrum belonging to the primary radical cations (holes) of cyclohexane. This identification was doubted in the papers of the Novosibirsk group [49]. The authors [49] considered the spectrum observed in [33] to belong to the radical cation of cyclohexene which resulted from abstraction of hydrogen from the primary radical cation of cyclohexane.

4.3 FAST TRANSPORT OF POSITIVE CHARGE IN CYCLIC HYDROCARBONS

Using the method of induced microwave absorption [2, 17, 18] it was shown that highly mobile positive charges with lifetimes of tens and hundreds of nanoseconds are formed under ionizing irradiation of liquid six-membered cyclic hydrocarbons such as cyclohexane, methylcyclohexane, *trans*-decalin. The authors [17, 18] identified them as the radical cations (holes) of the above solvents. The mobility of these charges was more than ten times higher than the mobility of molecular ions. The high mobility of the holes in these hydrocarbons was assigned to a close matching of nuclear coordinates between cations and initial molecules of hydrocarbons. In other hydrocarbons no highly mobile charges were detected.

The highly mobile charges are identified as holes by correlating the efficiency of their capture by an acceptor to the efficiency of the hole capture by the same acceptor.

The capture efficiency usually increases with increasing difference in the ionization potentials of the solvent and acceptor. The correlation is most explicit for a hole acceptor such as cyclopropane. Although this molecule has a high ionization potential (11.66 eV [55]), it efficiently reacts with the cyclohexane radical cation via the reaction:

$$c\text{-}C_6H_{12}^+ + c\text{-}C_3H_6 \rightarrow c\text{-}C_6H_{10}^+ + n\text{-}C_3H_8 \tag{13}$$

In liquid cyclohexane this reaction is verified by OD ESR results. In X-ray irradiated cyclohexane solution of 10^{-3} M *para*-terphenyl-d_{14} only the line of the (*para*-terphenyl-d_{14})$^+$/(*para*-terphenyl-d_{14})$^-$ pair is observed the width of which is 3 G (FWHM). Addition of 10^{-3} M cyclopropane to the solution gives a wide structureless line with a width which is almost the same as that of the envelope of the ESR spectrum of the cyclohexene radical cation. (Note that this line is observed at room temperature with a sufficiently strong mw field). Similar experiments in *trans*-decalin solutions fail to give wide lines. On the other hand, highly mobile charges are known to be captured by cyclopropane with a rate constant of about 10^{11} M^{-1} s^{-1} [2] and not captured in *trans*-decalin. Thus, the correlation between the hole reactivity and the identified charge is obvious. The best evidence for the existence of a highly mobile hole in cyclohexane and *trans*-decalin would be the presence of its very narrow OD ESR signal narrowed by a fast charge transfer. Such signals are, however, unobservable either in *trans*-decalin or cyclohexane, probably due to a short lifetime of the hole-anion pair prior to recombination related to the high hole mobility.

A hypothesis was reported [56] that $c\text{-}C_6H_{11}^+$ or $c\text{-}C_6H_{13}^+$ rather than $c\text{-}C_6H_{12}^+$ are the highly mobile particles, i.e.it is not the electron but H$^-$ or the proton which are transported in charge transfer:

$$c\text{-}C_6H_{11}^+ + c\text{-}C_6H_{12} \rightarrow c\text{-}C_6H_{12} + c\text{-}C_6H_{11}^+ \tag{14}$$

$$c\text{-}C_6H_{13}^+ + c\text{-}C_6H_{12} \rightarrow c\text{-}C_6H_{12} + c\text{-}C_6H_{13}^+ \tag{15}$$

The $c\text{-}C_6H_{13}^+$ particle is likely to emerge from the reaction involving an excited radical cation $c\text{-}C_6H_{12}^+$:

$$c\text{-}C_6H_{12}^{+*} + c\text{-}C_6H_{12} \rightarrow c\text{-}C_6H_{13}^+ + c\text{-}C_6H_{11} \tag{16}$$

In particular the above viewpoint was verified in [34], where using a time-resolved OD ESR technique (FDMR) the authors observed the formation kinetics of triethylamine radical cation directly in the irradiated cyclohexane solutions containing as additives the hole acceptor triethylamine and the scintillator PPO. If the radical cation of triethylamine $(TEA)^+$ results from the capture of fast holes by the molecules of triethylamine, the formation kinetics of $(TEA)^+$ must have a very short rise time corresponding to a high rate constant $K = 1.3 \times 10^{11}$ $M^{-1} s^{-1}$ determined from the experiments on mw-absorption [2]. However, the OD ESR experiments testify that the intensity of $(TEA)^+$ spectrum increases with time much more slowly [34]. A diffusion controlled rate constant is indicative of this increase.

Another argument in favour of a new interpretation of the results [17, 18] is the experiments on the temperature variations of the OD ESR spectrum of *trans*-decalin radical cation [57]. In *trans*-decalin and squalane matrices, at below liquid nitrogen temperature, this radical cation has no explicit OD ESR spectrum with a resolved hyperfine structure. However, with increasing temperature, its intensity sharply decreases and at temperatures above 80 K the spectrum is practically unobservable even when the matrix remains solid. Note that the spectrum of $(cis\text{-decalin})^+$ in squalane and 3-methylpentane solutions is observed even at room temperature. The disappearance of the spectrum is assigned [57] to the chemical instability of *trans*-decalin$^+$. In the frame of such an assumption, one might conclude that at room temperature the radical cation of *trans*-decalin lives less than 1 ns. At the same time, the lifetime of the highly mobile charges is known to be hundreds of nanoseconds [58]. From this contradiction it is concluded [57] that the observable highly mobile charges are not the holes.

Note, however, that not only the instability of $(trans\text{-decalin})^+$ but also, e.g. the shortening of the time T_1 of the spin-lattice relaxation of radical cation, with increasing temperature cause the temperature variations in the intensity of $(trans\text{-decalin})^+$ OD ESR spectrum.

It was shown in [59] that the degenerate ground state of a radical ion can give rise to enhanced electron spin-lattice relaxation. The results [60] show that in the radical cation of *trans*-decalin there is an excited electronic state close to the ground state. Different ESR spectra correspond to these states. For some matrices ESR lines are observed which belong to both spectra. Only one spectrum is observed for *cis*-decalin radical cation [60]. Therefore it is expected that T_1 for the *trans-* form is shorter than for the *cis-* one.

Thus, it is concluded that the question of the identification of the highly mobile positive charges is still open to discussion.

4.4 CATION-ANION PAIRS

The capture of electron and holes by acceptor molecules in reactions (2) and (3) leads to the appearance of cation-anion pairs D^+/A^- where a singlet spin orientation typical for the primary hole-electron pairs is preserved. The OD ESR spectrum of such a pair contains the lines belonging to both radical cations and radical anions. In many cases the OD ESR spectra bear the same information as the usual ESR spectra. The only difference is that due to the high sensitivity of the former, it is possible to observe the radicals with shorter lifetimes (in liquid as well).

Fig. 7 shows the temperature transformations of the OD ESR spectrum of $(9,10\text{-octalin})^+/(para\text{-terphenyl-d}_{14})^-$ in liquid hydrocarbon solution due to ring inversion of the 9,10-octalin radical cation [50]. Using the data in Fig. 7, the activation energy of the inversion was calculated [50] as 18.8 kJ mol^{-1}.

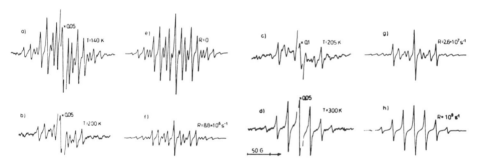

Fig. 7. Temperature change in the OD ESR spectra and their simulation for solutions of 10^{-3} M naphtalene-d$_8$ plus 10^{-2} M 9,10-octalin (a) in 3-methylpentane at 140 K; (b) in 1:1 mixture of squalane and 3-methylpentane at 200 K; (c) same as (b) at 205 K; (d) in squalane at 300 K. Simulation spectra (e), (f), (g) and (h) are to be compared with (a), (b) and (c), and(d) respectively. Spectra are simulated with the inversion frequencies R denoted [50].

4.5 SINGLET- AND TRIPLET-EXCITED RECOMBINATION PRODUCTS

As has been mentioned [61] the OD ESR spectra can be recorded by observing either recombination fluorescence or phosphorescence. In the latter case not singlet but triplet-excited recombination products are observed. Therefore an increasing intensity of sample luminescence is observed at resonance. Thus, the lines of the OD ESR spectrum detected by phosphorescence have an opposite sign and their derivatives have an opposite phase compared to the spectra detected by fluorescence. As a rule, the phosphorescence may be observed directly in the recombination in glasses. In the liquid, a sensibilized phosphorescence of a small quantity of biacetyl molecules purposely introduced into the solution is employed to detect ESR signals [61].

By comparing the line intensities in the OD ESR spectra of the pairs (naphthalene)+/(naphthalene)-, (biphenyl)+/(biphenyl)-, (*para*-terphenyl)+/(*para*-terphenyl)- detected by fluorescence and phosphorescence in liquid hydrocarbons, it was concluded that the ratio betwen the yields of singlet- and triplet-excited molecules on recombination of pairs with the corresponding multiplicity tends to unity [61].

4.6 SPIN-LOCKING PHENOMENON

The phenomenon of spin-locking is among the specific effects of OD ESR spectroscopy. This phenomenon consists in slowing down the S-T$_0$ transitions in the recombining pairs induced by either Δg- or *hfi*-mechanisms due to a high resonance microwave field. The presence of a high field H_1 causes such large splitting of the S and T$_0$ states that the transitions between them are impossible, if they are induced by either Δg- or *hfi*-mechanisms. At the same time, the microwave field cannot hamper the transitions due to phase relaxation.

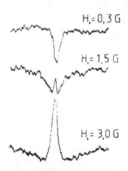

Fig. 8. Low-field OD ESR spectra of 5x10^{-3} M biphenyl-d$_{10}$ solution in squalane at various values of RF-field H$_1$.

In [37, 38] the spin-locking as applied to OD ESR spectroscopy was considered. When it is explicit (H_1 is larger than the full width of the OD ESR spectrum) the spectrum is represented by a single line with a sign which is opposite to the usual OD ESR one. Thus, under usual conditions, when the spectra are detected by fluorescence, the OD ESR line in the presence of spin-locking is consistent with an increase in the fluorescence intensity.

The phenomenon of spin-locking may be used to obtain information about the recombining pairs.

In squalane solution of 5×10^{-3} M biphenyl-d_{10} at low H_1 a single line with unresolved hyperfine structure with the width $\Delta H \approx 3$ G (FWHM) was observed to belong to the (biphenyl-d_{10})$^+$/(biphenyl-d_{10})$^-$ pair. With increasing H_1 the line begins to increase in intensity which is followed by the inversion of the central part of the line. At larger values of H_1 (at $H_1 \geq 3$ G) the line fully inverts (see Fig. 8). This means that S and T_0 sublevels are not mixed by hfi interaction. Hence, in the course of spin evolution starting from ionization up to recombination the electron spins interact with nuclei the hfi constants of which are not much higher than H_1^{inv}. More precisely, if hfi with values substantially higher than H_1^{inv} occurs at any stage of spin evolution, the time of this interaction is $t_p \ll 1/\gamma \Delta H_p$, where ΔH_p is the width of the ESR spectrum of a radical cation precursor. The measurements in glassy matrix show [41] that the spectrum width of the squalane radical cation is about 20 G. Consequently, $1/\gamma \Delta H_p \approx 3$ ns. Thus, if the hole width in squalane is close to that in a glassy matrix, the spin-locking at $H_1 \approx 3$ G is possible only if the time of spin evolution on the hole, t_p, is substantially shorter than three nanoseconds. As has been mentioned above, the squalane hole lives more than 10 ns, i.e. t_p is not limited to the hole lifetime. On the other hand, at the given concentration of 5×10^{-3} M biphenyl-d_{10} the condition of $t_p \ll 3$ ns will hold only for a very small part of the holes emerging from ionization in the vicinity of the biphenyl-d_{10} molecules. Since an explicit spin-locking is observed, it should be concluded that the hole spectrum in the liquid squalane is substantially narrowed by ion-molecular charge transfer. This spin-locking is also possible if most of the biphenyl radical cations are formed from ionization of its molecules on energy transfer from the superexcited solvent molecules, i.e. via the above mechanism (see Section 3.2).

The above considerations are also valid for n-pentadecane, cyclohexane, $trans$-decalin in which an explicit spin-locking is also observed at 10^{-3} M $para$-terphenyl-d_{14} and $H_1 \geq 3$ G [62].

In other words, in these solutions the radical cations with a wide ESR spectrum cannot be the precursors of the major part of the $para$-terphenyl-d_{14} radical cations.

In isooctane solutions of 10^{-3} M $para$-terphenyl-d_{14} at $H_1 \approx 3$ G, the line inversion of the ion radical pair ($para$-terphenyl-d_{14})$^+$/($para$-terphenyl-d_{14})$^-$ is practically unobservable (only a small central part of the line is inverted). Thus, the precursors of $para$-terphenyl-d_{14} radical cations in isooctane solutions have a wide ESR spectrum. They are represented by olefin radical cations, i.e. the products of the decay of isooctane holes. The absence of spin-locking proves the failure of the mechanism of the formation of the acceptor radical cation via energy transfer from superexcited isooctane molecules in isooctane. The latter is in agreement with the data in picosecond radiolysis [21] which show that the lifetime of excited isooctane molecules is very short due to their chemical dissociation.

4.7 GEMINATE RECOMBINATION IN POLAR MEDIA

Earlier [43] it was reported about the OD ESR signals of D^+/e^- pairs in liquid alcohols at low temperatures. The signals of cation-anion pairs were detected in many polar liquids even at room temperature [29, 43]. At first sight it is surprising. Indeed, to observe the OD ESR signal at the given mw field, it is necessary that the lifetime of the pairs prior to recombination was about $\pi/2\gamma H_1$, which for $H_1 = 1$ G is about 90 ns. At the same time, the so-called equivalent time of geminate recombination $\Gamma = 0.43\, r_c^2/D$ where r_c is the Onsager radius, D the coefficient of the mutual diffusion of ions [2] is very short in polar liquids. Thus, e.g. in ethanol at room temperature $\Gamma \approx 2$ ns. However, Warman claimed [2] that most of the ions escaping beyond the Onsager radius return to the region of Coulomb forces and recombine in a geminate way. These ions allow the observance of OD ESR signals. The estimates [2] show that about 10% of the ions escaping into the bulk, recombine in a geminate way in times exceeding $16.7\, r_c^2/D$ which for ethanol is 70 ns. Bearing in mind that for ethanol the fraction of ions escaping into the bulk is about 76% [43] it is concluded that a marked fraction of the pairs displays the time of geminate recombination equal to tens of nanoseconds.

As stressed in [43], the observance of OD ESR signals testifies also to the fact that in contrast to the previous opinion [63] the recombination of ions may induce the appearance of excited acceptor molecules in polar solutions.

5. Conclusion

The review of experimental results on the processes following the geminate pair recombination obtained using spin-dependent methods could be continued. For instance, interesting is the OD ESR technique for studying ion-molecular reactions involving proton transfer [13, 29, 57]. Abstraction of a proton from a radical cation causes the separation of charge and spin, and the disappearance of the OD ESR spectrum of a recombining pair. Highly convincing are the results obtained in [65] where the authors studied the reaction of the decay of alkane radical cations involving abstraction of an H_2 and formation of olefin radical cations. A number of papers has dealt with the processes which occur in concentrated acceptor solutions [48, 53, 65, 66] such as ion-molecular charge transfer leading to the broadening of components of OD ESR spectrum with subsequent collapse and narrowing. The formation of dimeric and multimeric radical cations has been observed in ion-molecular reactions. Some authors have studied the structure of radical ions [47, 50, 52]. For details one should refer to the original papers.

The spin-dependent methods have been under discussion quite recently and one may hope for their successful further development.

References

1. M. Szware, (ed.): *Ions and ion pairs in Organic Reactions*, Wiley-Interscience, A division of Joh Wiley and Sons Inc., London, Sydney, Toronto (1972).
2. J.M. Warman: "*The dynamics of electrons and ions in non-polar liquids*", IRI 134-81-23, Proceedings of the NATO advanced study institute, Capry, Italy. (1981).
3. K.M. Salikhov, Yu.N. Molin, R.Z. Sagdeev, and A.L. Buchachenko: *Spin Polarisation and Magnetic Effects in Radical Reactions*, Elsevier, Amsterdam, (1984).
4. B. Brocklehurst: *Nature* **221**, 921-923 (1969).
5. B. Brocklehurst: *International Reviews in Physical Chemistry* **4**, 279-306 (1985).
6. B. Brocklehurst, R.S. Dixon, E.M. Gardy, V.J. Lopata, M.J. Quinn, A. Singh, and F.P. Sargent: *Chem. Phys. Letters* **28**, 361-363 (1974).
7. R.S. Dixon, E.M. Gardy, V.J. Lopata, and F.P. Sargent: *Chem. Phys. Letters* **30**, 463-464 (1975).
8. F.P. Sargent, B. Brocklehurst, R.S. Dixon, E.M. Gardy, V.J. Lopata, and A. Singh: *J. Phys. Chem* **81**, 815 (1977).
9. J. Klein and R. Voltz: *Phys. Rev. Letters* **36**, 1214-1217 (1976).
10. O.A. Anisimov, V.L. Bizaev, N.N. Lukzen, V.M. Grigoryants, and Yu.N. Molin: *Chem. Phys. Letters* **101**, 131-135 (1983).
11. A.V. Veselov, V.I. Melekhov, O.A. Anisimov, and Yu.N. Molin: *Chem. Phys. Letters* **136**, 263-266 (1987).
12. A.V. Veselov, V.L. Bizyaev, V.I. Melekhov, O.A. Anisimov, and Yu.N. Molin: *Radiat. Phys. Chem.* **34**, No. 5, 567-573 (1989).
13. A.V. Veselov: *Magnetic modulation of recombination fluorescence in fast spur processes in liquid hydrocarbons*, Ph.D. Thesis, Novosibirsk, (1989).
14. V.O. Saik, O.A. Anisimov, A.V. Koptyug, and Yu.N. Molin: *Chem. Phys. Letters*, to be published (1989).
15. A.V. Il´jasov, Yu.M. Kargin, and I.D. Morozova: *Spektry EPR organicheskikh ion-radikalov*, Nauka, Moscow (1980).
16. A.I. Burshtein: *UFN* **143**, 553-600 (1984).
17. M.P. De Haas, J.M. Warman, P.P. Infelta, and A. Hummel: *Chem. Phys. Letters* **31**, 382-386 (1975).
18. J.M. Warman, P.P. Infelta, M.P. de Haas, and A. Hummel: *Can. J. Chem.* **55**, 2249-2257 (1977).

19. V.I. Melekhov, O.A. Anisimov, A.V. Veselov, and Yu.N. Molin: *Chem. Phys. Letters* **148**, 429-434 (1988).
20. L.H. Luthjens, H.C. de Leng, L. Woinarovits, and A. Hummel: *Radiat. Phys. Chem.* **26**, 509-511 (1985).
21. V.V. Lozovoy, V.M. Grigoryants, and O.A. Anisimov: Preprint No. 28, Institute of Chemical Kinetics and Combustion, Novosibirsk State University, Novosibirsk (1988).
22. S.I. Kubarev, S.V. Sheberstov, and A.S. Shustov: *Chem. Phys. Letters* **73**, 370-374 (1980).
23. Yu.N. Molin, O.A. Anisimov, V.M. Grigoryants, V.K. Molchanov, and K.M. Salikhov: *J. Phys. Chem.* **84**, 1853-1856 (1980).
24. E.L. Frankevich: *JETF* **50**, 1226-1234 (1966).
25. Y. Ruedin, P.-A. Schnegg, C. Jaccard, and M.A. Aegerter: *Phys. Stat. Solid.* (b) **54**, 565-576 (1972).
26. E.L. Frankevich, A.I. Pristupa, and V.I. Lesin: *Chem. Phys. Letters* **47**, 304-308 (1977).
27. O.A. Anisimov, V.M. Grigoryants, V.K. Molchanov, and Yu.N. Molin: *Chem. Phys. Letters* **66**, 265-268 (1979).
28. O.A. Anisimov, V.M. Grigoryants, V.I. Melekhov, V.I. Korsunskij, and Yu.N. Molin: *Doklady AN SSSR* **260**, 1151-1153 (1981).
29. A.V. Koptyug, V.O. Saik, O.A. Anisimov, and Yu.N. Molin: *Doklady AN SSSR* **297**, 1414-1417 (1987).
30. A.D. Trifunac and J.P. Smith: *Chem. Phys. Letters* **73**, 94-97 (1980).
31. J.P. Smith and A.D. Trifunac: *J. Phys. Chem.* **85**, 1645-1653 (1981).
32. J.P. Smith and A.D. Trifunac: *Chem. Phys. Letters* **83**, 195-198 (1981).
33. J.P. Smith, S. Lefkowitz, and A.D. Trifunac: *J. Phys. Chem.* **86**, 4347-4351 (1982).
34. S.M. Lefkowitz and A.D. Trifunac: *J. Phys. Chem.* **88**, 77-81 (1984).
35. S.I. Kubarev and E.A. Pshenichnov: *Chem. Phys. Letters* **28**, 66-67 (1974).
36. N.N. Lukzen, V.O. Saik, O.A. Anisimov, and Yu.N. Molin: *Chem. Phys. Letters* **118**, 125-129 (1985).
37. S.I. Kubarev, S.V. Sheberstov, and A.S. Shustov: *Khim. Fiz.* **6**, 784-793 (1982).
38. A.B. Doktorov, O.A. Anisimov, A.I. Burshtein, and Yu.N. Molin: *Chem.Phys.* **71**, 1-8 (1982).
39. S.N. Smirnov, V.A. Rogov, A.S. Shustov, S.V. Sheberstov, N.V. Panfilovich, O.A. Anisimov, and Yu.N. Molin: *Chem. Phys.* **92**, 381-387 (1985).
40. Yu.N. Molin, O.A. Anisimov, V.I. Melekhov, and S.N. Smirnov: *Faraday Discuss. Chem. Soc.*, No. 78, 289-301 (1984).
41. B.M. Tadjikov, V.I. Melekhov, O.A. Anisimov, and Yu.N. Molin: *Radiat. Phys. Chem.* **34**, 353-359 (1989).
42. O.A. Anisimov, Yu.N. Molin, S.N. Smirnov, and V.A. Rogov: *Radiat. Phys. Chem.* **23**, 727-729 (1984).
43. L.T. Percy, D.W. Werst, and A.D. Trifunac: *Radiat. Phys. Chem.* **32**, 209-213 (1988).

44. S.N. Smirnov, O.A. Anisimov, and Yu.N. Molin: *Chem. Phys.* **109,** 321-329 (1986).
45. O.A. Anisimov, V.M. Grigoryants, and Yu.N. Molin: *Chem. Phys. Letters* **74,** 15-18 (1980).
46. O.A. Anisimov, V.M. Grigoryants, and Yu.N. Molin: *Pis´ma v JETF* **30,** 589-592 (1979).
47. V.M. Grigoryants, O.A. Anisimov, and Yu.N. Molin: *Zhurnal strukturnoi khimii* **23,** 4-10 (1982).
48. O.A. Anisimov: *J. of Industr. Irradiation Tech.* **2,** 271-300 (1984).
49. V.I. Melekhov, O.A. Anisimov, V.O. Saik, and Yu.N. Molin: *Chem. Phys. Letters* **112,** 106-110 (1984).
50. A.V. Veselov, V.I. Melekhov, O.A. Anisimov, Yu.N. Molin, K. Ushida, and T. Shida: *Chem. Phys. Letters* **133,** 478-481 (1987).
51. V.I. Melekhov, O.A. Anisimov, A.V. Veselov, and Yu.N. Molin: *Chem. Phys. Letters* **127,** 97-100 (1986).
52. V.V. Lozovoy, V.M. Grigoryants, O.A. Anisimov, Yu.N. Molin, P.V. Schastnev, L.N. Schegoleva, I.I. Bilkis, and V.D. Shteingarts: *Chem. Phys.* **112,** 463-471 (1987).
53. V.O. Saik, O.A. Anisimov, V.V. Lozovoy, and Yu.N. Molin: *Z. Naturforsch.* **40a,** 239-245 (1985).
54. S.N. Smirnov, O.A. Anisimov, and Yu.N. Molin: *Chem. Phys.* **124,** 81-89 (1988).
55. B.C. Le Motais and C.D. Jonah: *Radiat. Phys. Chem.* **33,** 505-517 (1989).
56. A.D. Trifunac, Jr. M.C. Sauer, and C.D. Jonah: *Chem. Phys. Letters* **113,** 316-319 (1985).
57. A.D. Trifunac, D.W. Werst, and L.T. Percy: *Radiat. Phys. Chem.*, to be published (1989).
58. M.C. Sauer, K.H. Schmidt, and A.-D. Liu: *J. Phys. Chem.* **91,** 4836-4839 (1987).
59. H.M. McConnell: *J. Chem. Phys.* **34,** 13-16 (1961).
60. V.I. Melekhov, O.A. Anisimov, L. Sjöqvist, and A. Lund: to be published.
61. V.O. Saik, O.A. Anisimov, and Yu.N. Molin: *Chem. Phys. Letters* **116,** 138-141 (1985).
62. O.N. Antsutkin: Diploma thesis, Novosibirsk (1989).
63. J.H. Baxendale, D. Beaumond, and M.A.J. Rodgers: *Chem. Phys. Letters* **4,** 3-4 (1969).
64. D.W. Werst and A.D. Trifunac: *J. Phys. Chem.* **92,** 1093-1103 (1988).
65. V.O. Saik, N.N. Lukzen, V.M. Grigoryants, O.A. Anisimov, A.B. Doktorov, and Yu.N. Molin: *Chem. Phys.* **84,** 421-430 (1984).
66. M.F. Desrosiers and A.D. Trifunac: *Chem. Phys. Letters* **121,** 382--385 (1985).

II ANION RADICALS AND TRAPPED ELECTRONS

RADICAL ANIONS IN DISORDERED MATRICES

AKINORI HASEGAWA
Department of Chemistry, Kogakkan University, Ise-shi 516, Japan

1. Introduction

In this chapter we are concerned with the radical anions which were derived from guest molecules in disordered matrices by radiolysis or photolysis at low temperature and detected mainly by ESR spectroscopy. The matrix isolation technique in ESR studies has facilitated the selective detection of the highly reactive ionic radicals primarily formed from guest molecules by electron-gain or electron-loss, using suitable matrices.

Few review articles have been published in connection with the radical anions trapped in matrices. Symons has reviewed ESR studies on radicals in rare-gas matrices [1] and the effect of various matrices on the ESR parameters of trapped radicals [2]. Moreover, the electron-gain and electron-loss mechanisms caused in matrices by irradiation were discussed in his suggestive article [3]. The electron-gain mechanism directly relating to the formation and reaction of radical anions, was argued in terms of trapping electron, shape-change, bond-breaking, bond-making, protonation and solvation processes.

I myself, on the other hand, have reviewed our studies on the radical anions formed in tetramethylsilane (TMS) matrices [4], which were started in collaboration with Drs. Williams and Shiotani [5] and were developed in my group. In that review, the geometrical change caused by electron-capture was discussed for a series of radical anions having a central atom of the group IV elements.

Since the choice of suitable matrices is important for studies using this technique, the matrix itself and the radical anions generated in the matrix were outlined. The amount of literature to be reviewed is extremely large. However, rather than prune the quantity of papers in our interest too severely, it was attempted to cite as many of the papers as possible.

In some cases, molecular shape changes remarkably in the process of formation of radical anions. One of the most important driving force for the shape-change is the Jahn-Teller distortion and many examples of the Jahn-Teller distorted 'radical cations' have recently been observed using ESR [6]. Prior to these examples, it was pointed out that the shape-change of some 'radical anions' is caused by the Jahn-Teller distortion [4]. Some radical anions also undergo interconversion on annealing or photoillumination. These changes in molecular structure will be described in more detail. Dissociation processes of radical anions will also be discussed in connection with the structure of the radical anions and the distribution of spin density on the radical anions.

313

A. Lund and M. Shiotani (eds.), Radical Ionic Systems, 313–336.

2. Matrices and Radical Anions in Matrices

There are two types of matrices used in ESR investigations on radical anions. The first is rigid solids. In this category, we have non-hydroxylated polar matrices, nonpolar matrices, and hydroxylated polar matrices which are frozen at 77 K. 2-Methyltetrahydrofuran (MTHF) has been most frequently utilized as the non-hydroxylated polar matrices. 3-Methylpentane, methylcyclohexane, *etc.* belong to the nonpolar matrices. In addition to frozen alcohols, aqueous alkaline and neutral glasses are the hydroxylated polar matrices. Besides, rare gas matrices at *ca.* 4 K may be included in rigid solids.

The second is rotator solids. Matrix molecules in high symmetry can rotate freely before trapped radicals migrate to decay, so that they give the isotropic ESR spectra which allow easy spectral interpretation. For this purpose, SF_6, adamantane, neopentane, and tetramethylsilane (TMS) have been used. Even if a radical anion is too large to give isotropic spectra, and gives anisotropic spectra, the line-width observed for rotator matrices is much smaller than that for rigid matrices.

When a radical was investigated in different matrices, the radical was included in the section of the matrix predominantly used, along with the description of the other matrices.

2.1. 2-METHYLTETRAHYDROFURAN (MTHF) MATRIX

The ESR spectrum of a γ-irradiated pure MTHF glass recorded at 77 K, consists of the superimposition of an intense sharp singlet due to trapped electrons on a broad seven-line hyperfine structure with an average splitting of 20.5 G. This seven-line spectrum has been assigned to either or both the neutral radical (1) [7] or the protonated radical (2) [8]: the three methyl protons and one of the β-CH_2 protons giving a 20.5 G splitting and the other β-CH_2 proton a 41 G splitting. This assignment was, however, questioned because irradiation of MTHF-d_3 (3) gave a nine-line spectrum instead of the same seven-line spectrum. In order to obtain unambiguous results, Ling and Kevan exposed adamantane-d_{16} doped with MTHF to γ-rays, identifying two distinct radicals (1) and (4) [9]. These radicals are believed to be yielded upon irradiation of MTHF glasses, but Ling and Kevan could not explain the seven-line spectrum observed in the glass at 77 K using these radicals with the ESR parameters obtained at 240-270 K. Formation of radicals (1) and (4) was confirmed by analogous experiments with adamantane, by Dismukes and Willard [10], and by spin trapping of radiolysis products of MTHF in the glass and solution states, by Murabayashi *et al.* [11]. Dismukes and Willard found that the two equivalent β-protons in radical (1) at 230 K become inequivalent at 77 K, and they concluded that the seven-line spectrum in the MTHF glass was attributable to radical (1). This conclusion is inconsistent with the result obtained for the MTHF-d_3 (3) glass at 77 K by Ling and Kevan [9].

The geometrical structure of the electrons trapped in the MTHF solids has been studied by Kevan *et al.* using a spin echo modulation technique. It was concluded that the trapped electron has three MTHF molecules in the first solvation shell [12], as described in another chapter of this book. Positive holes, forming as the counterpart of the trapped electrons during irradiation, are generally considered to be stabilized to yield protonated species, either paramagnetic (2) or diamagnetic (5) [8].

When an MTHF glass contains guest molecules of electron scavengers, ESR spectra obtained after γ-irradiation of the glass have new spectral lines attributable to radical anions originating from the guest molecules but have no signal due to the trapped electrons. This means that the electrons formed upon irradiation migrate to be trapped by the guest molecules, yielding the radical anions. In this way, various radical anions were generated in MTHF matrices and identified through the analysis of their ESR srectra.

Some pioneer studies have been made concerning the ESR observation of the radical anions formed in MTHF matrices after γ-irradiation at 77 K. Ayscough et al. [13] reported that exposure of an MTHF glass containing small quantities of naphthalene or CCl_4 causes a diminishing of the singlet of trapped electrons and the appearance of signals attributable to naphthalene anion or $\cdot CCl_3$ formed by dissociation of CCl_4^-. The ESR spectrum of biphenyl radical anions was also observed by Smith and Pieroni [14]. ESR spectra of aromatic radical anions were observed after γ-irradiation of solid matrices of MTHF and DME (dimethoxyethane) containing benzene, biphenyl, naphtalene, styrene, nitrobenzene, and benzoquinone at 77 K [15]. In particular, annealing of the nitrobenzene anion gave the protonated $C_6H_5NO_2H$ radical which is stable even at room temperature in the liquid phase.

Analogous studies were applied to α-methylstyrene in MTHF [16] and maleic anhydride in MTHF [17] as well as to acrylic acid in MTHF, TEA (triethylamine) and MHX (3-methylhexane) [18], revealing formation of their radical anions, immediately after irradiation, and of the neutral radicals by protonation to the radical anions, on warming. Maleate radical anions were also detected [19]. Their irreversible isomerization will be mentioned in section 5. Radical anions of 1,3-butadiene and its derivatives were investigated and the conformation of the butadiene, isoprene and 2,3-dimethylbutadiene anions was proposed to be predominantly in the *trans* form [20].

An ESR spectrum for I_2^- formed in MTHF containing I_2 was reported, together with the electronic spectra for I_2^-, I_4^-, Br_2^-, and ICl^- in this matrix [21].

$SO_2Cl_2^-$ and Cl_2^- radical anions have been generated from sulfuryl chloride in MTHF, 3-methylpentane, methylcyclohexane, and n-butyronitrile [22]. In addition to these anions, SO_2^- was also detected in the MTHF matrix. The unpaired electron in $SO_2Cl_2^-$ was shown to be in a supramolecular orbital derived from the lowest antibonding orbitals of SO_2 and Cl_2. In contrast, $SO_2Cl_2^-$ is the only radical anion formed in irradiated crystalline sulfuryl chloride [23].

Irradiation of pure diethyl phosphite at 77 K gave two radical species. One of them was identified as the $C_2H_5P(O)O^-$ anion produced by electron attachment, because only this species was detected by ESR from diethyl phosphite in MTHF glass [24].

When phenyl dichlorophosphate and $PhP(S)Cl_2$ [25] as well as a series of chlorophosphate esters [26] were irradiated, the chlorophosphoranyl radical anions were formed by simple electron attachment to the parent molecules. In the dichlorophosphoranyl radicals, two equivalent chlorines with large Cl couplings have the apical positions of a trigonal bipyramidal structure, which is consistent with the result established in the studies of $POCl_3^-$ [27] and $\cdot PF_4$ [28] in single crystals. The anion radicals derived from $(MeO)_2P(S)Br$ [29] and $Me_2P(S)Br$ in organic matrices (MTHF, CH_3OH, CD_3OD, CH_3CN, CD_3CN) and in the pure solids [30], however, have tetrahedral configurations accommodating an unpaired electron in the P-Br σ* bond. Radiolysis of $(RO)_2P(X)SH$ (X=O,S) in MTHF and CD_3OD matrices or as powders yielded electron-gain species $(RSSR)^-$ of σ* type and $^-P(RO)_2(X)SSP(X)(OR)_2$ in phosphoranyl structure [31].

A triplet state species was detected by Konishi *et al.* [32] after irradiation of 1,3-dinitrobenzene and 1,3,5-trinitrobenzene in MTHF at 77 K and assigned to the radical pair, $(MTHF \cdots PhNO_2^-)$. Reinvestigation of this system by Symons *et al.* [33] has, however, revealed the formation of monoanion with an unpaired electron primarily on one nitro-group and of dianion in triplet state occupying each unpaired electron on one nitro-group. The monoanion arises at low γ-ray doses in MTHF and CD_3OD, while the dianion at high doses in MTHF. The effective separations between the two unpaired electrons in the triplet species was estimated to be in the region of 5-6Å.

The structure of the anions of C_6F_6 [34] and $(CF_3)_3CI$ [35] formed in MTHF matrices will be discussed in section 4.7 and 6.1, respectively.

Irradiation of 5-halogenouracils in neutral or alkaline aqueous glasses at 77 K gave the π^* anion (6) and, upon warming, the uracil-yl radical (7) [36]. However, when a range of 5-halogenouracil bases, ribo- and deoxyribo-sides were exposed to γ-rays or to photo-electrons in organic (MTHF, CH_3OH, CD_3OD) and neutral aqueous (12 M $LiCl-D_2O$) glasses, σ^* radical anions (8) were observed in addition to π^* anions (6), for bromo- and iodo-derivatives in all the matrices. In fluoro- and chloro-uracils, only π^* radicals were observed [37]. σ^* radical anions were also derived from C_6F_5I, $C_6F_4I_2$, C_6F_5Br, $C_6F_4Br_2$ and C_6F_5Cl in MTHF and CD_3OD matrices [38]. An unpaired electron in each radical anion is largely confined to a single C-X (X=Cl,Br,I) bond.

Moreover, 4-halogeno-, 5-halogeno-, and 4,5-dihalogeno-imidazole, halogeno being bromo and iodo, in MTHF and CD_3OD yielded similar σ^* radical anions [39]. As for 4,5-dihalogenoimidazole, the unpaired electron was concluded to be in the C_4-hal bond, from comparison of ESR parameters.

The radical anion of c-C_8F_8 was generated in an MTHF glsss. Its isotropic spectrum at 145 K was assigned to a simple aromatic π^* radical anion in planar D_{8h} structure [40]. Irradiation of methyl isocyanate and methyl isothiocyanate in MTHF or CD_3OD gave the corresponding radical anions which are bent at the cyano carbon atom, exhibiting h.f.c. of 7 G to a N nucleus and satellite coupling of 110 G to ^{13}C [41]. CS_2^- radical anion has been detected in MTHF, CD_3OD, CD_3OD-D_2O or CD_3OH-H_2O matrix [42]. Annealing the irradiated MTHF glass resulted in the formation of the dimer anion $(SC)S \doteq S(CS)^-$. In the course of this study, O_2^- radical anions were detected in CD_3OD and CD_3OD-D_2O matrices [42].

The radical-nucleophilic first order substitution ($S_{RN}1$) mechanism has been investigated with the following organo nitro-compounds in MTHF and CD_3OD matrices : $Me_2C(X)NO_2$ with B, Cl, SCN, NO_2 and so on [43,44], $Me_2C(N_3)NO_2$ [45], XCH_2PhNO_2 [46], and nitroimidazole derivatives [47]. Irradiation of these systems gave ESR spectra mostly attributable to the parent radical anions and the neutral radicals formed by dissociative electron capture or dissocation of the parent radical anions.

Various radical anions of transition-metal carbonyls and the related compounds have been generated by electron attachment in MTHF glasses containing $Co(\eta^5-C_5H_5)(CO)_2$ [48], $Co_2(CO)_6(ER_3)_2$ with ER_3=P-n-Bu_3, $P(OMe)_3$, and As-i-Bu_3 [49], $Mn_2(CO)_{10}$ [50], $Mn_2(CO)_8(PR_3)_2$ [50], $Re(CO)_5Br$ [51], $Re(CO)_5I$ [51], $Fe(CO)_5$ [52], $Co(\eta^5-$

C$_5$H$_5$)$_3$(μ_3-CPh)$_2$ [53], Mn$_3$(μ-H)$_3$(CO)$_{12}$ [54], Re$_2$(μ-H)$_2$(CO)$_8$ [55], and Re$_2$(μ-H)$_2$(CO)$_6$(μ-dppm) [55].

2.2. ALCOHOL MATRIX

Neutral radicals ·CH$_2$OH and CH$_3$ĊHOH have been identified in ESR spectra for the solid methanol and ethanol, respectively, irradiated at 77 K [56]. ESR evidence for trapped electrons in alcohol glasses was first given by Chachaty and Hayon [57,58]. Geometrical structure of an electron surrounded by four first-solvation-shell molecules has recently been given for the trapped electrons in ethanol from the analysis of electron spin echo modulation by Narayana and Kevan [59].

Exposure of solid methanol and ethanol containing benzene or naphthalene to X-rays at 77K revealed that upon irradiation, hydroxyl hydrogen is liberated and subsequently adds on the conjugated ring system [60,61]. For naphthalene, the hydrogen atom adds in the α position [61]. Similar hydrogen atom addition has been observed for benzene, hydroquinone, naphthalene, biphenyl, and styrene [15] as well as acrylic acid [18], in alcohol matrices. However, hydrogen atom addition or protonation to radical anion in alcohol matrices is remarkably reduced by using deuterated methanol, CD$_3$OD. Thus, various radical anions have been generated and characterized by ESR in CD$_3$OD as well as MTHF, as have been already described in the section of MTHF matrix.

The BrCN$^-$ anion was detected after irradiation of a CD$_3$OD or CD$_3$CN solution containing BrCN [62]. This anion was shown to be one of the three paramagnetic centers formed in γ-irradiated crystalline BrCN. A series of phosphites were irradiated in CH$_3$OH, and observed ESR spectra were attributed to the HṖ(OR)$_3$ radicals formed by protonation to P(OR)$_3{}^-$ anions [63]. However, trialkylphosphates in CD$_3$OD gave the parent phosphoranyl (RO)$_3$PO$^-$ anions and phosphoryl (RO)$_2$ṖO radicals [64]. The former was obtained in a relatively smaller yield in a CD$_3$OD matrix. The following scheme has been given for electron capture by (MeO)$_3$PO.

Scheme for electron capture by (MeO)$_3$PO. Reproduced with permission from *J. Chem. Soc., Perkin. Trans.* 2, 1977, 286. Copyright (1977) The Royal Society of Chemistry.

The phosphoranyl anion (9) having two MeO groups in the apical sites is thermodynamic favoured, while, in some cases of pure solid trialkylphosphates, the process of b and c occurs. The phosphoryl radical (10) is formed by process d rather than f. Process e occurs in pure solids.

Various alcohols and aqueous alcohols saturated with air or oxygen produced superoxide anions, O$_2{}^-$ after γ-irradiation at 77 K [65]. The anions gave low-field ESR features whose g values were solvent dependent. Well-defined ESR spectra of O$_2{}^-$ were also obtained for NaO$_2$ dissolved in these solvents. Solvation of O$_2{}^-$ anion was discussed.

Electron addition to 1,2-dihalogenoethanes in CD$_3$OD gave dihalide σ^* radical anions, Cl$_2{}^-$, Br$_2{}^-$, or I$_2{}^-$ [66]. Photolysis of CHBr$_3$, CDBr$_3$ and CBr$_4$ in matrices (alcohols, diethylether, benzene derivatives and KBr) brough about CBr_4^- anion and various neutral radicals [67].

Radical anions of 1,2,4,5-tetrazine and its derivatives were generated in CD_3OD. Their ESR spectra indicated that the SOMO is confined to the four ring nitrogen atoms [68]. Exposure of tetracyanonickelate (II) ions in CD_3OD-D_2O matrices produced electron-gain centers, $[Ni(CN)_4]^-$ (d^9) anions [69].

2.3. AQUEOUS MATRIX

Irradiation of aqueous glass is important particularly in biochemical aspects. However, the radical anions of biochemical interest which were generated in aqueous matrices, were omitted in this section because they may be included in another chapter of this book.

A series of radical anions of unsaturated carboxylic acid were generated in LiCl-D_2O or NaOD-D_2O glasses by photolysis of $Fe(CN)_6^{4-}$ ions at 77 K [70]. The sites of protonation to the radical anions were also investigated by warming the glasses. Radical anions of carboxylic acids, ketones and aldehydes were observed in LiCl-D_2O glasses and thermal hydrogen-atom abstraction by these radical anions were investigated [71]. Electron attachment at 77 K to esters and triglycerides in LiCl-D_2O glasses formed the radical anions which decay on annealing by β-scission [72].

A well-known species, formed by irradiation of aqueous alkaline glasses and having $g_\perp = 2.07$ and $g_{//} = 2.002$, was identified as O^-. This was supported by using H_2O enriched in $H_2^{17}O$, which gave $A_{//}(^{17}O) = 99$ G [73]. The spin density of ca. 0.69 deduced from this result suggests considerable delocalization on to the matrix. It is of interest to note that O^- is not normally detected in neutral aqueous glass, though well known in alkaline glasses. However, O^- in alkaline glasses was reported to form after photolysis of $K_4Fe(CN)_6$ in 8 M $NaClO_4$/D_2O [74], but it might have been generated by dissociation of ClO_4^- anion.

The follwing are fundamental and important ESR studies on the inorganic radical anions formed by γ-irradiation in frozen aqueous solutions at 77 K. All of them have been investigated entirely by Symons and his co-workers. Hydroxyl radicals formed by irradiation of ice crystals at 77 K were reported to be reversibly converted into O^- anions on cooling from ca. 30 K to ca. 4 K. Evaluation of hyperfine couplings to ^{17}O nuclei in $^{17}O^-$ and ^{17}OD radicals were achieved by using D_2O enriched in ^{17}O [75].

When aqueous solutions of potassium superoxide were frozen, broad g_z features appeared at $g = 2.11$, without irradiation. In this case, phase separation occurred on freezing so that the g_z value was not for the aquated O_2^- ion. Accordingly, a g_z value of 2.065 was acquired for the aquated ion by extrapolating the data obtained for aqueous methanolic glasses [76]. A list of g_z values for O_2^- in a range of environments is also given [76].

Irradiation of alkali-metal bromides or iodides solutions frozen at 77 K gave Br_2^- or I_2^-, ·OH, and the mixed species $BrOH^-$ or IOH^- [77]. Selenite ions in frozen solutions were attacked by ·OH radicals, forming the SeO_4^{3-} species weakly protonated in the solid [78]. Aqueous solutions of gallium(III), thallium(III), and mercury(III) salts were exposed to γ-rays at 77 K, giving ESR spectra attributable to Ga^{2+}, Tl^{2+}, and Hg^+, respectively [79]. A more detailed study was carried out with γ-irradiated aqueous and alcoholic solutions of mercurous(I) and mercuric(II) salts. With mercuric salts in frozen acidic solutions, the $^{199}Hg^+$ and $^{201}Hg^+$ species formed by electron capture were observed together with $HgOH^+$ species [80].

Irradiation of aqueous solutions of alkali-metal azides and cyanides gave $(CN)_2^-$ and HCN^- anions in addition to the hydroxyl, NH_2 and H_2CN radicals. The HCN^- anion was readily protonated to H_2CN even in strongly alkaline solid solutions. Some edivence was given for the formation of the $HCNO^-$ radical anion [81,82]. On the other hand, irradiation of aqueous solutions containing alkali-metal cyanates gave ESR spectra attributable to the $HCNO^-$ radical anions and trapped nitrogen atoms [83]. Similar experiments with alkali-metal thiocyanates [83] yielded the spectrum characteristic of $HCNS^-$, no nitrogen atom being detected. The unusually large proton couplings (73.0 G for $HCNO^-$ and 76.2 G for $HCNS^-$) were interpreted in terms of protonation on carbon. These anions are considered to have the unpaired electron primarily confined to the nitrogen p_z orbital.

Frozen aqueous glasses containing various oxyanions were exposed to γ-rays at 77 K. Among them, AsO_4^{3-}, SeO_4^{2-} and BrO_4^{2-} accepted an electron to give the radical anions which are thought to have an approximately trigonal bipyramidal structure [84].

Potassium carbonate enriched in ^{13}C was γ-irradiated in aqueous, alkaline aqueous, and CD_3OD glasses at 77 K. Aqueous glasses gave CO_2^- and CO_3^- anions in addition to the major primary products of ·OH and e^-. On the other hand, the alkaline solution produced O^- and e^-, but a new species, tentatively assigned to CO_4^{3-}, was detected on annealing. In CD_3OD, CO_2^{3-} radicals were observed together with other radical species, and signals due to the CO_3^- radical increased on warming. Analogous studies of aqueous glasses containing formates, nitrites and nitrates were reported [85]. Formates gave $H\dot{C}O$ and CO_2^- radicals and the species assigned to HCO_2^{2-} and $HCO(OH)^-$. Nitrites produced NO_2 and NO_2^{2-}, and annealing of alkaline glasses resulted in the formation of NO_3^{2-}. In neutral glasses, HNO_2^- was formed by protonation to nitrogen. Nitrates gave NO_3^{2-} in alkaline solutions, but NO_2 was formed with the decrease in pH. NO_2 is thought to be generated by electron addition to the system of NO_3^- and H_2O. These studies mentioned above were accomplished by Symons and his-coworkers.

When aqueous glasses containing silver salts were irradiated, Ag atoms were formed by electron attachment to Ag^+ ions. Drastic decrease in silver h.f. coupling was observed between the Ag atoms generated and observed at 4.2 K (site I) and those observed after brief thermal annealing at 77 K (site II) [86-88]. The geometrical model for the Ag atoms has been discussed on the basis of proton spin flip satellites observed for site II [87] and electron spin echo modulation [89,91], in addition to the Ag h.f. couplings [86-88,90]. Ultimately, it was concluded from the analysis of high power electron spin modulation for the Ag atoms in deuterated ice that the presolved Ag atom in site I is surrounded by eight D atoms, at *ca.* 3.1Å , of four water molecules, while the solvated Ag atom has one D atom at *ca.* 1.7Å and seven D atoms at *ca.* 3.1Å [92]. Thus, the change from site I to site II is brought about by rotation of one water molecule.

2.4. ACETONITRILE MATRIX

Irradiation of acetonitrile at 77 K gave two different electon-captured centers depending upon the nature of the crystalline phase: dimer anion radical $(MeCN)_2^-$ in metastable Crystal I [93] and monomer anion radical $MeCN^-$ in Crystal II [94]. A very weak interaction between methyl radicals and cyanide ions was observed for the methyl radicals formed by photobleaching of the irradiated Crystal II [94].

Alkyl radical-halide ion adducts were generated by dissociative electron capture in acetonitrile [95,96]. In contrast, N-chloro and N-bromo-amides in this matrix gave σ* radical anions without a loss of halide ion [97,98]. Two structures were detected for the NCBr⁻ anion in a CD₃CN matrix [99], which will be described in section 5.

The structure of the Ag atom center formed by elecrton-gain of an Ag⁺ ion in this matrix was discussed by both Symons' and Kevan's groups [100-102].

2.5. ARGON MATRIX

Kasai and his co-workers have developed a method of generating charged species in an argon matrix by co-deposition of suitable electron-donating and electron-accepting species at 4 K, followed by photolysis to transfer an electron between them [103]. Formation of molecular radical anions has been achieved by this method using Na atoms as electron donors and tetracyanoethylene (TCNE), diborane, and furan molecules, respectively, as the acceptors [103]. With TCNE, the ESR spectrum recorded prior to the photolysis, consisted of the Na quartet and a weak broad signal at g =2.0. Photoirradiation caused a complete diminishing of the Na signals and an increase in the signal at g =2.0 which had a structure of nine equally spaced lines and suggests the formation of the radical anion of TCNE. By a similar method, ESR spectra of $B_2H_6^-$ and $B_2D_6^-$ were observed [103,104], supporting the result by Palke and Lipscomb [105] that the unpaired electron in $B_2H_6^-$ possesses a b_{3g} orbital given by an antibonding combination of boron 2p orbitals perpendicular to the plane involving the four terminal hydrogen atoms.

The radical anions of phenol [106], pyrrole [107], pyrazole [107], imidazole [107], indole [107], furan [108], isoxazole [108], oxazole [108], dihydroxybenzenes [109], and hydroxypridines [109], have also been detected in argon matrices. Their structures which are of particular interest will be discussed in section 4.9.

Photolysis of acetylene co-condensed with alkali metals in an argon matrix gave the C_2^- radical anion which exhibited an ESR spectrum with small h.f. couplings to metal ions and orthorhombic g-tensors [110]. This indicates the formation of ion-pairs in C_{2v} symmetry. A stable reaction intermediate complex K⁺---HCl⁻ was reported to form after deposition of atomic potassium and HCl in argon matrix at 4 K [111]. It is of interest to compare this complex and the hydrogen halide radical anions formed in irradiated Me_3NSO_3 doped with Me_3NHX (X=F,Cl,Br,I) [112].

Radical anions weakly interacting with a sodium ion, $Na^+SO_2^-$ [113], $Na^+O_2^-$ [113,114], and $Na^+O_3^-$ [115], were generated by the reaction of sodium atoms and SO_2, O_2, or O_3 during deposition in an argon matrix at 4.2 K. A similar method produced the formation of $M^+Cl_2^-$ ion-pair as a result of the reaction of various metals with Cl_2 [116].

2.6. NEON MATRIX

Knight and his co-workers have established the neon matrix generation and trapping technique which allows ESR observation of various radical cations [117], as shown in a previous chapter of this book. This technique was shown to be applicable to ESR studies on radical anions. The anion radical F_2^- was generated by depositing neon matrix gas and F_2 gas under vacuum UV photolysis [118]. This radical anion is the first example of the anion trapped as a free ion in dilute rare gas matrices at 4 K, because all the radical anions formed in argon matrices interact weakly with metal cations. ESR parameters for this anion were compared with those for F_2^- in various crystalline environments.

The F_2CO^- radical was generated by two independent methods. The first being electron addition to F_2CO in neon matrix gas by photoionization and the second being co-deposition of F_2CO with Na into neon and argon matrices [119]. ESR spectra for this anion revealed that the radical anion is a carbon-central radical with a large h.f. coupling to the ^{13}C nucleus and that the anion deviates from C_{2v} symmetry, having two inequivalent F nuclei. In this experiment, the species tentatively assigned to F_2CO^+ was also detected by ESR.

Photoionization and electron bombardment of CH_4 during neon matrix deposition produced the CH_2^- radical anion which gave h.f. couplings to ^{13}C and H nuclei in its ESR spectrum [120]. Knight et al. have also described that they have made preliminary ESR assignments for other isolated anions including Cl_2^-, HF^-, DF^-, $^{16}O_4^-$, and $^{17}O_4^-$ [121].

2.7. KRYPTON MATRIX

Solid krypton was shown to dissolve certain transition-metal carbonyl compounds [122] and to generate mononuclear and polynuclear carbonyl radical anions as well as radical cations and neutral radicals, after irradiation. The following radical anions were detected in Kr matrices at 77 K: $HCo(CO)_4^-$ and $Co_2(CO)_8^-$ from $HCo(CO)_4$ [123,124] and $Ni_2(CO)_7^-$ from $Ni(CO)_4$ [125].

2.8. ADAMANTANE MATRIX

The adamantane matrix technique for the generation of free radicals and the observation of their solution-like isotropic ESR spectra has proven to be quite successful for the study of neutral free radicals [126]. This technique has also been developed for the study of radical anions by doubly doping the adamantane with a radical precursor and a strong electron donor, Me_3NBH_3, to avoid electron-hole recombination [127,128].

A series of fluorinated benzenes [127,128] and pyridines [129] were studied by this technique. σ^* radical anions were obtained from heavily fluorinated compounds, while π^* radical anions were detected from the other compounds. The details will be mentioned in section 4.8. The $SiCl_4^-$ anion [130] and alkyl radical-halid ion adducts [96,131] were detected in adamantane matrices.

Wang and Williams, however, suggested that the role of this additive may be simply that of a plasticizer in increasing the solubility of the anion precursor in the crystalline adamantane matrix [132]. In fact, we observed ESR spectra of $B_2H_6^-$ in adamantane matrices without the additive [133].

2.9. SF_6 MATRIX

The SF_6 matrix has a rotator phase above the solid-solid transition temperature at 94 K. Originally, SF_6^- and SF_5 radicals were prepared during continuous electron bombardment at 100 K of pure solid of SF_6 and their isotropic ESR spectra were observed [134]. In some cases, addition of fluorine atoms to guest molecules was observed after irradiation [134,135]. Since SF_6 itself behaves as an electron scavenger, the radical cations from guest molecules arise together with SF_6^- during radiolysis if the guest molecule has a lower ionization potential than SF_6 has [136].

However, if the guest molecule is a more effective electron scavenger, radical anions are generated from the molecule and no trace of or a remarkably reduced amount of SF_6^- is detectable. The following radical anions were observed in SF_6 matrices: AsF_5^- [137], SeF_6^- [138], TeF_6^- [138], F_3NO^- [139,140], and $FClO_3^-$ [141]. The AsF_5^- radical gave ESR spectra with h.f. couplings to four F nuclei, the fifth F nucleus exhibiting no splitting in a manner similar to the cases of SF_5 [142] and PF_5^- [137].

As well as other neutral hexafluoride radicals, the SeF_6^- and TeF_6^- anions have large h.f. interactions to the central atoms, indicating $^2A_{1g}$ ground state in O_h symmetry [138].

The F_3NO^- radical anion is the first example of a 33 valence electron radical composed only of first-row elements [139,140]. The isotropic ESR spectra of $FClO_3^-$ were observed in the SF_6 matrix, while anisotropic spectra were recorded in pure $FClO_3$ and in tetramethylsilane (TMS) matrices [141]. The structure and dissociation of these radical anions will be discussed in sections 4.1 and 6.2

2.10. NEOPENTANE MATRIX

Neopentane has a rotator phase suitable for giving isotropic ESR spectra of trapped radicals, above the solid-solid phase transition temperature of 130 K. This matrix is known to add hydrogen atoms to solute molecules and yield neutral radicals upon irradiation, as shown for the formation of a series of phosphoranyl radicals by Morton et al. [143]. On the other hand, Miyazaki et al. reported that the hydrogen atoms formed in this matrix extract hydrogen atoms from guest molecules, yielding neutral radicals, e.g. alkyl radicals [144].

Despite these facts, Morton et al. observed isotropic ESR spectra of the radical anions, SiF_4^- [145], SnH_4^- [145], and PF_5^- [146], generated in irradiated neopentane matrices containing these precursors. Williams and his co-workers, motivated by these results, started their studies on electron attachment to a series of fluorocarbons [5] and reported the ESR spectra of the perfluorocyclobutane anion c-$C_4F_8^-$ [147] and $C_6F_6^-$ [147] anion formed in this matrix. More details of these radical anions will be given in section 4.7.

2.11. TETRAMETHYLSILANE (TMS) MATRIX

Williams, Shiotani, and myself thought that since radical anions are trapped in neopentane matrices, they may be also trapped in tetramethylsilane (TMS) which is obtained by replacing the central C atom of neopentane by a Si atom, and that the radical anions may rotate more easily in TMS owing to the larger size of the TMS matrix molecule, giving better isotropic ESR spectra [5].

Shiotani and Williams proved this idea by detecting isotropic spectra for a series of perfluorocycloalkane radical anions: c-$C_3F_6^-$, c-$C_4F_8^-$, and c-$C_5F_{10}^-$ in TMS matrices at 130 K [5]. H.f. couplings to equivalent fluorines indicate that the unpaired electron is delocalized over the entire molecular framework in an orbital of high symmetry.

On the other hand, Williams and I observed well-defined anisotropic spectra for CF_3Cl^-, CF_3Br^-, and CF_3I^-, which were interpreted in terms of axial symmetric h.f. couplings to a halogen (Cl, Br, or I) and three equivalent ^{19}F nuclei [148,5]. This implies that the CF_3X^- radical has C_{3v} symmetry and that the CF_3 group is rotating rapidly about the symmetric C-X axis in the TMS matrix at 101 K (X=Cl), 121 K (X=Br) or 98 K (X=I). In contrast to the case of the c-$C_nF_{2n}^-$ radical anions, the unpaired electron is confined in the C-X σ^* orbital in these radical anions. Spectral resolution of the powder pattern in the TMS matrix is greatly superior to that observed in the rigid MTHF matrix. ESR spectra for $CF_2Cl_2^-$ and $CFCl_3^-$ were also observed [5]. Dissociation of these fluoromethyl halide radical anions will be mentioned in section 6.1.

Williams and his co-workers observed isotropic and anisotropic spectra for $C_2F_4^-$ in TMS-d_{12} at 120 K and in MTHF at 77 K, respectively [149]. In these two matrices, the novel cycloaddition reaction

$$C_2F_4^- + C_2F_4 \longrightarrow c\text{-}C_4F_8^-$$

was also observed to occur at 80 - 95 K [150]. Regarding the structure of the $C_2F_4^-$ radical, several papers were published thereafter. The details will be mentioned in section 4.6. Analogous radical anions, $C_2F_3Cl^-$ and $C_2F_3Br^-$, were also generated in TMS-d_{12} [151] and compared with the radical anions described above. It was concluded that the unpaired electron in these radicals occupies a C-X σ^* orbital in a manner quite similar to the case of the CF_3X^- radical anions.

My co-workers and I have extended the ESR studies on the CF_3X^- radical anions to those of the congeneric radical anions with a central Si atom, SiF_3X^- (X=Cl,Br,I) [152,153] and $SiF_nCl_{4-n}^-$ (n=0-2) [153]. As a result, in contrast to the CF_3X^-, these radical anions were revealed to have trigonal bipyramidal structures with two apical and two equatorial halogens. Moreover, generation of SiH_3X^- [154] and GeH_3X^- [155], where X =Cl,Br,I, was successful in TMS matrices. It is of interest that despite the fact that ·CH$_3$ radicals or ·CH$_3$---X^- adducts were produced from CH_3X, molecular radical anions arose from the congeneric compounds having a central Si or Ge atom, in TMS matrices. Analogous radical anions were also obtained from methyl derivatives of halogeno-silanes and -germanes [156], as well as other organo halogeno-silanes [157] and methyl derivatives of stannane [158]. The drastic geometrical change in these radical anions brought about by the difference of the central atom, will be discussed in sections 4.2 to 4.4.

We have also extended the studies on the radical anions of halides having a central atom of the group IV elements to those for $(CH_3)_2AlX^-$ (X=Cl,Br,I) anions having a central Al atom of group III [159]. In contrast to the result for the radical anions having a central atom of group IV, the unpaired electron is accommodated in the orbital composed largely of 3s and 3p orbitals of Al and slightly of halogen orbitals. This is consistent with the result reported for the BF_3^- radical anions generated in a TMS matrix [160]. These radical anions are considered to be nonplanar.

Recent ESR study on the CCl_4^- radical anions in TMS matrices by Muto *et al.* will be discussed in section 2.13 [161].

The following mechanism is thought to occur upon irradiation of TMS solid solutions containing guest molecules,

$$(CH_3)_4Si + G \xrightarrow{\gamma\text{-rays}} (CH_3)_4Si^+ + e^- + G \longrightarrow (CH_3)_4Si^+ + G^-$$

$$(CH_3)_4Si^+ \longrightarrow \cdot H + (CH_3)_3SiCH_2^+$$

$$\cdot H + (CH_3)_4Si \longrightarrow H_2 + (CH_3)_3Si\dot{C}H_2$$

Thus, since the positive hole generated by irradiation is considered to be stabilized in the $(CH_3)_3SiCH_2^+$ cation, the anion radical formed by electron attachment is trapped in the TMS matrix safely avoiding recombination with the positive hole, and detected by ESR together with the $(CH_3)_3Si\dot{C}H_2$ radical which masks the central region of observed ESR spectra.

2.12. HEXAMETHYLETHANE (HME) AND METHYLCYCLOHEXANE-d_{14} (MCHD) MATRICES

Hexamethylethane (HME) was shown to be a very suitable matrix for observing naturally abundant ^{13}C satellites in high-resolution isotropic spectra of $C_6F_6^-$ radical anions [162]. Compared with this matrix, neopentane and TMS gave some residual anisotropy so that the signal intensities were too low to detect ^{13}C satellites. Accordingly, this matrix may be the most suitable matrix to obtain isotropic spectra of trapped radical anions, since adamantane and SF_6 matrices can give excellent isotropic spectra but have some limitations for generation of radical anions.

On the other hand, methylcyclohaxene-d_{14} (MCHD) is also of interest, because the crystalline MCHD solution prepared by slow cooling gave partially ordered radical anions, $C_2F_4^-$ [163] and $F_2CCH_2^-$ [164], which gave well-defined ESR spectra with orientation dependent parallel and perpendicular features. Pronounced orientation effects similar to these cases had already been observed for $CF_2Cl_2^-$ in the TMS solid solution which was obtained by quick freezing at 77 K [5].

2.13. MISCELLANEOUS MATRICES

The ion pair $Li^+\cdot CO_2^-$ was obtained by condensation of a Li atomic beam in a CO_2 matrix at 77 K [165]. The results of g-tensor, $A_{//} < A_\perp$ for ^{13}C and ^{17}O coupling imply that the CO_2^- radicals are rotating about their long axes at 77 K.

To divert somewhat, it may be worthwhile noting the new method to investigate radical anions recently started by Bonazzola et al. They observed well-defined ESR spectra after exposure of $CFCl_3$ [166], CCl_4 [167] and CH_2Cl_2 [167] to γ-rays at 77 K in the presence of tetrahydropyran or some other impurities at very low concentration, and identified as $CFCl_3^-$, CCl_4^-, and $CH_2Cl_2^-$, respectively. As for $CFCl_3^-$, they observed well-defined anisotropic spectra at 157 K. In clear contrast, we have observed a broad featureless signal extending over 400 G for γ-irradiated solid solutions of $CFCl_3$ in TMS at 81 K and in MTHF at 77 K [5]. On annealing the TMS solution above 115 K, the broad spectrum decayed out irreversibly and the isotropic spectrum of the $\cdot CFCl_2$ radical grew in until only the latter was observable at 137 K [5]. Thus, in contrast to the spectrum reported by Bonazzola et al., the broad signal was assigned to the precursor of the $\cdot CFCl_2$ radical, namely the $CFCl_3^-$ radical anion.

Moreover, Muto et al. have quite recently observed well-defined anisotropic ESR spectra attributable to CCl_4^- radical anions in TMS containing $^{12}CCl_4$ or $^{13}CCl_4$ [161]. The observed spectra were nicely reproduced by computer simulation using the parameters: $(g_{//}, g_\perp) = 2.005, 2.016$, $(A_{//}, A_\perp) = 243, 183$ G for ^{13}C, $(A\gamma, A_\alpha = A_\beta) = 20, 4.5$ G for three equivalent Cl atoms, and $(A_{//}, A_\perp) = 9, 2$ G for the forth Cl atom. The direction of the principal $A\gamma$ axis inclines by $10°$ from the parallel axis. This result is clearly inconsistent with that assigned to randomly oriented CCl_4^- in CCl_4 by Bonazzola et al.: $(g_{//}, g_\perp) = 2.0131, 2.0084$, $(A_{//}, A_\perp) = 0, 13.3$ G for three equivalent Cl atoms, and $(A_{//}, A_\perp) = 0, -3.3$ G for the forth Cl atom.

The agreement between the observed spectra and the stick diagrams shown by Bonazzola et al. in both cases, may not always be unequivocal. Consequently, it may be desirable to carefully examine their method of generating radical anions.[#]

[#] Note added in proof: Computer simulation has recently suggested that the spectra for irradiated $CFCl_3$ favour the $\cdot CFCl_2$ formation, but that $CFCl_3^-$ radicals cannot be excluded. M. C. R. symons and J. L. Wyatt, J. Chem. Res. (S), 1989, 362.

3. Assignment to Radical Anions

It has been already shown that radical anions are generated by electron addition to guest molecules in such matrices as given in section 2, upon irradiation or photolysis. However, one must make sure whether or not the radical species were brought about by electron addition even if they actually arose in these matrices. There are several ways of confirming this. Photolysis of N,N,N',N'-tetramethyl-p-phenylenediamine (TMPD) in organic matrices [168,5,16] and potassium ferrocyanide, $K_4Fe(CN)_6$, in aqueous glasses [13], provide more direct evidence because the radical species formed therein are caused only by capture of the electrons released by these photolyses. Other ways are to investigate the effects of various efficient electron scavengers, *e.g.* CH_3Br [24], and to examine the changes which result from the use of different matrices.

4. Structure of Radical Anions

4.1. HYPERVALENCE IN RADICAL ANIONS

Let us consider the F_3NO molecule composed only of first-row elements. This molecule has already, perfectly enforced the Lewis octet rule, having 32 valence electrons and no low-lying atomic orbitals available for the capture of excess electrons. Thus, it is of much interest to know whether this molecule can gain one more electron than can be accommodated by valence-bond structures based on the Lewis octet rule, to yield the 'hypervalent' radical anions. Besides, it is of significance to reveal what kind of geometrical change is caused in the molecule by the capture of the electron if the hypervalent radical anion arises. These are essentially the differences when compared with radical cations.

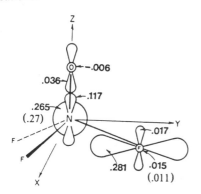

Fig. 1. Structure of F_3NO^- and spin densities calculated by the INDO method and those observed, in parentheses. Reprinted with permission from *J. Amer. Chem. Soc*, **103**, 3436 (1981). Copyright (1981) American Chemical Society.

In fact, the hypervalent F_3NO^- radical anion has successfully been generated in an SF_6 matrix [139,140]. The observed isotropic ESR spectra were interpreted in terms of h.f. couplings to the N and three equivalent F nuclei, indicating the two possibilities of the C_{3v} structure and the trigonal bipyramidal (C_s or C_{2v}) structure undergoing a rapid exchange of the F ligand between the axial and equatorial sites on the ESR time scale. However, careful observation of the spectra in the temperature range from 10 to *ca.* 110 K eliminated the latter possibility. Moreover, in accord with experimental results, INDO calculations for this radical anion suggested that the radical anion prefers the C_{3v} structure having the SOMO composed of largely the 2s orbital of the central N and 2p orbitals in the three equivalent F atoms [140], as shown in Fig. 1.

4.2. COMPARISON OF STRUCTURES BETWEEN HALOGENOMETHANE ANIONS AND HALOGENOSILANE ANIONS

A series of halogenomethanes in the gas-phase are known to immediately dissociate to neutral radicals and halide anions by dissociative electron capture [169]. Nevertheless, hypervalent molecular radical anions, CF_3Cl^-, CF_3Br^-, CF_3I^-, $CF_2Cl_2^-$, $CFCl_3^-$ [5], and CCl_4^- [161], were detected in TMS matrices. The CF_3X^- (X=Cl,Br,I) radical is in C_{3v} symmetry and accommodates an unpaired electron largely in the C-X σ^* orbital, having a streched C-X bond, as shown in Fig. 2(a). Spin density on halogen X increases as the electronegativity decreases with the change in X from Cl to I, in accord with the trend found in various radicals with an unpaired electron in antibonding orbitals. As for $CF_2Cl_2^-$, since the observed spectrum was interpreted in terms of two Cl nuclei with large and small spin densities and two equivalent F nuclei, this radical anion was concluded to have quasi C_{3v} symmetry having one unique Cl nucleus with the larger spin density in a manner similar to the case of CF_3Cl^-. Unfortunately, the observed spectrum for irradiated $CFCl_3$ in TMS has too broad a line-width to reveal the structure of $CFCl_3^-$. The CCl_4^- radical anion detected by Muto et al. in TMS has also a C_{3v} structure having a Cl nucleus with a smaller spin density in the unique (axial) site and the other three Cl nuclei with larger spin densities in the trigonal sites [161].

A congeneric radical anion, SiF_4^-, has been observed in a neopentane matrix by Morton et al. [145]. It was concluded from the presence of two different couples of equivalent F atoms that the SiF_4^- anion has a trigonal bipyramidal structure, possessing two F atoms with larger h.f. couplings in the apical positions and the other two F atoms with smaller couplings in the equatorial positions. The difference in structure found between the congeneric radical anions of CF_3X^- and SiF_4^- is of particular interest. Accordingly, it was attempted to generate SiF_3X^- (X=Cl,Br,I) and compare their structures with those of CF_3X^-.

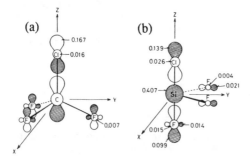

Fig. 2. Structure and spin density distribution of CF_3Cl^- (a) and SiF_3Cl^- (b) radical anions. Reprinted with permission from *J. Mag. Reson..* **38**, 391 (1980). Copyright (1980) Academic Press.

Approximate isotropic spectra were obtained for SiF_3Cl^- in a TMS matrix at 153 K [152,153]. The analysis of observed spectra revealed that the SiF_3Cl^- anion possesses a trigonal bipyramidal structure having the Cl nucleus and the F nucleus with a large spin density in the apical positions and the other two equivalent F nuclei in the equatorial positions, as shown in Fig. 2(b). Analogous results were also obtained for SiF_3Br^- and SiF_3I^-, indicating one F and Br or one F and I atom in apical positions.

As for $SiF_2Cl_2^-$, $SiFCl_3^-$, and $SiCl_4^-$, two equivalent Cl atoms have large spin densities, suggesting that these radical anions have trigonal bipyramidal structures having these Cl atoms in their apical positions [153].

Hence, it was concluded that halogenosilane radical anions have trigonal bipyramidal (C_s or C_{2v}) structures, in contrast to that the corresponding radical anions with a central C atom have C_{3v} or quasi C_{3v} structures.

4.3. STRUCTURE OF CH_3X^-, SiH_3X^-, AND GeH_3X^- ($X=Cl,Br,I$)

Methyl halides, CH_3X ($X=Cl,Br,I$), are well-known to be dissociated to $\cdot CH_3$ radicals and halides ions by dissociative electron capture in organic matrices such as CH_3OH. In contrast, $\cdot CH_3$---X^- adducts were observed in acetonitrile [95,96] and adamantane [96,131] matrices as well as under various conditions [170]. These adducts are considered to appear as a result of stabilization due to the cage effect which prevents a complete dissociation of CH_3X^- anions, and to have essentially planar CH_3 groups. Unfortunately, these adducts were not detected in TMS matrices.

Anisotropic ESR spectra obtained for irradiated SiH_3X ($X=Cl,Br,I$) in TMS were interpreted in terms of h.f. interactions to nuclei of the halogen X and one H: the interaction to the H nucleus being proved using SiD_3X [154]. These results imply that the SiH_3X^- anions have trigonal bipyramidal (C_{2v} or C_s) structures having the X and the H nucleus in the apical positions and the other two H nuclei in the equatorial positions. Small interactions to equatorial H nuclei were detected in the case of SiH_3Cl^-.

Analogous experiments were carried out with congeneric GeH_3X ($X=Br,I$) halides. As a result, observed were h.f. splittings due to three equivalent H nuclei in addition to the X nucleus. This clearly indicates that the radical anions GeH_3X^- have C_{3v} symmetry [155].

Consequently, remarkable changes in the stability and the geometric and electronic structures occur in a congeneric series of radical anions, CH_3X^-, SiH_3X^-, and GeH_3X^-, as shown in Fig. 3.

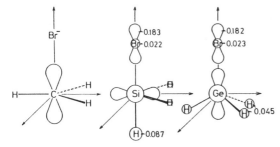

Fig. 3. Geometric and electronic structure of $\cdot CH_3$---Br^- adduct, SiH_3Br^- anion and GeH_3Br^- anion. Reprinted with permission from *Mol. Phys.*, **40**, 697 (1980). Copyright (1980) Taylor & Francis Ltd.

4.4. STRUCTURE OF SnH_4^- AND $SnH_n(CH_3)_{4-n}^-$

Radical anions derived from SnH_4 [145] and its methyl derivatives [158] have C_{2v} (or C_s) structures and methyl groups prefer equatorial sites [158]. This trend in methyl groups accords with that observed for a series of $SiH_n(CH_3)_{3-n}X^-$ radical anions. Thus, radical anions with a central Sn atom prefer the C_{2v} structure, similar to those with a central Si atom.

4.5. JAHN-TELLER DISTORTION

The radical anions formed by electron capture to molecules in high symmetry have C_{3v} or trigonal bipyramidal C_{2v} (or C_s) structure, as shown in sections 4.1 to 4.4. These two structures may be interpreted in terms of bond-stretching and bond-bending occurring in the relaxation process after electron capture, as proposed by Symons *et al.* for *e.g.* phosphorus compounds [64]. However, C_{3v} and C_{2v} structures have been predicted theoretically for CH_4^+ cations. The tetrahedral CH_4^+ formed in the initial stage of irradiation is deformed to have $^2A_1(C_{3v})$ and $^2E(C_{3v})$ states owing to the Jahn-Teller effect through the fundamental vibration $v_3(t_2)$. The ^2E state is not stable for the C_{3v} structure and it undergoes the Jahn-Teller distortion to have a C_{2v} structure [171]. Moreover, it has been stressed that the structure of radical ions is affected not only by electronic states but also by vibrational states [172].

Thus, I have concluded that the C_{3v} and C_{2v} structures of such radical anions as mentioned in sections 4.1 to 4.4, resulted from the Jahn-Teller effect [4]. Subsequently, many examples of Jahn-Teller distorted structures were detected for 'radical cations' by ESR [6].

4.6. STRUCTURE OF THE $C_2F_4^-$ RADICAL ANION

Williams and his co-workers reported in their original paper on the $C_2F_4^-$ radical anion that the anion has a planar σ^* (D_{2h}) structure rather than a planar π^* structure, judging from the values of ^{19}F and ^{13}C h.f. couplings [149]. In a subsequent paper with Morton and Preston [163], only the σ^* and π^* structures were discussed and it was concluded that there was insufficient evidence to decide in favor of either the σ^* or the π^* structure.

In contrast, Symons reported that the most probable structure for $C_2F_4^-$ has a pyramidal configuration and stressed that the small isotropic ^{13}C coupling and the large ^{19}F coupling, observed by Williams *et al.* [149], can be reasonably explained in terms of the spin-polarisation of the σ-electrons and an adjacent F atom effect, respectively, for the pyramidal configuration [173].

Three independent approaches were carried out with MO calculations to solve this problem. The first study was that of Merry and Thomson, using *ab initio* methods [174]. Subsequently, Paddon-Row *et al.* reported the result of calculations using a similar procedure, together with isotropic h.f. couplings to ^{13}C and ^{19}F calculated in an INDO level for the structure proposed by them [175]. Finally, in our study INDO methods were used [176]. The results obtained by these three groups are in agreement with one another, concluding that the chair (*anti*) structure in the pyramidal configuration is the most stable conformation for the $C_2F_4^-$ anion.

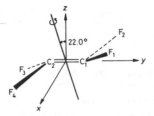

Fig. 4. Structure and rotation axis proposed for the $C_2F_4^-$ radical anion. Reprinted with permission from *J. Chem. Soc., Faraday Trans. 1*, **79**, 1565 (1983). Copyright (1983) The Royal Society of Chemistry.

Furthermore, we attempted to interpret anisotropic ESR spectra for $C_2F_4^-$ on the basis of the structure suggested by INDO calculations. As a result, the anisotropic ESR spectrum observed at 100 K in an MCHD matrix [163] was successfully interpreted under the assumption that the radical anion in the chair form undergoes rotation about the axis perpendicular to the plane containing the four F atoms (Fig. 4). The rigid chair form reasonably explained the spectra of the radical anion at 4 K.

4.7. STRUCTURE OF $C_6F_6^-$, c-$C_4F_8^-$, AND c-$C_8F_8^-$

There are arguments concerning the structure of $C_6F_6^-$ similar to those described already for $C_2F_4^-$. ESR spectra for $C_6F_6^-$ were observed in various matrices [127,128,147]. Wang and Williams observed naturally abundant ^{13}C satellite for $C_6F_6^-$ in HME and concluded that the observed h.f. coupling to ^{13}C suggests a σ^* rather than a π^* configuration in planar D_{6h} symmetry [162]. Moreover, the c-$C_4F_8^-$ radical anion has been discussed on the basis of D_{4h} symmetry, though the possibility of puckered D_{2d} structure was referred to [147].

On the other hand, Symons et al. observed rigid solution spectra for the $C_6F_6^-$ and c-$C_4F_8^-$ anions. From analysis and comparison of the outer features of the spectra for these two radical anions, it was concluded that the $C_6F_6^-$ anion does not have planar σ^* structure but has a puckered-ring structure with σ and seudo π delocalization [34].

There is one more radical anion to be discussed here. The octafluorocyclooctatetraene radical anion, c-$C_8F_8^-$, was generated from the parent molecule, in a tub conformation [177], in MTHF at 77 K. Its isotropic spectrum observed at 145 K is composed of nine lines suggesting the presence of eight equivalent F nuclei [40]. The observed isotropic ^{19}F splitting of 10.92 G for this radical anion is very small as compared with that reported for $C_6F_6^-$, 134.4 G [147], or for c-$C_4F_8^-$, 147 G [5,147]. Hence, the radical anion was concluded to be a simple aromatic π^* radical anion in a regular octagonal (D_{8h}) geometry. In accord with this conclusion, INDO calculations performed on the basis of this geometry, led the parameter Q^F_{eff} in the McConnell form to a reasonable value. This conclusion is surprising in view of the fact that radical anions derived from heavily fluorinated aromatic compounds prefer the σ^* structure to the π^* structure, as described in the following section.

4.8. σ^*- π^* ORBITAL CROSSOVER IN FLUORINATED BENZENE ANIONS AND PYRIDINE ANIONS

Mainly, σ^* radical anions were previously discussed in sections 4.1 to 4.7. However, σ^*-π^* orbital crossover was found for fluorinated benzene anions and pyridine radical anions generated in adamantane matrices [127,129]. Namely, hexa-, penta-, and 1,2,4,5-tetra-fluorobenzene anions were revealed to be σ^* radicals, while p-difluorobenzene radical anion was assigned to a π^* radical [127,128]. Similar results were obtained for a series of fluorinated pyridine radical anions: σ^* radicals were obtained from penta- and 2,3,4,6-tetra-fluoropyridine, π^* radicals from 2,6-difluoropyridine and 2-fluoropyridine [129]. The σ^*- π^* orbital crossover phenomenon observed for these fluorinated compounds was explained in terms of the combined effects of stabilization of σ^* orbitals and destabilization of π^* orbitals, on the basis of the results of INDO calculation.

4.9. STRUCTURE OF RADICAL ANIONS DERIVED FROM HETEROCYCLES

An ESR spectrum with a triplet-of-triplets structure was observed for the radical anions derived from phenol, in enol form, in an argon matrix [106]. The observed large triplet with the splitting of 44 G was attributed to the protons at C-2 projected above and below the skeletal plane and the small triplet with the splitting of 13 G to the protons at C-3 and C-5 situated within the plane. Thus, the phenol radical anion was concluded to be in the tautomeric keto form (11) having the negative charge on the oxygen atom and the spin density in the C-3 and C-5 allylic system. Similar tautomeric α-pyrrolenine form (12) was obtaind for the radical anion of pyrrole, giving a triplet with the splitting of 45 G for the two protons at C-2 and that with the splitting of 11 G for protons at C-3 and C-5 [107].

(11)   (12)   (13)

In contrast, the furan anion possesses a structure (13) resulting from a rupture of an oxygen bond, having the negative charge on the oxygen atom and the spin density in the broken σ orbital [108]. Other examples corresponding to these three types of radical anions were obtained from their related compounds [107-109]. Hence, electron capture by heterocycles in argon matrices was followed by H atom transfer or bond rupture even at *ca.* 4 K.

5. Isomerization or Interconversion of Radical Anions

The radical anions formed by γ-irradiation of maleates in MTHF at 77 K, have maleate forms (14) of their neutral states. Exposure to visible light at 77 K, however, caused irreversible isomerization from the maleate forms (14) to the fumarate forms (15), no isomerization being detected on irradiation of the parent molecules with visible light [19]. The isomerization of the radical anions was interpreted in terms of the considerable single-bond character which was brought about in the central C-C bond as a result of accommodation of *ca.* 40 % of the spin density on each the olefinic C atoms.

The NCBr⁻ radical anion generated in CD_3CN at 77 K has a linear structure having the unpaired electron almost equally shared by the C and Br orbitals, as indicated from the ^{13}C and Br couplings. After annealing at *ca.* 140 K, this species was converted irreversibly into a bent form having the unpaired electron in a pseudo-π* orbital delocalized over all three atoms [99].

(14)   (15)

These examples of isomerization and interconversion are of interest because these phenomena occurred between two disparate structures for the same radical anion.

6. Reaction of Radical Anions

Symons has extensively dealt with the reaction of radical anions, in his review [3]. Therefore, only fundamental problems for the reaction will be mentioned, avoiding an overlap with his review as far as possible, except section 6.1.

6.1. DISSOCIATION OF CH_3X^-, CF_3X^- (X=Cl,Br,I), AND $(CF_3)_3CI^-$

Molecular radical anions of methyl halides have never been detected by ESR, though $\cdot CH_3$ radicals and $\cdot CH_3$---X^- adducts were observed in various matrices and in such matrices as CD_3CN and adamantanes, respectively, as described in section 4.3. The adducts are considered to form probably due to the cage effect of the matrices. On the other hand, the CF_3X^- radical anions were observed in TMS and neopentane matrices and the $\cdot CF_3$ radicals formed by the dissociation of the radical anions were detected in neopentane matrices [5].

For these facts, Symons proposed the following [178]. There is no minimum in the potential energy curve in the dissociation process of the CH_3X^- anion because the CH_3 group undergoes a major continuous change in shape from a pyramidal form for the hypothetical radical anion to the planar structure of the $\cdot CH_3$ radical. In contrast, an energy minimum may arise in the dissociation of CF_3X^- because the CF_3 group needs not alter its pyramidal shape to any great extent, resulting in the formation of the molecular CF_3X^- anions. Thus, the tendency for the dissociation of RX^- was thought to be determined by the change in orbital hybridization and configuration of the central C atom, in the dissociation process.

Williams argued against this theory by showing evidence of the formation of $(CF_3)_3CI^-$ in MTHF, and stating that this radical anion is stable regardless of the nearly planar structure for the $\cdot C(CF_3)_3$ radical. Thus, he concluded that, in addition to the change in orbital hybridization and configuration, the distribution of spin density in RX^- was important for the stability of RX^- with respect to the dissociation [35].

Symons replied to this argument, stating that there are several other factors to be considered, a major one being the distribution of the three electrons in the C-X bond in these radical anions [179]. $\sigma*$ electron will be confined largely on C and the two σ electrons on the X atoms, for CH_3X^- and still even for CF_3X^-. However, the distribution of the electrons may be opposite in the case of $(CF_3)_3CI^-$, so that $(F_3C)C:^- + I\cdot$ may be yielded on the dissociation, regardless of that $(F_3C)C\cdot$ radicals might be formed in protic media. If this process is true, the anion $(F_3C)C:^-$ is pyramidal so that Symons' thought is strongly persuasive.

These arguments indicated that it was very important to understanding of the dissociation process not only to know the structure of radical anions but also to detect the radical species formed by dissociation. From this view point, there are two examples of particular interest.

6.2. DISSOCIATION OF F_3NO^- AND ClO_3F^-

The F_3NO^- radical anion has a C_{3v} structure, as shown in section 4.1. This radical anion was observed together with the neutral F_2NO radical, in SF_6 matrices. ESR observations before and after annealing at 102 K revealed that the F_2NO radical arose as a result of the dissociation of the F_3NO^- radical anion [140].

Similar results were obtained in the system of ClO_3F in SF_6, giving ClO_3F^- radical anions and neutral $\cdot ClO_3$ radicals [141]. Analysis of anisotropic ESR spectra for ClO_3F^- in the solids of ClO_3F and in TMS matrices revealed that the Cl and F h.f. tensors have their largest principal values along the same axis and that the unpaired electron occupies a σ^* orbital largely composed of the 3s and 3p of Cl and of the 2p of F, these two p orbitals lying along the same axis. These results clearly ruled out the possibility for a C_{2v} or C_s structure and supported a C_{3v} structure for the radical anion. During annealing, the neutral $\cdot ClO_3$ radical grew at the expense of this ClO_3F^- radical anion.

Comparison of the dissociations in both cases is of extreme interest because the neutral F_2NO and $\cdot ClO_3$ radicals were formed as a result of releasing one F ligand in the trigonal site and in the axial site, respectively, from their precursors in C_{3v} symmetry. In the F_3NO^- and ClO_3F^- precursors, F ligands have much higher spin densities than O ligands have, regardless of the difference in the geometrical site (trigonal or axial) [140,141]. Therefore, the dissociation of these radical anions is determined by electron distribution rather than by the forced withdrawal of the ligand from a particular geometrical site [140].

6.3. FORMATION OF DIMER ANIONS AND DIANIONS

These are a few examples of the dimer radical anions which arose as a result of dimerization of a radical anion and its parent molecule. $(MeCN)_2^-$ in metastable crystals of acetonitrile [93] and the dimer anion of $(RSSR)^-$ type derived from $(RO)_2PXSH$ ($X=O,S$) [31] were observed immediately after irradiation, while $(SC)S\dot{-}S(CS)^-$ [42] and $c\text{-}C_4F_8^-$ [150] were detected on annealing of CS_2^- and $C_2F_4^-$ radical anions, respectively.

In this connection, it may be of interest to refer to a dianion. ESR spectra attributable to a triplet state species were observed for the MTHF solutions containing 1,3-dinitrobenzene or 1,3,5-trinitrobenzene which were irradiated at high doses [33]. The triplet state species was assigned to an intramolecular dianion having each unpaired electron on one nitro-group. More details have already been given in section 2.1.

6.4. OTHER REACTIONS

Protonation is one of the important reactions which radical anions experience frequently, particularly in alcohol and aqueous matrices, as described in sections 2.2 and 2.3. However, the details have to be omitted. On the subject of protonation, Symons has discussed not only inorganic and organic radical anions but also metal complex anions, in his review [3].

Competition reactions and solvation for radical anions are also included in his review [3].

Acknowledgments The author wishes to express his thanks and gratitude to Professor M. C. R. Symons for his helpful advice and constant encouragement throughout this work, and to Professor M. Shiotani for his continuing support including computer survey of papers cited. The author also thanks Mrs. B. M. McGuckin for proof reading and Miss. A. Kohjiya for typing this manuscript.

References

1. M. C. R. Symons: *Matrix Isolation Spectroscopy*, A. J. Barnes, W. J. Orville-Thomas, A. Müller, and R. Graufres, Ed., Reidel, Dordrecht (1981), p.69.
2. M. C. R. Symons: *Matrix Isolation Spectroscopy*, A. J. Barnes, W. J. Orville-Thomas, A. Müller, and R. Graufres, Ed., Reidel, Dordrecht (1981), p.369.
3. M. C. R. Symons: *Pure Appl. Chem.*, **53**, 223 (1981).
4. A. Hasegawa: *Radia. Chem.*, **14**, 2 (1979).
5. A. Hasegawa, M. Shiotani, and F. Williams: *Faraday Discuss. Chem. Soc.*, **63**, 157 (1978).
6. For example L. B. Knight, Jr., J. Steadman, D. Feller, and E. R. Davidson: *J. Amer. Chem. Soc.*, **106**, 3700 (1984); M. Iwasaki, K. Toriyama, and K. Nunome: *J. Amer. Chem. Soc.*, **103**, 3591 (1981); K. Toriyama, K. Nunome, and M. Iwasaki: *J. Chem. Phys.*, **77**, 5891 (1982); M. Iwasaki, K. Toriyama, and K. Nunome: *J. Chem. Soc., Chem. Commun.*, **1983**, 202; A. Hasegawa, S. Kaminaka, T. Wakabayashi, M. Hayashi, and M. C. R. Symons: *J. Chem. Soc., Chem. Commun.*, **1983**, 1199.
7. F. S. Dainton, J. P. Keene, T. J. Kemp, G. A. Salmon, and J. Teply: *Proc. Chem. Soc. London*, 265 (1964); F. S. Dainton and G. A. Salmon: *Proc. R. Soc. London, Ser. A*, **285**, 319 (1965).
8. D. R. Smith and J. J. Pieroni: *Can. J. Chem.*, **43**, 876 (1965).
9. A. C. Ling and L. Kevan: *J. Phys. Chem.*, **80**, 592 (1976).
10. G. C. Dismukes and J. E. Willard: *J. Phys. Chem.*, **80**, 1435 (1976).
11. S. Murabayashi, M. Shiotani, and J. Sohma: *Chem. Phys. Lett.*, **51**, 568 (1977); **48**, 80 (1977).
12. L. Kevan, M. K. Bowman, P. A. Narayana, R. K. Boeckman, V. F. Yudanov, and Yu. D. Tsvetkov: *J. Chem. Phys.*, **63**, 409 (1975).
13. P. B. Ayscough, R. G. Collins, and F. S. Dainton: *Nature*, **205**, 965 (1965).
14. D. R. Smith and J. J. Pieroni: *Can. J. Chem.*, **43**, 2141 (1965).
15. C. Chachaty: *J. Chim. Phys.*, **64**, 614 (1967).
16. J. Lin, K. Tsuji, and F. Williams: *Trans. Faraday Soc.*, **64**, 2896 (1968).
17. M. Fukaya, H. Muto, K. Toriyama, and M. Iwasaki: *J. Phys. Chem.*, **80**, 728 (1976).
18. M. Iwasaki, M. Fukaya, S. Fujii, and H. Muto: *J. Phys. Chem.*, **77**, 2739 (1973).
19. A. Torikai, T. Suzuki, T. Miyazaki, K. Fueki, and Z. Kuri: *J. Phys. Chem.*, **75**, 482 (1971).
20. T. Shiga and A. Lund: *Ber. Bunsenges. Phys. Chem.*, **78**, 259 (1974).
21. T. Sida, Y. Takahashi, H. Hatano, and M. Imamura: *Chem. Phys. Lett.*, **33**, 491 (1975).
22. C. M. L. Kerr and F. Williams: *J. Amer. Chem. Soc.*, **94**, 5212 (1972).
23. C. M. L. Kerr and F. Williams: *J. Amer. Chem. Soc.*, **93**, 2805 (1971).
24. C. M. L. Kerr, K. Webster, and F. Williams: *Mol. Phys.*, **25**, 1461 (1973).
25. S. P. Mishra and M. C. R. Symons: *J. Chem. Soc., Dalton Trans.*, **1973**, 1494.
26. C. M. L. Kerr, K. Webster, and F. Williams: *J. Phys. Chem.*, **79**, 2663 (1975).
27. T. Gillbro and F. Williams: *J. Amer. Chem. Soc.*, **96**, 5032 (1974).
28. A. Hasegawa, K. Ohnishi, K. Sogabe, and M. Miura: *Mol. Phys.*, **30**, 1367 (1975).
29. M. C. R. Symons: *Chem. Phys. Lett.*, **40**, 226 (1976).
30. J. C. Evans and S. P. Mishra: *J. Inorg. Nucl. Chem.*, **43**, 481 (1981).
31. B. C. Gilbert, P. A. Kelsall, M. D. Sexton, G. D. G. McConnacchie, and M. C. R. Symons: *J. Chem. Soc., Perkin Trans. 2*, **1984**, 629.
32. S. Konishi, M. Hoshino, and M. Imamura: *J. Phys. Chem.*, **85**, 1701 (1981).
33. M. C. R. Symons, S. P. Maj, D. E. Pratt, and L. Portwood: *J. Chem. Soc., Perkin. Trans. 2*, **1983**, 191.
34. M. C. R. Symons, R. C. Selby, I. G. Smith, and S. W. Bratt: *Chem. Phys. Lett.*, **48**, 100 (1977).
35. J. T. Wang and F. Williams: *J. Amer. Chem. Soc.*, **102**, 2860 (1980).
36. L. D. Simpson and J. D. Zimbrick: *Internat. J. Radiat. Biol.*, **28**, 461 (1975); M. D. Sevilla, B. Failor, and G. Zorman: *J. Phys. Chem.*, **78**, 696 (1974).
37. H. Riederer, J. Hüttermann, and M, C, R, Symons: *J. Chem. Soc., Chem. Commun.*, **1978**, 313.
38. M. C. R. Symons: *J. Chem. Soc., Faraday Trans. 1*, **77**, 783 (1981).
39. M. C. R. Symons, W. R. Bowman, and P. F. Taylor: *Tetrahedron Lett.*, **30**, 1409 (1989).
40. B. W. Walther, F. Williams, and D. M. Lemal: *J. Amer. Chem. Soc.*, **106**, 548 (1984).
41. M. C. R. Symons and P. M. R. Trousson: *Radiat. Phys. Chem.*, **23**, 127 (1984).

334

42. J. S. Lea and M. C. R. Symons: *J. Chem. Soc., Faraday Trans. 1*, **84**, 1181 (1988).
43. W. R. Bowman and M. C. R. Symons: *J. Chem. Soc., Perkin Trans. 2*, **1983**, 25.
44. W. R. Bowman and M. C. R. Symons: *J. Chem. Res. (S)*, **1984**, 162.
45. S. I. Al-Khalil, W. R. Bowman, and M. C. R. Symons: *J. Chem. Soc., Perkin Trans. 1*, **1986**, 555.
46. M. C. R. Symons and W. R. Bowman: *J. Chem. Soc., Chem. Commun.*, **1984**, 1445; *J. Chem. Soc., Perkin Trans. 2*, **1988**, 583.
47. M. C. R. Symons and W. R. Bowman: *J. Chem. Soc., Perkin Trans. 2*, **1988**, 1077.
48. M. C. R. Symons and S. W. Bratt: *J. Chem. Soc., Dalton Trans.*, **1979**, 1739.
49. S. Hayashida, T. Kawamura, and T. Yonezawa: *Inorg. Chem.*, **21**, 2235 (1982).
50. M. C. R. Symons, J. Wyatt, B. M. Peake, J. Simpson, and B. H. Robinson: *J. Chem. Soc., Dalton Trans.*, **1982**, 2037
51. M. C. R. Symons, S. W. Bratt, and J. L. Wyatt: *J. Chem. Soc., Dalton Trans.*, **1982**, 991.
52. B. M. Peake, M. C R. Symons, and J. L. Wyatt: *J. Chem. Soc., Dalton Trans.*, **1983**, 1171.
53. S. Enoki, T. Kawamura, and T. Yonezawa: *Inorg. Chem.*, **22**, 3821 (1983).
54. T. Sowa, T. Kawamura, and T. Yonezawa: *J. Organometal. Chem.*, **284**, 337 (1985).
55. T. Sowa, T. Kawamura, T. Yamabe, and T. Yonezawa: *J. Amer. Chem. Soc.*, **107**, 6471 (1985).
56. R. S. Alger, T. H. Anderson, and L. A. Webb: *J. Chem. Phys.*, **30**, 695 (1959); H. Zeldes and R. Livingston: *J. Chem. Phys.*, **30**, 40 (1959).
57. C. Chachaty and E. Hayon: *Nature*, **200**, 59 (1963).
58. C. Chachaty and E. Hayon: *J. Chim. Phys.*, **61**, 1115 (1964).
59. M. Narayana and L. Kevan: *J. Amer. Chem. Soc.*, **103**, 1618 (1981).
60. J. A. Leone and W. S. Koski: *J. Amer. Chem. Soc.*, **88**, 224 (1966).
61. J. A. Leone and W. S. Koski: *J. Amer. Chem. Soc.*, **88**, 656 (1966).
62. S. P. Mishra, G. W. Neilson, and M. C. R. Symons: *J. Chem. Soc., Faraday Trans. 2*, **70**, 1280 (1974).
63. B. W. Fullam and M. C. R. Symons: *J. Chem. Soc., Dalton Trans.*, **1975**, 861.
64. D. Nelson and M. C. R. Symons: *J. Chem. Soc., Perkin Trans. 2*, **1977**, 286.
65. G. W. Eastland and M. C. R. Symons: *J. Phys. Chem.*, **81**, 1502 (1977).
66. S. P. Mishra and M. C. R. Symons: *J. Chem. Soc., Perkin Trans. 2*, **1975**, 1492.
67. R. Stösser, B. Pritze, W. Abraham, B. Dreher, and D. Kreysig: *J. prakt. Chem.*, **327**, 317 (1985).
68. H. Fischer, I. Umminger, F. A. Neugebauer, H. Chandra, and M. C. R. Symons: *J. Chem. Soc., Chem. Commun.*, **1986**, 837; H. Fischer, T. Müller, I. Umminger, F. A. Neugebauer, H. Chandra, and M. C. R. Symons: *J. Chem. Soc., Perkin Trans. 2*, **1988**, 413.
69. M. C. R. Symons, M. M. Aly, and D. X. West: *J. Chem. Soc., Dalton Trans.*, **1979**, 1744.
70. C. van Paemel, H. Frumin, V. L. Brooks, R. Failor, and M. D. Sevilla: *J. Phys. Chem.*, **79**, 839 (1975).
71. M. D. Sevilla, S. Swarts, R. Bearden, K. M. Morehouse, and T. Vartanian: *J. Phys. Chem.*, **85**, 918 (1981).
72. M. D. Sevilla, K. M. Morehouse, and S. Swarts: *J. Phys. Chem.*, **85**, 923 (1981).
73. S. Schlick and L. Kevan: *J. Phys. Chem.*, **81**, 1093 (1977).
74. M. D. Sevilla and J. B. D'Arcy: *J. Phys. Chem.*, **82**, 338 (1978).
75. M. C. R. Symons: *J. Chem. Soc., Faraday Trans. 1*, **78**, 1953 (1982).
76. M. C. R. Symons, G. W. Eastland, and L. R. Denny: *J. Chem. Soc., Faraday Trans. 1*, **76**, 1868 (1980).
77. I. Marov and M. C. R. Symons: *J. Chem. Soc., (A)*, **1971**, 201; I. S. Ginns and M. C. R. Symons: *J. Chem. Soc., Dalton Trans.*, **1972**, 143.
78. K. V. S. Rao and M. C. R. Symons: *J. Chem. Soc., Dalton Trans.*, **1972**, 147.
79. M. C. R. Symons and J. K. Yandell: *J. Chem. Soc., (A)*, **1971**, 760.
80. R. J. Booth, H. C. Starkie, and M. C. R. Symons: *J. Chem. Soc., (A)*, **1971**, 3198.
81. I. S. Ginns and M. C. R. Symons: *J. Chem. Soc., Dalton Trans.*, **1972**, 185.
82. I. S. Ginns and M. C. R. Symons: *J. Chem. Soc., Chem. Commun.*, **1971**, 893.
83. I. S. Ginns and M. C. R. Symons: *J. Chem. Soc., Dalton Trans.*, **1973**, 3.
84. I. S. Ginns and M. C. R. Symons: *J. Chem. Soc., Dalton Trans.*, **1975**, 514.
85. M. C. R. Symons and D. N. Zimmerman: *Internat. J. Radiat. Phys. Chem.*, **8**, 595 (1976).

86. B. L. Bales and L. Kevan: *J. Phys. Chem.*, **74**, 1098 (1970); *J. Chem. Phys.*, **55**, 1327 (1971); **57**, 1813 (1972).
87. L. Kevan, H. Hase, and K. Kawabata: *J. Chem. Phys.*, **66**, 3834 (1977).
88. D. R. Brown and M. C. R. Symons: *J. Chem. Soc., Faraday Trans. 1*, **73**, 1490 (1977).
89. P. A. Narayana, D. Becker, and L. Kevan: *J. Chem. Phys.*, **68**, 652 (1978).
90. M. C. R. Symons: *J. Chem. Phys.*, **69**, 3443 (1978).
91. L. Kevan: *J. Chem. Phys.*, **69**, 3444 (1978).
92. T. Ichikawa, L. Kevan, and P. A. Narayana: *J. Chem. Phys.*, **71**, 3792 (1979).
93. T. Gillbro, K. Takeda, and F. Williams: *J. Chem. Soc., Faraday Trans. 2*, **70**, 465 (1974).
94. E. D. Sprague, K. Takeda, J. T. Wang, and F. Willams: *Can. J. Chem.*, **52**, 2840 (1974).
95. E. D. Sprague and F. Williams: *J. Chem. Phys.*, **54**, 5425 (1971).
96. M. C. R. Symons: *J. Chem. Soc., Perkin Trans. 2*, **1981**, 1180.
97. G. W. Neilson and M. C. R. Symons: *J. Chem. Soc., Faraday Trans. 2*, **68**, 1582 (1972).
98. S. P. Mishra and M. C. R. Symons: *J. Chem Soc., Perkin Trans. 2*, **1973**, 391.
99. M. C. R. Symons and S. P. Mishra: *J. Chem. Soc., Faraday Trans. 1*, **78**, 3019 (1982).
100. M. C. R. Symons, D. R. Brown, and G. Eastland: *Chem. Phys. Lett.*, **61**, 92 (1979).
101. T. Ichikawa, H. Yoshida, A. S. W. Li, and L. Kevan: *J. Amer. Chem. Soc.*, **106**, 4324 (1984).
102. M. C. R. Symons, D. Russell, A. Stephens, and G. Eastland: *J. Chem. Soc., Faraday Trans. 1*, **82**, 2729 (1986).
103. P. H. Kasai: *Accounts Chem. Res.*, **4**, 329 (1971).
104. P. H. Kasai and D. McLeod, Jr.: *J. Chem. Phys.*, **51**, 1250 (1969).
105. W. E. Palke and W. N. Lipscomb: *J. Chem. Phys.*, **45**, 3948 (1966).
106. P. H. Kasai and D. McLeod, Jr.: *J. Amer. Chem. Soc.*, **94**, 6872 (1972).
107. P. H. Kasai and D. McLeod, Jr.: *J. Amer. Chem. Soc.*, **95**, 27 (1973).
108. P. H. Kasai and D. McLeod, Jr.: *J. Amer. Chem. Soc.*, **95**, 4801 (1973).
109. P. H. Kasai and D. McLeod, Jr.: *J. Amer. Chem. Soc.*, **96**, 2342 (1974).
110. D. M. Lindsay, M. C. R. Symons, D. R. Herschbach, and A. L. Kwiram: *J. Phys. Chem.*, **86**, 3789 (1982).
111. W. R. M. Graham, K. I. Dismuke, and W. Weltner, Jr.: *J. Chem. Phys.*, **61**, 4793 (1974).
112. J. B. Raynor, I. J. Rowland, and M. C. R. Symons: *J. Chem. Soc., Dalton Trans.*, **1987**, 421.
113. F. J. Adrian, E. L. Cochran, and V. A. Bowers: *J. Chem. Phys.*, **59**, 56 (1973).
114. D. M. Lindsay, D. R. Herschbach, and A. L. Kwiram: *Chem. Phys. Lett.*, **25**, 175 (1974).
115. F. J. Adrian, V. A. Bowers, and E. L. Cochran: *J. Chem. Phys.*, **61**, 5463 (1974).
116. J. V. Martinez de Pinillos and W. Weltner, Jr.: *J. Chem. Phys.*, **65**, 4256 (1976).
117. L. B. Knight, Jr. and J. Steadman: *J. Chem. Phys.*, **77**, 1750 (1982); *J. Amer. Chem. Soc.*, **106**, 900 (1984).
118. L. B. Knight, Jr., E. Earl, A. R. Ligon, and D. P. Cobranchi: *J. Chem. Phys.*, **85**, 1228 (1986).
119. L. B. Knight, Jr. and J. Ott: *Faraday Discuss. Chem. Soc.*, **86**, 71 (1988).
120. L. B. Knight, Jr., M. Winiski, P. Miller, and C. A. Arrington: *J. Chem. Phys.*, **91**, 4468 (1989).
121. L. B. Knight, Jr., to be published; Reference 11 in [120].
122. J. J. Turner and M. Poliakoff: *ACS Symp. Ser.*, **211**, 35 (1983).
123. S. A. Fairhurst, J. R. Morton, and K. F. Preston: *Organometallics*, **2**, 1869 (1983).
124. S. A. Fairhurst, J. R. Morton, and K. F. Preston: *J. Magn. Reson.*, **55**, 453 (1983).
125. J. R. Morton and K. F. Preston: *Inorg. Chem.*, **24**, 3317 (1985).
126. S. DiGregorio, M. B. Yim, and D. E. Wood: *J. Amer. Chem. Soc.*, **95**, 8455 (1973) and references therein.
127. L. F. Williams, M. B. Yim, and D. E. Wood: *J. Amer. Chem. Soc.*, **95**, 6475 (1973).
128. M. B. Yim and D. E. Wood: *J. Amer. Chem. Soc.*, **98**, 2053 (1976).
129. M. B. Yim, S. DiGregorio, and D. E. Wood: *J. Amer. Chem. Soc.*, **99**, 4260 (1977).
130. M. C. R. Symons: *Chem. Phys. Lett.*, **60**, 418 (1979).
131. I. G. Smith and M. C. R. Symons: *J. Chem. Soc., Perkin Trans. 2*, **1980**, 1362.
132. J. T. Wang and F. Williams: *Chem. Phys. Lett.*, **71**, 471 (1980).
133. K. Sogabe, A. Hasegawa, and M. Miura: *Bull. Chem. Soc. Japan*, **48**, 1643 (1975).
134. R. W. Fessenden and R. H. Schuler: *J. Chem. Phys.*, **45**, 1845 (1966).

336

135. For example K. Nishikida, F. Williams, G. Mamantov, and N. Smyrl: *J. Amer. Chem. Soc.*, **97**, 3526 (1975); A. R. Boate, J. R. Morton, and K. F. Preston: *Inorg. Chem.*, **14**, 3127 (1975).
136. For example M. Iwasaki, K. Toriyama, and K. Nunome: *J. Amer. Chem. Soc.*, **103**, 3591 (1981); K. Toriyama, K. Nunome, and M. Iwasaki: *J. Chem. Phys.*, **77**, 5891 (1982).
137. A. R. Boate, A. J. Colussi, J. R. Morton, and K. F. Preston: *Chem. Phys. Lett.*, **37**, 135 (1976).
138. A. R. Boate, J. R. Morton, and K. F. Preston: *J. Magn. Reson.*, **29**, 243 (1978).
139. K. Nishikida and F. Williams: *J. Amer. Chem. Soc.*, **97**, 7166 (1975).
140. A. Hasegawa, R. L. Hudson, O. Kikuchi, K. Nishikida, and F. Williams: *J. Amer. Chem. Soc.*, **103**, 3436 (1981).
141. A. Hasegawa and F. Williams: *J. Amer. Chem. Soc.*, **103**, 7051 (1981).
142. A. Hasegawa and F. Williams: *Chem. Phys. Lett.*, **45**, 275 (1977).
143. A. J. Colussi, J. R. Morton, and K. F. Preston: *J. Phys. Chem.*, **79**, 1855 (1975).
144. T. Miyazaki, T. Wakayama, M. Fukaya, Y. Saitake and Z. Kuri: *Bull. Chem. Soc. Japan*, **46**, 1030 (1973); T. Miyazaki and T. Hirayama: *J. Phys. Chem.*, **79**, 566 (1975); T. Miyazaki, K. Kinugawa, and J. Kasugai: *Radiat. Phys. Chem.*, **10**, 155 (1977).
145. J. R. Morton and K. F. Preston: *Mol. Phys.*, **30**, 1213 (1975).
146. A. R. Boate, A. J. Colussi, J. R. Morton, and K. F. Preston: *Chem. Phys. Lett.*, **37**, 135 (1976).
147. M. Shiotani and F. Williams: *J. Amer. Chem. Soc.*, **98**, 4006 (1976).
148. A. Hasegawa and F. Williams: *Chem. Phys. Lett.*, **46**, 66 (1977).
149. R. I. McNeil, M. Shiotani, F. Williams, and M. B. Yim: *Chem. Phys. Lett.*, **51**, 433 (1977).
150. R. I. McNeil, M. Shiotani, F. Williams, and M. B. Yim: *Chem. Phys. Lett.*, **51**, 438 (1977).
151. R. I. McNeil, F. Williams, and M. B. Yim: *Chem. Phys. Lett.*, **61**, 293 (1979).
152. A. Hasegawa, S. Uchimura, K. Koseki and M. Hayashi: *Chem. Phys. Lett.*, **53**, 337 (1978).
153. A. Hasegawa, S. Uchimura, and M. Hayashi: *J. Magn. Reson.*, **38**, 391 (1980).
154. S. Uchimura, A. Hasegawa, and M. Hayashi: *Mol. Phys.*, **38**, 413 (1979).
155. A. Hasegawa, S. Uchimura, and M. Hayashi: *Mol. Phys.*, **40**, 697 (1980).
156. A. Hasegawa, S. Uchimura, and M. Hayashi: *J. Chem. Soc., Perkin Trans. 2*, **1980**, 1690.
157. A. Hasegawa, A. Nagayama, and M. Hayashi: *Bull. Chem. Soc. Japan*, **54**, 2620 (1981).
158. A. Hasegawa, T. Yamaguchi, and M. Hayashi: *Chem. Lett.*, **1980**, 611.
159. A. Hasegawa and M. Hayashi: *Chem. Phys. Lett.*, **77**, 618 (1981).
160. R. L. Hudson and F. Williams: *J. Chem. Phys.*, **65**, 3381 (1976).
161. H. Muto and K. Nunome: *Proceeding of 2nd Japan-China Bilateral ESR Symposium*, Kyoto, Japan (1989), p. 140.
162. J. T. Wang and F. Williams: *Chem. Phys. Lett.*, **71**, 471 (1980).
163. J. R. Morton, K. F. Preston, J. T. Wang, and F. Williams: *Chem. Phys. Lett.*, **64**, 71 (1979).
164. J. T. Wang and F. Williams: *J. Amer. Chem. Soc.*, **103**, 2902 (1981).
165. J. -P. Borel, F. Faes, and A. Pittet: *J. Chem. Phys.*, **74**, 2120 (1981).
166. L. Bonazzola, J. P. Michaut, and J. Roncin: *Chem. Phys. Lett.*, **149**, 316 (1988).
167. L. Bonazzola, J. P. Michaut, and J. Roncin: *Chem. Phys. Lett.*, **153**, 52 (1988).
168. W. H. Hamill: *Radical Ions*, E. T. Kaiser and L. Kevan, Ed., Interscience-Wiley, New York (1968), p.321.
169. W. E. Wentworth, R. George, and H. Keith: *J. Chem. Phys.*, **51**, 1791 (1969).
170. S. P. Mishra and M. C. R. Symons: *J. Chem. Soc., Perkin Trans. 2*, **1973**, 391; Y. Fujita, T. Katsu, M. Sato, and K. Takahashi: *J. Chem. Phys.*, **61**, 4307 (1974); E. D. Sprague: *J. Phys. Chem.*, **83**, 849 (1979); O. Claesson and A. Lund: *Z. Naturforsch.*, **39a**, 1056 (1984).
171. A. D. Liehr: *J. Chem. Phys.*, **27**, 476 (1957).
172. K. Fueki: *Houshasen to Genshi Bunshi (Radiation, Atom, and Molecule)*, Kyoritsu, Tokyo (1966), p.119.
173. M. C. R. Symons: *J. Chem. Research (S)*, **1981**, 286.
174. S. Merry and C. Thomson: *Chem. Phys. Lett.*, **82**, 373 (1981).
175. M. N. Paddon-Row, N. G. Rondan, K. N. Houk, and K. D. Jordan: *J. Am. Chem. Soc.*, **104**, 1143 (1982).
176. A. Hasegawa and M. C. R. Symons: *J. Chem. Soc., Faraday Trans. 1*, **79**, 1565 (1983).
177. B. B. Laird and R. E. Davis: *Acta. Cryst.*, **B38**, 678 (1982).
178. D. J. Nelson and M. C. R. Symons: *Chem. Phys. Lett.*, **47**, 436 (1977).
179. M. C. R. Symons: *Chem. Phys. Lett.*, **72**, 559 (1980).

TRAPPED ANIONS IN ORGANIC CRYSTALS

HACHIZO MUTO

Government Industrial Research Institute, Nagoya, Hirate, Kita, Nagoya 462, Japan

1. Introduction

This chapter describes the structure and reaction of the radical anions of organic molecules studied mainly by single crystal ESR and ENDOR spectroscopy. The study makes it possible to precisely determine ESR parameters such as g, hyperfine, and quadrupole tensors. The geometrical and electronic structures of the radicals can be deduced from these parameters by comparing the tensor axes with the molecular orientation in the crystals. Analysis of the matrix ENDOR spectra yields information on the crystalline environment of the radicals which provides for a better understanding of the radical reactions in crystals. The radical anions can be formed along with the counter cations by irradiating the crystal with ionizing radiation. The anions are usually unstable but can be trapped in low temperature solids. Firstly, a characterization of the anions is given in terms of simple molecular orbital theory. Secondly, the details of the structures and reactions of the anions reflecting the electronic state, conformation, and crystalline environment of the radical are discussed.

2. Simple Molecular Orbital Description of Radical Anions

For the past three decades, a great number of organic radical anions have been studied by paramagetic resonance spectroscopy. Firstly, the electronic characterization of anions will briefly be described.

The stabilization of the anions is characterized by the "electron affinity(EA)", which is the energy released by adding an excess electron to the parent molecule. The values of EA of organic anions are listed in TABLE 1[1-3]. The excess electron occupies an antibonding molecular orbital (SOMO), which is the lowest available orbital. Therefore, except for large aromatics and freons, most organic anions are unstable and have a negative EA value. For the anion of the molecules to be formed, it must have a low-lying antibonding MO. The radical anions are classified into two categories, one having a σ^* SOMO and the other a π^* SOMO. The electronic state of molecules is described by a determinant composed of the energy matrix elements based on a linear combination of

A. Lund and M. Shiotani (eds.), Radical Ionic Systems, 337–359.

atomic or valence orbitals. The elements are approximated by the following eqs.(1-2) in a simple MO theory of Hueckel type[4],

$$\alpha(Y) = -(IP_Y + EA_Y)/2 , \qquad \alpha(X) = -(IP_X + EA_X)/2 \qquad (1)$$

$$\beta_{YX} = (1/2)[\beta_{YY}(0) + \beta_{XX}(0)] \; S(YX) \qquad (2)$$

$$\varepsilon(\pm) = [\alpha(Y)+\alpha(X)]/2 \pm \beta_{YX}\{ 1+[\alpha(Y)-\alpha(X)]^2/(4\beta_{YX}^2\}^{1/2}$$

$$\approx [\alpha(Y)+\alpha(X)]/2 \pm \beta_{YX} \qquad (3)$$

$$c(X)^2 \approx \{ 1- [\alpha(Y)-\alpha(X)]/(2\beta_{YX})\}/2 \qquad (4-1)$$

$$c(Y)^2 \approx \{ 1+ [\alpha(Y)-\alpha(X)]/(2\beta_{YX})\}/2 \qquad (4-2)$$

$$\text{if } |\alpha(Y)| < |\alpha(X)|, \quad \text{then } c(Y)^2 > c(X)^2 \qquad (5)$$

where $\alpha(Y)$, $\alpha(X)$, and β_{YX} are the Coulomb(core) integrals of the Y and X atomic orbitals, and the resonance integral of the YX bond, respectively. IP and EA are the the ionization potential and the electron affinity of the atomic orbitals. The resonance integral $\beta(YX)$ is approximated to be the average of the resonance parameters of the homonuclear bonds $(\beta_{XX}(0)+\beta_{YY}(0))/2$ multiplied by the overlap integral S(XY) (eq.2). The parameters α and $\beta(0)$ are listed in TABLE 2[3-5]. Some compounds with a functional bond Y-X form anions whose unpaired electron is localized in the bond. In such cases, the energies of antibonding and bonding MOs of the bond($\varepsilon(-)$ and $\varepsilon(+)$) are given by eq.(3). The unpaired electron densities are defined by the square of MO coefficients of the SOMO ($c(X)^2$, $c(Y)^2$). Since anions have an electron spin in an antibonding MO(the lowest one), the unpaired electron density is expected to be higher on the atom with the smaller core integral. This relationship is easily obtained from eq.(4) and is represented by

TABLE 1. Electron affinity of organic anions

compound	EA (eV)	compound	EA (eV)
-C≡C- ; CH≡CH	-2.6	tetrahydrofuran	-1.76
-CH=CH-	-1.78~-2.27	ketone ; CH_3COCH_3	-1.51
(-CH=CH-)$_2$	-0.62	aldehyde ; CH_3COH	-1.19
(-CH=CH-)$_3$	>0	nitril ; $CH_3C≡N$	-2.84
benzene	-1.17	alkyl halide; CH_3I	0.2
pyridine	-0.62	CF_3Br	0.91,
napthalene	-0.19	CF_3I	1.4-1.57
p-quinone	1.89		

Cited from Ref. 1-3.

TABLE 2. Coulomb and resonance parameters for atomic orbitals(in eV).

orbital	H	C	N	O	S	F	Cl	Br	I
s;(IP+EA)/2	7.18	14.05	19.32	25.39	17.65	32.27	21.59	19.63	--
p;(IP+EA)/2		5.57	7.27	9.11	6.99	11.08	8.71	7.60	6.76
$-\beta_{\mu\mu}(0)$	9	21	25	31	18.15	39	22.33	19.6[a]	17.4[a]

Cited from Ref. 3-5. a; estimated from $\beta(XX)$ values in Ref. 5.

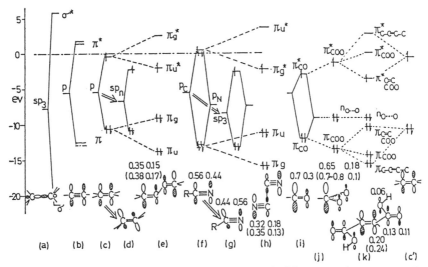

Figure 1. Energy diagram for π* anions. The numbers with and without parentheses represent the observed values and the calculated unpaired electron densities using the Hueckel type MO approximation, respectively. Dotted lines show the correlation of MOs (see text).

eq.(5), the trend being that the atom with the lower electronegativity has the higher unpaired electron density. This behavior is opposite to the counter cation, which has an electron spin in a bonding MO.

As typical examples the energy diagrams for a →C-C← σ single bond and a >C=C< π bond are illustrated in Figure 1a and 1c. Although the former bond distance is longer(1.54 Å) than the latter(1.32 Å), the overlap integral of the former sp3-sp3 bond is much larger(S(σ)=0.64) than that of the latter p-p bond(S(π)=0.28). Therefore the resonance integral for the former bond is larger(β(σ)=-13.45 eV) than the latter (β(π)=-6.40eV). Thus the energy levels of σ(C-C) bonding and σ(C-C)* antibonding MOs are widely split as compared with those for π(C-C) and π(C-C)* MOs.

The following three kinds of compounds have small σ overlap integrals and hence small resonance integrals; halogenated compounds such as alkyl halide(sp3-p for →C-X), N-halogenated imide(sp2-p for >N-X), and disulfide (p-p for -S-S-). Consequently their σ* antibonding molecular orbitals are low-lying, as is illustrated in Figure 2e-g. Their σ* radical anions are metastably trapped in low temperature solids.

Since σ overlap integrals for →C-H bonds(sp3-1s), →C-N< and →C-O-bonds(sp3-sp3) have also large values, their σ and σ* molecular orbitals are widely split just as for the →C-C← bond. The energy diagrams are shown in Figure 2a-d. Molecules composed of these bonds have no low-lying σ* antibonding MO. They form no anions by irradiation, except for a small number of trapped electrons in defects of the low temperature solids[6], which involve alkyl amines, alcohols, ethers, and furans, and saccharose in addition to alkanes.

The second category of anions are the π* radicals. The radical

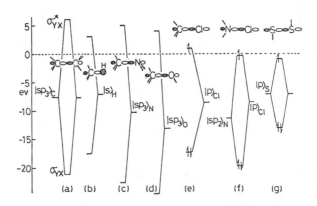

Figure. 2   Comparison of energy splittings of σ and σ* molecular orbitals for fundamental σ bonds calculated using the Hueckel type MO approximation.   Molecules composed of groups a-d have high-lying antibonding MO, giving no formation of their anions.   Molecular anions with group e-g are formed(see text).

anions   of   unsaturated   hydrocarbons   are   typical.   Their   electron affinities   change   as   follows;   C≡C(-2.6 eV),   C=C(-1.78~ -2.3 eV),   (-C=C-)$_2$ (-0.62eV),   and   (-C=C-)$_3$ (> 0   eV).   Alkyne anions have large negative   EA   values   that   have   never   been detected by ESR or optical measurements[103].   The   high-lying π* SOMO of alkyne anions is due to a large   overlap   integral   of p-p π orbitals caused by the short C≡C bond distance.   A comparison of the energy diagrams of C≡C and C=C π bonds is shown   in   Figure   1b   and   1c.   Monoene(alkene)   radical   anions have a slightly   smaller   negative   EA   than   that   of alkynes because of their longer CC distances.   They have been recently observed in mixed crystals of n-alkene/n-alkane   irradiated   at   4.2 K.   Precisely speaking, the alkene anions have a slightly bent radical structure shown in Figure 1d. The details will be described in section 4.   Polyene anions (-CH=CH-)$_n$ •−

with n≥2 have a relatively low-lying SOMO due to an extent of π electron delocalization.   The   correlation of their molecular orbitals to monoene MOs is shown in Figure 1c and 1e.

The energy level of the π* antibonding MO is lowered by substitution of   groups such as COOH which increase the extent of π delocalization or by   replacement   of   carbon atoms   by N, O, or S atom with a negative large   core   integral.   This   category   involves unsaturated carboxylic acids,   organic   bases,   and   many   substituted aromatics.   The energy diagram for the former anions is shown in Figure 1k.

Organic compounds with the carbonyl group(-C=O) form another kind of anion with   π* SOMO, including anions of ketones, aldehydes, amino acids,   carboxylic   acids,   esters,   peptides, and amides.   These anions have been studied most precisely in crystals.   Their energy diagrams are illustrated   in   Figure   1i   and 1j.   These carbonyl and carboxyl anions have   a   limiting   molecular formula -C·(O−)R (or -C·(O−)OH) with a high unpaired   electron   density   on   the   carbon atom, which can be easily expected   from   eq.(5),   since   the carbon 2p orbital has a smaller core integral(-5.57   eV)   than   the   oxygen   2p orbital(-9.11 eV).   Next, the nitrile(-C≡N) anion is formed in neat crystals in spite of its high-lying π* antibonding MO originating from the short CN bond. An interesting observation is the trapping of a dimer anion of CH_3CN, stabilized by π delocalization. Their energy diagrams are shown in Figure 1f-h. The details will be described in section 5.

As briefly described above, radical anions are classified to σ* and π* radicals. The organic compounds whose anions have been observed are fewer than those whose cations have been detected[7-9], since the anions have an excess electron in the antibonding MO and are more unstable.

3. The Carboxyl Anion and Its Protonated Form

3.1. HISTORY

Unstable radicals of carboxylic acids and amino acids, which differ from the stable radicals at room temperature, were first observed by Gordy et al. by irradiation at 77 K(1955)[10]. The unstable species were suggested by Miyagawa and Gordy(1961) to be the carboxyl anion[11]. Box et al. studied the ^{13}C enriched compound and confirmed the suggestion that the species was the carboxyl(-C·OO) positive or negative radical ion(1965). [12] Bennett and Gale created the anions by codeposition of carboxylic acids and sodium atoms into inert gas matrices using a rotating cryostat and confirmed the assignment as anions (1968)[13]. The protonated form of the alanine anion ($^+NH_3CH(CH_3)C·OO^2-H^+$) was found independently by three groups; Sinclair(1967)[14], Minegishi(1967)[15], and Ayscough (1968)[16]. The occurrence of a stereospecific protonation to only one of the carboxyl oxygen atoms was proposed by Miyagawa et al.(1969)[17]. The protonation reaction was precisely studied by Muto, Iwasaki, and Nunome using the single crystal ENDOR technique(1973-74)[18-21]. They detected superhyperfine couplings of all hydrogen bonded protons in some amino acids and carboxylic acids. They found that protonation reaction occurs through a hydrogen bond and specified the controlling factors for the specific proton transfer. Protonation is now known to be a common charge neutralization reaction for many organic and inorganic anions with unsaturated bonds.

3.2. STRUCTURE OF THE ANION AND β PROTON COUPLINGS

The hyperfine coupling tensors of ^{13}C α carbon and CH and OH β protons in carboxyl -CH(β)-C(α)·(O-)OH(β) and carbonyl π anions -CH(β)-C(α)·(O-)-R reported are listed in TABLE 3-5[18-46]. In order to summarize the reported data, the anisotropic term for β-protons is plotted in Figure 3 as a function of the isotropic coupling. The coupling is a good measure for the conformational angle(θ) of OH and CH β-bonds with respect to the lobe axis of unpaired electron orbital. The CH β-bonds in anions with two or three CH β-protons have conformational angles of θ=~0º and ~(±)120º as deduced from this plot and the $\cos^2\theta$ rule to be described later. The anisotropy of the OH β-proton couplings is larger than that of CH protons, mainly because the former proton has a shorter

TABLE 3. ^{13}C and ^{14}N hyperfine tensors(in MHz) for carboxyl and carbonyl anions, estimated electron spin density (ρ) on the atoms, and angle of deviation for the RCOX group from the plane(δ)

anion	a(iso)	B(max)	ρ_p(C)	ρ_s (C)	$\Sigma\rho$	d_s^a/ρ_p	$\delta(^0)$	Ref.
$CH_3 C\cdot OO-(1/2Zn^{2+})$	295	126	0.71	0.095	0.80	0.100	12.1	22
$CH_3 C\cdot OOH-$	318	141	0.78	0.102	0.88	0.098	11.9	13
$CH_3 C\cdot (O-)NDCH_2 COOD$	226	129	0.71	0.072	0.78	0.068	10.1	23
$CH_3 CONDCH_2 C\cdot OOD-$	291	130	0.72	0.094	0.81	0.096	11.8	24
$+NH_3 CH_2 C\cdot OO^2-(H+)$	256	120	0.66	0.082	0.74	0.091	11.5	25
$HOOCCH_2 CH_2 C\cdot (OH)O-H+$	314	134	0.74	0.101	0.84	0.103	12.2	12,26
$DOOCCHODCHOD-C\cdot OOD-$	304	133	0.73	0.098	0.83	0.100	12.1	27
$+NH_3 CH(CH_3)C\cdot OO^2-H+$	251	128	0.71	0.081	0.79	0.079	10.8	28
$+NH_3 C(CH_3)_2 C\cdot OO^2-H+$	280	110	0.61	0.090	0.70	0.114	12.7	29

	a(iso;N)	B(N)			ρ(N;p)	ρ(N;s)		
$ND_2 C\cdot (O-)CH_2 CH_2 COOH$ (succinamic acid)	63.8	(18.5 -5.3 -12.9)			0.19	0.04		30

a) d_s:Unpaired electron density in the carbon 2s orbital estimated by subtracting the

contribution of spin polarization.

TABLE 4. Hyperfine tensor of CH β-protons in carbonyl and carboxyl anions determined by single crystal ENDOR spectroscopy. (in MHz)

	CH $\beta1$		CH $\beta2$ and CH $\beta3$		Ref.
anion	a(iso)	B	a(iso)	B	
$CH_3 C\cdot OOO^2-(1/2Zn^{2+})$	87.9	(7.3,-2.7,-4.5)	15.8	(10.5,-4.2,-6.4)	22
			6.0	(10.8,-4.0,-6.8)	
$CH_3 C\cdot (O-)NDCH_2 COOD$	85.3	(6.7,-2.7,-3.9)	15.8	(9.6,-4.4,-5.2)	23
			7.5	(9.5,-4.1,-5.5)	
$CH_3 CONDCH_2 C\cdot OOD-$	76.4	(6.0,-1.8,-4.2)	19.7	(10.1,-4.5,-5.7)	23
$+NH_3 CH_2 C\cdot OO^2-(H+)$	76.2	(6.5,-1.6,-4.8)	12.7	(10.2,-4.1,-6.2)	19
$HOOCCH_2 CH_2 C\cdot (OH)O-H+$	76.3	(7.0,-2.9,-4.1)	23.6	(11.2,-5.2,-6.0)	21,31
$DOOCCHODCHOD-C\cdot OOD-$	62.5	(7.1,-2.2,-4.9)			27
$+NH_3 CH(CH_3)C\cdot OO^2-H+$	53.7	(7.7,-2.9,-4.7)			18

anion	a(iso); $\beta1$	$\beta2$	Ref.	anion	a(iso); $\beta1$	$\beta2$	Ref.
$HOOCCH_2 SCH_2 C\cdot OOH-$	76.9	18.4	30	citric acid	83.5	13.7	37
$ClNH_3 CH_2 C\cdot (O-)NHCH_2 COOH$	83.9	25.7	31	aspartic acid HCl	81.9		38
$ClNH_3 CH_2 CONHCH_2 C\cdot OOH-$	78.5	10.8	31	hydroxyproline	61.0		39
$D_2 NC\cdot (O-)CH_2 CH_2 COOH$	65	21	32	histidine HCl	63.7		41
$D_2 NCOCH_2 CH_2 C\cdot OOH-$	77.3	11.7	33	DL-serine	63.2		42
$D_2 NC\cdot (O-)CH_2 CH(ND_3+)COO-$	54.2	36.6	34	cysteine HCl	60.9		43
$D_2 NCOCH_2 CH(ND_3+)C\cdot OO^2-$	59.6		34	glycine HCl	78.8	32.1	44
$+ND_3 CH_2 CH_2 C\cdot OO^2-$	75.7	15.1	35	L-tyrosine HCl	9.1		45
$K+-OOCCH_2 C\cdot OOH-$	70.7	10.9	36	hydroxy prolineHCl	25.4		46

$C(\alpha)...H(\beta)$ distance than the latter. It is found from the ^{13}C tensors that the C-COO group of the anion has a slightly bent structure [12,13, 22-29], as is shown in Figure 3. The angle of deviation for the C-COO group from the plane(δ) and the spin density on the α carbon atom are estimated to be $\delta = 10\sim13^0$, and $\rho(2s) = 0.08\sim0.1$, $\rho(2p) = 0.66\sim0.78$, respectively. The spin densities on the two oxygen atoms and the C(β)

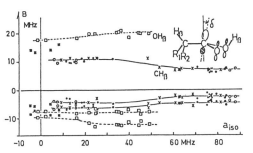

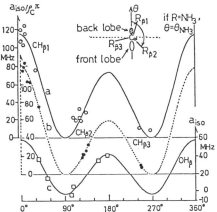

Figure 3. Anisotropic hyperfine coupling tensor B plotted as a function of the isotropic coupling a(iso) for the OHβ (open squares) and the CHβ protons in the carboxylanions. X, filled and open circles indicate data for the CHβ protons of the anions with one, two, and three CHβ protons, respectively. Filled squares are for quasi transferred protons(see text).

Figure 4. The Conformational dependence of (a,b) a(iso)/$\rho(\pi;C)$ of the CHβ protons and (c) a(iso) of the OHβ protons in the carboxyl and carbonyl anions(see text).

atom were estimated to be $\rho(2p;O)=0.1$ and $\rho(2p; C)=-0.05$, respectively, from a comparison of the calculated and observed dipolar hyperfine tensors for β protons[20, 21, 25]. These anions may be stabilized by mixing the [2s;C> orbital with a large Coulomb integral to the [2p;C> orbital. A similar deformation is reported for olefinic and nitrile anions, which will be described in sections 4 and 5.

Since these anions have a bent structure, the isotropic β proton couplings must have a dihedral angle(θ) dependence with a 2π period. This differs from that found for the usual planar π radicals[25,47]. Figure 4 shows the θ dependence for the CH and the OH β-proton couplings. The dihedral angles for the β bonds plotted in Figure 4a and 4c were estimated from the principal directions of the ^{13}C- and the β-proton hyperfine tensors by comparison with the crystallographic data. This dependence gives an experimental confirmation for the above prediction and results in the following rules[25].

$$a(CH\beta) = B_2 \rho *\cos^2 \theta \qquad (6)$$

B_2 = 120±6, 60 MHz for CHβ protons in the back and front lobe sides with the lower and higher spin densities, respectively.

$$a(OH\beta) = -4 + B_2 '*\cos^2 \theta \qquad (7)$$

B_2 '= 61, 43 MHz for back and front lobe sides.

The dihedral angles estimated from the tensor axes were supposed to include a maximum error of 10⁰. The B_2 value in eq.(6) is a more

TABLE 5. Hyperfine and superhyperfine tensors of hydrogen bonded protons for carboxyl anions determined by single crystal ENDOR spectroscapy. Calculated and crystallographic distances between α-carbon and the protons are given.

anion	a(iso)	B	$r(C(\alpha)...H)$ (Å)		assign.[c]	Ref.
proton	(MHz)		dipol.[a]	X-ray[b]		
succinic acid	$HOOCCH_2\ CH_2\ C\cdot(OH)O{-}H{+}$					21
OH(1)	-1.4	(17.3,-7.6,-9.7)	2.09	1.97	orig.	
OH(2)...O	2.4	(17.5,-7.6,-9.9)	2.08	2.6	transf.	
L-valine HCl	$(CH_3)_2\ CH(NH_3{+}Cl{-})C\cdot(OH)O{-}H{+}$					20
OH(1)	16.1	(18.1,-8.3, -9.7)	2.05	1.97	orig.	
OH(2)...N	36.6	(20.9,-8.1,-12.7)	1.96	2.9	transf.	
L-alanine	$+NH_3\ CH(CH_3)C\cdot OO^2{-}H{+}$					18
OH(1)...N	39.8	(18.3,-6.9,-11.4)	2.05	2.97	transf.	
O...H(2)N	2.1	(9.8,-4.8, -5.0)	2.52	2.57	neigh.	
O...H(3)N	-0.14	(9.7,-4.5, -5.2)	2.53	3.00	neigh..	
glycine	$+NH_3\ CH_2\ C\cdot(O{-}..H^{\delta+})_2$					19
O..H(1)..N	4.3	(13.9,-4.8, -9.1)	2.25	2.84	quasi tra.	
O..H(2)..N	-4.7	(14.1,-5.4, -8.7)	2.24	2.46	quasi tra.	
O...H(3)N	0.2	(7.8,-3.8, -4.0)	2.73	2.75	neigh.	
α-aminoisobutyric acid	$(CH_3)_2\ CHCH(NH_3{+})C\cdot(O{-}...H^{\delta+})_2$					19
O..H(1)..N	8.2	(16.2,-7.2, -9.0)	2.14	2.78	quasi tra.	
O..H(2)..N	-2.8	(12.7,-5.8, -7.4)	2.29	2.62	quasi tra.	
O...H(3)N	0.	(5.0,-2.5, -2.5)	3.15	2.96	neigh.	
L-cysteic acid	1) $-SO_3\ CH_2\ CH(NH_3{+})C\cdot OOH{-}$		[CHβ=68.6; at 4 K]			40
	2) $-SO_3\ CH_2\ CH(NH_3{+})C\cdot(OH)O{-}H{+}$		[CHβ=48.7; at 77K]			

	proton	a(iso)	B (MHz)	dipol.[a]	assign.	Ref.
	2)OH...N	35.3	(20.3,-7.9,-12.5)	1.98 Å	transf.	40
histidine HCl	OH...N	44.4	(19.9,-7.3,-12.6)	1.99	transf.	41
DL-serine	OH...N	not determined			transf.	42
cysteine HCl	OH	45.2	(19.7,-8.7,-11.0)	2.00	orig.	43
glycine HCl	OH	34.3	(20.3,-7.9,-12.5)	1.98	orig.	44
L-tyrosine HCl	OH	49.1	(19.7,-7.7,-12.0)	2.00	orig.	45
hydroxyproline	1)OH	21.8	(19.7,-8.0,-11.7)	2.00		46
HCl	2)OH	24.5	(19.7,-6.7,-13.0)	2.00		

a;distance estimated from the dipole tensor. b;crystallographic data. c;orig., transf., quasi tra., and neigh. represent originally bonded, transferred, quasi transferred protons and proton in neighboring molecule, respectively.

reliable value than that obtained from Figure 4a, It is estimated by a least square procedure, using isotropic couplings of the CH β-protons and the α-carbon spin density in the anions with two or three CH β-protons (TABLES 3 and 4) where assuming that the dihedral angles for each of the two CH β bonds is 120^0. Figure 4b shows the comparison of the $\cos^2\theta$ rule (eq.(6)) with the observed data.

The B$_2$ values for CH and OH ß-protons in the front lobe side are smaller than those in the back lobe side. The overlap integral between the 2p component orbital of the α-carbon atom and the hydrogen 1s orbital is smaller for the proton on the front lobe side(S(2p-1s)= 0.054 for CH ß-proton with θ=180⁰) than that in the back lobe side(0.11 for proton with θ=0⁰)(see insert in Figure 4). The cause of the smaller B$_2$ value in the front side can be attributed to the difference in the overlap integrals. The B$_2$ values on both sides are smaller (120, 60 MHz for CHß protons) than the value commonly used for π radicals (150 MHz). The reason is not because of the overlap integral, since the integral for the back lobe side proton is larger(S(2pπ-1s)= 0.11) than the corresponding integral for planar π radicals(0.081). The reason can not also be attributed to the decrease of effective nuclear charge caused by electron capture, because the protonated anions also have similar small couplings. One possible reason may be the lower-lying SOMO(rather than the pure pπ orbital), which is more widely separated from C-H pseudo orbital and results in smaller hyper-conjugation. The B$_2$ ρ value for carboxyl anions has been assumed to have the value ~126 MHz for π neutral radicals or 76 MHz for succinic acid anion for both sides[21]. These values have created something of a dilemma in estimating the electron spin density and the radical conformation [23]. A value of 28-33 G and the half value should be used for CH ß-protons on the back and front lobe sides, judging from the 2p(π) spin density on α-carbon atom (0.66-0.78). The cos² θ rule for anions obtained here may be valuable in estimating the conformation of radicals of other acids, peptides, and macromolecules such as proteins. Similar small ß proton couplings are found for alkene anions with a bent structure, which is described in section 4.

3.3. REACTIONS

3.3.1. *Selective proton Transfer Through One of The Hydrogen Bonds*
The selective proton transfer to radical anions is an interesting reaction, which was studied in amino acids by Muto et al. as follows[18-20]. Amino acid molecules usually exhibit the zwitterion form (+NH$_3$ CR$_1$ R$_2$ COO-) in crystals. The COO- group usually makes three hydrogen bonds with NH protons of three neighboring molecules, as shown in Figure 5a for L-alanine. The superhyperfine tensors of the three hydrogen bonded protons for the alanine anion were determined by single crystal ENDOR spectroscopy[18]. The angular dependence of the ENDOR spectra is shown in Figure 5b. The superhyperfine tensors are listed in TABLE 5. Those couplings have been assigned to specific NH protons by comparing the orientation of the maximum dipolar tensor element A(max) with the crystal structure. The projection of A(max) orientations is shown in Figure 5a. The neighboring NH proton H(1), which is positioned out of the COO plane, has about two times as large dipolar tensor as the other two NH protons. The distance between the radical carbon and the neighboring protons (r(C...H)) are estimated from their dipolar tensors and listed in TABLE 5. Proton H(1) has a distance shorter by 1 Å as compared to its original position in the neighboring molecule,

indicating the occurrence of a selective transfer of the H(1) proton to the anion side.

$$O^-\ldots H(1)-NH_2\overset{+}{\underset{}{-}} \qquad\qquad O-H(1)\ldots NH_2-$$

$$^+NH_3\,CH(CH_3)\,C \cdot \qquad\longrightarrow\qquad ^+NH_3\,CH(CH_3)\,C \cdot$$

$$\hspace{8cm} (r1)$$

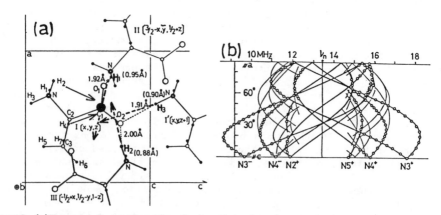

Figure 5. (a)The crystal structure of L-alanine. Three dotted arrows indicate the projection of the maximum hyperfine and superhyperfine tensors of exchangeable protons (see text); (b)The angular dependence of matrix ENDOR signals for alanine anion in single crystals. The circles indicate data for exchangeable protons. From H. Muto and M. Iwasaki, J.Chem.Phys. **59**, 4821 (1973); copyright © 1973; American Institute of Physics; reproduced here by kind permission of American Institute of Physics.

The large isotropic coupling of the H(1) proton(40 MHz) is character-istic of OH β-protons with small dihedral angles; supporting the above argument. Thus, the specific proton transfer via one of the three hydrogen bonds following electron capture has been unequivocally deduced. Similar studies were made for glycine, α-aminoisobutyric acid, and valine hydrochloride(TABLE 5). The valine anion is also found to be protonated from a position out of the COO plane[20]. For the former two anions, two exchangeable protons have intermediate dipolar tensors between OH β-protons and NH protons in the neighboring molecules(filled squares in Figure 3). These results suggest that the two protons have moved to an intermediate position of each hydrogen bond or are jumping (or tunneling) between the two minima of the double minimum potential [19]. The controlling factor for selective proton transfer was deduced from the following results[19].

    (1) The proton located out of the COO plane and has a relatively short hydrogen bond length is the one most favored for transfer.

    (2) The proton with the shortest hydrogen bond length in the COO plane favors transfer, when the out of plane proton has a long hydrogen bond.

The latter deduction(2) indicates that the driving force for protonation is electrostatic attraction. The attraction is large in the p lobe direction with a large electron density and may result in the preferred proton transfer(1), which is an orbital steering reaction. These selective proton transfers might have a similarity to the specific charge transfer in biological systems.

Box and Budzinski suggested direct observation of proton transfer to the anions in L-cysteic acid single crystals by warming to 48 K after 4K irradiation[40]. They observed an appearance of an ENDOR signal assigned

$$-SO_3 CH_2 CH(NH_3 +)C \cdot (O-)OH + H+ \longrightarrow -SO_3 CH_2 CH(NH_3 +)C \cdot (OH)OH$$
$$(r2)$$

to an OH β-proton together with a change of the CH β-proton coupling induced by the warming(TABLE 5). Protonation of the anions is also reported for other amino acids such as histidine and serine (TABLE 5)[41,42].

3.3.2. Deamination and Dissociation of Peptide Linkage

Radical anions of amino acids and peptides produced by irradiation at low temperatures undergo deamination and dissociation of peptide linkage (β-elimination) by warming[24,48-50].

$$+NH_3 CR_1 R_2 C \cdot OO^2 -(or H+) \longrightarrow NH_3 + R_1 R_2 C \cdot COO-(H+) \quad (r3)$$

$$CH_3 C \cdot (O-)NHCH_2 COOH \longrightarrow CH_3 CONH- + C \cdot H_2 COOH \quad (r4)$$

Shields and Hamrick,Jr. found that the deamination process is influenced by the conformational angle(θ) of the C(β)-NH_3 bond with respect to lobe axis of the unpaired electron orbital(see insert in Figure 4)[48]. They studied the deamination temperatures T(deam) for the amino acid radical anions and found a linear correlation of temperature with $\sin^2 \theta$;eq.(8).

$$T(deam) = 5 + 155 \sin^2 \theta \quad (8)$$
$$\Delta E = 8.8 - 10\cos^2 \theta \approx A\sin^2 \theta \quad (9)$$
$$k = k_0 \exp(-\Delta E/k_B T) \quad or \quad = k_0 \exp(-A\sin^2 \theta /k_B T) \quad (10)$$
$$T = A\sin^2 \theta /(k_B \ln(k_0 /k)) \quad (11)$$

They showed that the activation energy also has a correlation with $\sin^2 \theta$ (eq.(9)) and have interpreted these results as follows: There is a repulsion between the unpaired electron and the bonding pair of electrons of the C(β)-N bond. Suppose the repulsion increases as the dihedral angle(θ) for the C(β)-N bond to the lobe axis of the unpaired electron orbital decreases. Then the activation energy of the deamination process will be expressed by $A\sin^2 \theta$, giving an explanation for the experimental result(eq.(9)). The specific rate of deamination is given by the Arrehnius equation (10). The equation (10) is rewritten as eq.(11), which is the same form as eq.(8), if the small constant is neglected. Thus they have beautifully interpreted the deamination reaction originating from the repulsive interaction between the pair of electrons in C(β)-N bond and the unpaired electron.

β-alanine $+NH_3 CH_2 CH_2 COO-$ has no C(β)-N bond. Its carboxyl radical anion decomposes to the acyl radical(RC·=O) and does not deaminate [51], supporting the above deamination mechanism.

3.3.3. Dehydroxylation and Water Elimination

The radical anion of carboxylic acids undergoes dehydroxylation or water elimination reaction. Muto et al. found that the radical anion of succinic acid is trapped in a protonated form RC·(OH)O-H+ by irradiation at 77 K.[21] They showed that the anion converts to the acyl radical R-C·=O at an elevated temperature from 77 K.[52] The acyl radical had been found by Kwiram and other groups[11,51-55] in formic acid, malonic acid, and succinic acid by irradiation at room temperature or by warming after 77 K irradiation. Thus the overall reactions of radical anions of saturated carboxylic acids may be as follows:

$$R-C·OOH- \ + \ H+ \longrightarrow \ R-C·(OH)_2 \ \xrightarrow{\Delta} \ R-C·=O \ + \ H_2O \qquad (r5).$$

Before the ENDOR observation of the protonated radical anion of succinic acid, it had been supposed that carboxyl anions dehydroxylate(R-C·OOH⁻ ⟶ R-C·=O + OH-). It is hard to detect the transferred or originally bonded OH protons by ESR, since they usually are located in the COO plane and have a small hyperfine coupling. A protonated anion similar to those in the carboxylic acids is found by Rao et al. in oxalic acid/urea cocrystals by an analysis of matrix ENDOR spectra[56]. The following water elimination reaction (r6) is reported by Samskog et al. [57] for the tartaric acid anion with OH β bonds. This is a β-bond elimination and may occur by the same mechanism as the deamination from amino acid anions described in the previous section.

$$K+ \ -OOCCHOH-CHOHC·OO^2 - \ Na+ \ + \ H+ \longrightarrow$$
$$\longrightarrow \ K+ \ -OOCHOH-CHOHC·OOH- \ Na+$$
$$\longrightarrow \ K+ \ -OOCCHOH-C·HCOO- \ Na+ \ + \ H_2O \qquad (r6)$$

3.3.4. Miscellaneous

Carbonyl and carboxyl radical anions are simultaneously trapped in irradiated acetyl amino acids and peptides[23,32-34]. In acetyl L-alanine the carbonyl radical anion is formed primarily by irradiation at 77K. It converts to a carboxyl radical anion only during illumination by λ>350 nm light and returns back to the carbonyl anion upon switching off the lamp[58].

$$CH_3C·(O-)NHCH(CH_3)COOH \ \xrightarrow{h\nu} \ CH_3CONHCH(CH_3)C·OOH- \qquad (r7)$$

This fact indicates the EA value of the latter form is slightly higher than that of the former carbonyl radical anion. Cyanoacetic acid NCCH₂-COOH forms a carboxyl anion[59], and does not form a -CN·- anion reflecting the occurence of a high-lying N≡C π* orbital as discussed in section 2.

The following interesting electron transfer by warming to 77 K after 4K irradiation is reported in glycine hydrochloride crystals whose exchangeable protons are partially deuteriated[42]. (The reaction might be better understandable by H-D exchange reaction rather than the electron transfer.)

$$Cl^{-+}NX_3 CH_2 C \cdot OOH^- + Cl^{-+}NX_3 CH_2 COOD \quad \longrightarrow \Delta \longrightarrow$$

$$\longrightarrow Cl^{-+}NX_3 CH_2 COOH + Cl^{-+}NX_3 CH_2 C \cdot OOD^- \quad (X=D \text{ or}H) \quad (r8)$$

4. The Monoene and Diene Anions of Olefines

Hexene $\pi*$ radical anions have been observed recently together with the corresponding cations in mixed crystals of hexane and hexene isomers (-1, -2, or -3) irradiated at 4.2 K by the present author[60]. Both ions of hexene-3($CH_3 CH_2 CH=CHCH_2 CH_3$) have three pairs of equivalent proton hyperfine couplings(TABLE 6), reflecting the molecular symmetry C_2.

TABLE 6. Isotropic and anisotropic hyperfine couplings(in gauss) and spin density on carbon atoms(ρ)for monoene and diene anions.

molecule	proton	a(iso)	(B_1	B_2	B_3)	[cation]^a	ρ(C;π)	Ref.
$CH_3 CH_2 -CH=CH-CH_2 CH_3$	CH	2	(5	0	-5)	[13]	0.32	60
	CH_2	a(iso)= 13, 5.5				[45, 33]		
$CH_2 =CH-CH=CH_2$	CH_2	-7.8	(-6.0	0.	6.0)		0.38	61,62
		-7.7	(-5.4	-0.2	5.6)			
	CH	-2.7	(-1.3	-1.2	2.6)		0.17	

a; isotopic hyperfine couplings for the counter cation.

One pair of couplings with a large anisotropy associated to the anions is assigned to α-protons and the other two to β-protons. These couplings are very small, and only about one third the size of those for the counter cations. The dipolar tensor of two α protons, however, is not small (B=5,0,-5 G), indicating that each half of the unpaired electron spin density resides on equivalent C=C α-carbon atoms. Their isotropic coupling (a=2 G) is very small as compared with α-proton couplings for planar π radicals. Such a small α-proton coupling can be expected only for bent radicals(Figure 1d), in which spin polarization and spin delocalization to the hydrogen orbital are cancelled with the opposite signs. The small β-proton couplings observed also support the bent structure of olefinic anions in a similar way as is discussed for carboxyl radical anions in section 3.2. The monoene anion decays by recombination with the counter cation upon illumination or warming. The reaction may be initiated by releasing the excess electron from the anion, since the anion has high-lying $\pi*$ SOMO as discussed in section 2.

$$(>C=C<) \cdot - \longrightarrow >C=C< + e-, \quad e- +(>C=C<) \cdot + \longrightarrow >C=C< \quad (r9)$$

ESR parameters for diene anions are listed in TABLE 6[61,62]. The large α-proton couplings for the diene anions(-7.8 G) indicate that the radical has a planar structure. As is easily seen from the phase of the unpaired electron orbital of the diene($H_2 C=CH-CH=CH_2$) anion, shown in Figure 1e, the orbital energy is lowered by introducing π bonding in the interior C(2)-C(3) bond. The anion exhibits a higher π spin density on the end carbon atoms(ρ(1)=ρ(4)=0.38) than the interior atoms(ρ(2)= ρ(3)=0.17), since the interior π bond is weaker. The monoene anions without any groups to extend the π electron delocalization may be stabilized by mixing [2s;C> atomic orbital with a large core integral to the [2pπ;C> orbital. The mixing makes both π and π* orbitals lower, as shown in Figure 1c and 1d.

5. The Monomer and Dimer Anions of Nitriles

The π* radical anion of acetonitriles was studied by Williams et al.[63-65] and Symons[66]. The monomer and dimer anions are trapped in the low temperature crystalline phase and a metastable phase by irradiation at 77 K, respectively. Their ESR parameters are listed in TABLE 7. The former group reported that the monomer anion has a ^{13}C isotropic coupling(61.4 G) similar to that for the HCN·⁻ anion(74.4 G)[67], suggesting a R-C-N bent radical structure. The HCN·⁻ anion has the unpaired electron orbital; ψ= 0.52[1s;H> -0.47[sp3 ;C> +0.65[p;N>, whose MO coefficients are estimated from the unpaired electron density. The higher spin density on the N atom(ρ(N)=0.42) as compared to that on the carbon atom(ρ(C)= 0.22) can be reasonably understood from eq.(5), since the absolute value of the Coulomb integral of nitrogen 2p orbital is smaller(-7.27 ev) than that of carbon sp3 valence orbital(-7.70 ev). Unfortunately the details of the ESR parameters for the acetonitrile monomer anion are still reserved for publication except for the preliminary report where the following information is reported for the dimer anion by the former research group[64]. The anion has a much smaller isotropic ^{13}C coupling (4.3 G) and the unpaired electron orbital is composed of nearly pure 2p orbitals of a pair of C and N atoms;(ρ(C)= 0.13, ρ(N)=0.34), indicating the molecules in the dimer are more nearly linear than the monomer radical anion. A greater fraction of the unpaired electron resides on N atoms in the dimer (0.34 x2) rather than in the monomer radical anion(0.28). Both research groups suggested the following dimer structure [I] as the most probable form.

CH3 -C≡N   CH3 -C≡N   CH3 -C≡N...N≡C-CH3   CH3 ┐C≡N
    ┆ ┆   ┆   ┆
[I] N≡C-CH3   [II] N≡C-CH3   [III]   [IV] N≡C-CH3

TABLE 7. Hyperfine coupling tensors of ^{14}N, ^{13}C, and H(in gauss) and spin densities in nitrile anions.

molecule	atom	A_1	A_2	A_3	a(iso)	B(//)	ρ_p	ρ_s	ρ_s/ρ_p	Ref.
CH3 C≡N	N						0.28			64
	C				61.4		0.32			
H-C≡N	N	21.1	-0.2	-0.6	6.8	14.3	0.42	0.012	3 %	67
	C	88.7	65.3	69.2	74.4	14.3	0.22	0.067	23	
	H	141.7	135.3	134.4	137.1	4.6	-	0.27		
(CH3 CN)2	Nx2	17.3	0.	0.	5.8	11.5	0.34	0.011	2.9	64
	Cx2				4.3		~0.13			
(N≡C-C≡N)	Nx2	14.8	-2.5	-2.5	3.3	11.5	0.34	0.006	1.7	68
	Cx2	67.	56.	56.	60.	7.3	0.11	0.05	30	

However, MO calculations for I-III do not predict such a large
density on N atoms, since the nitrogen 2p atomic orbital has a larger
negative core integral(-7.27 ev) than the carbon 2p orbital(-5.57 ev).
Only conformation IV gave densities in agreement with the observation
(Figure 1h). The high density on the end N atoms for IV is similar to
that on the end C atoms in diene anion >C=C-C=C<·-(Figure 1e), which
is discussed in the previous section. The smaller s nature of the
carbon orbital in the nitrile dimer compared to that in the monomer
resembles the change from alkene to diene anions. The spin density
distribution in nitrile dimer anions resembles that in the (N≡C-C≡N)·-
anion except for the s nature of carbon orbitals(TABLE 7)[68]. The
similarity might support conformation IV for the dimer anion. Nitriles
may be the only compound whose dimer anion has been observed. The cause
may be the high-lying SOMO described in section 2. The formation of the
dimer anion is stabilized by the electron delocalization. Both anions
dissociate by photoillumination[65] and become protonated upon thermal
annealing[66].

$$CH_3 CN\cdot-, (CH_3 CN)_2 \cdot- \xrightarrow{-h\nu} CH_3\cdot + CN^- \text{ (or adduct)} \qquad (r10)$$

$$CH_3 CN\cdot-, + H^+ \xrightarrow{\Delta} CH_3 HCN\cdot \qquad (r11)$$

6. The Unsaturated Carboxylic Acid Anions

Since unsaturated carboxylic acids have both COOH groups and a C=C bond,
their anions have a lower-lying antibonding molecular orbital(SOMO) than
their constituents. The energy diagram and the correlation of molecular
orbitals to the MOs of their constituents are illustrated in Figure 1j,
k, and c'. The ESR parameters reported for their anions are listed in
TABLE 8[69-74]. An interesting radical is the fumaric acid anion, which
was observed in a single crystal of succinic acid doped with fumaric
acid [69]. The radiolysis of the mixed crystals was studied first at
room temperature by Miyagawa and Itoh[75]. They found the formation of
the DOOC-CHD-C·H-COOD radical(S1) in the mixed crystals where the
acidic protons were deuterated. Irradiation of pure succinic acid
crystals yields a similar radical HOOC-CH2-C·H-COOH (S2), formed by a
C-H bond scission. The latter radical has two CH β-protons with
nonequivalent isotropic couplings, a=27.3 and 8.4 G[76]. The β
deuterium atom in S1 radical selectively occupies the one position of
the two β protons, which exhibits the larger isotropic coupling. They

TABLE 8. Hyperfine tensor of protons and p(π) spin density
on carbon atoms in unsaturated carboxylic acid anions.

molecule	A_1 (G)	A_2	A_3		ρ(C)	Ref.
HOOCCH=CHCOOH	-2.7	-6.8	-9.9	(2H)	0.24	69
maleic acid anhydride	-2.1	-7.0	-10.5	(2H)	0.24	70
HOOCC=CCOOH	2-4			(2H)		71
K+[-OOCCH=CHCOOH]	-5.4	-11.1	-15.1	(1H)	0.39	72
maleic acid	-3.2	-8.5	-13.0	(1H)	0.30	73
	-1.9	-5.2	-6.4	(1H)	0.17	
CH2=CHCOOH	a(iso)=13 G(2H; CH2=)					74

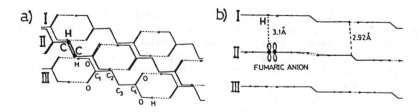

Figure 6. Projection of the crystal structure of succinic acid to (a) the molecular plane and (b)a plane perpendicular to (a). Fumaric acid (guest) molecules have a similar orientation with those of the host molecules(see text).; From M.Iwasaki, H.Muto, and K.Toriyama, J. Chem. Phys. 55, 1894 (1971); copyright © 1971; American Institute of Physics; reproduced here by kind permission of American Institute of Physics.

interpreted this behavior as a stereospecific D atom addition to fumaric acid molecules. Iwasaki et al. irradiated the mixed crystals containing 0.5 mole % fumaric acid at 77 K[69]. They found a selective formation of fumaric acid radical anion instead of the succinic acid anion. It was further observed that the fumaric radical anion converted to the D addition radical S1 when warmed, indicating that the latter radical is formed by protonation of the anion.

$$DOOC-CH=CH-COOD \cdot - + D^+ \longrightarrow_\Delta\longrightarrow DOOC-CHD-C\cdot H-COOD \qquad (r12)$$

A comparison of the ESR parameters with the crystal structure indicates that an acidic proton of a succinic acid molecule in the upper molecular layer is located just above the C=C bond of fumaric molecules(Figure 6.). It was suggested that the D addition radical S1 was formed as a result of the transfer of this proton to the C=C bond from the vertical direction. This transfer from the p lobe direction of the unpaired electron orbital would result in a smaller dihedral angle for the CH β-bond and hence a larger isotropic coupling than for the originally bonded proton. A similar orbital steering for proton transfer from the p lobe direction is discussed for carboxyl anions in section 3.3.1.

Budzinski and Box studied the radiolysis of mixed crystals with a lower content of fumaric acid(~0.1 mol %) at 4.2 K[77]. They found about equal amounts of succinic acid and fumaric acid anions were formed at 4 K but no conversion occurred from the former anion to the latter by warming to 77 K. From these results they supported the interpretation offered by Miyagawa and Fikes[78] that selective reduction of fumaric acid is the result of a more efficient trapping of radiation induced electrons by solute molecules with a higher EA.

Maleic acid anhydride and acetylene dicarboxylic acid form radical anions with a symmetric π electron spin distribution(TABLE 8)[70,71], while maleic acid and acrylic acid anions and the fumaric acid anions in potassium hydrogen fumarate have nonequivalent C-H α proton couplings reflecting their molecular asymmetries[72-74]. However, the sum of spin densities on the two carbon atoms of the C=C bond is the same for each anion($\sim$0.48). These anions, trapped at 77 K, also undergo a protonation by the acidic protons to the C=C bond upon warming[70-74,79]. In summary, the unsaturated carboxylic acid anions are charge-neutralized by protonation just as is the case for other saturated carboxyl radical anions, although the atom attacked differs.

7. The Disulfide Anions

The radical anion of disulfides was first found and precisely studied in cysteic acid single crystals irradiated at 77 K by Akasaka et al.[80-82]. The g tensor and hyperfine tensors of ^{33}S and the CH β-protons are listed in TABLE 9. The combined data showed that the principal directions of the maximum ^{33}S tensor element and of the minimum (parallel) g element are parallel to the S-S bond. The spin densities on the two sulfur atoms are equal and estimated to be $\rho(3p\sigma;S) = 0.63$ and $\rho(3s;S)=0.05$. These results indicate that the unpaired electron resides in the σ(S-S) bond and that the orbital is mainly composed of the two sulfur $3p\sigma$ orbitals, as is shown in Figure 2g. They assigned the radical to the $\sigma*$ anion based on a consideration of g anisotropy in addition to the above results.

Many disulfide anions have been observed in the crystalline phases and in a variety of matrices[83, 84]. Box et al. studied the dipolar couplings of the CHβ protons of those anions and showed that the anisotropic terms could be reconstructed with a radical model such that a quarter of the unpaired electron density resides in each of the four $p(\sigma)$ orbital lobes of two sulfur atoms[84]. The disulfide anion is stable, since it has a low-lying $\sigma*$ SOMO as is discussed in section 2. It decays by geminate recombination with the counter hole detrapped by illumination[82].

$$(-S-S-)\cdot + \xrightarrow{h\nu} \text{mobile hole} \xrightarrow{+ (-S-S-)\cdot -} 2 \ (-S-S-) \quad (r11)$$

TABLE 9. g tensor and hyperfine coupling tensor for the L-cystine anion.

g and A	principal value			direction	Ref.
[g_1 g_2 g_3]=	2.002 2.018 2.018			g(min) // S-S	80
	a(iso)	B	[MHz]		
^{33}S	76	(104 -52	-52)	B(max) // S-S	81
H(1)	20.1	(3.98 -1.21	-2.77)		82
H(2)	20.1	(4.03 -1.22	-2.80)		
H(3)	28.8	(4.25 -1.29	-2.96)		
H(4)	28.9	(4.34 -1.31	-3.03)		
electron spin density on S; $\rho(s)=0.028$, $\rho(p)=0.63$					

8. The Anions of Halogenated Compounds

The radical anions of N-halogenated succinimide were first studied by Symons et al.[85,86], who assigned the anions to $\sigma*$ radicals, in which the unpaired electron is localized in the >N-X σ bond. Pace et al. and Muto and Kispert confirmed this assignment by single crystal ESR spectroscopy and MO calculations[87,88]. The ESR parameters for those anions are listed in TABLE 10. The unpaired electron orbital is composed from nitrogen sp_2 and halogen pure $p\sigma$ orbitals. The spin density on the halogen atom is higher for anions containing the heavier X atom; $\rho(X)=0.51$, 0.61, and 0.80 for Cl, Br, and I, respectively. The energy levels of the >N-X group calculated using eqs.(1-3) are shown in Figure 7. The energy of the $\sigma(N-F)*$ SOMO of the fluorosuccinimide anion is much higher than those of the other halogenated imides, because the F atom has a very large resonance parameter $\beta(0)=39$ eV. The formation of this anion has not been reported. Since the p orbital of the heavier atoms has the smaller Coulomb integral as compared with the nitrogen sp_2 orbital, the MO calculation predicts a higher unpaired electron density on the heavier halogen atom [eq.(5)]; $\rho(X)= 0.52$, 0.61, 0.65, and 0.68 for X= F, Cl, Br, and I. This trend predicted from the calculations agrees with that obtained experimentally, as shown in Figure 8. The agreement supports the above simple MO description for the nature of the $\sigma(Y-X)*$ antibonding orbital.

The ESR parameters reported for the anions with carbon-halogen bonds are summarized in TABLE 10[89-101]. The unpaired electron densities on the halogen atoms as estimated from the parameters are plotted in Figure 10. The anions are classified into three types: (a)≡C-X(sp-p), (b)=C-X(sp_2-p), and (c)>C-X(sp_3-p). Their energy diagrams, obtained by a simple MO calculation, are shown in Figure 9 for the case X=I. Since the core integral of carbon valence orbital has a lower value in the order c, b, a, the $\sigma(C-X)$ bonding and the $\sigma(C-X)*$ antibonding orbitals

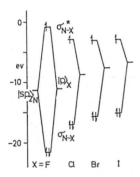

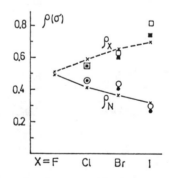

Figure 7. Energy diagram for the the >N-X group of N-halogenated succinimide anions.

Figure 8. Unpaired electron densities on halogen and nitrogen atoms in halogenated succinimide anions. x and open marks (circles and squares) represent the calculated and observed values, respectively. Filled circles and squares indicate the relative fraction of observed values on N and X atoms.

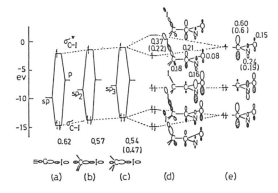

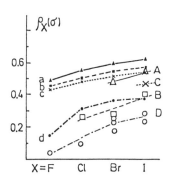

Figure 9. Energy diagram for the a)≡C-I, b)=C-I, c)→C-I group of corresponding anions and d)halogenated amide and e) amide anions, respectively. Numbers without and with parentheses represent the calculated and observed unpaired electron densities, respectively.

Figure 10. Unpaired electron density on halogen atoms in anions with carbon-halogen bond. (a,A), (b,B), (c,C), and (d,D) and their corresponding marks represent the densities for ≡C-X, =C-X, →C-X, and halogenated acetamide anions. The capitals (A, B,...) and the letters (a,b...) represent the observed and calculated data, respectively.

become lower-lying in this order, as is expected from eq.(3). Therefore the former two types of anions(a,b) are relatively stable. The last type of anions(c) have relatively high-lying σ(C-X) and σ(C-X)* orbitals and are unstable. Simple alkyl halides such as CH_3X easily undergo dissociative electron attachment. This type of anions has been reported only for halogenated acetamides and acids[93-101] and multihalogenated alkanes. The latter alkane anions are described in another Chapter.

In every type of anion(a-c) the unpaired electron density is higher on the heavier halogen atom just as in the halogenated succinimides anions described previously. A comparison of the calculated and observed spin densities is shown in Figure 10. The calculated spin density for type (a) anions(with ≡C-X) agrees with the observation[89, 90], supporting the above MO description. The type(b) anions(=C-X·⁻) have a lower unpaired electron density than the corresponding calculated value, probably suggesting a delocalization to the C=O β-bond in the 5-halogenated uracil anions and to the β-sulfur atom in the thiophene anion(TABLE 10 and Figure 10)[91,92]. The $(CF_3)_3C$-I·⁻ anion has a high unpaired electron density on the iodine atom; ρ(I)=0.47[93]. It may be the only reported alkyl halide anion whose unpaired electron is localized in a σ(C-X) bond. The type (c) anions such as halogenated acetamide or acid anions have a much lower spin density on the halogen atom(0.04-0.28)[94,101]. Their densities are also lower than those calculated for simple alkyl halide anions but have nearly the

356

TABLE 10. g tensor and hyperfine coupling tensors of ^{14}N, ^{19}F, [35]Cl,[81]Br, and [127]I atoms for halogenosuccinimide and halogenated organic anions and estimated unpaired electron densities.

molecule	g(//, 1)		$A_{//}$ A_1 (G)		a_{iso}	$B_{//}$ (MHz)	ρ_p^a	ρ_s	Ref.
halogenated succinimides									
>N-Cl	2.003	2.026	Cl 118	26.5	156	173	0.51	0.03	85,87
			N 68	49.5	170	122	0.35	0.10	
>N-Br	1.989	2.101	Br 603	125.5	727	950	0.61	0.03	85,88
			N 61	41.5	170	173	0.34	0.09	
>N-I	1.96	2.41	I 560	145	478	1057	0.80	0.02	86
			N 45	27	102	21	0.22	0.07	
phC≡C-I	1.98	2.09	I 639	263	1049	720	0.55		89
N≡C-Br	2.00	2.065	Br 567	187	850	736	0.47		90
Cl—thiophene	2.002	2.006	Cl 84	35	143	92	0.27		91
5-BrUracil	1.997	2.030	Br 373	143	608	434	0.28	[other atoms]	92
5-I Uracil	1.991	2.044	I 505	220	869	537	0.41	(G)	92
$(CF_3)_3 C-I$	1.963	2.165	I 467	162	671	612	0.47		93
$ICH_2 CONH_2$	1.990	2.042	I 242	96	397	276	0.21	[18-23(2H)]	94-96
$ICH_2 COOH$	1.98	2.04	I 360	163	631	366	0.28		97
$BrCH(CONH_2)_2$	1.998	2.043	Br 289	84	416	392	0.25	[14 (1H)]	98
$BrCH_2 COOH$	--	--	Br 277	90	417	357	0.23	[13-20(2H)]	99
$BrCF_2 CONH_2$	2.002	2.029	Br 238	87	372	269	0.17	[108(//), <3(1)(2F)]	100
$ClCF_2 CONH_2$			Cl ~28	~12	17	11	0.11	[215(//),~38(1) (2F)]	100
$FCF_2 CONH_2$	2.003	2.005	F 139	71	261	128	0.04	[12 (2F)]	101
$[F-..CH_2 COOH]$ (adduct)			F 13	6.5				[(-8 -21 -34)x1H] [(-15 -21 -35)x1H]	47

a;Estimated using the B(//) values for X_2 •⁻ centers[102] assuming that one half of the unpaired density resides on each of the atoms.

values calculated for $XCRRCONH_2$ •⁻ anions. These results indicate that these are carbonyl anions(X1) whose unpaired electron is delocalized to a halogen p orbital by σ(C-X)-π(COR) hyperconjugation. These anions have a C-X(β) conformational angle of nearly θ=0⁰ just as the parent molecules in the crystals[95,100,101]. Therefore the conformational angles for the CH β-bonds may be nearly ±120⁰ for the XCH_2 -$CONH_2$ •⁻ anions, predicting a small CH β-proton coupling(less than ~10 G). On the other hand, large values in the range of 14-22 G are observed(TABLE 10). Symons et al. assigned these couplings to α protons and proposed that the radical is a β halo-radical and is formed by halogen atom migration to a carbonyl carbon atom[92,99]. Kispert et al. showed from single crystal analyses that the maximum principal axes of the hyperfine and quadrupole coupling tensors of the halogen atom are parallel to the C-X bond of the parent molecules, thereby excluding the halogen migration model[95]. They proposed that the species is the σ(C-X)* anion whose unpaired electron is localized in the σ bond. However, the unpaired electron density on the halogen atom is so low(0.04-0.28) as compared with the value for the $(CF_3)_3C-I \cdot^-$ anion(0.47) that the radicals still seem to favor the carbonyl anion. The large proton couplings should be ascribed to the superposition of hyperconjugation from the carbonyl group and spin delocalization from the C-X carbon orbital.

9. Concluding Remarks

Organic radical anions are unstable and metastably trapped in low temperature solids because an excess electron resides in an antibonding MO. They are classified as π^* and σ^* radicals. π^* anions have a trend to be neutralized by proton addition to the unsaturated bond such as C=O, C=C, C≡C, and C≡N, from a neighboring molecule. The protonated carboxyl anions and σ^* anions of the halogenated compounds undergo decomposition reactions by warming(β and α eliminations, respectively). In some anions such as the monoene anion with high-lying SOMO, the excess electron is ejected by photoillumination or by elevating the temperature and subsequently recombines with the counter cation. The disulfide anions with a low-lying SOMO decay by recombination with the photo-activated counter mobile hole.

References

1. V. I. Vedereyev, L. V. Gurvich, V. N. Kondra'yev, and V.A. Medvedev: Bond Energies, Ionization Potential and Electron Affinities, Edward Arnold Ltd., London (1966).
2. Hand Book of Chemistry, Vol. Fundamentals II , 3-rd Ed. (Japanese) Ed. by Chem. Soc. Jpn., Maruzen Publisher, Tokyo, p. II-588 (1984).
3. CRC Handbook of Chemistry and Physics, 58th ed. Ed. by R. C. Weast: CRC Press Inc., NY, pp. E67-68 (1978).
4. J. A. Pople: and D. L.Beveridge, Approximate Molecular Orbital Theory, McGraw-Hill Inc., NY, pp. 69-79 (1970).
5. M. A. Whitehead:'Semiempirical-All-Valence-Electron SCF-MO-CNDO Theory(III-2)' in O. Sinanoglu and K. B. Wiberg (eds.), Sigma Molecular Orbital Theory, Yale University Press, New Haven and London, pp. 49-80 (1970).
6. H. Muto, K. Nunome, K. Toriyama, and M. Iwasaki: *J.Chem. Phys.* **93**, 4898 (1989), and references cited therein.
7. H. C. Box: Radiation Effect; ESR and ENDOR Analysis, Academic, New York, (1977).
8. M. Shiotani: *Magn. Reson. Rev.* **12**, 333 (1987).
9. M.C.R. Symons: *Chem. Soc. Rev.* pp. 393-439 (1984).
10. W. Gordy, W. B. Ard, and H. Shield: *Proc. Nat. Aca. Sci.* **41**, 996 (1955).
11. I. Miyagawa and W. Gordy: *J. Am. Chem. Soc.* **83**, 1036 (1961).
12. H. C. Box, N. G. Freund, and K. T. Lilga: *J. Chem. Phys.* **42**, 1471 (1965).
13. J. E. Bennett and L. H. Gale: *Trans. Faraday Soc.* **64**, 1174 (1968).
14. J. W. Sinclair and M. H. Hanna: *J. Phys. Chem.* **71**, 84 (1967).
15. A. Minegishi, Y. Shinozaki, and G. Meshitsuka: *J. Chem. Soc. Jpn.* **40**, 1549 (1967); ibid **41**, 3035 (1968).
16. P. B. Ayscough and A. K. Roy: *Trans. Faraday Soc.* **64**, 582 (1968).
17. I. Miyagawa, N. Tamura, and J. W. Cook,Jr.: *J. Chem. Phys.* **51**, 3520 (1969).
18. H. Muto and M. Iwasaki: *J. Chem. Phys.* **59**, 4821 (1973).

358

19. M. Iwasaki and H. Muto: *J. Chem. Phys.* **61**, 5315 (1974).
20. H. Muto, K. Nunome, and M. Iwasaki: *J. Chem. Phys.* **61**, 5311 (1974).
21. H. Muto, K. Nunome, and M. Iwasaki: *J. Chem. Phys.* **61**, 1075 (1974).
22. R. LoBrutto, E. E. Budzinski, and H. C. Box: *J. Chem. Phys.* **73**, 6349 (1980).
23. H. C. Box, E. E.Budzinski, and K. T. Lilga: *J. Chem. Phys.* **57**, 4295 (1972).
24. J. Sinclair and P. Codella: *J. Chem. Phys.* **59**, 1569 (1973).
25. H. Muto: Ph. D. Thesis, ESR and ENDOR study on the structure and reaction of radicals in amino acids and carboxylic acids irradiated, Nagoya University(Jpn) (1978).
26. D. J. Whelan: *Chem. Rev.* **69**, 179 (1969).
27. G. C. Moulton and G. McDearmon: *J. Chem. Phys.* **72**, 1665 (1980).
28. J. Sinclair and M. Hanna: *J. Phys. Chem.* **50**, 2125 (1969).
29. H. C. Box, H. G. Freund, K. T. Lilga, and E. E.Budzinski: *J. Chem. Phys.* **63**, 2059 (1975).
30. D. C. Straw and G. C. Moulton: *J. Chem. Phys.* **60**, 1231 (1974).
31. J. W. Wells: *J. Chem. Phys.* **52**, 4062 (1970).
32. J. Y. Lee and H. C. Box: *J. Chem. Phys.* **61**, 428 (1974) .
33. C. L. Ko and H. C. Box: *J. Chem. Phys.* **55**, 2446 (1971).
34. G. C. Moulton and J. M. Coleman: *J. Chem. Phys.* **80**, 4748 (1984).
35. H. C. Box and E. E. Budzinski: *J. Chem. Phys.* **55**, 2446 (1971).
36. H. C. Box, E. E. Budzinski, and W. Potter: *J. Chem. Phys.* **55**, 315 (1971).
37. L. L. Finch, J. E. Johnson, and G. C. Moulton: *J. Chem. Phys.* **70**, 3662 (1979).
38. S. M. Adams, E. E. Budzinski, and H. C. Box: *J. Chem. Phys.* **65**, 998 (1976).
39. W. H. Nelson and C. R. Nave: J. Chem. Phys. **74**, 2710 (1981); W. H. Nelson: *J. Phys. Chem.* **92**, 554 (1988).
40. H. C. Box and E. E. Budzinski: *J. Chem. Phys.* **60**, 3337 (1974).
41. F. Q. Ngo, E. E. Budzinski, and H. C. Box: *J. Chem. Phys.* **60**, 3373 (1974).
42. J. Y. Lee and H. C. Box: *J. Chem. Phys.* **59**, 2509 (1972).
43. W. W. H. Kou and H. C. Box: *J. Chem. Phys.* **64**, 3060 (1976).
44. a)H. C. Box, H. G. Freund, and E. E. Budzinski: *J. Chem. Phys.* **57**, 4290 (1972); b)E. E. Budzinski, K. T. Lilga, and H. C. Box, *J. Chem. Phys.* **59**, 2899 (1973).
45. H. C. Box, E. E. Budzinski, and H. G. Freund: *J. Chem. Phys.* **61**, 2222 (1974).
46. C. L. Ko and H. C. Box: *J. Chem. Phys.* **68**, 5357 (1978).
47. O. Awadelkarim, A. Lund, and P.-O. Samskog: *Radiat. Phys. Chem.* **27**, 353 (1986).
48. H. Shields and P. J. Hamrick, Jr.: *J. Chem. Phys.* **64**, 263 (1976).
49. a)H. C. Box and H. G. Freund: *J. Chem. Phys.* **44**, 2345 (1966); b)E. Westhof , W. Flossmann, and A. Mueller: *Mol. Phys.* **28**, 151(1974); c)G. Saxebl and E. Sagstuen: *Int. J. Radiat. Biol.* **26**, 373 (1974).
50. a)E. Westhof , W. Flossman, H.-D. Luedemann, and A. Mueller: *J. Chem. Phys.* **60**, 3376 (1974); b)W. H. Nelson: *J. Phys. Chem.* **92**, 554 (1988).
51. D. M. Close, G. W. Fouse, W. A. Bernhard, and R. S. Andersen: *J. Chem. Phys.* **70**, 2131 (1979).
52. H. Muto, T. Inoue, and M. Iwasaki: *J. Chem. Phys.* **57**, 3220 (1972).
53. R. W. Holmbrg: *J. Chem. Phys.* **51**, 3255 (1969).
54. R. C. McCalley and A. L. Kwiram: *J. Chem. Phys.* **53**, 2541 (1970).
55. M. B. Yim and R. E. Klinck: *J. Chem. Phys.* **60**, 538 (1974).
56. a)M. V. V. S. Reddy, K. V. Lingam, and T. K. G. Rao: *J. Chem. Phys.* **76**, 4398 (1982); b)T. K. G. Rao, and K. V. Lingam: *Mol. Phys.* **54**, 999 (1985).
57. a)P.-O. Samskog: *Acta. Chem. Scand.* **A35**, 559 (1981); b)P.-O. Samskog, A. Lund, G. Nilsson, and M. C. R. Symons: *Chem. Phys. Lett.* **66**, 199 (1979).
58. D. J. T. Hill, J. H. O'Donnell, P. J. Pomery, and A. K. Whittaker: *Radiat. Phys. Chem.* **26**, 191 (1985).
59. K. Toriyama and W. C. Lin: *J. Phys. Chem.* **76**, 3377 (1972).
60. H. Muto and K. Nunome: "25th Symmposium of Radiation Chemistry, Japan", Hiroshima University

(1989).

61. T. Shiga and A. Lund: *Ber. Buns. Gesellsh.* **78**, 259 (1974).
62. D. H. Levy and R. J. Myers: *J. Chem. Phys.* **44**, 4177 (1966).
63. M. A. Bonin, K. Tsuji, and F. Williams: *Nature*, **218**, 946 (1968).
64. a)E. D. Sprague, K. Takeda, and F. Williams: *Chem. Phys. Lett.* **10**, 299 (1971); b)T. Gillbro, K. Takeda, and F. Williams: *J. Chem. Soc. Faraday Trans. 2*, **70**. 465 (1974); and references cited therein.
65. E. D. Sprague, K. Takeda, J. T. Wang, and F. Williams: *Can. J. Chem.* **52**, 2840 (1974).
66. R. J. Egland and M. C. R. Symons: *J. Chem. Soc.* (A) 1326 (1970).
67. I. S. Ginns and M. C. R. Symons: *J. C. S. Dalton*, 185 (1972).
68. K. D. J. Root and M. C. R. Symons: *J. Chem. Soc. (A) Inorg. Phys. Theor.* 21 (1968).
69. M. Iwasaki, H. Muto, and K. Toriyama: *J. Chem. Phys.* **55**, 1894 (1971).
70. M. Fukaya, H. Muto, K. Toriyama, and M. Iwasaki: *J. Phys. Chem.* **80**, 728 (1976).
71. H. Muto, K. Toriyama, and M. Iwasaki: *J. Chem. Phys.* **57**, 3016 (1972).
72. M. Iwasaki, K. Minakata, and K. Toriyama: *J. Chem. Phys.* **55**, 1472 (1971).
73. B. Eda and M. Iwasaki: *Mol. Phys.* **24**, 589 (1972).
74. M. Iwasaki, F. Fukaya, S. Fujii, and H. Muto: *J. Phys. Chem.* **77**, 2739 (1973).
75. I. Miyagawa, K. Itoh: *Bull. Am. Phys. Soc.* **9**, 488 (1964); *Nature*, **209**, 504 (1966).
76. P. Pooley and D. H. Wiffen: *J. Am. Chem. Soc.* **84** 366 (1962).
77. E. E. Budzinski and H. C. Box: *J. Chem. Phys.* **63**, 4927 (1975).
78. I. Miyagawa and J. W. Fikes: *Bull. Am. Phys. Soc.* **15**, 166 (1970).
79. K. Toriyama, H. Muto, and M. Iwasaki: *J. Chem. Phys.* **55**, 1885 (1971).
80. K. Akasaka, S. Ohnishi, T. Suita, and I. Nitta: *J. Chem. Phys.* **40**, 3110 (1964).
81. A. Naito, K. Akasaka, and H. Hatano: *J. Magn. Reson.* **24**, 53 (1976); *Chem. Phys. Letters*, **32**, 247 (1975).
82. K. Akasaka, S. Kominami, and H. Hatano: *J. Phys. Chem.* **75**, 3746 (1971).
83. a)R. Franzi, M. Geoffroy, M. V. V. S. Reddy, and J. Weber: *J. Phys. Chem.* **91**, 3187 (1987); b)L. Bonazzola, J. P. Michaut, and J. Roncin: *J. Chem. Phys.* **83**, 2727 (1985); c)D. N. R. Rao, M. C. R. Symons, and J. M. Stephenson: *J. Chem. Soc. Perkin Trans.* 2, 727 (1983).
84. M. J. Colaneri and H. C. Box: *J. Chem. Phys.* **84**, 2926 (1986).
85. G. W. Neilson and M. C. R. Symons, *J. C. S. Faraday Trans.* 2, **68**, 1582 (1972).
86. G. W. Neilson and M. C. R. Symons: *Mol. Phys.* **27**, 1613 (1974).
87. D. Pace, K. Ezell, and L. D. Kispert: *J. Chem Phys.* **71** 3971 (1979).
88. H. Muto and L. D. Kispert: *J. Chem. Phys.* **72**, 2300 (1980).
89. D. J. Neilson and M. C. R. Symons: *Chem. Phys. Letters*, **47**, 436 (1977).
90. S. P. Mishra, G.W. Neilson, and M.C.R. Symons: *J.Chem.Soc. Faraday Trans.* 2, **70**, 1280 (1974).
91. N. Nagai and T. Gillbro: *J. Phys. Chem.* **81**, 1793 (1977).
92. M. C. R. Symons: Radiat. *Radiat. Phys. Chem.* (1980).
93. J. T. Wang and F. Williams: *J. Am. Chem. Soc.* **102** 2860 (1980).
94. R. F. Picone and M. T. Rogers: *J. Magn. Reson.* **14**, 279 (1974).
95. P.-O. Samskog, L. D. Kispert, and B. Kayanaraman: *J. Chem. Soc. Faraday Trans.* 2, **80**, 267 (1984).
96. G. W. Neilson and M. C. R. Symons: *J. C. S. Chem. Commun.* 717 (1973).
97. A. Lund, P.-O. Samskog, and G. Nilsson: *Chem. Phys.* **57**, 399 (1981).
98. R. F. Picone and M. T. Rogers: *J. Chem. Phys.* **61**, 4814 (1974).
99. R. J. Booth, S. P. Nishra, G.W. Neilson, and M.C.R. Symons: *Tetrahedron Lett.* **34**, 2949 (1975).
100. L. D. Kispert, R. Reeves, and T. C. S. Chen: *J. Chem. Soc. Faraday Trans.* 2, **74**, 871 (1978); P.-O. Samskog, and L. D. Kispert: *J. Phys. Chem.* **88**, 1385 (1984).
101. P.-O. Samskog and L. D. Kispert: *J. Chem. Phys.* **78**, 2129 (1983).
102. D. Shoemaker: *Phys. Rev.* **149**, 693 (1966).
103. The alkyne radical anion has been recently observed by ESR. It has a non-linear molecular structure. H. Muto, K. Matsuura, and K. Nunome:to be submitted for publication.

Anion Radicals in Inorganic Crystals

JØRGEN R. BYBERG
Department of Chemistry, Aarhus University, DK-8000 Aarhus C, Denmark

1. Introduction.

When the Electron Spin Resonance (ESR) technique became available to chemists some thirty years ago it was soon noted, that many inorganic ionic solids with polyatomic ions yielded well-resolved ESR signals after exposure to ionizing radiation. The ESR signals were initially assigned to paramagnetic 'centers' in analogy with the results of previous studies of radiation damage in alkali halides. It eventually became apparent, however, that the observed properties of many of these centers could be accurately represented in terms of well-defined chemical species having an odd number of valence electrons (radicals), 'matrix-isolated' in specific orientations in an otherwise passive host lattice. The first clear-cut case seems to be that of ClO_2 trapped in $KClO_4$ [1]. Over the years, a large number of inorganic radicals have been detected and characterized by ESR in irradiated crystals. Moreover, ESR signals proving difficult to associate with reasonable radical structures have in several cases been successfully assigned to molecular complexes arising from fragmentation of the anions [2]. The study of radiation damage in this type of solids by ESR and other spectroscopic techniques has therefore come to be considered a major source of molecular parameters of inorganic radicals and radical-complexes.

The observation of well-defined molecular products has naturally led to a description of the processes generating the radiation damage in terms of radiation-induced chemical transformations (radiolysis) of the polyatomic ions. This picture, emphasizing the chemical properties of the isolated ions and, hence, corresponding to a solid-solution model of the ionic crystal, has helped to establish a link to the radiochemistry of the ions as studied in aqueous solution. It represents an oversimplification, though, because the detailed mechanism of formation and trapping of radicals and other products in the irradiated crystal, and also the subsequent reactions and the dynamics of the products, often involve structural and dynamical properties of the host lattice as well as the chemistry of isolated species.

In the present chapter I shall discuss the characteristics of anion radicals in inorganic crystals as revealed by spectroscopic studies of the radiation damage in alkali halates and perhalates, which may serve to illustrate also the intimate connection between the identification and characterization of radicals and the unravelling of radiolytic processes in solids. The discussion is clearly anything but comprehensive. For a convenient guide to the very extensive literature on inorganic radicals the reader is referred to the series of reviews by Martyn Symons, published in *Specialist Periodical Reports* [3].

Anion radicals, usually defined as molecular systems with an odd number of electrons in the valence shell and a net negative charge, arise in ionic crystals by modification of the anions and, hence, occupy anion sites. However, as the neutral radicals formed by one-electron oxidation of

A. Lund and M. Shiotani (eds.), Radical Ionic Systems, 361–383.
© 1991 *Kluwer Academic Publishers. Printed in the Netherlands.*

monovalent anions also remain at the anion sites and play the same roles in the radiolytic processes as their negatively charged analogs derived from polyvalent anions, I shall here consider as anion radicals in crystals all radicals derived from the anions.

It may be noted, that the labels *radical* and *paramagnetic defect* are currently used interchangeably in radiation-damage studies, the choice merely placing the emphasis either on the molecular system to which the odd electron is confined, or on the host crystal. Hence the labeling of a molecular system embedded in a crystal as a radical does not imply that the system may exist outside the actual lattice.

2. Radiation-induced processes in ionic solids.

The processes induced by ionizing radiation (X- or γ-rays, fast electrons) in ionic crystals may be divided in three classes, each having its characteristic time scale:

i. *The electronic processes* (10^{-14} s) in which the radiation energy is transferred to electronic degrees of freedom in the crystal while the nuclei retain the initial configuration, commence by ejection of fast electrons, which become *moderated* or slowed down by interaction with the stationary electrons of the crystal. In the process ejection of secondary electrons and electronic excitations occur along the path of the fast electron. The excitations transfer electrons from filled to empty bands of the crystal, thus creating holes in the filled bands. About 25 eV is expended per excitation. The secondary electrons similarly cause electronic transitions and ejections of new secondary electrons, the process continuing until the energies of all electrons have sunk below the lowest excitation potential of the crystal. The cross section for interaction between a stationary electron and a moving electron increases as the kinetic energy of the latter approaches the binding energy of the former, implying that many excitations occur in dense clusters arising from a relatively slow electron and its descendants of still slower electrons, which interact strongly with the lattice electrons.

The holes created by ejection of secondary electrons may have energies exceeding the lowest excitation potential. In that case an Auger process is likely to occur, creating two holes with smaller energy plus an electron [4]. As the process continues, the resulting Auger cascade produces a cluster of *sub-excitation* holes and electrons. Thus the spatial distribution of the excitations arising from a single photon or fast electron is strongly nonuniform, comprising low-density regions along the path of fast initial electrons and dense clusters arising from the final moderation of low-energy electrons and from Auger cascades. This pattern of excitations is referred to as the *track* associated with the primary particle.

ii. *The relaxations* ($10^{-13} - 10^{-11}$ s) transfer the electronic energy to vibrational degrees of freedom, in part under formation of defects in the lattice. Due to the very efficient electron-phonon coupling in ionic crystals, the sub-excitation electrons and holes very rapidly attain the lowest available states, becoming stabilized by delocalization and polarization *via* local distortion of the lattice. If local distortion prevails, the electron or hole is said to be *self-trapped*. At this stage the processes, by which a part of the excitation energy bound in the (thermalized) electrons and holes is utilized to generate defects, depend critically on how the stabilization is obtained and hence on the type of material. In the alkali halides, where only the holes become self-trapped whereas the electrons become delocalized, immediate recombination of electrons and holes *via* self-trapped excitons, acting as precursors for defects, are important [5]. In crystals with polyatomic anions, self-trapping of both electrons and holes in the form of localized electron-excess and hole defects is generally observed, implying that self-trapping competes successfully with immediate recombination. In fact an investigation of the radiolysis of $KClO_4$, discussed below, indicates that the self-trapping is a fast process compared to all other processes involving electrons and holes, including recombination.

Although some dissipation of the initial spatial distribution of excitations must be expected to accompany the relaxation, the resulting distribution of self-trapped electrons and holes still comprises dense clusters and low-density regions, which have profound effects on the subsequent reactions. In particular, spontaneous reactions may occur within the clusters during irradiation even at very low temperature, thus creating secondary defects which may in turn trap electrons or holes.

iii. *The thermally activated processes* ($> 10^{-11}$ s) comprise local transformations of defects as well as reactions occurring when defects become thermally mobile in the lattice.

3. Radiolytic formation of radicals in $KClO_4$

In this section the results of a series of spectroscopic studies of radiation damage produced by 50 kV X-rays in $KClO_4$ are summarized. A coherent picture of the formation and trapping of radicals and their reactions has gradually evolved in the course of these studies. Accordingly, some of the interpretations given here deviate from those proposed originally.

$KClO_4$ has several merits as a model substance: The radiolysis is fairly simple because addition of radiolytic fragments to the 'saturated' anion does not occur. Furthermore, the good resolution of the ESR signals from radicals in $KClO_4$, exhibiting resolved hyperfine splittings from the involved chlorine nuclei only, has allowed an unambiguous determination of the nuclear quadrupole interaction and, hence, the electric field gradient at the chlorine nuclei, which has proved crucial for the identification of the radicals. The quality of the ESR signals has also facilitated the quantitative study of accumulation and reactions of the radicals.

3.1. ESR SIGNALS AND PARAMAGNETIC SPECIES

The ESR spectrum of a $KClO_4$ crystal irradiated with X-rays below 30K consists of intense signals in the spectral region around g ~ 2 (i.e. corresponding to radicals with S = 1/2), which have been assigned [6,7] to the radicals ClO_4^{2-} and $(ClO_4)_2^-$, and to the complex $[ClO_2,O_2]$. Moreover, several extremely anisotropic signals have been assigned to oxygen molecules in the triplet ground state (S = 1) trapped as part of the complex $[ClO_2^-,O_2]$ [8].

The observed anisotropic splittings of the ESR signals from radicals are compactly expressed in terms of the conventional spin Hamiltonian

$$\mathcal{H}^s = \beta\, S \cdot \mathbf{g} \cdot B_o + \sum_i I^i \cdot \mathbf{A}^i \cdot S + \sum_i I^i \cdot \mathbf{Q}^i \cdot I^i - \sum_i \gamma^i\, I^i \cdot B_o$$

Here β is Bohr magneton, S is the electron spin operator, B_o is the static magnetic field, while I^i, A^i, Q^i, γ^i are the nuclear spin operator, the magnetic hyperfine tensor, the nuclear quadrupole tensor, and the gyromagnetic ratio of nucleus i, respectively. The symmetric 3 × 3 matrices $\mathbf{g}$, $\mathbf{A}^i$, and $\mathbf{Q}^i$, which constitute the parameters representing the individual properties of the radical, may be specified by their principal values and the orientation of the corresponding principal axes. The principal values in the spin Hamiltonians for ClO_4^{2-}, $(ClO_4)_2^-$, and $[ClO_2,O_2]$ are shown in Table 1.

TABLE 1. Principal values in the spin Hamiltonians of radicals in $KClO_4$.

Species	g_x	g_y	g_z	$A_x{}^a$	A_y	A_z	$Q_x{}^a$	Q_y	Q_z
$(ClO_4)_2^-$ [b]	2.0029	2.0062	2.0152	9.6[e]	10.3[e]	11.7[e]	-0.5	-0.5	1.0
				11.9[e]	12.8[e]	14.8[e]	-0.6	-0.6	1.2
$[O^-,ClO_3]$ (A)[c]	2.0543	2.0506	2.0013	158.2	153.5	181.8	-5.3	-5.1	10.4
$[O^-,ClO_3]$ (A')[c]	2.0415	2.0423	2.0014	185.8	180.9	213.8	-5.1	-4.5	9.6
$[O^-,ClO_3]$ (B)[c]	2.0382	2.0395	2.0019	194.9	196.5	225.8	-5.4	-4.1	9.5
$[ClO_2,O_2]$ (C)[b]	2.000	1.999	2.0063	-8	-11	70.0	9	-1	-8
$[ClO_2,O_2]$ (C')[b]	1.999	2.001	2.0065	-8	-13	66.0	9	-1	8
$[ClO_2,O_2]$ (C")[b]	2.000[d]	1.999	2.0064	-8[d]	-15[d]	68	9[d]	-1[d]	-8[d]
$[ClO_2,O_2]$ (C‴)[b]	2.000	2.001[d]	2.0059	-8[d]	-13[d]	68	9[d]	-1[d]	-8[d]

[a] Hyperfine- and quadrupole tensors (in MHz) refer to ^{35}Cl nuclei.
[b] Data from reference 6.
[c] Data from reference 13.
[d] Estimated value.
[e] Relative sign undetermined.

The structural information conveyed by the parameters in the spin Hamiltonians is conveniently discussed within the LCAO picture:

i. The dimer $(ClO_4)_2^-$ represents the self-trapped hole [⊕]. The hyperfine tensors (Table 1) indicate that the hole, shared almost equally between the two ClO_4^-, is largely confined to the oxygen atoms, the spin density at the chlorine nuclei corresponding to about 0.01 electron. The molecular orbital containing the hole may be visualized as a combination of a component of the highest occupied nonbonding t_1 orbitals on either ClO_4^-. The nuclear quadrupole tensors indicate a very small electric field gradient at the chlorine nuclei. The Townes-Dailey model [9], attributing the field gradient at a nucleus solely to the population of the atomic valence p orbitals centered at that nucleus, has been shown to work well for oxygen-halogen compounds [10]. For $(ClO_4)_2^-$, the model predicts a vanishing contribution to the field gradient from the hole. Hence the observed field gradients indicate that the distortion of the individual ClO_4^- tetrahedra associated with the self-trapping of the hole is very limited. Rather, the dominant distortion appears to be a reduction of the distance between the two ions as observed for the analogous V_K-type defects (e.g. Cl_2^-) in alkali halides [11].‡ A thermally activated migration of [⊕], observed above 30K, is probably induced by a lattice mode that modulates the spacing between adjacent anions, thus enabling the hole to jump to new, equivalent positions in the lattice.

‡ Dimeric species of the type $(ClO_4)_2^-$ have been termed σ* radicals [12] on the assumption that bonding and antibonding combinations of the highest occupied molecular orbital on either component are responsible for the bond, as observed in noble gas ions like $(He)_2^+$.

$(ClO_4)_2^-$ has a broad optical absorption band centered at 660 nm. Excitation within this band leads to migration of the holes in the lattice, which strongly indicates that the absorption arises from a charge transfer transition, in which an electron is transferred from the medium to $(ClO_4)_2^-$ thereby displacing the self-trapped hole one lattice distance. This interpretation suggests in turn that the singly occupied electronic level of $(ClO_4)_2^-$ lies 2 eV above the upper edge of the valence band, and, as the band gap in $KClO_4$ is ~ 6 eV, $(ClO_4)_2^-$ may therefore have an 'oxidation potential' of approximately 4 eV.

ii. The electron-excess species ClO_4^{2-} is the self-trapped electron [e⁻]. The Q tensor of the chlorine nucleus indicates a very substantial distortion of ClO_4^- associated with the self-trapping of the electron. In fact, the field gradient at chlorine is the same as that observed in ClO_3^-. This, combined with the observation of non-parallel principal axes of the terms in $\mathcal{H}^s$ and of a g tensor consistent with O⁻ in an axial crystal field, have led to the conclusion [13] that ClO_4^{2-} should be represented as the complex $[O^-,ClO_3^-]$. The orientation of the principal axes in relation to the geometry of the host lattice suggest that O⁻ is displaced by 0.6 - 0.8 Å. The principal distortion of ClO_4^{2-} is, hence, a rather drastic elongation of one Cl–O bond. The model implies that the electron spin resides in a 2p orbital on O⁻. The g tensor is accounted for in terms of an axial crystal field, splitting the singly occupied $2p_z$ level from the degenerate $2p_x$, $2p_y$ levels by an amount ΔE. The crystal field is a simplified one-parameter representation of the interaction (including electronic overlap) of O⁻ with the associatedClO_3^-, which gives rise to the observed hyperfine splitting, as well as with the adjacent section of the 'cavity wall' consisting of one ClO_4^- and three K⁺. The interaction with the cavity wall results in a largely unresolved superhyperfine splitting. The formulation [e⁻] = $[O^-,ClO_3^-]$ has recently been further substantiated by ¹⁷O labeling: Only one ¹⁷O hyperfine splitting is observed, and the apparent spin density at the ¹⁷O nucleus is the same as seen in O⁻ trapped in NaOH glass [14].

In accordance with the large internal distortion, [e⁻] is stationary in the lattice below 270K.

$[O^-,ClO_3^-]$ has an intense optical absorption band at 252 nm, polarized along the z axis of g, which has been assigned to the charge-transfer transition [15]

$$O^-(^2\Sigma_u^+) + ClO_3^- \rightarrow O(^1D, \Sigma_g^+) + ClO_3^{2-}, \tag{1}$$

where the symbol attached to oxygen indicates the state in the axial crystal field. This assignment implies that the charge-transfer configuration $[O,ClO_3^{2-}]$ contributes significantly to the ground state of the complex.

iii. The identification of the complex $[ClO_2,O_2]$ also rests on the field gradient at the chlorine nucleus. It was noted that with one choice of the common arbitrary sign of Q and A, Q became virtually identical to the Q tensor of ClO_2. However, A would then correspond to a *negative* spin density at the chlorine nucleus, being related to the A tensor of ClO_2 simply as $A = -\frac{1}{3} A(ClO_2)$. The complex $[ClO_2,O_2]$ accounts for these observations if it is assumed that the spin doublet component (ClO_2) and the spin triplet component (O_2) are coupled to an overall doublet ground state by an isotropic exchange interaction [7]. It may be noted that the g tensor of the complex is also correctly represented in terms of this model. The interaction between ClO_2 and O_2 is too weak to modify the molecular orbitals significantly, only the spin distribution is governed by the exchange.

iv. The structure of the complex $[ClO_2^-,O_2]$ was obtained indirectly *via* that of the bromine analog $[BrO_2^-,O_2]$ [8]. The ESR signals of these complexes reveal the presence of an immobilized, but otherwise almost unperturbed oxygen molecule in its spin triplet ground state, but convey no information on the associated fragment. However, the extreme anisotropy of the signals (sweeping the field range 1.5 - 11 kG at 9 GHz) makes the signals very sensitive to minor differences in the environment of O_2. It was noted [8] that very similar ESR signals from O_2 could be produced by photolysis of $KClO_4$ (at $\lambda = 185$ nm), of $KBrO_4$ (at $\lambda = 265$ nm), and of $KBrO_4$ embedded in $KClO_4$ ($KClO_4$: $KBrO_4$) (at $\lambda = 229$ nm). These signals also closely resembled those obtained from $KClO_4$ and $KBrO_4$ after radiolysis below 30 K. It was concluded that the same type of complex was responsible for the signal in all cases. In the photolyzed samples of $KBrO_4$ (and $KClO_4$: $KBrO_4$) IR bands at 713 (717) cm^{-1} and at 1552 (1554) cm^{-1} were seen to respond in exactly the same way as the ESR signal from O_2 to thermal and optical bleaching, thus linking the IR bands to the complex. The bands at 713-717 cm^{-1} were assigned to the v_1 vibration of BrO_2^- while the bands at 1552-1554 cm^{-1} were taken to represent an oscillation of charge between BrO_2^- and O_2, induced by the vibration of O_2. Thus the structure associated with the ESR signal from O_2 is the charge-transfer complex [BrO_2^-,O_2]. By the similarity of the ESR signals, the O_2 complex in $KClO_4$ was identified as [ClO_2^-,O_2], also a charge-transfer complex.

3.2. INFLUENCE OF THE HOST LATTICE ON THE SPECTROSCOPIC PROPERTIES OF RADICALS

3.2.1. *Equivalent and Inequivalent Configurations.*

In general, the most tangible effect of the host lattice is that the radical or complex is formed in one or several well-defined orientations. Depressions and protrusions in the wall of the cavity in which the radical is trapped make the potential sufficiently nonuniform to orient the radical. The orientational energy may have several minima, however, resulting in the formation of distinct configurations. *Equivalent configurations* arise from sets of such minima related by the symmetry operations of the point group of the site. Moreover, magnetically distinct sites in the unit cell may exist, implying (regardless of the symmetry of the radical) the presence of sets of equivalent configurations related by symmetry operations involving translations. The principal values of the spin Hamiltonians corresponding to equivalent configurations are identical, the only distinction residing in the orientation of the principal axis.

The spin Hamiltonians corresponding to *inequivalent configurations* of a radical or complex have similar rather than identical principal values. Moreover, the thermal stabilities of inequivalent configurations are different. Inequivalent configurations may arise whenever the nominal symmetry of the parent anion is higher than both the symmetry of the radical and the site symmetry. Then the distortion associated with the formation of the radical may take several directions, which are equivalent with respect to the anion itself, but are inequivalent with respect to the lattice.

The $KClO_4$ crystal is orthorhombic, belonging to the space group Pnma [16], with four molecules in the unit cell. Two magnetically distinct anion sites exist, while the site symmetry is C_s. Hence asymmetric radicals have four equivalent configurations, two for each distinct site, whereas symmetric radicals conforming to the site symmetry have only the two equivalent configurations that arise from the distinct sites.

The self-trapped electrons in $KClO_4$ exhibit eight distinct ESR signals for a general orientation of B_o [6], corresponding to three inequivalent configurations labeled A, A' and B. As noted above, the distortion associated with the self-trapping may be visualized as an elongation of one Cl–O bond. Distortion along the bonds lying in the mirror plane through ClO_4^- leads to the symmetric configurations A and A', while distortions along the out-of-plane bonds, which are equivalent by the site symmetry, lead collectively to the asymmetric configuration B. The configurations A' and B are metastable with respect to configuration A, the relaxations A' → A and B → A occuring at 50K and 100K, respectively. A, A', and B are produced in the proportion 2:1:3. Hence the yields of symmetric and asymmetric configurations are equal, suggesting that the Cl–O bond to be stretched is chosen at random. The unequal yields of A and A' probably result from a radiation-induced partial relaxation of the metastable A'.

The complexes $[ClO_2,O_2]$ are formed in two symmetric and two asymmetric configurations labeled C, C'', and C', C''', respectively [6]. C', C'', and C''' are metastable with respect to C, the relaxations occuring between 30 and 45K.

Inequivalent configurations imply different external perturbations acting on the radical just as if the radical were embedded in different host lattices. Accordingly, the difference between corresponding principal values of $\mathcal{H}^s$ for inequivalent configurations may be taken to indicate the sensitivity to varying environments.

In the three configurations of $[e^-] = [O^-,ClO_3^-]$, the observed hyperfine splitting, associated with a spin density on ClO_3, varies linearly with the splitting ΔE of the 2p levels of O^- [17]. This corroborates the assumption that both arise from the interaction between O^- and ClO_3^- and, hence, depend on their distance. As the metastable configurations A' and B both have larger hyperfine splittings (and values of ΔE) than has the stable configuration A (Table 1), the principal interaction between O^- and ClO_3^- is repulsive so that their separation is determined by the walls of the cavity. The observed spin Hamiltonian may therefore be taken as a measure of the space available to the complex in the actual configuration.

While the spin Hamiltonians of the electron-excess system $[O^-,ClO_3^-]$, which appear to exert a pressure on the cavity walls, depend markedly on the orientation within the cavity, the spin Hamiltonians of the four configurations of $[ClO_2,O_2]$ are very similar (Table 1). This reflects that the precise magnitude of the exchange interaction between O_2 and ClO_2 has no influence on the ESR signal. Accordingly, as long as the two components are not squeezed so hard together that the orbitals of ClO_2 become significantly deformed, the spin Hamiltonian does not depend on the distance between the components or, equivalently, on the space available to the complex. The similarity of the spin Hamiltonians indicates, therefore, that the components of $[ClO_2,O_2]$ have sufficient room in the cavity to maintain their integrity regardless of the orientation.

The complexes $[O^-,ClO_3^-]$ and $[ClO_2,O_2]$ exemplify a difference generally observed between electron-excess and hole species: The excess electron resides in a diffuse antibonding orbital which tends to increase substantially the effective size of the anion, thus giving rise to a rather strong interaction with the cavity walls, whereas the hole tends to decrease the effective size of the anion and to decouple it from the lattice. In agreement with this picture, superhyperfine splittings arising from the nuclei of adjacent ions are frequently observed in the ESR signals from electron-excess species.

3.2.2. Relaxation of Metastable Configurations and Photochemical Conversions

The roughness of the cavity wall prevents the spontaneous relaxation of metastable configurations by rotation of the entire trapped radical or complex. Hence, the relaxation requires some thermally induced flexibility of the trapped species itself. The observed relaxations of $[ClO_2,O_2]$ involve rotation of the individual components and shifts in their relative position, whereas the relaxations of

$[O^-, ClO_3^-]$ appear essentially to be changes in the direction of the distortion. Vibrational excitation deforms the nuclear geometry of the metastable configuration, producing momentarily a 'transition geometry' which allows the distribution of the excess electron and the direction of the distortion to switch to those characterizing the stable configuration. According to this description, it is highly unlikely that metastable configurations relaxing below 100K could survive if they were formed in vibrationally excited states. Thus the mere existence of such configurations demonstrates the extreme efficiency of the 'cooling' by energy transfer to the lattice: Most of the excess energy released in the formation process is dissipated while the nuclei of the radical or complex are still moving on to their new equilibrium positions.

A set of photochemical conversions, induced by optical excitation of the complexes $[O^-, ClO_3^-]$ within the charge-transfer band ($\lambda_{max} = 252$ nm), has provided a further demonstration of the near-instant dissipation of excess energy [15]. Bleaching of the irradiated $KClO_4$ crystal at $\lambda = 254$ nm caused a rapid decay of the ESR signal from the configuration B while the signals from the configurations A and A' increased. The total amount of $[O^-, ClO_3^-]$ decreased very slowly during the bleaching, indicating that the main relaxation path of the charge-transfer state $[O(^1D), ClO_3^{2-}]$ leads back to $[O^-, ClO_3^-]$. Thus the excitation energy (5 eV) is absorbed by $[O^-, ClO_3^-]$ and dissipated again without producing any net chemical change. However, the final configuration of the complex may be different from the initial configuration. Irrespective of the initial relative amounts of the three configurations their proportion approached A : A': B = 1:1: 0.1, demonstrating the occurrence of the efficient photolytic conversions B $\rightarrow$ A, B $\rightarrow$ A', A $\rightarrow$ A', and the less efficient conversions A $\rightarrow$ B and A' $\rightarrow$ B. The conversions are attributed to electronic rearrangement within the charge-transfer state: $O(^1D)$ becomes bound to ClO_3^{2-} from which another oxygen atom is expelled as O^-, carrying with it most of the excess energy. The energy must be very rapidly dissipated since the observed formation of the metastable configuration A' in high yield is inconsistent with vibrational excitation of the final configuration. The low relative rates of the conversions of the symmetric configurations A and A' into the asymmetric configuration B are interpreted as the result of a symmetry-based selection, occurring in the excitation process, of the oxygen atom to become the new O^-. Accordingly, the expulsion of O^- from ClO_3^{2-} must begin immediately upon the transfer of the electron since vibrations of ClO_3^{2-} would otherwise obliterate the memory of the selective excitation.

3.3. THERMAL ELECTRON-HOLE REACTION.

The self-trapped holes disappear rapidly at 32K in a reaction with the self-trapped electrons [8,18]

$$[\oplus] + [e^-] \rightarrow \text{products with integral spin.} \tag{2}$$

The products include the complex $[ClO_2^-, O_2]$ since the ESR signal from this complex is seen to increase at a rate proportional to the rates of decay of $[\oplus]$ and $[e^-]$ at any temperature in the range 15-32K. Moreover, the complex $[ClO_3^-, O_2, ClO_3^-]$ is probably formed in reaction (2), which suggests that the initial product is an oxygen atom that may react either with the cognate ClO_3^- or with a ClO_4^-

in the wall of the cavity, abstracting in both cases an additional oxygen atom:[†]

$$[\oplus] + [e^-] \rightarrow [ClO_3^-,O] + 2ClO_4^- \rightarrow \begin{cases} [ClO_2^-,O_2] + 2ClO_4^- & (2a) \\ \\ [ClO_3^-,O_2,ClO_3^-] + ClO_4^- & (2b) \end{cases}$$

Reaction (2) occurs when $[\oplus]$ performing thermally activated migration in the lattice encounters $[e^-]$ the migration of $[\oplus]$ being the rate-determining process. By annealing at any fixed temperature between 15 and 23K, the signal heights of $[e^-]$ and $[\oplus]$ decrease rapidly for the first few minutes and then very slowly, whereas raising the temperature a few degrees causes a new rapid decrease. This indicates that $[\oplus]$ has varying barriers to migration, the Coulomb interaction causing certain $[\oplus]$ - $[e^-]$ configurations to collapse at temperatures at which configurations with larger separations are virtually stable. The fraction of $[\oplus]$ decaying below 23K is 0.3 at low X-ray dose and increases with increasing dose, in agreement with this interpretation. 60-70% of the initial amount of $[e^-]$ survives the total decay of $[\oplus]$.

3.3.1. Kinetics

Experimental information on the spatial distribution of self-trapped electrons and holes has been obtained indirectly from a quantitative study of the electron-hole reaction [18]. Reaction (2) may be divided in two parts:

i. Reaction between $[e^-]$ and $[\oplus]$ formed simultaneously along the same track (i.e. descended from the same X-ray photon), and

ii. Reaction between $[e^-]$ and $[\oplus]$ belonging to different tracks, which may arise at different times.

The two parts can be distinguished experimentally because the metastable and stable configurations of $[e^-]$, being spectroscopically distinct but chemically identical, offer a labeling of $[e^-]$ according to the time of formation. In a series of irradiations and intervening annealings at 32 and 100K of the same crystal it was noted, that the metastable configurations $[e^-]$ A' and $[e^-]$ B (transformed into the stable configuration A by annealing at 100K) were in each step formed and consumed in reaction (2) in the proportion 1:3. It was assumed, therefore, that the new $[e^-]$ were in each step formed and consumed in the proportion A : A' : B = 2:1:3 as observed after the first irradiation. The proportion of A (old and new) to A' and B (new only) consumed by reaction (2) after the n'th irradiation was seen to increase with n, indicating that reactions of $[e^-]$ and $[\oplus]$ belonging to different tracks competed more successfully with reactions occurring within the newly formed tracks. However, the new $[\oplus]$ maintained a marked preference for those $[e^-]$ belonging to the same track, indicating a strong spatial correlation between $[\oplus]$ and $[e^-]$ arising simultaneously. Analysis of these proportions allowed determination of the probability, P_{dif}, for reaction in the n'th step between $[e^-]$ and $[\oplus]$ belonging to different tracks as a function of the macroscopic concentration C^0 of $[e^-]$ at the end of the n'th irradiation.

[†] The postulated intermediate $[ClO_3^-,O]$ merely serves to rationalize the formation of the observed products in terms of reactions of well-defined chemical species. A more realistic description might run as follows: During the relaxation following the recombination, the system moves on a hypersurface that encompasses two troughs leading to $[ClO_2^-,O_2]$ and $[ClO_3^-,O_2,ClO_3^-]$, respectively, and possibly also a third trough leading back to (3) ClO_4^-.

3.3.2. Model of the electron-hole reaction

The information contained in the function $P_{dif}(C^o)$ was employed to determine the values of two adjustable parameters in a simple model of the spatial distribution of $[e^-]$ and $[\oplus]$, treating reaction (2) as a diffusion-controlled bimolecular process: The migration of $[\oplus]$ is considered as a diffusion in an isotropic, continuous medium. The diffusion is assumed to be unaffected by the presence of other defects when the distance r to the nearest $[e^-]$ exceeds a critical value r_o, whereas spontaneous reaction occurs for $r \leq r_o$. The complex spatial distribution of $[e^-]$ and $[\oplus]$ along the tracks is represented by assuming that $[e^-]$ and $[\oplus]$ are formed in pairs (as they are indeed) and taking the probability $\rho(r)$, that $[e^-]$ and $[\oplus]$ belonging to a pair have initially the distance r, as

$$\rho(r) = \begin{cases} 4\pi r^2 M \exp[-(r/\lambda r_o)^2], & r > r_o \\ 0, & r \leq r_o \end{cases} \tag{3}$$

where M is a normalization constant and λ is a parameter determining the spatial extent of the distribution. Except for the correlation between $[e^-]$ and $[\oplus]$ belonging to the same pair, a random distribution is assumed. From a modified version of Waite's treatment of diffusion-controlled reactions [19] the probability α for reaction between $[e^-]$ and $[\oplus]$ which *do not* belong to the same pair was calculated as a function of C^o with different values of r_o and λ. α was taken to represent, within the model, the probability of reaction between $[e^-]$ and $[\oplus]$ belonging to different tracks. Agreement with the experimentally determined P_{dif} was obtained with $\lambda = 0.8$ and $r_o = 62$ Å. The resulting pair-distribution function, shown in Figure 1, suggests that the collapse of $[e^-] - [\oplus]$ configurations having separations in the range 50-60 Å accounts for the thermal decay observed below 23K, which corresponds, then, to only one or two jumps of $[\oplus]$.

FIG.1. Pair-distribution function $\rho(r)$ calculated from Eq.(3) with $\lambda = 0.8$ and $r_o = 62$Å.
(Reproduced with permission from reference 18).

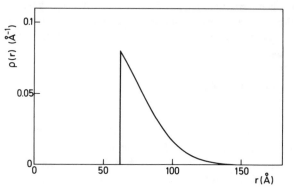

3.4. SPONTANEOUS ELECTRON HOLE-REACTION

The spatial distribution of $[e^-]$ and $[\oplus]$ existing at the end of the irradiation indicates that a substantial fraction of electrons and holes have recombined spontaneoulsy, in accordance with the observation that 80% of the total amount of the recombination product $[ClO_2^-, O_2]$ arise *during* the irradiation at 4K [8]. At low dose most of the recombination takes place within the dense clusters belonging to the individual tracks. However, at increasing dose spontaneous reactions of new electrons and holes with preexisting species of opposite effective charge become important. The latter type of spontaneous electron-hole process was studied by observing the accumulation of paramagnetic species during the irradiation [8]. In Fig. 2 the yields obtained at 26K are given in units of C_1, the concentration of [e⁻] after the first minute of irradiation. C_1, determined from the intensity of the ESR signal, was 2.4×10^{-6} mol [e⁻]/mol ClO_4^-. The differential yield (per unit X-ray dose) depends for each species on the dose in a characteristic way. Most remarkably, the differential yield of [⊕] becomes negative after irradiation for 45 minutes, indicating that the rate of formation of new [⊕] has by then sunk below the rate of recombination of electrons with already existing [⊕]. The maximum amount of [⊕] corresponds to the concentration $C_+ = 7 \times 10^{-6}$ mol [⊕]/mol ClO_4^-. This is in qualitative agreement with the value of the critical distance between [⊕] and [e⁻] (Figure 1), which defines a reaction zone of volume 10^6 Å³ around each [e⁻]. At 45 minutes, the combined reaction zones of all [e⁻] comprise 75% of the volume of the crystal if overlap is neglected, leaving little space to formation of new, stable [⊕]. The differential yield of [e⁻] (A) similarly becomes negative at high dose (Fig. 2) if the concentration is made abnormally high by thermal relaxation of the metastable configurations A' and B. Hence the rate of reaction of new holes with already existing [e⁻] at this stage exceeds the rate of formation of new [e⁻] in the configuration A, representing one-third of the new [e⁻].

FIG.2. Concentrations C_i of paramagnetic species in $KClO_4$ at 26K vs irradiation time t. After irradiation for 135min, the crystal was annealed at 110K until reaction (2) and the relaxation of metastable configurations of [e⁻] and [ClO₂,O₂] were complete, and then reirradiated at 26K. The concentrations are given in units of C_1, the yield of [e⁻] after the first minute of irradiation. Circles ● represent [e⁻](B), triangles ▼ represent $0.5 \times$ [e⁻] (A), squares ■ represent $2 \times$ [⊕], and triangles ▲ represent $4 \times$ [ClO₂,O₂](C'). B and C' are metastable configurations. Curves are calculated from the model outlined in section 3.4.
(Reproduced with permission from reference 18).

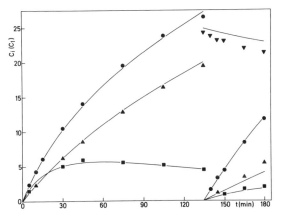

The spontaneous reactions of new electrons and holes with preexisting species of opposite effective charge during irradiation may conceivably take place before as well as after self-trapping. Thus the reactions should be written:

$$e^- + [\oplus] \rightarrow P \tag{4a}$$

$$\oplus + [e^-] \rightarrow P \tag{4b}$$

$$[\oplus] + [e^-] \rightarrow P, \qquad\qquad (4c)$$

where e^- and $\oplus$ are the radiation-induced 'free' electrons and holes. The process (4c) here represents the reaction of new self-trapped holes with preexisting self-trapped electrons and *vice versa*, which is analogous to the thermal reaction. The three reactions (4a)-(4c) cannot be distinguished experimentally. However, their relative importance was inferred from the simulation of the differental yields of all species shown in Figure 2. The simulation was based on the assumptions that electrons and holes are formed at constant rate and with equal probability per unit volume independently of preexisting radiation-induced species, and that their probability of reaction is described by the concentration of the pertinent species and an associated reaction volume. It turned out that the combined reaction volume of the processes (4a)-(4c) fitting the observed yields was 1.3×10^6 Å^3 which, the crudeness of the model considered, agrees with the volume of reaction (4c) alone (1.0×10^6 Å^3). Thus it was concluded that the dominant part of the reaction of new electrons and holes with preexisting species is the collapse of unstable configurations of $[e^-]$ and $[\oplus]$, whereas free electrons and holes have only a low reactivity toward $[\oplus]$ and $[e^-]$, respectively. This result was furthermore taken to suggest that the probability of recombination of electrons and holes prior to self-trapping is also low, so that the spontaneous recombination occurring within the dense clusters is similarly a collapse of $[e^-]$ - $[\oplus]$ configurations.

3.4.1. Mechanism of formation of [ClO₂,O₂] and [ClO₃,O₂]

The complex $[ClO_2,O_2]$ is formed in high yield during irradiation at 26K (Figure 2), whereas no change of the amount of $[ClO_2,O_2]$ has been observed to accompany the thermal decay of $[\oplus]$ at 32K. Therefore, in the model for accumulation of radiation-induced defects outlined above, the complex $[ClO_2,O_2]$ was considered as an alternative, immobile form of the self-trapped hole, which influenced the electron-hole processes in an indirect way. However, the presence of copious amounts of $[ClO_2^-,O_2]$ combined with the observed formation of the complex $[BrO_2,O_2]$ in a reaction of $[BrO_2^-,O_2]$ with $[\oplus]$ eventually led to the conclusion [8] that $[ClO_2,O_2]$ may arise as a tertiary species by a similar oxidation of $[ClO_2^-,O_2]$:

$$[\oplus] + [ClO_2^-,O_2] \rightarrow [ClO_2,O_2] \qquad\qquad (5)$$

However, most of the observed $[ClO_2,O_2]$ apparently stem from the spontaneous collapse of unstable configurations consisting of two $[\oplus]$ and one $[e^-]$

$$2[\oplus] + [e^-] \; (=[O^-,ClO_3^-]) \rightarrow [O,ClO_3] \rightarrow [ClO_2,O_2], \qquad\qquad (6)$$

rather than from the consecutive reactions (2a) and (5). The importance of reaction (6) was inferred from the observed formation of the complex $[ClO_3,O_2]$, one of the minor radiolytic products in $KClO_4$ [20]. $[ClO_3,O_2]$ is structurally similar to $[ClO_2,O_2]$, but unlike $[ClO_2,O_2]$, $[ClO_3,O_2]$ is formed almost exclusively in a thermal reaction coinciding with the decay of $[\oplus]$ at 32K. Hence $[ClO_3,O_2]$ is taken to arise by oxidation of the complex, $[ClO_3^-,O_2,ClO_3^-]$ formed in reaction (2b). As $[ClO_3^-,O_2,ClO_3^-]$ competes successfully with $[e^-]$ and, in particular, with $[ClO_2^-,O_2]$ for the holes surviving in the low-density regions, reaction (2b) probably accounts also for a substantial fraction of the recombination occurring within the clusters. The negligible spontaneous yield of $[ClO_3,O_2]$ indicates, then, that the relaxation time associated with reaction (2b) exceeds the life time of $[\oplus]$ in the clusters, which precludes the subsequent oxidation of $[ClO_3^-, O_2, ClO_3^-]$ to $[ClO_3, O_2]$. Accordingly, the formation of $[ClO_3, O_2]$ is confined to the low-density regions where the necessary time-lapse between recombination and oxidation is ensured.

With the reasonable assumption that the relaxation times of reactions (2a) and (2b) are similar, both being associated with oxygen abstraction and formation of O_2, it follows that the life time of $[\oplus]$ in the clusters is also too short to allow the formation of $[ClO_2, O_2]$ via $[ClO_2^-, O_2]$. Hence the high spontaneous yield of $[ClO_2, O_2]$ is attributed mainly to reaction (6), in which the time-consuming formation of O_2 occurs as the final relaxation.

The investigations discussed above indicate, that the formation and accumulation of paramagnetic species in $KClO_4$ irradiated with X-rays at low temperature may be represented in terms of generation and subsequent reactions of self-trapped electrons and holes, $[e^-]$ and $[\oplus]$. Thus recombination (reaction (2)) accounts for the formation of secondary species with integral electron spin, whereas subsequent oxidation by reaction with $[\oplus]$ produces tertiary radical species (reaction (5)). Both recombination and oxidation are taken to comprise a thermal part as well as a spontaneous part, associated with low-density regions and clusters of excitations, respectively. However, electron-transfer processes leading to the collapse of unstable configurations of $[\oplus]$ and $[e^-]$ in the clusters are too fast to allow the occurrence of genuine consecutive reactions involving several $[\oplus]$. Instead, recombination and oxidation may occur simultaneously as the reaction of two $[\oplus]$ with one $[e^-]$ (reaction (6)).

4. Comparison with the Radiolysis of $KBrO_4$, $KClO_3$, and $KBrO_3$.

A brief discussion of the low-temperature radiolysis of solid $KBrO_4$, $KClO_3$, and $KBrO_3$, formulated in accordance with the scheme developed for $KClO_4$, may serve to illustrate the applicability of this scheme as well as to indicate the kinds of modifications that may arise from the chemical characteristics of the actual anion and from the geometry of the actual lattice.

4.1. $KBrO_4$

Like $KClO_4$, $KBrO_4$ belongs to the space group Pnma with four molecules in the unit cell [21]. The lattice geometries of $KBrO_4$ and $KClO_4$ are so similar that any structural differences between corresponding radiation-induced species in the two crystals may be attributed to the electronic properties of the halogen atoms.

The self-trapped hole $(BrO_4)_2^-$ and the secondary and tertiary complexes $[BrO_2^-, O_2]$ and $[BrO_2, O_2]$ are structurally very similar to their chlorine analogs [7,8,22]. In fact, the self-trapped electron BrO_4^{2-} exhibits the only significant deviation: The distortion of this species with respect to the tetrahedral shape of the parent BrO_4^- is much less pronounced than the distortion of ClO_4^{2-}. Moreover, the excess electron in BrO_4^{2-} is located on two oxygen atoms and the bromine atom rather than on one oxygen atom, indicating that two Br–O bonds are stretched to yield a C_{2v} geometry [13,23]. Two inequivalent configurations of BrO_4^{2-}, one symmetric and one asymmetric, are produced in equal amounts. At 50K the asymmetric configuration relaxes into the symmetric configuration. In spite of the structural difference between BrO_4^{2-} and ClO_4^{2-}, both react with O_2 to form $[O_3^-, XO_3^-]$ by annealing above 240K [13]. Hence BrO_4^{2-} as well as ClO_4^{2-} are chemically equivalent to O^-.

The thermal recombination of self-trapped electrons and holes occurs in $KBrO_4$ at 70K. Formation of the complexes $[BrO_2^-, O_2]$ and $[BrO_2, O_2]$ in thermal processes corresponding to reactions (2a) and (5) has not been observed in $KBrO_4$, indicating that both arise predominantly from spontaneous processes within the dense clusters of excitations. The electron-hole recombination probably produces also the complex $[BrO_3^-, O_2, BrO_3^-]$. However, as BrO_3^- in $KBrO_4$ does not trap the migrating holes, no tertiary complex analogous to $[ClO_3, O_2]$ is formed.

4.2. SELF-TRAPPED ELECTRONS AND HOLES IN $KClO_3$ AND $KBrO_3$

$KClO_3$ is monoclinic, belonging to the space group $P2_1/m$ with two molecules in the unit cell [24]. As in $KClO_4$, the symmetry of the anion site is C_s. The anions are arranged in strings along a with the oxygen atom in the mirror plane acting as a bridge between adjacent chlorine atoms.

The radiation-induced hole becomes self-trapped in the form of the radical ClO_3 [25], which retains the orientation of ClO_3^- in the lattice. In spite of the low symmetry of the site, the ESR signal reveals no deviation from the C_{3v} symmetry of the free ClO_3 radical, in which the hole occupies an antibonding orbital belonging to the representation A_1. Hence the bond angle and bond length of ClO_3 are sufficiently different from those of ClO_3^- to allow self-trapping of the hole on a single anion without further distortion. ClO_3 exhibits a broad optical absorption band with $\lambda_{max} = 560$ nm, which has been attributed to charge transfer from the medium in agreement with the observation of a photo-induced electron-hole recombination [15].

The electron becomes self-trapped as the complex $[O^-, ClO_2^-]$, formed like the analogous complexes in $KClO_4$ by stretching of one Cl–O bond [17]. In the asymmetric configuration B, resulting from stretching of one of the bonds inclined to the mirror plane, O^- interacts almost exclusively with the associated ClO_2^-, whereas in the symmetric configuration C, with the bridging oxygen atom carrying the excess electron, a marked electronic overlap with the adjacent anion is observed. Accordingly, configuration C is represented as the complex $[ClO_2^-, O^-, ClO_3^-]$. In addition to B and C, a related, symmetric complex $D = [ClO_3^-, O^-, ClO_3^-]$ is formed during irradiation at 26K in a yield equal to that of C. At 35K the metastable configuration B relaxes into C

(B)   (C)

and at 110K the further relaxation C → D occurs, which closely resembles the relaxation $[e^-]$ (A')
→ $[e^-]$ (A) in $KClO_4$,

(C)   (D)

The observation of the latter process and the initial yield of the secondary product D suggest that the stretching of the Cl–O bond lying in the mirror plane may induce the relaxation C → D. The relaxations of metastable configurations, continuing with an additional transfer of bridging oxygen atoms between adjacent anions at 125K, leads to migration of the excess charge along the crystallographic axis a, a phenomenon that was exploited to study the process $O^- + BrO_3^- \rightarrow BrO_4^{2-}$ [26].

The electronic and optical properties of the O^- complexes are analogous to those of $[O^-,ClO_3^-]$ in $KClO_4$ [15]. All configurations in $KClO_3$ absorb strongly near 254 nm, which is attributed to charge-transfer from O^- to ClO_3^- according to Eq. (1). Moreover, the configurations B and C exhibit strong absorption bands at 383 and 386 nm, respectively, which are similarly assigned to charge transfer from O^- to ClO_2^-. Excitation within these bands induces the relaxation processes B → C (366 nm), C → D (254, 366 nm) as well as the regeneration of metastable configurations by the reverse processes D → C (254 nm) C → B (254, 366 nm). As in $KClO_4$, the excitation energy is dissipated without causing genuine chemical transformations or vibrational excitation of the complexes. However, in $KClO_3$ the photochemical conversions imply a limited migration of charge.

The $KBrO_3$ crystal is rhombohedral, belonging to the space group R3m, with one molecule in the unit cell [27]. The point group of the anion site is C_{3v}. As in $KClO_3$, the self-trapped hole in $KBrO_3$ is confined to a single anion [28], but unlike ClO_3 the resulting radical has a very small hyperfine splitting from the halogen nucleus and has no symmetry, which indicates that the hole enters an orbital that is degenerate in the symmetric configuration. Due to poor resolution of the ESR signal no detailed structural information has been obtained for this radical in $KBrO_3$. However, as BrO_3^*, the oxidation product of BrO_3^- embedded in $KClO_4$, has a similar hyperfine splitting and is also asymmetric, the structural data for BrO_3^* probably pertain also to the radical in $KBrO_3$. In BrO_3^* the hole occupies a nonbonding orbital located on two oxygen atoms, while only a small distortion of the nuclear geometry with respect to that of BrO_3^- is indicated by the observed shift of the field gradient at the bromine atom [28].

The radiation-induced electron also becomes self-trapped on a single anion. The resulting radical, BrO_3^{2-}, has approximately the shape of BrO_3^- although the symmetry is C_s rather than C_{3v}, the dominant distortion being a moderate elongation of one Br–O bond [29]. Hence BrO_3^{2-} is structurally similar to the self-trapped electron BrO_4^{2-} in $KBrO_4$.

4.3. ELECTRON-HOLE REACTIONS AND SECONDARY PRODUCTS

A simultaneous decrease of the ESR signals from [e⁻] and [⊕] above 90K in $KClO_3$ and above 50K in $KBrO_3$ indicates the occurrence of electron-hole recombination as a thermal reaction in both crystals [30]. The thermal recombination in $KBrO_3$ has been directly shown to yield the charge transfer complex $[BrO^-,O_2]$. The analogous complex $[ClO^-,O_2]$ is formed in $KClO_3$. Other paramagnetic radiolytic products identified by ESR in both crystals are the radical XO_2 [29,31] and the exchange-coupled complex $[XO,O_2]$, X = Cl,Br [30]. The formation of the complexes may be represented by the reactions

$$[\oplus] + [e^-] \rightarrow [XO^-,O_2] \tag{7}$$

$$[\oplus] + [XO^-,O_2] \rightarrow [XO,O_2] \qquad (8)$$

The oxidation (reaction (8)) has been observed in $KBrO_3$ as a thermal process when the precursor $[BrO^-,O_2]$ was generated by photolysis. However, $[XO,O_2]$ and also XO_2 arise normally in spontaneous processes during the irradiation which, in analogy with the spontaneous formation of $[ClO_2,O_2]$ in $KClO_4$ (reaction (6)), may be formulated

$$2[\oplus] + [e^-] \rightarrow [O,XO_2] + 2XO_3^- \rightarrow \begin{cases} [XO,O_2] + 2XO_3^- & (9a) \\ XO_2 + XO_4^- + XO_3^- & (9b) \end{cases}$$

ClO_2 formed in reaction (9b) has a symmetric position in the $KClO_3$ lattice with the molecular plane perpendicular to the mirror plane, indicating that it is the bridging oxygen which is transferred to the adjacent anion.

In $KClO_3$ a thermal reaction occurs that corresponds closely to the formation of $[ClO_3,O_2]$ in $KClO_4$: During the decay of $[\oplus]$ at $\sim 90K$, the complex $[ClO_2,O_2]$ is formed in two inequivalent, but very similar configurations, both with ClO_2 moieties having the symmetric position observed for the isolated ClO_2 [30]. The obvious interpretation is that $[ClO_2,O_2]$ arises from oxidation of $[ClO_2^-,O_2]$, formed in one of the processes

$$[\oplus] + [e^-] \, (C) \, (=[ClO_2^-,O^-,ClO_3^-]) \rightarrow [ClO_2^-,O_2,ClO_2^-] \qquad (10a)$$

$$[\oplus] + [e^-] \, (D) \, (=[ClO_3^-,O^-,ClO_3^-]) \rightarrow [ClO_2^-,O_2,ClO_3^-] \qquad (10b)$$

which are both analogous to reaction (2a) accounting for the formation of $[ClO_2^-,O_2]$ in $KClO_4$. The observed symmetric configurations of $[ClO_2,O_2]$ indicate that the precursors are also symmetric, which restricts the oxygen-abstraction involved in the formation of O_2 to the bridging oxygens lying in the mirror plane.

The results summarized here show that the principal features of the radiolysis of KXO_3 closely resemble those of the radiolysis of KXO_4, the formation of XO_2 being an exception attributable to the obvious fact that XO_3^-, but not XO_4^-, may accept an additional oxygen atom. The observation, that the self-trapped hole in KXO_3 is localized on one anion rather than shared equally by two anions as in KXO_4, similarly reflects a difference between the electronic structures of XO_3^- and XO_4^-.

The more subtle differences between the electronic properties of the halogen atoms, expressed in the order and spacing of the molecular electronic levels in XO_3^- and XO_4^-, are reflected in the dissimilar shapes of the self-trapped holes in $KClO_3$ and $KBrO_3$, and in the distortions associated with self-trapping of the electron: In $KClO_3$ and $KClO_4$, O^- is spontaneously expelled from the electron-excess species, while moderate distortions of BrO_3^{2-} and BrO_4^{2-} sufffice to stabilize the excess electron.

Features attributable to the structure of the lattice are clearly seen in the radiolysis of $KClO_3$: The 'delocalization' of excess charge over several anions, resulting in O^- complexes similar to those observed in $KClO_4$, as well as the associated formation of the complexes $[ClO_2^-,O_2]$ and $[ClO_2,O_2]$, and the thermal migration of charge at low temperature, all arise from the unique arrangement of anions with bridging oxygen atoms.

5. Reaction of [⊕] with Impurities in KClO₄

The formation of $[ClO_2, O_2]$ in reaction (5) may be considered as the one-electron oxidation of the radiation-induced impurity $[ClO_2^-, O_2]$. Similar oxidations by reaction with [⊕] have been observed for a range of anionic impurities in $KClO_4$ including ClO_3^- [32], BrO_3^- [28] and BrO_4^- [8] as well as products obtained from these ions and also from IO_4^- by selective photolysis with light of wavelengths at which $KClO_4$ does not absorb [8,33]. The photolyzed samples all exhibit ESR signals from O_2, indicating the formation of the charge-transfer complexes $[XO^-, O_2]$ and $[XO_2^-, O_2]$ from XO_3^- and XO_4^-, respectively. The complexes $[XO_2^-, O_2]$ dissociate to $XO_2^- + O_2$ by annealing above 240K, which enables O_2 to escape by diffusion through the lattice while XO_2^- is retained at the site of formation [34,35].

In each case the oxidation of the impurity R^- to the corresponding radical R occurs partly as a spontaneous process during the irradiation with X-rays at 10 or 26K and partly as a thermal post-irradiation reaction when the surviving [⊕] become mobile at 30-32K. As illustrated in Figure 3 for the impurities XO_3^- and $[XO^-, O_2]$, X = Cl, Br, the thermal part of the oxidation

$$[\oplus] + R^- \rightarrow R \tag{11}$$

progresses linearly with the electron-hole recombination (reaction (2)), which shows that also for reaction (11) the migration of [⊕] determines the rate of reaction. Hence reaction (11) has no significant energy of activation, in accord with its character as a genuine electron transfer process. If reaction (11) is taken to represent also the spontaneous part of the oxidation, the description becomes similar to that given above of the recombination process: To each impurity R^- is assigned a reaction zone of radius r_0^R. A hole becoming self-trapped within the reaction zone reacts spontaneously with R^-, thus contributing to the part of the oxidation occurring during the irradiation, whereas [⊕] formed at a distance $r > r_0^R$ from the nearest R^- cannot react until thermal migration has brought it within the reaction zone. The sizes of the reaction zones have not been determined. However, an estimate of the initial relative yields at 26K of the impurity radical ClO_3 and of [⊕] in $KClO_4$ containing 0.6 mol % $KClO_3$ suggests that the combined reaction zones of ClO_3^- occupy 40% of the volume of the crystal. Hence r_0^R for ClO_3^- is of the order 10 Å, indicating that [⊕] and ClO_3^- must be separated by at least two layers of anions to escape spontaneous reaction. Although detectable ESR signals have been observed from ClO_3 produced by oxidation of ClO_3^- in concentrations as low as 10^{-4} mol %, the concentrations of impurities have typically been chosen in the range 0.1 - 0.3 mol %, thus vastly exceeding the maximum concentration of [⊕] (7×10^{-4} mol %). Such high concentrations of R^- imply that even with a small reaction zone like that of ClO_3^-, a few thermally activated jumps suffice to bring [⊕] in contact with the nearest R^-. Hence the thermal part of reaction (11) might be appropriately described as the collapse of metastable configurations of R^- and [⊕] rather than as resulting from diffusion of [⊕] in a homogeneous medium. This observation helps to reconcile the assumption of a single process to represent the spontaneous as well as the thermal formation of radicals with a result derived from Figure 3: When two different impurities are present simultaneously, the relative thermal yields of the corresponding radicals may be very different from the relative spontaneous yields at 26K. Since the spontaneous yields of XO_3 and $[XO, O_2]$ shown in Figure 3 are similar, so are the total reaction volumes of their precursors according to the model. However, the concentration of XO_3^- is probably an order of magnitude higher than that of $[XO^-, O_2]$, so that on the average fewer jumps of [⊕] are required to reach XO_3^- which, therefore, wins the competition for [⊕] at the thermal stage.

FIG.3. Heights h (in arbitrary units) of the ESR signals from $[XO,O_2]$ (circles), XO_3 (triangles), and $[e^-] = [O^-,ClO_3]$ (squares) plotted vs the heights of the ESR signal from $[⊕] = (ClO_4)_2^-$. Signals were measured at 26K between successive annealings of $KClO_4:KXO_3$ crystals in the temperature range 28-35K after photolysis at 214 or 229 nm and subsequent irradiation with X-rays. The reactions proceeds from right to left. The linear plots show that the rates of formation of $[XO,O_2]$ and of XO_3 and the rate of decay of $[e^-]$ are all proportional to the rate of decay of [⊕] throughout the range of temperatures. The thermal relative increments in the amounts of XO_3 and $[XO,O_2]$ are ~130% and ~20%, respectively, indicating that $[XO^-,O_2]$ competes much more successfully with XO_3^- for [⊕] in the spontaneous part of reaction (11) than in the thermal part.
(Reproduced with permission from reference 33).

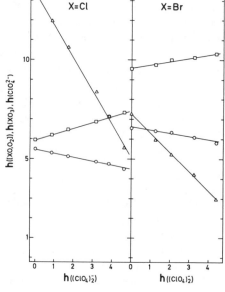

5.1. SHAPE AND POSITION OF OXIDATION PRODUCTS

The oxidation products of the anions XO_3^- and XO_2^- are the radicals XO_3 and XO_2, respectively, while the exchange-coupled complexes $[XO,O_2]$ and $[XO_2,O_2]$ arise by oxidation of the charge-transfer complexes $[XO^-,O_2]$ and $[XO_2^-,O_2]$, respectively. Hence the oxidation represented by reaction (11) is a gentle transfer of an electron from the impurity to the lattice, occurring without breaking or making of bonds. The resulting radical has in general approximately the same orientation in the lattice as the anion precursor and, moreover, metastable configurations of the precursor are often reflected in corresponding metastable configurations of the radical, which shows that the oxidation, just like the self-trapping processes, proceeds without significant vibrational excitation of the resulting species. It appears, therefore, that the equilibrium geometry of the radical R, which may differ considerably from that of the precursor R^-, is attained almost simultaneously with the electron transfer to the medium rather than by a relaxation following a vertical electronic process.

No significant spread of electron spin density from R to adjacent ions is observed, implying that the hole is strongly localized. Hence the singly occupied electronic level in R lies well above the level containing the hole in $(ClO_4)_2^-$ which, as noted above, presumably lies 2 eV above the upper edge of the valence band.

The parameters in the spin Hamiltonians for the trapped radicals resemble closely those inferred or known from other sources for the 'free' radicals. Thus the assumption, that the influence of the host lattice is largely limited to orientation of the trapped species, is generally justified. However, the radicals produced by oxidation of BrO_3^- present an interesting exception: The asymmetric radical BrO_3^* obtained in $KClO_4$ bears no resemblance to the symmetric analog of ClO_3 formed similarly by oxidation of BrO_3^- embedded in KNO_3. Hence the host lattice determines in this case the electronic ground state of the oxidation product. An interpretation in terms of the different shapes of the cavities available to BrO_3 in $KClO_4$ and in KNO_3 has been proposed [28].

The photolytical formation of $[XO^-,O_2]$ and $[XO_2^-,O_2]$ and the subsequent oxidation to $[XO,O_2]$ and $[XO_2,O_2]$, respectively, have offered a unique opportunity to study these complexes and also the radicals XO_2 in a convenient, 'inert' matrix. In addition, three significant results concerning the radiolytic and photolytic processes have emerged:

i. The structural correspondence observed between precursors and products in the one-electron oxidations strongly suggests that the formation of some radical R in a process correlated with the thermal decay of [⊕] implies the presence of a precursor R^- of similar structure. (The discussion of the formation of $[ClO_3,O_2]$ in $KClO_4$ (section 3.4.1.) and of $[ClO_2,O_2]$ in $KClO_3$ (section 4.3.) was based on this observation).

ii. The formation of $[XO,O_2]$ and $[XO_2,O_2]$ *can* occur in two steps *via* $[XO^-,O_2]$ and $[XO_2^-,O_2]$ which, combined with observed formation of these precursors in the radiolysis, corroborate the interpretation of $[XO,O_2]$ and $[XO_2,O_2]$ as tertiary species formed by reaction of two [⊕] with one $[e^-]$.

iii. The photolysis of XO_3^- and XO_4^- embedded in $KClO_4$ yields products obtained also by electron-hole recombination in the radiolysis of KXO_3 and KXO_4. This observation pertains also to the photolysis of KXO_3 and KXO_4, as shown by a detailed study of $KClO_3$ [36,37], identifying $[ClO,O_2]$ as the main photolytic product and $[ClO_2^-,O_2,ClO_2^-]$ as a minor product, and by the detection of the photolytic products $[BrO_2^-,O_2]$ and $[BrO^-,O_2]$ in $KBrO_4$ [8] and $KBrO_3$ [30], respectively. Hence the relaxations of excited species apparently follow the same path irrespective of the mechanism of excitation.

The equivalence of optical excitation and electron-hole recombination has an interesting corollary: The stable tertiary radicals and the corresponding number of $[e^-]$ surviving the thermal decay of [⊕] are the only distinctive radiolytic products. As the tertiary radicals arise primarily in the spontaneous collapse of configurations consisting of two [⊕] formed close to one $[e^-]$, the difference between radiolysis and photolysis is to a large extent determined by the initial yield of such configurations in the radiolysis.

6. Reversible Processes

Although radicals trapped in a crystal constitute singularities in the periodic lattice, they are of course coupled to the vibrational modes of the lattice that modulate the potential at the trapping sites or the distance to the adjacent ions. Thus the excitation of lattice modes leads to irreversible thermal processes like the relaxation of metastable configurations, discussed in section 3.2.2., and the electron-transfer reactions that occur when the migrating self-trapped holes enter the reaction zones of other species. In addition to these processes, the lattice modes induce various kinds of motion of the radicals, each reflected in a characteristic dependence on the sample temperature of the position and shape of the lines in the ESR signals. This motional distortion may at 77K be sufficiently severe to shroud the true symmetry and shape of the trapped species, and significant molecular motion has been shown to persist in some cases even at 4K [38]. Hence the identification of unknown species and the determination of static molecular properties may require observations at several sample temperatures. In this context, the ESR signals observed at high temperature are considered merely as distorted versions of those observed at low temperature. However, the motional effects are often a worthwhile study in their own right since they may convey valuable information (albeit implicitly) about the structure and dynamics of the radicals and, perhaps more important, also about the dynamics of the host lattice. Thus ESR signals from radical probes have been widely used to monitor phase transitions, particularly in ferroelectrics.

Rigid-body motion as well as intramolecular motion may contribute to the observed motional distortion of ESR signals. Apart from the ubiquitous libration, which modulates the orientation of the molecular axes with respect to B_o and, hence, reduces the apparent anisotropy of the spin Hamiltonian, the rigid-body motion may comprise jumps of the radical between equivalent positions in the cavity, rotation about a fixed axis and, ultimately, free tumbling in the cavity. The intramolecular motion comprises switching of the direction of distortions, modulation of the relative positions of components of complexes, and switching of secondary interactions (e.g. hydrogen bonds in ferroelectrics). Excitation of internal vibrations in the radicals may be involved in many of these processes, but do not in general affect the observed spectrum directly except at high temperatures.

6.1. JUMPS BETWEEN EQUIVALENT CONFIGURATIONS

Most of the spectacular motional effects observed in the ESR signals from radicals in KXO_3 and KXO_4 are associated with equivalent configurations of the type arising when the symmetry of the radical is lower than the symmetry of the site. Radicals in equivalent configurations yield distinct ESR signals at low sample temperature except for orientations of B_o that are symmetric with respect to the site. However, as the temperature is raised, these signals tend to merge into a single signal displaying the symmetry of the site rather than that of the radical. This transition may often be represented in terms of the 'jump-model': In analogy with the thermal relaxation of metastable, inequivalent configurations, the radical is assumed to jump rapidly across the barrier between the equivalent configurations with a probability that increases with increasing temperature. At low temperature, when the reciprocal life time Ω of the radical in a given configuration is small compared to the splittings Δv between the signals from different configurations, these signals are observed individually without distortion. However, when Ω at higher temperatures becomes comparable to Δv, the anisotropic terms in the spin Hamiltonians, which fluctuate with the jumps, induce rapid transitions between the electron spin states, thus strongly broadening the lines of the ESR signal. At still higher temperatures Ω becomes much larger than Δv and the distinction between the individual spin Hamiltonians is averaged out. Accordingly, the set of equivalent configurations exhibit a single signal with narrow lines arising in the process known as motional narrowing. The spin Hamiltonian representing the signal in the high temperature limit ($\Omega \to \infty$) is the algebraic mean of the spin Hamiltonians of the low-temperature configurations. The principal features of the motional effects discussed below all conform to the predictions of the jump model.

6.1.1. $[O_3^-, BrO_3^-]$ in $KBrO_4$

The complex $[O_3^-, BrO_3^-]$, formed in $KBrO_4$ by addition of O_2 to BrO_4^{2-} at 240K, has no symmetry[13]. Accordingly, two distinct ESR signals are observed for each anion site at 26K, corresponding to a pair of equivalent configurations connected by reflection in the mirror plane through the site. These signals merge as the temperature is raised, and at 220K a single, well-resolved signal displaying C_s symmetry is observed. At intermediate temperatures the signals are broadened beyond detection except for B_o lying in the mirror plane, where also the positions of the lines depend only slightly on the temperature. The Q tensors of the BrO_3^- moiety in the otherwise asymmetric low-temperature configurations almost conform to the C_s symmetry, indicating that BrO_3^- is stationary in a symmetric position throughout, while O_3^- at high temperature jumps as a rigid body between the two asymmetric positions.

6.1.2. BrO_3^* in $KClO_4$: $KBrO_3$

A formally analogous example is provided by the radical BrO_3^* formed by oxidation of BrO_3^- in $KClO_4$ [28]: Two asymmetric configurations of BrO_3^* for each anion site are observed at 26K, while the ESR signal at 80K represents a single, symmetric configuration with a spin Hamiltonian very close to the algebraic mean of the spin Hamiltonians determined at 26K. However, in contrast to the rigid-body motion of O_3^- in $[O_3^-, BrO_3^-]$, the 'jump' performed by BrO_3^* is merely a shift of the direction of the distortion with respect to the shape of BrO_3^-, as indicated by the near-symmetrical orientation of the pair of Q tensors observed at 26K.

6.1.3. BrO_3^{2-} in $KBrO_3$

The qualitative description of motional effects in terms of jumps between equivalent configurations helps to establish a link between observations pertaining to the low and high temperature limits. However, to describe the spectral changes in the transition region ($\Omega \approx \Delta\nu$) a quantitative representation of the jump model in terms of the stochastic Liouville equation must be employed [39,40]. The dynamic behaviour of the self-trapped electron BrO_3^{2-} in $KBrO_3$ was studied in this manner [41].

The relationship between the point groups of radical and site (C_s and C_{3v}, respectively) implies three equivalent configurations of BrO_3^{2-} connected by 120° rotations, as observed at 26K. Below 80K inhomogeneous broadening and partially resolved superhyperfine structure dominate the line shape of the signals. In the region 80-115K a marked homogeneous broadening increases rapidly with temperature, whereas the positions of the hyperfine lines are virtually unchanged. In the region 115-160K the lines are broadened beyond detection except for B_o directed along the C_3 axis of the site or along a few other special axes. Above 160K a single, averaged spectrum emerges, the linewidth generally decreasing with increasing temperature. The stochastic Liouville equation combined with the spin Hamiltonians determined at 77K represent adequately the observed lineshapes in the region 87.5-117K in terms of a temperature dependence of Ω given by $\Omega(T) = 4.8 \times 10^{11} \exp(-1015/T)$. Moreover, the equation represents satisfactorily the entire, strongly anisotropic temperature dependence of the signals up to 160K except for $B_o \parallel C_3$, showing that the dominant motion may be described in terms of the single parameter $\Omega(T)$. At higher temperatures additional adjustable parameters are needed, however, to account quantitatively for the observations, thus indicating the onset of a more complicated motion.

The description in terms of the stochastic Liouville equation does not refer to a specific physical process and, hence, yields no clue to the type of motion performed by BrO_3^{2-}. The moderate distortion suggests, however, that the motion of BrO_3^{2-} (like that of BrO_3^* in the previous example) is a shift in the direction of the distortion rather than a rotation of the entire radical. The temperature dependence of Ω corresponds to a barrier of 705 cm^{-1}. As BrO_3^- in aqueous solution has the IR bands $v_1 = 806$ and $v_3 = 836$ cm^{-1}, it seems plausible that the observed motion is a consequence of excitation of a corresponding vibration in BrO_3^{2-}.

7. Summary

The investigations of the reactions and accumulation of paramagnetic species in $KClO_4$ irradiated with X-rays at low temperature indicate, that whenever a sequence of events can be established, self-trapping on anions precedes all other processes involving the radiation-produced electrons and holes. On this basis, the formation of the identified radiolytic products in solid alkali halates and perhalates is formulated in terms of generation of self-trapped electrons and holes, [e$^-$] and [⊕], which constitute the *primary anion radicals*, and their subsequent reactions resulting from the mobility of [⊕] in the lattice. *Secondary products*, arising from the recombination of [e$^-$] and [⊕], consist of charge-transfer complexes of anions with O_2. The formation of O_2 apparently occurs in two steps: The one-electron oxidation of [e$^-$] leads to ejection of an oxygen atom, which in turn abstracts a second oxygen atom from an adjacent anion in a 'thermal' process. The secondary products are identical to those obtained by photolysis of the halate and perhalate ions. *Tertiary products* comprise exchange-coupled complexes of radicals with O_2 and (in halates) the isolated radical XO_2. They arise from the reaction of two [⊕] with one [e$^-$] occurring either as consecutive processes (one-electron oxidation of a secondary product) or as a two-electron oxidation of [e$^-$].

In accordance with the nonuniform spatial distribution of primary electronic excitations, comprising clusters and low-density regions, the initial spatial distributions of [e$^-$] and [⊕] are strongly correlated. Therefore, the reactions of [⊕] occur partly as the spontaneous collapse of unstable configurations of [⊕] and [e$^-$] formed within the clusters, and partly as electron-transfer processes governed by the thermally activated migration of those [⊕] surviving in the low-density regions. This applies also to the oxidation of anionic impurities by reaction with [⊕]. The formation of the immobile tertiary radicals, which prevents that recombination annihilates all radicals, occurs primarily as spontaneous processes. Hence the nonuniform spatial distribution not only governs the kinetics of the reactions, but also greatly enhances the survival of radicals, thus distinguishing the radiolysis from the photolysis.

The structures of all radiation-induced products are adequately represented in terms of well-defined molecular species trapped in cavities in a solid matrix. The specific properties of the crystalline host are expressed in the electron-transfer processes associated primarily with the mobility of [⊕], in reactions and structures involving pairs of adjacent anions, and in the pattern of equivalent and inequivalent configurations, arising from various types of mismatch of the symmetries of trapped species and trapping site and leading to reversible as well as irreversible thermal transformations.

Self-trapping of electrons and holes is probably the initial step in the radiolytic formation of anion radicals in all inorganic crystals with polyatomic anions. However, the diversity of structures and mobilities, displayed even in the closely related substances discussed above, indicates that generalizations concerning the subsequent reactions of the self-trapped species are unwarranted.

References

1. T. Cole: *Proc. Natl. Acad. Sci. US.* **46**, 506 (1960).
2. J.R. Byberg, in *Magnetic Resonance of the Solid State*, edited by J.A. Weil (The Canadian Society for Chemistry, Ottawa 1987) p. 413.
3. M.C.R. Symons, in *Specialist Periodical Reports, Electron Spin Resonance* (Royal Society of Chemistry, London), Vol. **3** (1976) p. 134, Vol. **4** (1977) p. 84, Vol. **5** (1979) p. 134, Vol. **6** (1981) p. 96, Vol. **7** (1982) p. 124, Vol. **8** (1983) p. 166, Vol. **9** (1985) p. 87, Vol. **10B** (1987) p. 198, Vol. **11B** (1989) p. 175.
4. T.A. Carlson: *Radiat. Res.* **64**, 53 (1975).
5. M.N. Kabler, in *Proceedings of the Nato Advanced Study Institute on Radiation Damage Processes in Materials, Corsica 1973*, edited by C.H.S. Dupuy (Noordhoff, Leyden, 1975) p. 171.
6. J.R. Byberg and S.J.K. Jensen: *J. Chem. Phys.* **52**, 5902 (1970).
7. J.R. Byberg and J. Linderberg: *Chem. Phys. Lett.* **33**, 612 (1975)
8. N. Bjerre and J.R. Byberg: *J. Chem. Phys.* **82**, 2206 (1985).
9. C.H. Townes and B.P. Dailey: *J. Chem. Phys.* **17**, 782 (1949).
10. J.R. Byberg and J. Spanget-Larsen: *Chem. Phys. Lett.* **23**, 247 (1973).
11. W. Känzig: *Phys. Rev.* **99**, 1890 (1955).
12. M.C.R. Symons and S.P. Mishra: *J. Chem. Research (S)* **1981**, 214.
13. J.R. Byberg: *J. Chem. Phys.* **75**, 2663 (1981).
14. S. Schlick and L. Kevan: *J. Phys. Chem.* **81**, 1093 (1977).
15. N. Bjerre and J.R. Byberg: *J. Chem. Phys.* **75**, 4776 (1981).
16. G.B. Johansson and O. Lindqvist: *Acta Crystallogr.* **B33**, 2918 (1977).
17. J.R. Byberg: *J. Chem. Phys.* **75**, 2667 (1981).
18. N. Bjerre and J.R. Byberg: *Phys. Rev. B* **20**, 3597 (1979).
19. T.R. Waite: *Phys. Rev.* **107**, 463, 471 (1957).
20. J.R. Byberg: *Chem. Phys. Lett.* **56**, 130 (1978).
21. S. Siegel, B. Tani, and E. Appelman: *Inorg. Chem.* **8**, 1190 (1969).
22. J.R. Byberg: *J. Chem. Phys.* **55**, 4867 (1971).
23. J.R. Byberg: *J. Chem. Phys.* **86**, 6065 (1987).
24. J. Danielsen, A. Hazell, and F.K. Larsen: *Acta Crystallogr.* **B37**, 915 (1981).
25. R.S. Eachus and M.C.R. Symons: *J. Chem. Soc.* **1968**, 2433.
26. J.R. Byberg: *J. Chem. Phys.* **76**, 2179 (1982).
27. W.H. Zachariasen: *Norske videnskapsselsk. Skr. (Oslo)* **4**, 90 (1928).
28. J.R. Byberg: *J. Chem. Phys.* **83**, 919 (1985).
29. J.R. Byberg and B.S. Kirkegaard: *J. Chem. Phys.* **60**, 2594 (1974).
30. J.R. Byberg: *J. Chem. Phys.* **84**, 6204 (1986).
31. J.-C. Fayet and B. Thieblemont: *C.R. Acad. Sci. Paris* **261**, 5420 (1965).
32. J.R. Byberg: *Chem. Phys. Lett.* **56**, 563 (1978).
33. J.R. Byberg: *J. Chem. Phys.* **84**, 4235 (1986).
34. J.R. Byberg: *J. Chem. Phys.* **85**, 4790 (1986).
35. J.R. Byberg: *J. Chem. Phys.* **88**, 2129 (1988).
36. N. Bjerre: *J. Chem. Phys.* **76**, 2881 (1982).
37. N. Bjerre and J.B. Bates: *J. Chem. Phys.* **78**, 2133 (1983).
38. N. Bjerre: *J. Chem. Phys.* **76**, 3347 (1982).
39. R. Kubo: *Adv. Chem. Phys.* **15**, 101 (1969).
40. R. Kubo: *J. Phys. Soc. Jap. Suppl.* **26**, 1 (1969).
41. J.R. Byberg, S.J.K. Jensen, and B.S. Kirkegaard: *J. Chem. Phys.* **61**, 138 (1974).

Trapped Electrons in Disordered Matrices

LARRY KEVAN

Department of Chemistry, University of Houston Houston, Texas 77204-5641, USA

1. Introduction

Electrons generated by ionizing radiation or photoionization can be trapped in a variety of solids. Extensive studies have been carried out on trapped electrons in alkali halide crystals in which the electron is trapped in an anion vacancy in the crystal lattice [1]. These trapped electrons have an intense optical absorption in the visible and are thus called color centers or farben centers (F-center). Electrons may also be trapped in molecular crystals such as crystalline ice and a variety of polyhydroxy compounds such as alcohols and carbohydrates [2]. Electron trapping in molecular crystals containing hydroxy groups is qualitatively different than in alkali halides since facile rearrangements of hydrogen positions in the presence of an electron can occur in such molecular crystals. Thus, although there may be some vacancies or sites in molecular crystals with hydroxy groups oriented to favor trapping of an electron, the detailed geometry of the trapping site can be changed by the presence of the electron. This is analogous to the solvation of an electron in which the surrounding molecular groups are reoriented by the presence of the electron compared to their orientation in the absence of the electron.

In this chapter the focus is on trapped electrons in disordered matrices rather than crystalline matrices. In general, the yields of trapped electrons are higher in disordered matrices compared to crystalline ones. Various kinds of direct spectroscopic evidence show that considerable local rearrangement of the molecule surrounding the electron occurs in the process of localization and trapping [3,4,5]. This rearrangement is perhaps best called solvation and trapped electrons in disordered matrices are probably best described as trapped, solvated electrons. Trapped, solvated electrons are generated by ionizing radiation or photoionization in a variety of aqueous and organic glassy matrices. A variety of alcohols, ethers, amines and alkanes readily form glassy matrices when rapidly frozen and most of these trap electrons relatively efficiently. Water however crystallizes rapidly when frozen and few electrons are trapped in polycrystalline ice. However, the addition of a sufficient concentration of a highly soluble ionic salt in water causes the aqueous solution to freeze readily to a glassy matrix. Examples of such salts are sodium hydroxide, sodium perchlorate, and potassium carbonate among others. In fact, the first report of a trapped electron in a disordered matrix is probably that in 1962 of a trapped electron in a aqueous glass composed of 10 M sodium hydroxide [6]. This trapped electron was characterized by a strong optical absorption spectrum in the visible and an intense electron spin resonance line near $g = 2.002$.

A. Lund and M. Shiotani (eds.), Radical Ionic Systems, 385–408.
© *1991 Kluwer Academic Publishers. Printed in the Netherlands*

Also in 1962 a solvated electron in liquid water was first reported and detected by a transient optical absorption in the visible following pulsed radiolysis [7]. The optical absorption spectra of the solvated electron in liquid water and the trapped, solvated electron in the alkaline aqueous glass at 77 K are very similar.

There is a strong parallelism between solvated electrons generated in liquids by radiolysis or photoionization and in glassy matrices of the same or analogous liquids at low temperature. In both liquids and glasses one can usefully consider that solvated electrons are formed kinetically in a two step process. The initially produced electrons are mobile or quasi-free and move in a conduction band. They are localized in a first step which depends on the relative energies of the conduction electron level of the disordered medium and the energy of the localized electron state in the medium. A second step, after initial localization occurs, is solvation in which the electron induces rearrangements in the surrounding solvent molecules to generate an equilibrium geometry characterized by the forces between the electron and the solvent molecules. Thus, by a trapped, solvated electron we mean one that is equilibrated energetically with its surroundings. Perhaps it is better called a fully solvated or equilibrated electron, but following common usuage we will call it a solvated electron. A localized electron becomes partially solvated as soon as some surrounding molecular rearrangement occurs different from that of the bulk medium. A partially solvated electron is characterized by a range of local geometries which have not yet reached equilibrium. The word 'presolvated' has been used to describe this range of partially solvated electrons. The term localized electron would seem to refer to both partially and fully solvated electrons and is thus not discriminatory.

In glassy matrices there is ample evidence for this solvation process in all matrices from aqueous through alcohol to alkanes. The evidence is most clear in alcohol matrices in which time dependent optical spectral shifts as well as time dependent shifts in the electron spin resonance linewidths are observed which are indicative of the solvation process. However, in a disordered glassy matrix there are a number of varying molecular arrangements with a distribution of local energies. Thus the concept of preexisting traps has been invoked to describe various aspects of the yields and other features of trapped electrons in disordered matrices [8]. The variation of the molecular potential throughout a glassy matrix is undoubtedly important but it is probably most important for the initial stages of electron localization. The glassy matrix potential itself does not control the ultimate electronic or geometric structure of the equilibrated trapped, solvated electron or even of many of the partially solvated electron states. The width and shape of the optical absorption spectra of trapped electrons in glassy matrices have been variously interpreted as being compatible with a distribution of largely preexisting, trap depths. However, once initial localization occurs, the local potential around the electron causes molecular rearrangement which modifies the potential of the matrix itself. This molecular rearrangement does not necessarily reach equilibrium; it may be arrested at sufficiently low temperature. It is also clear from continued studies to the present day in both liquids [9,10] as well as glasses [11, 12] that there is not a continuous, smooth progression in the process of electron solvation to a final equilibrium solvated species. It seems that the molecular structure plays a role and that intermediate geometries of varying stability exist.

The two primary spectroscopic characteristics of trapped, solvated electrons in disordered matrices are the optical absorption spectrum and the electron spin resonance spectrum. We regard the optical absorption spectrum as being primarily related to the electronic structure of the trapped, solvated electron and the electron spin resonance spectrum as being primarily related to the geometrical structure of the trapped, solvated electron. Of course, this division is overly simplistic and the optical absorption spectrum is also related to the geometrical structure and vice versa. In the following sections we will discuss the relationship of the optical spectrum to the electronic structure of trapped electrons in matrices of varying polarity. This information will be related to current theoretical treatments of solvated electrons, which formally apply to the liquid phase, as well as to femtosecond studies of electron solvation in the liquid phase.

The electron spin resonance spectrum will be discussed with respect to the specific information that can be obtained about the geometrical structure in matrices of different polarities. The same information is not available in the liquid phase because the electron spin resonance spectrum is transient and gives little direct structural information [13]. The structural information in the electron spin resonance spectrum is carried largely by electron nuclear dipolar interactions which are averaged out in the liquid phase. The conclusions about the geometrical structure of trapped, solvated electrons will be related to current theoretical descriptions of the geometry of solvated electrons in liquids. In particular, the widely utilized cavity model, which is substantiated by the geometrical structure deduced from electron magnetic resonance data, is found to be compatible with current quantum statistical mechanical calculations.

2. Optical Spectra - Relation to Electronic Structure

The optical spectra of trapped electrons in aqueous and organic matrices are broad and structureless with an asymmetric shape extending toward higher energy when plotted on a scale linear in energy. The optical absorption maxima qualitatively correlate with matrix polarity and extend from 580 nm for the typical aqueous glass, 10 M NaOH/H_2O, to about 530 nm for ethanol glass, 1200 nm for 2-methyltetrahydrofuran (MTHF) glass and 1700 nm for 3-methylpentane (3MP) glass [14]. Similar spectra are observed for solvated electrons in liquids. The structureless spectra alone do not give much information about the geometrical structure of these trapped electrons. The only approach utilized to date is to compare a calculated spectrum based on some theoretical model with an experimental spectrum to infer something indirectly about the geometrical structure.

2.1 BOUND-BOUND AND BOUND-CONTINUUM TRANSITIONS

A major question is what is the nature of these absorption spectra in terms of the optical transitions involved. In the liquid phase it has been difficult to experimentally distinguish between bound-bound and bound-continuum transitions or a mixture of these contributing to the spectra. Some information could probably be given by clever hole burning or photobleaching experiments, but to date no definitive results have been reported.

In contrast, for trapped electrons in solid disordered matrices the nature of the optical transitions contributing to the absorption spectra can be determined by a comparison of the optical spectra and the photoconductivity or photobleaching spectra versus wavelength. The most direct approach to study bound-continuum or bound-free transitions is to carry out photoconductivity spectral measurements.

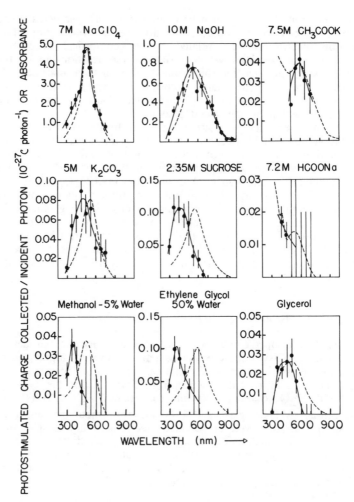

Fig. 1. Photoconductivity spectra at 85 K, ——, and optical absorption spectra at 77 K, -----, for trapped electrons produced by gamma irradiation in various aqueous and alcoholic glassy matrices. Error bars are shown for the photoconductivity measurements. The optical spectra peaks were normalized to the maxima of the photoconductivity spectra. The matrices are arranged to illustrate the two types of behavior found; the top view shows matrices where the optical and photoconductivity spectra nearly coincide while the other two rows show matrices where the photoconductivity spectra are blue shifted from the optical spectra. [From reference 15 with permission.]

Any noncoincidence of the photoconductivity and optical absorption spectra allows one to roughly assign portions of the optical spectra to bound-bound and bound-continuum transitions. Similar information can be obtained by a comparson of optical absorption spectra with photobleaching spectra; however, photobleaching only indirectly measures the promotion of electrons to a conduction state and may be misleading if the excited electrons can tunnel efficiently to an acceptor species in the matrix without actually being promoted to a conduction state. In contrast, photoconductivity measurements do measure the direct promotion of the electron to a conduction state, but they are also sensitive to the range of electrons in the conduction state.

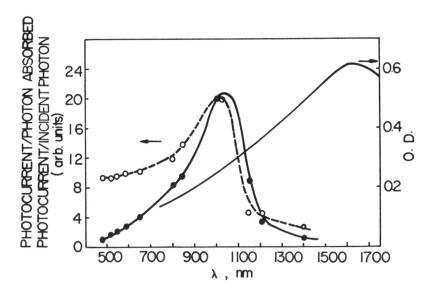

Fig. 2. The wavelength dependence of photocurrent in γ–irradiated 3-methylhexane at 77 K: (o) photocurrent/incident photon; (o) photocurrent quantum efficiency curve; and (- - - -) optical absorption curve. [From reference 16 with permission.]

In general, it has been found for trapped, solvated electrons in alcohol, ether and alkane glasses that the photobleaching and photoconductivity spectra at 77 K are rather similar and distinctly different in wavelength range from the optical absorption spectrum. This clearly indicates the occurrence of both bound-bound and bound-continuum transitions. In aqueous glasses there is generally a closer correspondence between the photoconductivity and optical absorption spectra for trapped electrons which suggests the dominance of bound-continuum transitions. Fig. 1 shows a comparison of photoconductivity and optical absorption spectra for trapped, solvated electrons in various alcoholic glassy matrices [15]. It can been seen that even among aqueous glasses there is a range of behavior where there is close correspondence between these two types of spectra for 7 M sodium perchlorate and 10 M sodium hydroxide glasses whereas there is a distinct difference between the two spectra for 5 M potassium carbonate glasses. In general, the alcohol glasses all show a distinct shift between the photoconductivity and the optical absorption spectra indicating larger relative contributions of bound-bound transitions. The results in Fig. 1 may be contrasted with those in Fig. 2 showing photoconductivity and optical absorption spectra for trapped, solvated electrons in 3-methylhexane glass which is essentially a nonpolar matrix [16]. Here the photoconductivity spectrum indicates that only the high energy tail of the optical absorption spectrum corresponds to bound-continuum transitions. This is also found for glassy ether matrices such as 2-methyltetrahydrofuran [17].

2.2. PRESOLVATED VERSUS SOLVATED ELECTRON SPECTRA

In general, when electrons are generated by radiolysis or photoionization in glassy matrices at 77 K the optical absorption spectrum seems to be characteristic of an equilibrium species in that little change is observed upon increasing the temperature. In contrast, if the electrons are generated in the same way at 4 K the optical spectrum is generally much broader and shifted toward lower energy.

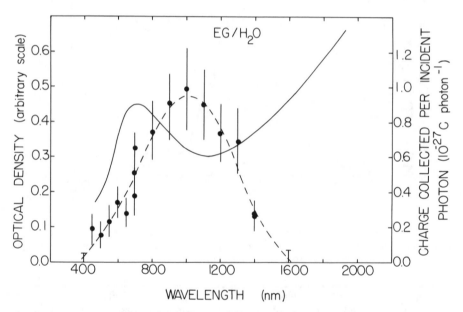

Fig. 3. Optical spectrum of ethylene glycol/50 % D_2O glass at 4.2 K irradiated with γ-rays to a dose of 0.15 Mrad, ——. Photoconductivity spectrum of the same system at an x-ray dose of 25 krad; experimental points, o, and errors are shown together with a suggested spectrum, - - - -. [From reference 18 with permission.]

This spectrum is unstable and slowly shifts toward higher energy as the temperature is raised toward 77 K. This has been interpreted as localization of presolvated electrons at 4 K in a nonequilibrium state in which the electron has not been able to fully reorient the adjacent molecules of the matrix to an equilibrium configuration of the fully solvated state.

Fig.3 shows the optical and photoconductivity spectra for trapped electrons generated in an ethylene glycol-water glass at 4 K [18]. It can be seen that both the optical absorption and the photoconductivity spectra extend significantly toward longer wavelengths than for solvated electrons produced in matrices at 77 K as shown in Fig. 1. However, it still appears that both bound-bound and bound-continuum transitions contribute to the total optical spectrum of the trapped, presolvated electrons at 4 K. It is also interesting to look at Fig. 4 which shows the optical and photoconductivity spectra for 10 M sodium hydroxide glass with the electrons generated at 4 K. Here there is a distinct difference between the photoconductivity and the optical spectra indicating some contribution from bound-bound transitions for the presolvated electron in this aqueous glass by comparison with the analogous spectrum in Fig. 1 in which there is good coincidence between the optical absorption and the photoconductivity spectra. There seem to be real changes in the relative contributions of the bound-bound and bound-continuum transitions for a presolvated versus a solvated electron trapped in aqueous glasses.

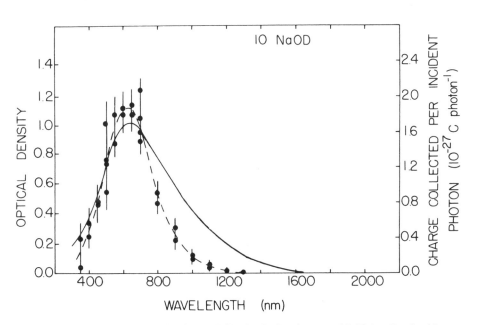

Fig. 4. Optical spectrum of 10 M NaOD in D_2O glass at 4.2 K irradiated with x-rays to a dose of 0.06 Mrad, ———. Photoconductivity spectrum of 10 M NaOD in D_2O at 4.2 K irradiated ato an x-ray dose of 25 krad; experimental points, o, and errors are shown together with a suggested spectrum, - - - -. [From reference 18 with permission.]

2.3. ENERGY LEVEL STRUCTURE

In less polar organic matrices such as ethers and alkanes a combination of optical, photoconductivity and photobleaching experiments has lead to an energy level diagram for trapped, solvated electrons as shown in Fig. 5. In addition to the bound-bound and bound-continuum transitions shown from the ground state, another set of energy levels is shown which corresponds to population of an intermediate state. The evidence for this is shown most clearly in photobleaching experiments [16,17]. This intermediate state must be optically forbidden with respect to the ground state and is considered to be a relaxed 2S state in which the nuclei have rearranged in response to the new charge distribution of the optically excited 2P state. These features are seen in both alkane glasses [16] as well as in ether glasses [17] with slightly different values of the energy level differences.

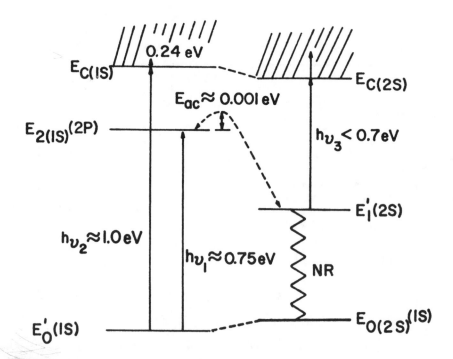

Fig. 5. The energy level diagram for trapped electrons in 3-methylhexane glass at 77 K. 1s, 2s and 2p refer to hydrogenic type wavefunctions; for vertical transitions to unrelaxed states the orbital in the subscript refers to the wavefunction which determines the orientational polarization of the matrix. Primes denote relaxed states. NR refers to a possible nonradiative transition. [From reference 16 with permission.]

The energy level structure in Fig. 5 and the energy level structure in more polar matrices based on the data in Figs.1-4 has been semiquantitatively rationalized and accounted for by a semicontinuum potential model for solvated electrons in liquids and solids [19]. In the semicontinuum model of electrons in polar media the electron is considered to interact with a number of specifically oriented matrix dipoles in the first solvation shell by short range attractive and repulsive potentials and with the rest of the matrix molecules beyond the first solvation shell by a long range average polarization potential. The total energy of the electron in the medium is given by the sum of the electronic energy of the electron and the energy necessary to rearrange the medium due to its interaction with the electron. This model can account well for *changes* in the optical absorption maxima and for the presence of both bound-bound and bound-continuum transitions [20] in amines, ethers, alcohols and aqueous systems. The semicontinuum potential is like a particle-in-a-box potential near the electron origin and has a Coulombic tail from the charge-matrix polarization interaction. The Coulombic tail gives rise to the bound excited states. Thus solvated electrons typically have both bound-bound and bound-continuum transitions.

In pure aqueous glasses predominantly bound-continuum transitions for trapped, solvated electrons are indicated by the experimental data in which the photoconductivity spectrum almost exactly superimposes on the optical absorption spectrum as shown in Fig. 1. This is inconsistent with predictions of the semicontinuum model *unless* the Coulombic tail of the semicontinuum potential is truncated. In aqueous glasses containing a large concentration of ionic solute it is probable that the long range polarization interactions involved in the semicontinuum model are not operable for the solvated electron because of the intense short range orientation of the water molecules throughout the matrix by the high ion concentration. To account for this, the Coulomb tail of the potential may be justifiably truncated. In this case the potential does not then sustain a bound excited state and thus the dominance of bound-continuum transitions found experimentally is accounted for.

The semicontinuum model has also been adapted for nonpolar systems such as alkanes by emphasizing charge-polarizability interactions and taking into account some aspects of the molecular structure of the solvent molecules [21]. The electron-polarizability interaction is made dependent on molecular structure by considering the electron interaction with the polarizability of each C-C bond of the first solvation shell molecules. In addition, the effect of hydrogen-hydrogen repulsion between first solvation shell molecules is incorporated. This model accounts well for the variety of experimental data on solvated electrons in alkane glasses at 77 K.

2.4. RELATION OF CURRENT THEORY TO OPTICAL SPECTRA

The detailed understanding of the optical spectrum of trapped, solvated electrons in matrices as well as that of solvated electrons in liquids is still rather imperfect. A better and improved understanding depends upon an intimate interplay between theory and experiment. Until the mid 1980's the most successful theoretical studies of trapped, solvated electrons were based on a dielectric continuum picture of the solvent which necessarily averages over the molecular structure of the solvent [22]. The so-called semicontinuum model of solvated electrons, which has been mentioned above, has been quite successful as far as it goes. The two main variations of the semicontinuum model by Jortner, Kestner and coworkers [23,24] and by Fueki, Feng and Kevan [19-21,25] differ in some calculational details but both give the same general picture of the trapped, solvated electron energy level structure.

Another approach was to carry out *ab initio* quantum chemical calculations of a negatively charged molecular cluster [22]. These studies are oriented toward determination of the geometrical details of the structure of the ground state of the solvated electron. This approach is limited by the number of molecules that can be included and doesn't take into account the overall effect of the solvent. However, this approach can give useful insight into the geometrical structure of solvated electrons and will be discussed to a greater extent below after the geometrical structure of solvated electrons based on experimental data is described.

Beginning in 1984 a quantum statistical mechanical approach to theoretical studies of solvated electrons was introduced by Chandler and coworkers [26-30]. This built upon recent developments of the statistical mechanics of liquids. Following this general quantum statistical mechanical approach, at least five groups have carried out detailed calculations on various aspects of solvated electrons [31-49]. The Feynman path integral formulation of quantum statistical mechanics has been used which provides a method to obtain an approximate expression for the partition function of a quantum particle moving in a potential. This method is combined with molecular dynamics or Monte Carlo computer simulations. The implementation of this approach depends upon the development of an effective and realistic interaction potential between a solvent molecule and an external electron. One approach is to specify a plausible interaction potential and determine the consequences by Monte Carlo computer simulations [31-36]. Another approach is to use current electron-molecule density functional theory to obtain a potential [39]. In all the formulations utilized the effective potential is treated as a so-called pseudopotential which smooths out the potential near the nuclei of the atoms in a molecule. The essential differences between the formulations of the various groups that have carried out detailed quantum statistical mechanical calculations on solvated electrons, mainly in water, lie in somewhat different formulations of the electron-molecule pseudopotential. However, in all cases it must be recognized that the uncertainty in the final potential relative to the "true" potential probably remains substantial.

Rossky and Schnitker [41,40] and Berne and coworkers [44] have both presented calculations of the optical absorption spectrum of the solvated electron in water. Without going into detail, the results of Rossky and Schnitker give a smooth optical spectrum which is asymmetric toward higher energy as observed experimentally, but with an absorption maximum at 2.2 eV rather than at the experimental value of 1.7 eV. In contrast, with a different pseudo-potential Berne and coworkers find a calculated optical absorption maximum of 1.7 eV which agrees well with experiment, but their calculated spectrum is not asymmetric to high energy which conflicts with experiment.

In both cases the interpretation from the quantum statistical mechanical simulation is that the optical absorption spectrum is composed primarily of bound-bound transitions with very little if any contribution from bound-continuum transitions. Even though there isn't any direct experimental evidence to distinguish between bound-bound and bound-continuum transitions of electrons in liquids, recent work has involved the analysis of the high energy side of the optical absorption spectrum of solvated electrons in several liquid alcohols and water within an Urbach tail scheme [50,51]. In this scheme direct bound-continuum transitions are assumed and this assumption seems to fit the observed spectra well. Thus, this question remains an open one in liquid systems.

Unfortunately, the quantum statistical mechanical calculations can not be adapted directly to the experimental data available in disordered matrices in aqueous systems because these aqueous glasses contain large concentrations of added salts. However, when it is possible to carry out quantum statistical mechanical calculations of similar accuracy for electrons in methanol or ethanol solvents, a more direct comparison can be made between the liquid phase and the solid disordered matrix in which there is direct evidence for bound-continuum transitions.

3. Electron Spin Resonance Spectra - Relation to Geometrical Structure

Trapped, solvated electrons in aqueous and organic glassy matrices are characterized by an electron spin resonance (ESR) spectrum in addition to their optical absorption spectrum. Proper analyses of the ESR spectrum has potential for giving the detailed nuclear geometry around the trapped electron which, in turn, provides a molecular geometrical picture of these trapped electrons. However, the electron spin resonance spectrum is only a single Gaussian-shaped line for trapped electrons in the various aqueous and organic glasses studied. The essential structural information is hidden within the unresolved electron-nuclear hyperfine structure that composes the substructure of the single observed ESR line. Thus it was necessary to develop new techniques to isolate the weak hyperfine interactions in order to solve the problem of the geometrical structure of the trapped, solvated electrons.

These new methods include second moment electron spin resonance lineshape analysis in selectively deuterated solvents [57-54], analysis of both the intensity and energy separation of allowed and forbidden "proton spin flip" electron spin resonance transitions [55,56], and analysis of electron spin echo modulation patterns [52,57-60]. The second moment and "spin-flip" transition analyses are extensions of older techniques which are limited in their effectiveness to certain experimental conditions. However, the electron spin echo modulation (ESEM) analysis represents a much more general technique which is broadly applicable.

Several reviews of electron spin echo methods and analysis of the associated nuclear modulation observed in disordered systems are available and the details will not be repeated here [57,61-63]. Fig. 6 illustrates the essential idea. The two-pulse experiment is perhaps the simplest. The first resonant microwave pulse phases the spins. They then precess and dephase, in part due to electron-nuclear interactions, over a short distance range in the vicinity of the unpaired electron. This dephasing period is called τ in Fig. 6. Then, a second microwave pulse causes the electron spin systems to start to rephase and after a *second* time τ the spins do rephase and generate a burst of microwave energy called an echo. The interpulse time τ is typically varied from 50 nanoseconds to several microseconds. As τ is swept an echo intensity pattern versus τ is generated which is often modulated due to weak electron-nuclear, largely dipolar, interactions. Thus, analysis of this modulation pattern can lead to information about the electron-nuclear hyperfine interaction and hence to information about the number and distance of surrounding magnetic nuclei. In practice this distance varies from about 0.2 to 0.6 nm. It should be emphasized that in a glassy matrix an average structure is determined by this method. Also there are some approximations made in the analysis of the

396

modulation pattern [57, 61-63] and this means that fine details of the molecular structure around a trapped, solvated electron will not be delineated by this method.

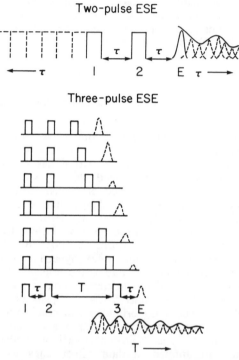

Fig. 6. Illustration of modulation of two- and three-pulse electron spin echo decay. In the two-pulse experiment microwave pulses 1 (90°) and 2 (180°) separated by time τ produce the echo signal at time τ after pulse 2. At τ is increased, the echo amplitude changes and traces out an echo envelope which may be modulated. In the three-pulse experiment the second pulse is split into two 90° pulses separated by time T. As T is swept, the echo amplitude changes and traces out an echo envelope which may be modulated.

3.1. PRESOLVATED ELECTRON STRUCTURE

Little experimental evidence has yet been obtained to describe the geometrical structure of presolvated electrons in a specific matrix. The general picture is that the nonequilibrium solvation shell consists of disoriented molecules or molecular dipoles around the electron. This has been supported theoretically in polar systems by semicontinuum model calculations of the ground and excited state energy levels of an excess electron in ethanol as a function of the point molecular dipole orientation around the electron [64]. The results show that the ground state energy is strongly affected by the average dipole orientation angle and that this accounts nearly quantitatively for the magnitude of the spectral shift observed between the presolvated and solvated electron configurations in glassy ethanol.

However this does not give a detailed geometrical picture of the presolvated electron since it is based on a point molecular dipole model. Analysis of electron magnetic resonance data has enabled the measurement of average changes in the electron-proton distances to first solvation shell molecules between the presolvated and solvated electron states in alcohol and ether matrices. Analysis of electron spin resonance lineshapes by second moment methods [65] and electron nuclear double resonance linewidths [66] has shown *independently* that the average distance differences correspond to only 0.02 to 0.03 nm. Thus the structure of the presolvated electron is relatively close to that of the equilibrium trapped,solvated electron which will now be discussed.

3.2. STRUCTURE IN AQUEOUS GLASSES

The detailed geometrical structure deduced for the trapped, solvated electron in 10 M sodium hydroxide aqueous glass is shown in Fig.7. Six water molecules solvate the electron in an approximately octahedral structure, in which the molecules are approxi- mately equivalent, in which each water molecule has its O-H bond oriented toward the electron and the distances are as shown. It should be stressed that the conclusion of six first solvation shell molecules from an analysis of the electron spin echo data [67] has been independently verified by a second moment lineshape analysis of electrons trapped in oxygen-17 enriched water [68]. The deduced isotropic coupling constants to the protons and oxygen atoms of the molecules in the first solvation shell correspond to a total unpaired electron spin density of about 4 %. This would be even far less for second solvation shell molecules. Thus the trapped, solvated electron is described as having a wavefunction that is fairly compact with about 96 % of the spin density concentrated near the center of the solvated structure. Hence a solvated electron looks very much like a solvated anion. It is also noteworthy that a completely independent study of this trapped, solvated electron geometry has been carried out by analysis of forbidden proton spin-flip satellite lines which confirms the structure shown in Fig. 7 [54].

There are two important aspects of this structure. One is the O-H bond orientation of the first solvation shell water molecules. The second is the number of water molecules being six. Of course, an average structure is determined in the glassy matrix and hence there may be a distribution of numbers of first solvation shell water molecules with an average of about six.

Since this structure was published in 1976 it has been pleasing that a variety of recent quantum statistical mechanical calculations have confirmed the O-H bond dipole oriented structure of the first solvation shell water molecules [31.40,44]. It is also noteworthy that these recent theoretical calculations have concluded that a cavity model is the best description for a solvated electron in water. This is consistent with the charge distribution deduced from the experimental results shown in Fig. 7. The general size of the calculated electron distribution of the solvated electron in water also is in semiquantitative agreement with the geometry shown in Fig. 7. And finally the total coordination number of six seems consistent with the calculation of Rossky and Schnitker [40].

It should also be emphasized that the experimental data exemplified in Fig. 7 seem quite incompatible with alternative structures based upon the solvation of a water anion (H_2O^-) which has been suggested in some analyses and some quantum chemical calculations based primarily on optical spectra [69,70]. However, recently the electron spin echo modulation data of solvated electrons in glassy matrices has been analyzed in a different way by Astashkin and coworkers [71]. In this analysis they attributed a significant amount of the modulation pattern as arising from beyond the first solvation shell and this part of the total modulation pattern was filtered out to obtain a residual modulation pattern which was analyzed in terms of an excess electron coordinated by only one or two water molecules. We believe that this analysis method overestimates the nuclear modulation effect from the nuclei in the solvation shells beyond the first one.

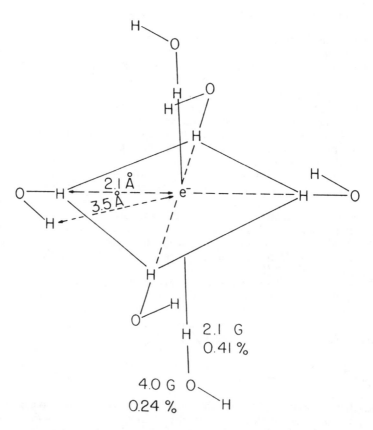

Fig. 7. Geometrical structure of the solvated electron in aqueous glasses determined from electron spin-echo modulation analysis. The isotropic couplings to H and ^{17}O are given along with the corresponding percentage of solvated electron spin density on these atoms. [From reference 78 with permission.]

It should be pointed out that a free nuclear Zeeman frequency component, which can be associated with second and further solvation shell molecules, will also be given by nuclei in the first solvation shell when the polar angle of the first solvation shell nuclei with respect to the external magnetic field is close to the magic angle of 54°. It should also be stressed that a completely independent study of oxygen-17 enriched water by second moment lineshape analysis of a trapped, solvated electron in aqueous glasses [68] gave six first solvation shell molecules and is incompatible with one or two. A third independent study based on proton spin-flip satellite analysis [54] also gave six first solvation shell molecules. So, although there may be some argument about the best analysis of the electron spin echo modulation data, the sum total of the experimental data based on several different types of magnetic resonance experiments still seems to strongly support the structure shown in Fig 7.

Quantum chemical calculations of solvated electron molecular clusters probably do not account adequately for the role of the entire solvent. However, it is interesting that recent *ab initio* calculations by Clark and Illing [72] for an aqueous solvation structure of an electron with six water molecules is consistent with the bond dipole orientation of the first solvation shell molecules. This also supports the structure shown in Fig. 7.

3.3. STRUCTURE IN ETHANOL GLASS

The trapped, solvated electron in ethanol glassy matrices provides a distinct contrast in its geometrical structure compared to the trapped, solvated electron in aqueous matrices. The analysis of electron spin echo modulation data for this system [73] shows that the electron is solvated by four ethanol molecules, in contrast to the six found for water, and perhaps more importantly that the O-H bond dipole of the ethanol molecule is *not* oriented toward the electron as was found in the aqueous system, but instead the molecular dipole of the entire ethanol molecule is oriented toward the electron. Fig. 8 shows the orientation of one first solvation shell molecule with respect to the solvated electron in ethanol glassy matrix. In this matrix it has been possible to study three different specifically deuterated ethanols, C_2H_5OD, CD_3CH_2OH, and CH_3CD_2OH. Separate spin echo modulation data to the deuterons located in these different positions have been analyzed and give an unambiguous determination of the orientation of an individual ethanol molecule with respect to the solvated electron. It is found that the electron is located approximately on the bisector of the C-O-H angle which is the direction of the molecular dipole for ethanol. This is consistent with a charge-molecular dipole interaction as the dominant interaction determining the solvation shell geometry.

It is reassuring that recent *ab initio* quantum chemical calculations of an electron solvated by four ethanol molecules indicate that the energy is minimized when the position of the localized electron *departs* from the O-H bond orientated configuration found for the hydrated electron and moves toward a molecular dipole oriented configuration consistent with the experimental data in Fig. 8 [74]. The number of solvating molecules being four in ethanol is also suppor-ted by recent independent experimental results on the optical absorption spectra by Ichikawa and Yoshida [75]. They carefully analyze the shape of the optical absorption spectra in ethanol/2-methyltetrahydrofuran glassy mixtures at 77 K. In the mixed matrices the trapped, solvated electron shows five spectral components as the mole fraction of ethanol is changed and the intensities of the components depend upon the fraction of ethanol. One component is the same as the absorption spectrum of trapped, solvated electrons in pure 2-tetramethylhydrofuran glass. The other four components are assigned to trapped electrons solvated by 1, 2, and 4 ethanol molecules respectively. It is also concluded in that work that most of the electrons are solvated by ethanol molecules with their molecular dipoles pointing toward the electron. Thus the optical work fully supports the originally published geometrical structure of the trapped, solvated electron in ethanol shown in Fig. 8.

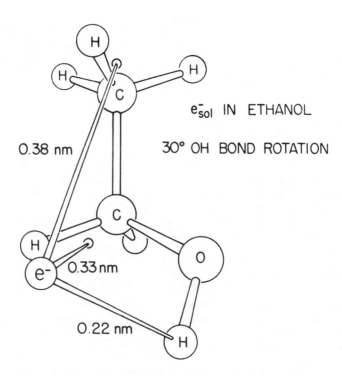

e^-_{sol} IN ETHANOL

30° OH BOND ROTATION

0.38 nm

0.33 nm

0.22 nm

Fig. 8. Orientation of an individual first solvation shell molecule with respect to the solvated electron in ethanol. The OH bond is rotated 30° around the OC bond from the CCO plane in ethanol. The distances are averages to the OH, CH_2, and CH_3 proton groups. The entire solvation shell includes four ethanol molecules so oriented and arranged approximately tetrahedrally about the electron. The molecular dipole of ethanol approximately bisects the COH angle and can be seen to be oriented approximately toward the electron. [From reference 18 with permission.]

3.4. STRUCTURE IN ETHYLENE GLYCOL GLASS

Ethylene glycol readily forms a glassy matrix when frozen to 77 K and also traps solvated electrons with relatively high efficiency. It represents a very interesting system for the geometry of the trapped, solvated electron in that each molecule has two hydroxyl groups. Thus, it is unlikely that each hydroxyl group or the molecular dipole associated with the ethylene glycol molecule can be optimally oriented toward the electron. One might then expect that this glassy matrix would not solvate electrons very efficiently. Nevertheless, this molecule seems to be a very excellent solvent for electrons in glassy matrices.

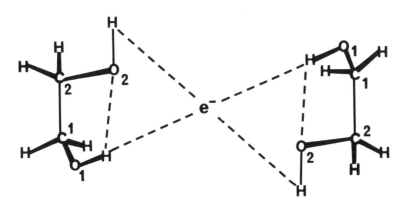

Fig. 9. Perspective drawing of a suggested solvation structure for electrons in ethylene glycol glass. The C_2, O_1, $H(O_1)$, O_2 atoms form a plane. The electron to $H(O_1)$ and $H(O_2)$ distances are 0.27 and 0.32 nm, respectivvely. The $H(C_2)$ and $H(C_1)$ hydrogens closest to the electrons are out of the plane and are both about 0.30 nm from the electron. The $H(C_2)$ and $H(C_1)$ hydrogens farthest from the electron are about 0.42 nm from the electron. Two additional molecules above and below the plane formed by the dashed lines, with their OH groups oriented away from the electron, are postulated. [From reference 76 with permission.]

Electron spin echo modulation data was analyzed on the specifically deuterated ethylene glycols, $DOCH_2CH_2OD$ and $HOCD_2CD_2OH$ [76]. The geometrical structure based upon the analyses of that data is shown in Fig. 9. This structure is consistent with the fact that different deuteron-electron distances are found for two sets of two deuterons from the analyses of the electron spin echo modulation data. Thus, two molecules have one of their O-H bonds oriented toward the electron whereas the other O-H bond in each of these ethylene glycol molecules is not oriented toward the electron. This accounts for the two different electron-deuteron distances found. The analysis of the data with deuterated alkyl groups in the ethylene glycol molecule indicates that four alkyl deuterons are closer to the electron and 12 alkyl deuterons are further from the electron. These results indicate that more than two ethylene glycol molecules are involved in the overall solvation shell. The closest alkyl deuterons, are interpreted as belonging to the four non-trans deuterons, relative to the electron, in the two solvating molecules shown in Fig. 9. The average distance found is consistent with the hydroxyl-deuteron distances as deduced from molecular models. The other four alkyl deuterons in the two solvating molecules shown in Fig. 9 are at a greater distance which is about the same as the distance to eight additional alkyl deuterons which must be from at least two additional molecules. It was suggested that these two additional molecules are above and below the approximate plane shown in Fig. 9 with their O-H groups oriented away from the electron such that the alkyl deuterons are the closest interacting deuterons. Thus, it has been concluded that a total of four first solvation shell molecules exist for the trapped, solvated electron in ethylene glycol glass, but only two of these have their O-H groups oriented toward the electron to provide the dominant potential of the solvated configuration.

3.5. STRUCTURE IN 2-METHYLTETRAHYDROFURAN GLASS

Electrons are also readily solvated in nonhydroxylic solvents and glassy matrices which are consequently less polar. These systems include ethers and alkanes. The cyclic ether, 2-methyltetrahydrofuran (MTHF), has been studied extensively as a glassy matrix that is effective for solvating and trapping electrons. The geometrical structure of trapped, solvated electrons in MTHF glass is shown in Fig. 10 which is based upon electron spin echo modulation data of a variety of selectively deuterated MTHF molecules in which the deuterium is located in different positions on the MTHF ring as well as in the methyl group [58].

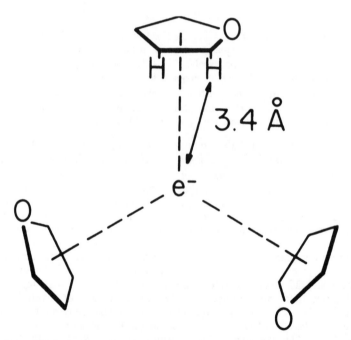

Fig. 10. Geometrical structure for solvated electrons in 2-methyltetrahydrofuran glass determined from electron spin-echo modulation and other electron spin resonance data. The isotropic hyperfine coupling to the nearest protons is 0.14 G which corresponds to 0.03% of the solvated electron spin density on each of the nearest first solvation shell protons. [From reference 78 with permission.]

These spin echo results show that the trapped, solvated electron interacts equivalently at the same interaction distance with deuterium located near and away from the polar oxygen end of specifically deuterated MTHF molecules. This suggests that the electron is located on a line perpendicular to the carbon atom plane of the surrounding MTHF molecules in the first solvation shell. Redundant analyses of both two-pulse and three-pulse electron spin echo modulation results show that the first solvation shell consists of three MTHF molecules oriented as shown in Fig. 10. It was also deduced that each MTHF molecule has an equal chance of being oriented with either side of its carbon plane toward the electron. The methyl group on carbon atom number 2 is axial and hence labels the side of the molecule that is oriented toward the electron for which the number of nearest deuterons differs. This statistical configuration of two possible orientations with respect to the electron is also consistent with second moment data analysis of the electron spin resonance lineshapes for the trapped, solvated electron in the same specifically deuterated matrices [58]. The methyl groups are intentionally not shown in Fig. 10 to imply that either side of the MTHF ring can be oriented perpendicular to a line toward the electron.

The semicontinuum model of electrons solvated by large nonhydroxylic molecules such as ethers and alkanes does indicate that the solvation geometry is dominated by maximizing the electron-molecular polarizability interaction. This means geometrically that the solvation geometry is dominated by the optimal orientation of CH_2 or CH_3 groups toward the electron. However no quantum chemical calculations have been carried out to assess the detailed molecular structure in the first solvation shell as shown in Fig. 10. With increased computer power such calculations should become feasible and it will be interesting to see what will be predicted in comparison with the experimental results.

3.6. STRUCTURE IN 3-METHYLPENTANE GLASS

Trapped, solvated electrons can also be formed in alkane glassy matrices such as 3-methylpentane and 3-methylhexane. In general, the electronic energy level structure deduced for trapped, solvated electrons in these alkane glassy matrices is quite analogous to that deduced for trapped, solvated electrons in ether matrices such as MTHF. Based on the geometrical structure found for the trapped, solvated electrons in MTHF it is anticipated that the geometrical structure of trapped, solvated electrons in alkane glasses will depend on optimizing the electron to C-H bond interactions. This is theoretically equivalent to maximizing the electron-alkyl polarizability interactions.

The experimental results on the geometry of trapped, solvated electrons in 3-methylpentane glassy matrix are not as complete as those for the MTHF glassy matrix. Electron spin echo modulation data has been analyzed for completely deuterated 3-methylpentane-d_{14} for which the number of equivalent interacting deuterons is found to be 18 to 21 [60]. Additional information is needed about the possible location of the electron with respect to the 3-methylpentane molecule in order to deduce the number of first solvation shell molecules.

A series of specifically deuterated 3-methylpentanes has also been studied, but it was not possible to obtain good electron spin echo modulation data on these systems because of interference from an alkyl radical which was generated simultaneously with the trapped, solvated electron. However, it was possible to detect changes in the electron spin resonance linewidth between the different specifically deuterated matrices, and a second moment lineshape analysis gave

404

information about the contributions from different positions in the 3-methypentane molecule. The analysis of the second moment data gave electron-to-proton group distances from 0.35 to 0.43 nm. These distances do not seem compatible with any *single* orientation of the 3-methylpentane molecules in the first solvation shell of the electron. However, they are compatible with a statistical distribution of different orientations as was found for electron solvation in MTHF glass. Fig. 11 shows two such orientations. In both these orientations the electron is in the carbon skeletal plane of 3-methylpentane, excluding the carbon of the methyl group at the 3 position, and is on a line approximately perpendicular to the terminal carbon-carbon bond and passing through the terminal carbon. In configuration 11a the electron interacts with five nearest protons in configuration 11b with seven nearest protons, for an average carbon interaction number of six protons per solvating 3-methylpentane molecule. So for a total of 18 interacting protons in the first solvation shell, as determined from the electron spin echo modulation data for the completely deuterated 3-methylpentane molecule, this corresponds to six 3-methylpentane molecules in the first solvation shell of the electron.

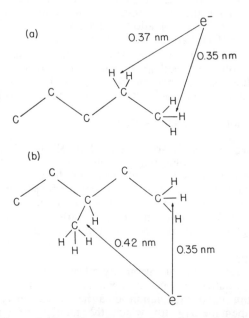

Fig. 11. Two equally probable orientations of one first solvation shell 3-methylpentane molecule with respect to the solvated electron viewed perpendicular to the carbon skeletal plane of 3-methylpentane. The scale is approximate. The distances shown are average distances to the proton groups shown. The solvated electron geometry in 3-methylpentane is suggested to be composed of three first solvation shell 3-methylpentane molecules each oriented statistically with configurations (a or b) and with the three molecules arranged trigonally around the electron. [From reference 78 with permission.]

A semicontinuum theoretical model with a structured first solvation shell for trapped, solvated electrons in liquid and glassy alkane matrices has been developed [21]. In this model the charge-molecular polarizability energy is dependent upon the molecular structure by considering the electron interaction with the polarizability of each C-C bond of the first solvation shell molecules. This model accounts for a number of features of the experimental data known for trapped, solvated electrons in 3-methylpentane glass and implies that the charge-molecular polarizability interaction with nearest neighbor CH_2 and CH_3 groups is probably dominant for controlling the solvation structure. It is also interesting that an *ab initio* quantum chemical calculation of an electron interacting with one molecule of propane for a particular configuraton shows some bonding [72]. This suggests that solvation of electrons by alkanes can be rationalized theoretically. It is anticipated that in the near future more realistic calculations of electron solvation in alkanes can be attacked.

4. Conclusions

In general, the geometrical structures of trapped, solvated electrons in aqueous and organic glasses determined primarily from electron spin echo modulation analysis with supporting data from second moment lineshape analysis as well as proton spin-flip satellite analysis of electron spin resonance spectra, as shown in Fig. 7 to 11, [78,79] have been supported by subsequent experimental and theoretical work. This support has dealt primarily with the solvated electron in aqueous systems and to a lesser extent with the solvated electron in ethanol systems. It is particularly significant that the rather detailed quantum statistical mechanical simulations of solvated electron structure in aqueous systems have given full support to the bond-dipole oriented model of the solvated electron which was advanced first on the basis of electron spin echo modulation data. This holds true even for the differences in pseudopotentials used in the quantum statistical mechanical calculations by several different groups. The concept of the electron having a compact spin density or wavefunction in a cavity of solvent molecules has also been fully supported by these detailed simulations. The number of first solvation shell molecules is at least consistent with the quantum statistical mechanical simulations, but this aspect is not as well supported as the bond dipole orientation. It is still possible that some average distribution of solvated molecules for the first solvation shell may occur in a dynamic aqueous system as well as in the frozen glassy matrices.

No quantum statistical mechanical calculations have been carried out for the structure of solvated electrons in ethanol, but perhaps this will be a focus of work in the near future. It is quite interesting that *ab initio* quantum chemical calculations support the difference between a bond-dipole orientation of first solvation shell molecules around a electron in water to a molecular dipole orientation of solvating molecules around an electron in ethanol [72]. As for the number of solvating molecules being reduced from six in the case of water to four in the case of the ethanol, this is supported by totally independent experimental evidence based upon an analysis of changes in the structure of the optical spectra in mixed matrices of ethanol and methyltetrahydrofuran as a function of the ethanol mole fraction [75].

The detailed energy level structure of trapped, solvated electrons has not advanced much beyond the experimental data with some semicontinuum model theoretical support of more than a decade ago. In particular, there is still some uncertainty in a complete interpretation for the solvated electron optical absorption spectrum, but this is clearly an area of active experimental and theoretical interest. The interest has mainly been on the theoretical side where attention is also focused on the process of localization and solvation of electrons in water and alcohols. The current experiments on picosecond studies of this process in both water and alcohol solvents [9,10] should lead to improved understanding of this process in the near future when coupled with detailed quantum statistical mechanical calculations.

References

1. W.B. Fowler, ed.: *Physics of Color Centers*, Academic Press, New York (1968).
2. A. Lund and S. Schlick: *Research Chem. Intermed.* , **11**, 37-66 (1989).
3. Larry Kevan: *J. Phys. Chem.*, **84**, 1232 - 1240 (1980).
4. Larry Kevan: *J. Phys. Chem.*, **82**, 1144 - 1148 (1978).
5. David C. Walker: *J. Phys. Chem.*, **84**, 1140 - 1144 (1980).
6. D. Schulte-Frohlinde and K. Eiben: *Z. Naturforsch A*, **17a**, 445 (1962).
7. E.J. Hart and J.W. Boag: *J. Am. Chem. Soc.*, **84**, 4090 (1962).
8. M. Hilczer, W.M. Bartczak and J. Kroh: *J. Chem. Phys.*, **89**, 2286 - 2291 (1988).
9. A. Migus, Y. Gaudel, J.L. Martin and A. Antonetti: *Phys. Rev. Lett.*, **58**, 1559 - 1562 (1987).
10. F.H. Long, H. Lu and K.B. Eisenthal: *J. Chem. Phys.*, **91**, 4413 (1989).
11. M. Ogasawara and L. Kevan: *J. Phys. Chem.*, **82**, 378 (1978).
12. M. Ogasawara, H. Yoshida and L. Kevan: *Chem. Phys. Lett.*, **66**, 295 (1979).
13. H. Shiraishi, K. Ishigare and K. Morokuma: *J. Chem. Phys.*, **88**, 4637 - 4649 (1988).
14. L. Kevan: In *Advances in Radiation Chemistry*, Vol. 4, M. Burton and J.L. Magee, eds., Wiley-Interscience, New York, 1974; pp. 181 - 305.
15. Stephen A. Rice and Larry Kevan: *J. Phys. Chem.*, **80**, 847 - 850 (1977).
16. T. Huang and L. Kevan: *J. Am. Chem. Soc.*, **95**, 3122 - 3128 (1973).
17. T. Huang, I. Eisele, D.P. Lin and L. Kevan: *J. Chem. Phys.*, **56**, 4702 - 4710 (1972).
18. Stephen A. Rice, G. Dolivo and Larry Kevan: *J. Chem. Phys.*, **70**, 18 -25 (1979).
19. K. Fueki, D.-F. Feng and Larry Kevan: *J. Am. Chem. Soc.* **95**, 1398 - 1407 (1973).
20. K. Fueki, D.-F. Feng and Larry Kevan: *J. Chem. Phys.* **59**, 6201 - 6208 (1973).
21. Toyoaki Kimura; Kenji Fueki, P.A. Narayana and Larry Kevan: *Can. J. Chem.*, **55**, 1940 - 1951 (1977).
22. Da-Fei Feng and Larry Kevan: *Chemical Reviews*, **80**, 1-20 (1980).
23. D. Copeland, N.R. Kestner and J. Jortner: *J. Chem. Phys.*, **53**, 1189 (1970).
24. N.R. Kester: In *Electron-Solvent and Anion-Solvent Interactions*, L. Kevan and B. Wedster, eds., Elsevier, Amsterdam, 1976; Chapter 1
25. K. Fueki, D. F. Feng and L. Kevan: *J. Phys. Chem.*, **74**, 1976 (1970).
26. David Chandler; Yashwant Singh and Diane M. Richardson: *J. Chem. Phys.*, **81**, 1975 - 1982 (1984).
27. Albert L. Nichols III; David Chandler; Yashwant Singh and Diane M. Richardson: *J. Chem. Phys.*, **81**, 5109 - 5115 (1984).
28. Michiel Sprik; Michael L. Klein and David Chandler: *J. Chem. Phys.*, **83**, 3042 - 3049 (1985)
29. Albert L. Nichols III and David Chandler: *J. Chem. Phys.*, **84**, 398 - 403 (1986).
30. Daniel Laria and David Chandler: *J. Chem. Phys.*, **87**, 4088 - 4092 (1987).
31. Charles D. Jonah, Claudio Romero and Annesur Rahman: *Chem. Phys. Lett.*, **123**, 209 - 214 (1986).
32. M. Sprik; R.W. Impey and M.L. Klein: *J. Statistical Phys.*, **43**, 9967 - 972 (1986).
33. Michiel Sprik and Michael L. Klein: *J. Chem. Phys.*, **87**, 5987 - 5999 (1987).

34. Michiel Sprik and Michael L. Klein: *J. Chem. Phys.*, **90**, 7614- (1989).
35. Michiel Sprik and Michael L. Klein: *J. Chem. Phys.*, **89**, 1592 1601 (1989).
36. Michiel Sprik and Michael L. Klein: *J. Chem. Phys.*, **91**, 5665 - 5671 (1989).
37. Jürgen Schnitker;Peter J. Rossky and Geraldine A. Kenney-Wallace: *J. Chem. Phys.*, **85**, 2986 - 2998 (1986).
38 Peter J. Rossky, Jürgen Schnitker and Robert A. Kuharski: *J. Statistical Phys.*, **43**, 949 - 965 (1986).
39. Jürgen Schnitker and Peter J. Rossky: *J. Chem. Phys.*, **86**, 3462 -3470 (1987).
40. Jürgen Schnitker and Peter J. Rossky: *J. Chem. Phys.*, **86**, 3471 - 3485 (1987).
41. Peter J. Rossky and Jürgen Schnitker: *J. Phys. Chem.*, **92**, 4277 - 4285 (1988).
42. Kazi A. Motakabbir, Jürgen Schnitker and Peter J. Rossky: *J. Chem. Phys.*, **90**, 6916 - 6924 (1989).
43. D.F. Coker; B.J. Berne and D. Thirumalai: *J. Chem. Phys.*, **86**, 5689 -5701 (1987).
44. A. Wallqvist; G. Martyna and B.J. Berne: *J. Phys. Chem.*, **92**, 1721-1730 (1988).
45. Uzi Landman; R.N. Barnett; C.L. Cleveland; Dafna Scharf and Joshua Jortner:*J. Phys. Chem.*, **91**, 4890 - 4899 (1987).
46. R.N. Barnett, Uzi Landman; C.L. Cleveland and Joshua Jortner: *Phys. Rev. Letts.*, **59**, 811 - 814 (1987).
47. R.N. Barnett, Uzi Landman; C.L. Cleveland and Joshua Jortner: *J. Chem. Phys.*, **88**, 4421 - 4427 (1988).
48. R.N. Barnett, Uzi Landman; C.L. Cleveland and Joshua Jortner: *J. Chem. Phys.*, **88**, 4429 - 4447 (1988).
49. R.N. Barnett, Uzi Landman and Abraham Nitzan:*J. Chem. Phys.*, **88**, 5567 - 5580 (1989).
50. C.Houée-Levin and J.-P. Jay-Gerin: *J. Phys. Chem.*, **92**, 6454-6456 (1988).
51. C.Houée-Levin; C. Tannous and J.-P. Jay-Gerin: *J. Phys. Chem.*, **93**, 7074-7077 (1989).
52. L. Kevan, M.K. Bowman, P.A. Narayana, R.K. Boeckman, V.F. Yudanov and Yu.D. Tsvetkov: *J. Phys. Chem.* **93**, 409 (1975).
53. B.L. Bales, J. Helbert and L. Kevan:*J. Phys. Chem.*, **78**, 221 (1974).
54. B.L. Bales, M.K. Bowman, L. Kevan and R.N. Schwartz: *J. Chem. Phys.*, **63**, 3008 (1975).
55. M. Bowman, L. Kevan and R.N. Schwartz: *Chem. Phys. Lett.*, **30**, 208 (1975).]
56. B.L. Bales and E. Lesin: *J. Chem. Phys.*, **65**, 1299 (1976).
57. L. Kevan: In *Time Domain Electron Spin Resonance*; L. Kevan, R.N. Schwartz, Eds.; Wiley-Interscience: New York, 1979; Chapter 8.
58. T. Ichikawa, L. Kevan, M.K. Bowman, S.A. Dikanov and Yu.D. Tsvetkov: *J. Chem. Phys.*, **71**, 1167 (1979).
59. W.B. Mims and J.L. Davis:*J. Chem. Phys.*, **65**, 4836 (1976).
60. P.A. Narayana and L. Kevan:*J. Chem. Phys.*, **65**, 3379 (1976).
61. J.R. Norris, M.C. Thurnauer and M.K. Bowman: In *Advances in Biological and Medical Physics*, J.H. Lawrence, J.W. Grofman and T.L. Hayes, Eds., Vol. 17, Academic Press, New York, 1980, p. 365.
62. S.A. Dikanov, V.F. Yudanov and Yu.D. Tsvetkov: *J. Struct. Chem.*, **18**, 370 (1977).
63. P.A. Narayana and L. Kevan: *Magn. Reson. Rev.*, **1**, 234 (1983).
64. K. Fueki, D.F. Feng and L. Kevan: *J. Chem. Phys.*, **56**, 5351 (1972).
65. D.P. Lin and L. Kevan: *Chem. Phys. Lett.*, **40**, 517 (1976).
66. H. Hase, F.Q.H. Ngo and L. Kevan: *J. Chem. Phys.*, **62**, 958 (1975).
67. P.A. Narayana, M.K. Bowman, L. Kevan, V.F. Yudanov and Yu.D. Tsvetkov: *J. Chem. Phys.*, **23**, 385 (1976).
68. S. Schlick, P.A. Narayana and L. Kevan: *J. Chem. Phys.*, **65**, 4836 (1976).
69. T.R. Tuttle, S. Golden, S. Lwenje and C.M. Stupak: *J. Phys. Chem.*, **88**, 3811 (1984).
70. H.F. Hameka, G.W. Robinson and C.J. Marsden: *J. Phys. Chem.*, **91**, 3150 - 3157 (1987).
71. A.V. Astashkin, S.A. Dikanov and Yu.D. Tsvetkov: *Chem. Phys. Lett.*, **144**, 258 (1988).
72. Timothy Clark and Gerd Illing: *J. Am. Chem. Soc.*, **109**, 1013 - 1020 (1987).

73. M. Narayana and L. Kevan: *J. Am. Chem. Soc.*, **103,** 1618 (1981).
74. Alan R. Reed and Timothy Clark: *Faraday Discuss. Chem. Soc.*, **85,** 365 - 372 (1988).
75 Tsuneki Ichikawa and Hiroshi Yoshida: *J. Phys. Chem.*, **93,** 5943 - 5947 (1989).
76. M. Narayana; Larry Kevan; P.O. Samskog; A. Lund and L.D. Kispert: *J. Chem. Phys.*, **91,** 2297 - 2299 (1984).
77. L. Kevan, T. Ichikawa and T. Ichikawa: *J. Phys. Chem.*, **84,** 4360 (1980).
78. Larry Kevan: *Accts. Chem. Research*, **14,** 138 - 145 (1981).
79. Larry Kevan: *Radiat. Phys. Chem.*, **17,** 413 - 423 (1981).

TRAPPED ELECTRONS IN CRYSTALS

HAROLD C. BOX
Biophysics Department
Roswell Park Memorial Institute
Buffalo, New York 14263
U.S.A.

INTRODUCTION

This chapter has a very specific topic - the trapping of electrons in single crystals. To be even more specific, consideration is limited to electrons trapped in single crystals by dipolar forces. Electrons trapped in vacancies, for example, as in F centers, are outside the scope of this review.

Several crystal structures of polyhydroxy compounds have been shown to trap electrons when subjected to ionizing radiation at low temperature [1-10]. This phenomenon provides new insights into the stabilization of charged species in an environment of electric dipoles. These insights derive from the fact that the matrix in which electron trapping occurs is a regular structure. It may be useful at the outset to enumerate those facets of the phenomenon which make electron trapping in single crystals so interesting.

1. Trapping occurs at preexisting trapping sites in the crystal structure. The mechanism of trapping is clearly different from the polaron mechanism of trapping where the charged particle itself purportedly initiates the trapping process by polarizing the surrounding medium [11].

2. Trapping occurs at a unique site in the crystal structure. All of the data derived to date from ESR-ENDOR measurements on single crystals show that a specific site in each crystal structure is utilized to trap electrons. The inference that only one specific site is occupied can be drawn unequivocally even though it may be difficult to deduce exactly where in the crystal structure the site is located. Implicit also in this conclusion is that electron trapping is not associated with imperfections or irregularities in the crystal structure.

3. The electric potential which makes trapping possible is generated by a favorable configuration of electric dipoles associated with hydroxy groups in the vicinity of the trap. The architecture of the preexisting configuration of dipoles can be examined in atomic detail. This information is available from the crystal structure as determined from X-ray or neutron diffraction studies.

4. The structure of the trap after electron accession is different from that of the preexisting trap. It is clear that after accession of an electron, the protons constituting the positive poles of the hydroxy dipoles forming the trap reposition somewhat. The protons in question are usually involved in hydrogen bonding and repositioning is expected to be facile. Experimental data on the repositioned protons comes from ESR or ENDOR studies which measure the hyperfine couplings between the electron and interacting protons. From the anisotropic component of a hyperfine coupling an electron-proton vector can be deduced giving the distance and orientation of the proton with respect to the electron [12].

5. Because of repositioning it may be difficult to correlate proton positions in the occupied trap, deduced from magnetic resonance measurements, with proton positions in the preexistant trap, available from crystal structure data. Data from non-exchangeable protons more remote

A. Lund and M. Shiotani (eds.), Radical Ionic Systems, 409–426.

110

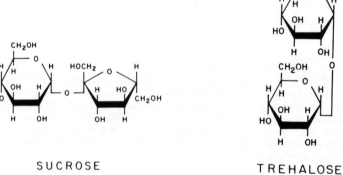

HOCH₂-(CH₂)₄-CH₂OH

1,6-HEXANEDIOL

HOCH₂-(CH₂)₆-CH₂OH

1,8-OCTANEDIOL

DULCITOL   D-SORBITOL   XYLITOL   PENTAERYTHRITOL

α-L-RHAMNOSE   β-L-ARABINOSE   GLUCOSE PHOSPHATE
di K SALT

SUCROSE   TREHALOSE

Fig. 1.   Polyhydroxy compounds for which ESR data on trapped electrons in single crystals are available [1-10].

from the trapping site may eventually prove more helpful in identifying the preexisting trapping site and relating it to the restructured site following electron accession. These protons are associated with carbon-bound hydrogen atoms. Being covalently bound to a relatively bulky polyhydroxy molecule, such protons are likely to remain fixed in their positions. Recently it has been shown that ENDOR measurements of hyperfine couplings can be obtained for a number of covalently bound hydrogen nuclei [13]. It is premature to judge how useful these additional data will prove, but the possibility should be explored that structural information can be obtained which is at once relevant to the occupied trap as well as to the preexistant trap.

Electron trapping has been observed in several crystals of polyhydroxy compounds exposed to ionizing radiation at low temperature. The simplest are the diols, 1,6-hexanediol and 1,8-octanediol [10]. The simple polyhydroxy compounds, xylitol [5], sorbitol [5], dulcitol [2] and pentaerythritol [9] also exhibit the phenomenon. There is evidence from ESR spectroscopy that inositol traps electrons but an analysis of the data has not been published [14]. Another class of compounds represented among the single crystal electron trappers are the monosaccharides. Arabinose [1,2], rhamnose [3] and the dipotassium salt of glucose-1-phosphate [4] represent this class. The crystal structures of the disaccharides sucrose and trehalose also trap electrons [3,8]. The molecular structures of these compounds are displayed in Fig. 1. Their obvious common denominator is having multiple hydroxy substituents. It is also noteworthy that none has a substituent that imparts a significant electron affinity to the molecule. It has been the general experience that crystalline organic compounds subjected to ionizing radiation at sufficiently low temperature yield free radical products that can be related to radiation-induced oxidation and reduction processes. On the reduction side electron addition or dissociative electron attachment are typical processes yielding free radical products. The compounds shown in Fig. 1 have little electron affinity and electron trapping may be viewed as the energetically favorable alternative to molecular reduction.

2. Observations and Measurements

In this section we review the experimental evidence concerning electrons trapped in single crystals. Data is available from two spectroscopies, namely electron spin resonance (ESR) and optical absorption. ESR spectroscopy and its adjunct, electron-nuclear double resonance (ENDOR) spectroscopy, have provided detailed information about trapping sites. Optical absorption spectroscopy and related photobleaching phenomenon provide data on the depth of the traps. The results of studies on the reaction of electrons with adjacent trapping molecules are included in this section.

2.1. ESR Absorption

Compelling evidence that single crystals of polyhydroxy compounds can trap electrons comes from ESR studies. Representative ESR absorption spectra obtained from an X-irradiated single crystal of rhamnose are shown in Fig. 2. The crystal was maintained at 4.2°K during irradiation and at 1.6K for the ESR measurements. The absorption pattern contains three components arising from different free radical products. The overall pattern varies markedly with orientation of the crystal with respect to the applied magnetic field. However the g value of the trapped electron absorption shows only a small anisotropy and is always less than 2.0023 (the g value of the free electron); consequently the trapped electron absorption is the high field component in each of the spectra shown in Fig. 2. This characteristic of the ESR absorption is consistent with a trapped electron interpretation, a g value of less than the free spin value being indicative of an excess electron [15]. Another feature of the ESR absorption consistent with a trapped electron attribution is the effect of deuteration on the hyperfine pattern. The overall width of the absorption is greatly reduced when the crystal is grown out of heavy water. The trapping site is constituted of hydroxy groups and the hyperfine pattern arises mainly from magnetic interaction between the electron and the protons of these hydroxy groups. Replacement of the exchangeable hydroxy hydrogen atoms by deuterium atoms reduces the hyperfine interaction because of the smaller magnetic moment of the deuteron compared with the proton. The effect of deuteration is shown on the right hand side of Fig. 2.

412

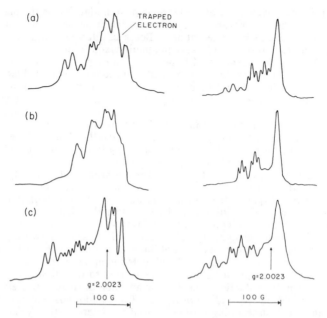

Figure 2.   V band (70 GHz) ESR absorption spectra from single crystals of rhamnose x-irradiated at 4.2°K. Spectra on the left are from crystals grown out of H_2O; those on the right from crystals grown out of D_2O. The applied magnetic field was parallel to the a, b and c axis in (a), (b) and (c) respectively. The trapped electron absorption occurs at a g value close to 2.0023.

A full description of the ESR absorption of a free radical is embodied in the spin hamiltonian

$$H = \beta H \cdot g \cdot S - g_N \beta_N \sum_i H \cdot I_i + \sum_i I_i \cdot A_i \cdot S \qquad (1)$$

where S and I_i are, respectively, the spin angular momenta for an electron and interacting nuclei. The first term is the Zeeman energy for the electron. The effective magnetic moment of the electron is given by $\beta g \cdot S$, where β is the Bohr magneton and g is a tensor quantity. Eq. (1) serves equally well to describe the trapped electron. For the trapped electron the g value is so nearly isotropic the tensor g can be replaced by a scalar quantity without significantly compromising the description of the absorption. The second term in (1) accounts for the Zeeman energy of magnetic nuclei. In the case of the trapped electron this term refers to protons in the environment which interact with the electron. The interaction between the ith proton and the electron is described by the third term where A_i is the hyperfine coupling tensor for the ith proton. The tensor quantities A_i can be measured experimentally and provide the principal insights into the nature of the electron trap. For example, Samskog et al. [8], identified the trapped electron in trehalose crystals from the g value of the ESR absorption and the fact that the protons responsible for the hyperfine structure were exchangeable. From ESR measurements of the hyperfine splittings with the magnetic field applied parallel to three orthogonal planes, these investigators deduced the hyperfine coupling tensors listed in Table 1. The tensors are described in terms of their three principal values and the direction cosines of the principal axes.

Table 1.   Exchangeable proton couplings associated with the trapped electron absorption in single crystals of irradiated trehalose as deduced from ESR and from ENDOR measurements. Principal values are in MHz.

Tensor	Principal Values	a	b	c
	ESR Determination			
A_1	47.6	0	0	1
	0	0	1	0
	0	1	0	0
A_2	86.8	1	0	0
	47.6	0	1	0
	47.6	0	0	1
	ENDOR Determination			
A_1	51.77	-0.250	-0.219	0.942
	8.17	0.018	0.972	0.231
	6.95	0.967	-0.075	0.239
A_2	78.57	0.822	0.449	0.350
	36.63	-0.377	0.890	-0.254
	34.45	0.425	0.076	0.901

Thus, ESR studies sufficed for recognizing the trapped electron phenomenon in trehalose. However, ENDOR is able to provide more accurate data on hyperfine couplings.

2.2. ENDOR Studies: Exchangeable Protons.

The widths of individual lines in the hyperfine patterns of ESR spectra from single crystals are invariably broadened due to weak unresolved couplings from a multiplicity of weakly coupled nuclei. Additional difficulties in interpreting ESR spectra may arise due to overlapping of the trapped electron spectrum with the spectra of other paramagnetic species. These difficulties limit the accuracy of hyperfine coupling tensors obtainable from ESR spectra. The difficulties are largely overcome in the ENDOR method of deducing hyperfine couplings [12]. The nuclear resonances, detected indirectly by the ENDOR method, have much narrower linewidths compared with ESR resonances. For this reason our discussion of the hyperfine couplings between the trapped electron and protons of the surrounding hydroxy groups will rely mainly on ENDOR data. The coupling tensors for the protons interacting with the trapped electrons in trehalose, as determined from ENDOR measurements (previously unreported), are listed in Table 1. These data are in agreement with the results of Samskog et al. (Table 1) to within the accuracy of the ESR data. The trehalose structure belongs to the $P2_12_12_1$ space group and the asymmetric unit is one molecule [16,17]. The tensors listed in Table 1 refer to one of four symmetry-related sites in the unit cell of the trehalose structure. It is typical that ENDOR spectroscopy is able to distinguish among the absorptions of symmetry-related sites, whereas ESR spectroscopy often does not.

Table 2 contains a compilation of the principal values of hyperfine coupling tensors obtained from ENDOR measurements on electrons trapped in single crystals. The hyperfine coupling tensor defining the magnetic interaction between the electron and a proton consists of two contributions. The Fermi contact interaction gives rise to an isotropic component, A_{ISO}, which is the mean of the three principal values of the hyperfine coupling tensor. The spin density of the electron at the proton can be calculated from A_{ISO} using the expression,

$$\rho = 3A_{ISO}h/8\pi g_N \beta_N \tag{2}$$

where h is Planck's constant and the other quantities were defined in (1). The magnetic dipole-dipole interaction between the electron and nuclear spins gives rise to an anisotropic contribution. The listing A_{MAX}-A_{ISO} in Table 2 provides a measure of the anisotropic interaction. Assuming a point dipole model for the electron distribution, the electron-proton distance can be calculated from A_{MAX}-A_{ISO} using the expression,

$$R = [2g\beta g_N \beta_N / h(A_{MAX} - A_{ISO})]^{1/3} \tag{3}$$

The validity of (3) improves as R becomes large compared with the mean radius of the electron distribution. The calculation of the expectation value of R for a distributed electron presents no problem provided the wavefunction for the trapped electron is known [18-20]. However, rather than introduce wavefunctions of questionable validity, the simple approximation (3) was used to compile the values of R listed in Table 2. In this approximation it is seen that the electron-hydroxy proton distances range from 1.59Å upward. Allowing for delocalization of the electron would yield a smaller calculated value of R.

Table 2.   Exchangeable proton couplings. A_{MAX}, A_{INT} and A_{MIN} refer to principal values of the hyperfine coupling. A_{ISO} is the isotropic component. The quantities ρ and R were calculated using Eqs. (2) and (3) respectively.

	A_{MAX}	A_{INT}	A_{MIN}	A_{ISO}	A_{MAX}-A_{ISO}	ρ	R
Dulcitol	132.34	79.94	73.14	93.14	39.20	0.065	1.59
Arabinose	88.30	31.12	30.24	49.89	38.41	0.035	1.60
Dulcitol	132.22	76.88	74.54	94.55	37.67	0.066	1.61
Arabinose	71.26	16.54	14.96	34.25	37.01	0.024	1.62
Arabinose	50.14	-2.06	-3.28	14.93	35.21	0.011	1.65
Rhamnose	42.98	-8.08	-9.48	8.47	35.51	0.006	1.66
Sorbitol	76.02	26.28	24.64	43.21	33.70	0.030	1.67
Sorbitol	65.96	16.74	16.02	32.90	33.05	0.023	1.68
Sucrose	85.00	37.66	34.56	52.41	32.59	0.037	1.69
Rhamnose	88.50	40.32	39.88	56.23	32.27	0.039	1.70
Sucrose	78.66	33.08	29.66	47.13	31.53	0.033	1.71
Rhamnose	72.44	26.46	24.96	41.29	31.15	0.029	1.72
Xylitol	75.94	31.84	27.80	45.19	30.74	0.032	1.73
Glucose phosphate	86.40	41.42	40.18	56.00	30.40	0.039	1.73
Glucose phosphate	49.76	5.56	3.68	19.67	30.09	0.014	1.74
Sucrose	20.86	-7.87	-8.88	1.37	29.74	0.001	1.74
Xylitol	65.72	22.66	19.98	36.12	29.60	0.025	1.75
Trehalose	51.77	8.17	6.93	22.28	29.49	0.016	1.75
Trehalose	78.57	36.63	34.45	49.88	28.69	0.035	1.76

2.3. ENDOR Studies: Non-exchangeable Protons.

Recently it has been demonstrated that the interactions of trapped electrons with non-exchangeable protons can be detected by ENDOR spectroscopy [13]. These protons belong to carbon-bound hydrogen atoms and are more distant from the electron. Consequently the hyperfine couplings are considerably smaller than those of exchangeable hydroxy protons. The first measurements of non-exchangeable proton couplings were made for electrons trapped in rhamnose crystals. The crystals are monoclinic belonging to the space group P2$_1$ with two molecules per unit cell [22,23]. Plots of the ENDOR resonances obtained from the trapped electron signal in crystals of rhamnose are shown in Figs. 3 and 4. The actual laboratory data shows some shifting of ENDOR resonances as a function of magnetic field orientation due to changes in the g value of the absorption. The resonances plotted in Figs. 3 and 4 were adjusted to a magnetic field strength corresponding to a resonance frequency of 106 MHz for free protons.

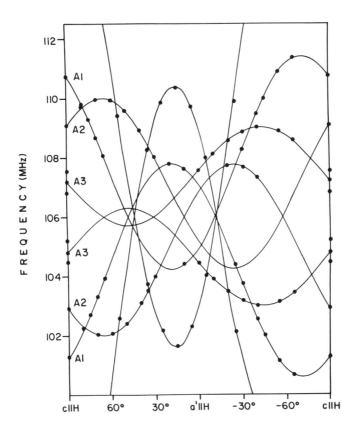

Figure 3. ENDOR resonance frequencies plotted as a function of magnetic field direction applied in the plane perpendicular to the b axis of rhamnose. These data, from the trapped electron component of the ESR absorption of irradiated rhamnose, are adjusted to an applied magnetic field strength corresponding to a free proton resonance frequence of 106 MHz.

Each proton coupling (A1, A2 and A3) gives rise to two resonances symmetrically disposed above and below the free proton resonance frequency. Furthermore the two symmetry-related sites in the unit cell give rise to distinct resonances (Fig. 4) except when the magnetic field is applied parallel or perpendicular to the b axis. When the magnetic field is parallel or perpendicular to b the two sites are equivalently oriented with respect to the field. The extraneous and highly anisotropic resonances appearing in Figs. 3 and 4 are due to an exchangeable proton. Other exchangeable proton resonances associated with the trapped electron signal in rhamnose are off scale in these figures [3].

The limited data available on carbon-bound proton coupling tensors are tabulated in Table 3. The spin density at the proton and the electron-proton distance calculated from Eqs. (2) and

416

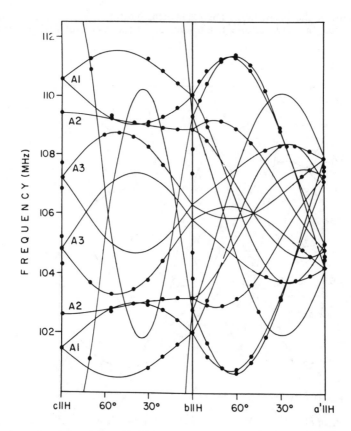

Figure 4.   ENDOR resonance frequencies plotted as a function of magnetic field direction applied in planes perpendicular to the a' and c axes of rhamnose. These data, from the trapped electron component of the ESR absorption of irradiated rhamnose, are adjusted to an applied magnetic field strength corresponding to a free proton resonance frequency of 106 MHz. Except when b is parallel or perpendicular to the field, site splitting occurs.

(3) are also listed. The most interesting characteristic of these tensors is the fact that the isotropic components, and therefore the spin densities, are all negative which is in marked contrast with the hydroxy proton spin densities listed in Table 2. The basis for judging the sign of trapped electron spin densities merits comment. ESR and ENDOR measurements do not directly yield the overall sign of the hyperfine coupling. Thus, a priori, either branch of the observed ENDOR frequencies for a given coupling in Figs. 3 and 4 might be followed in deducing a coupling tensor. Alternative choices yield principal values opposite in sign. The appropriate choice rests on the fact that in the trapped electron milieu, proton-electron dipole-dipole couplings are axially symmetric or nearly so. The dipole-dipole coupling tensor is axially symmetric if the electron has a spherical distribution or if the average electron-proton distance is large compared with the dimensions of the electron distribution. Both conditions are approximately satisfied in the case of trapped electrons. The principal axis of the hyperfine coupling tensor must correspond with maximum positive dipole-dipole coupling. The approximately equal transverse principal values correspond to negative dipole-dipole coupling. These considerations determined the overall choice of sign for the hyperfine coupling and consequently for the isotropic component for both

the hydroxy proton couplings (Table 2) and the carbon-bound proton couplings (Table 3). This feature, that the dipole-dipole coupling is axially symmetric, has been a helpful feature of trapped electron studies. The implications of the positive and negative spin densities on the hydroxy and carbon-bound protons respectively are discussed below.

Table 3. Non-exchangeable proton couplings. A_{MAX}, A_{INT} and A_{MIN} refer to principal values of the hyperfine coupling. A_{ISO} is the isotropic component. The quantities ρ and R were calculated using Eqs. (2) and (3) respectively.

	A_{MAX}	A_{INT}	A_{MIN}	A_{ISO}	$A_{MAX}-A_{ISO}$	ρ	R
Sucrose	4.46	-15.56	-16.75	-9.36	14.02	-0.0066	2.24
Sucrose	5.33	-10.64	-13.02	-6.11	11.44	-0.0043	2.40
Rhamnose	5.89	-10.32	-11.20	-5.21	11.10	-0.0036	2.42
Sucrose	5.63	-10.19	-11.44	-5.34	10.97	-0.0037	2.43
Rhamnose	4.12	-5.84	-8.63	-3.46	7.50	-0.0024	2.75
Rhamnose	4.06	-4.02	-6.52	-2.16	6.22	-0.0015	2.94

2.4. Photobleaching.

From the outset of studies on electrons trapped in crystals, the photobleachability of trapped electron absorptions was noted. Photobleachability of the optical absorption is a basic characteristic of the trapped electron phenomenon, not only in crystals, but in other media as well. The loss of trapped electrons upon exposure of the crystal to light can be observed by monitoring either the ESR spectrum or the optical absorption spectrum. The electrons absorb sufficient energy to either escape the trap or undergo a reaction. The optical absorption spectra of trapped electrons, considered in the following section, are a more refined observation of light absorption by trapped electrons.

2.5. Optical Absorption Spectra.

Electrons trapped in single crystals have optical absorption properties similar to those for electrons trapped in alcohols or in aqueous glasses. The latter matrices have been employed in numerous studies of trapped electrons. Ershov and Pikaev [24] have tabulated the wavelength at maximum absorption, λ_{MAX}, reported from many such experiments. For alcohols λ_{MAX} ranges between 485 and 585 nm. For glassy aqueous solutions λ_{MAX} ranges between 510 and 617 nm . For electrons trapped in crystals λ_{MAX} is shifted toward shorter wavelengths. Buxton and Salmon [25] were the first investigators to measure the trapped electron absorption in crystals and reported the "bluest electrons so far observed in molecular solids or liquids". The value of λ_{MAX} was 476 nm. Their measurements were made on sucrose crystals maintained at 6K.

In rhamnose crystals the trapped electron absorption has a λ_{MAX} = 500 nm at 278K, whereas at 4K, λ_{MAX} = 400 nm [26,27]. A similar shifting of λ_{MAX} to longer wavelengths at higher temperatures is also observed for electrons trapped in glasses where the phenomenon has been attributed to lattice expansion [27]. The λ_{MAX} value is not affected by replacing the exchangeable hydrogen atoms of rhamnose by deuterium. The optical absorption spectrum of electrons trapped in crystals of deuterated rhamnose is shown in Fig. 5. The width of optical absorption spectra of electrons trapped in crystals are broad. In rhamnose, for example, the width at half maximum is 250 nm (Fig. 5).

The optical absorption spectrum of δ-irradiated pentaerythritol crystals could be resolved into two components [9]. The shorter-lived component is associated with trapped electrons and has a λ_{MAX} = 425 nm [9]. The optical absorptions of electrons trapped in 1,6-hexanediol and 1,8-octanediol are unusual because the spectra exhibit resolved peaks [28]. In 1,8-octanediol these peaks occur at 740, 590 and 490 nm. This feature has been attributed to variation in the density of states in the continuum. Also, these spectra do not change on warming the crystals from 4K to 77K.

418

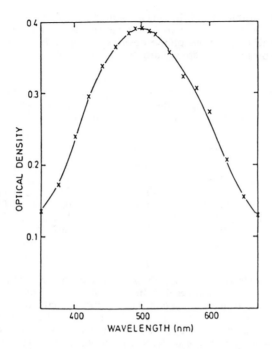

Fig. 5.   The absorption spectrum of a transient species from a deuterated single crystal of
rhamnose at 5.5°C. This result is from a pulse radiolysis study by Samskog et al. [26].

It is usually assumed that the optical absorption of trapped electrons is indicative of the depth of the trap. Thus, the shorter wavelengths associated with the trapped electron absorption in crystals compared with other matrices indicate a more deeply trapped electron. This conclusion is consistent with proton hyperfine couplings which typically are larger in crystals compared with other media. The larger couplings reflect smaller electron-proton distances. These protons are the positive poles of the hydroxy dipoles that constitute the trap. Since the trapping potential generated by the protons varies inversely as the electron-proton distances, the traps in crystals exhibiting larger couplings are expected to be correspondingly deeper.

2.5. Hydroxyalkyl Radical Formation.

The initial ESR-ENDOR studies of electrons trapped in single crystals were carried out on rhamnose crystals X-irradiated at 4.2K [3]. Warming the crystals to 77K resulted in irreversible loss of the trapped electron signal. Conversion of the trapped electron absorption upon warming to that of another paramagnetic species was not detected. However, subsequent investigations by Samskog et al. [26] using rhamnose crystals partially deuterated by replacement of exchangeable hydrogen atoms demonstrated that on warming the electrons react to form a free radical product. The product was identified as an hydroxyalkyl radical. The transformation is envisaged as a two step process:

$$>CHOD + e^- \rightarrow >CHO^- + D$$

$$>CHOD + D \rightarrow >COD + HD$$

3. Computational Studies

Computational studies have been aimed at identifying the features of crystal structures that favor electron trapping. Intuitively two structural features seem necessary for electron trapping to occur. Firstly, there should exist void volumes in the crystal structure that will accommodate the electron. The electron must be able to delocalize somewhat, else the momentum of the electron mitigates against trapping. Secondly the electric potential within the void must favor trapping.

3.1. Search for Voids

The electron trapping phenomenon focuses our attention on a usually ignored aspect of crystal structures, namely the space between molecules. A program was written to search for and quantitate voids. The imput data consisted of the coordinates of all atoms in the reference unit cell. The space within the unit cell was examined at a network of points spaced approximately Å/4 along the three crystal axes. Each point was evaluated as to whether it was a distance R_{MIN} or greater from all atoms in the structure. Atoms in the unit cell and in the 26 unit cells which share a face, an edge or a point with the reference unit cell were included in making the evaluation. At each point where this criterion was satisfied, the electric potential was calculated by summing the contributions from each polar hydroxy group in the vicinity of the point. The potential was calculated in terms of the quantity A,

$$V = \sigma A / l \tag{4a}$$

$$A = \sum_{N} \left(\frac{1}{R(H)_N} - \frac{1}{R(O)_N} \right) \tag{4b}$$

where V is the potential, σ is the electric dipole moment of an hydroxy group, l is the OH bond length, $R(H)_N$ and $R(O)_N$ are the distances of the point from the hydrogen and oxygen atoms of the Nth hydroxy group. For most space groups the electric potential calculated from the contributions of dipoles from the 27 unit cells does not reflect exactly the inherent crystal symmetry. In order to maintain exact symmetry the summation over hydroxy atoms was curtailed to those within a distance R_0 of the point in question, where R_0 was chosen such that for any point within the reference cell, the sphere defined by R_0 lay within the 27 unit cells employed in the calculation. To compensate for truncation of hydroxy groups resulting in an excess of either hydrogen or oxygen atoms being included in the potential calculation, a correction plus or minus n/R_0 was added to the calculation of A, where n is the number of hydrogen atoms either deficient or excess due to truncation.

The crystal structure of dulcitol was analyzed for likely electron trapping sites. Crystals of dulcitol are monoclinic belonging to the space group $P2_1/c$. There are 4 molecules per unit cell [29]. For this calculation R_{MIN} was set at 1.25Å, R_0 at 5Å and A>0.25(Å)$^{-1}$. The void outlined in the projections of Fig. 6 was identified as a likely trapping site. This void is centered 0.75 along $\underline{a}$, 0.36 along $\underline{b}$ and 0.24 along $\underline{c}$ [30]. There are, of course, four such symmetry-related sites per unit cell.

Electron trapping in the dulcitol structure is particularly interesting from another point of view. This trap is composed of two hydroxy groups as indicated by the observed exchangeable proton couplings (see Table 2). The principal values of the two proton coupling tensors are nearly equal. Moreover the axes of axial symmetry for the two tensors point in the same general direction suggesting a trapping site having an approximate center of symmetry. Examination of the crystal structure of dulcitol reveals a pseudo center of symmetry. This pseudo center is located at the midpoint between two hydrogen bonds that connect O5 and O6 of one molecule with O1 and O2 of an adjacent molecule (see Fig. 6). In terms of fractional coordinates this point is located 0.76 along $\underline{a}$, 0.38 along $\underline{b}$ and 0.27 along $\underline{c}$. Thus, both symmetry considerations and a calculational analysis of the structure suggest the same trapping site. It can be asserted with

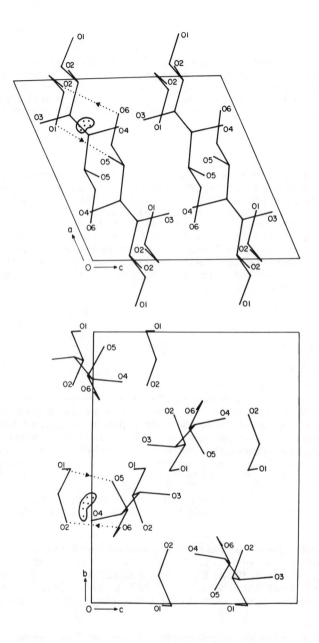

Figure 6. The crystal structure of dulcitol projected along the b and a' = b x c axes. The void wherein the electron is trapped is outlined. The hydrogen bonds formed by the two hydroxy groups of the trap are indicated by dotted lines.

some confidence that two hydroxy groups in apposition, as in Fig. 6, constitute a structural motif capable of trapping electrons.

3.2. Best Fit Calculations

An implicit assumption in the preceding discussing of electron trapping in dulcitol is that the center of symmetry retains symmetry after electron accession, i.e. whatever rearrangement of protons takes place, occurs symmetrically. Intuitively, the assumption seems reasonable. Unfortunately, this argument is not generally available for locating the trapping site in other crystal structures. Consider the crystal structure of rhamnose, shown in projection in Fig. 7. Calculation suggests two possible trapping sites. The larger void is probably the actual trapping site but it would be of interest to further substantiate this conclusion. It will be recalled that trying to match electron-proton distances determined from ENDOR measurements with electron-proton distances determined from crystal structure data seems an unpromising exercise for exchangeable protons due to their likely repositioning upon accession of the electron. Carbon-bound protons are less subject to repositioning. Taking cognizance of this an additional parameter has been incorporated into the void calculation. G in Eq. (5) is a measure of the agreement between electron-proton vectors, R_i, deduced from hyperfine coupling tensors and electron-proton vectors, r_i, calculated from crystal structure for an electron located at point X, Y and Z.

$$G = \sum_i [(R_{ix}-r_{ix})^2 + (R_{iy}-r_{iy})^2 + (R_{iz}-r_{iz})^2] \qquad (5)$$

Three ambiguities are encountered in the evaluation of G: (1) The sense of electron-proton vectors is not determined from ENDOR measurements. (2) A correlated set of ENDOR tensors can be referred to either one of two symmetry-related sites in the crystal structure. (3) The vectors obtained from ENDOR data can be paired with the vectors obtained from crystal structure in different ways. The best (lowest) value of G was obtained by testing all possibilities allowed by these ambiguities. For the rhamnose structure, using the carbon-bound proton data given in Table 3, the best G value obtained within the larger void outlined in Fig. 7 is 1.51. The best fit utilized atoms H1, H6 and H6', all on the same molecule. For the smaller void the best value that can be obtained for G is 2.72.

4. DISCUSSION

4.1. Trap Depths.

The most basic parameter that describes an electron trap is the trap depth. In single crystals, as in frozen alcohols and glasses, the optical absorption can be taken as an indicator of trap depth. Shorter λ_{MAX} values are generally observed for electrons trapped in crystals than for electrons trapped in frozen alcohols or glasses indicating the traps are deeper in single crystals. The λ_{MAX} value is not a precise indicator of trap depth because of the breadth of the absorption. The fact that the absorption in single crystals is ~200 nm wide is interesting in itself. In frozen alcohol and glasses it is unlikely there is a well defined trap structure and it was thought possible the breadth of the optical absorption spectra in these media could be due to the variety of trap depths. Since in single crystals the electrons occupy identical traps this explanation of line breadths cannot apply. It seems likely that the breadth of optical absorptions of trapped electron arises from the continuum of states above the true trap depth.

4.2. Trap Dimensions.

Most of the hyperfine couplings between hydroxy protons and the trapped electron measured in single crystals to date have been markedly larger than those observed in frozen alcohols and glasses. In the latter media the width of the ESR absorption serves as an overall measure of the hyperfine coupling even though individual couplings are not resolved. The larger couplings observed in crystals reflect smaller electron-proton distances and consequently deeper potential

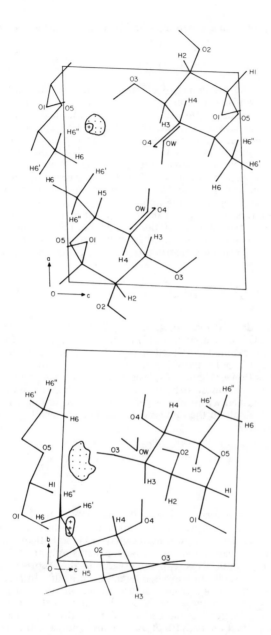

Fig. 7. The crystal structure of rhamnose projected along the b and a' = b x c axes. Voids considered as possible electron trapping sites are outlined.

wells. No systematic measurements of hyperfine couplings versus $(\lambda_{MAX})^{-1}$ have been attempted in single crystals but a close correlation is to be expected.

In an interesting study, Muto et al. [31] measured the width of the ESR absorption spectra associated with trapped electrons in mixed crystals of binary n-alkanes. In these systems electrons are trapped in cavities that result from chain length differences. Trapping is not observed in neat alkanes. Cavity size is proportional to the difference in chain length between the components of the binary system. The width of the ESR absorption in these systems is found to be inversely related to cavity size, a result consistent with the preceding discussion of linewidths and consistent with early views of the origin of linewidths associated with the ESR of trapped electrons [32].

4.3. Trap Geometries.

An additional experimental tool could be brought into play for studying trapped electrons when the scope of these studies was extended to single crystals. Electron-nuclear double resonance or ENDOR spectroscopy provides detailed information concerning electron-proton couplings in single crystals. New insights into the trapped electron phenomenon have come from ENDOR studies. Complete determinations of the hyperfine couplings associated with the trapped electron absorption have been reported from ENDOR studies on eight different crystal structures. In five cases two hydroxy groups constitute the trap and in three cases three hydroxy groups form the trap. From the anisotropic component of the hyperfine coupling tensors, electron-hydroxy proton distances have been inferred. The distances range from 1.59 and 1.76Å. However within a given structure the distances show a pleasing consistency. The difference between electron-proton distances in traps formed by two hydroxy groups do not exceed 0.02Å. In traps formed by three hydroxy groups the differences do not exceed 0.05Å. This consistency is probably achieved by repositioning of the protons upon accession of an electron.

The extent to which polar protons rearrange in crystals upon accession of an electron has been an open question since the earliest observations of trapping in crystals of polyhydroxy compounds [2]. An extreme view would be that the preexistant trap remains unchanged by accession of the electron. The other extreme view would be that the vestige of trap structure after electron accession bears little resemblance to the preexistant trap. Nature's response lies somewhere between these extremes. It has long been known that the protons of hydrogen bonds in single crystals subjected to ionizing radiation can relocate over significant distances. Even at the temperature of liquid helium protons of hydrogen bonds rearrange in response to the oxidation or reduction of a molecule involved in the bond [33-36]. For this reason, reliance on exchangeable proton coupling data to locate the electron is unwise. Identification of the trapping site in the crystal structure shown in Fig. 6, based on symmetry considerations and electric potential calculations, is reasonably certain. More remote structural features of the structure may contribute to the trapping potential but clearly the pair of dipoles is the primary feature of this trapping site. It may be preferable to describe trapping geometries in terms of the hydrogen bonds formed by the participating hydroxy groups. Thus, in Fig. 6 the trap is delineated by a pair of hydrogen bonds in apposition. Whereas the proton constituents of these bonds relocate upon addition of the electron, some form of bonding between the donor and receptor oxygen atoms likely persists.

4.4. Spin Density Distributions.

Another insight that derives from ENDOR studies concerns the distribution of spin densities in the vicinity of the trapped electron. The accurate determination of hyperfine couplings allows the sign of spin densities at the exchangeable and non-exchangeable protons to be inferred. The spin density at hydroxy protons is positive whereas the spin density at protons of carbon-bound hydrogens is negative. This interesting result can be interpreted in terms of a valence bond model. In Fig. 8 the three orbitals of interest are shown schematically, namely the orbital of the trapped electron and the bonding orbitals of an OH fragment. The two dispositions of three electron spins corresponding to formation of an hydroxy bond are shown. On the basis of Hund's rule disposition (a) should be favored. This preference places a net positive spin on the hydroxy

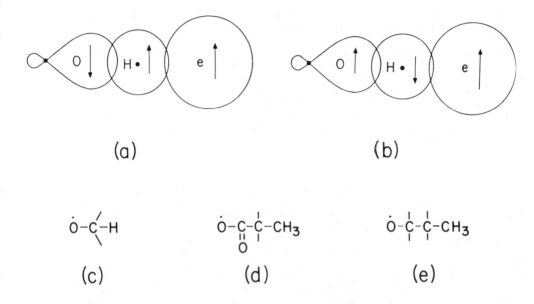

(a)   (b)

(c)   (d)   (e)

Figure 8.   (a) and (b) Disposition of electron spins in a valence bond depiction of the orbital of
a trapped electron and the bonding orbitals of a nearby hydroxy group.

proton and a negative spin density on the oxygen atom. The negative spin density on oxygen is,
of course, not experimentally observable since the ^{16}O nucleus does not have a magnetic moment.
It seems likely that the negative spin densities appearing on the carbon-bound protons are the
result of the intramolecular transmission of spin density from the oxygen atom. The expectation
is based on results from studies of oxygen centered free radicals. Alkoxy radicals, Fig. 8(c), in
which the unpaired electron is localized primarily on the oxygen atom, exhibit large spin densities
at β protons. The isotropic component of the β-proton hyperfine coupling is given approximately
by

$$A_{ISO} = Bcos^2\Theta \tag{6}$$

where B is 336 MHz [37]. The dihedral angle Θ is between two planes passing through the O-C
bond axis; one plane contains the axis of the oxygen p orbital bearing the unpaired electron, the
other contains the β-proton. The spin density at the proton is the same sign as on the oxygen
atom. Couplings between the unpaired electron of alkoxy radicals and γ-protons have also been
observed but are generally of much less magnitude [37]. Relevant to our discussion are the
substantial δ-proton couplings observed in alkoxy radicals containing methyl groups, Figs. 8(d)
and 8(e) [38,39]. Hyperfine couplings of up to 70 MHz have been reported between the unpaired
electron and a δ-proton of a methyl group. In these radicals also, the spin density on the δ-
proton is the same sign as on the oxygen atom. Returning to a consideration of the negative
spin densities anticipated on the oxygen atoms of hydroxy groups constituting the electron trap,
we may expect significant negative densities to be transmitted to protons β and δ relative to
oxygen. This point of view explains the consistently negative spin densities observed at proton
carbon-bound hydrogen atoms in the vicinity of a trapped electron.

4.5. Trap Models.

Various arrangements of water or alcohol molecules for the trapping or solvation of electrons have been evaluated theoretically [40,43]. The subject has been reviewed by Feng and Kevan [44]. The theoretical methods used in these studies range from molecular orbital calculations of the energy of specific atomic arrangements to calculations based on continuum models. Some studies use a mix of both approaches. Most of the basic concepts used in the analysis of electron solvation can be adapted to analyzing the stabilization of electrons in crystals. Kevan et al. have applied their semicontinuum model of electron solvation [45] to an analysis of electron trapping in single crystals of sorbitol [5]. The main conclusion of this study was that considerable local disturbance of the lattice, specifically rearrangement of nearby protons, must occur in order for the electron to be in a bound state.

Improved definition of trapping sites in crystals should encourage further theoretical analyses of trapping and solvation phenomena. The several theoretical studies referenced have not specifically considered the geometry of Fig. 6 in which two hydroxy groups form the electron trap. This configuration of dipoles is the trapping geometry best verified experimentally. On the experimental side, one anticipates additional trapping geometries will be recognized and characterized.

5. ACKNOWLEDGEMENTS

Through the years this laboratory has had the benefit of V band (70 GHz) ESR-ENDOR spectroscopy. V band spectroscopy has particular advantages for single crystal studies because good cavity filling factors can be achieved with relatively smaller crystals. Harold G. Freund is responsible for making V band technology effective. Edwin E. Budzinski utilizes this technology to the fullest. The author is also indebted to George Potienko and Gloria Ford for their help in drafting and composition of this report. This work is dedicated to Professor Oddvar F. Nygaard.

This research is supported by grant CA25027 from the National Cancer Institute.

426

References

1. H.C. Box, E.E. Budzinski and H.G. Freund: J. Chem. Phys. **69**, 1309-1311 (1978).
2. H.C. Box, E.E. Budzinski, H.G. Freund and W.R. Potter: J. Chem. Phys. **70**, 1320-1325 (1979).
3. E.E. Budzinski, W.R. Potter, G. Potienko and H.C. Box: J. Chem. Phys. **70**, 5040-5044 (1979).
4. S.E. Locher and H.C. Box: J. Chem. Phys. **72**, 828-832 (1980).
5. E.E. Budzinski, W.R. Potter and H.C. Box: J. Chem. Phys. **72**, 972-975 (1980).
6. H.C. Box and H.G. Freund: Applied Spectroscopy **34**, 293-295 (1980).
7. G. Nilsson, A. Lund and P-O. Samskog: J. Phys. Chem. **86**, 4144-4148 (1982).
8. P-O. Samskog, L.D. Kispert and A. Lund: J. Chem. Phys. **78**, 5790-5794 (1983).
9. G. Nilsson and A. Lund: J. Phys. Chem. **88**, 3292-3295 (1984).
10. P-O. Samskog, A. Lund and G. Nilsson, Chem. Phys. Lett. **79**, 447-451 (1981).
11. L.D. Landau: Phys. Z. Sowjetunion **3**, 664-665 (1933).
12. H.C. Box: Radiation Effects: ESR and ENDOR Analysis, Academic Press, New York, 1977.
13. H.C. Box, E.E. Budzinski and H.G. Freund: Radiat. Res., in press.
14. A. Lund, G. Nilsson and P-O. Samskog: Radiat. Phys. Chem. **27**, 111-121 (1986).
15. A.J. Stone: Proc. R. Soc. London **A271**, 424 (1963).
16. T. Taga, M. Senma and K. Osaki: Acta Crystallogr. Sect. **B28**, 3258-3263 (1972).
17. G.M. Brown, D.L. Rohrer, B. Berking, C.A. Beevers, R.O. Gould and R. Simpson: Acta Crystallogr. Sect. **B28**, 3145-3158 (1972).
18. H.M. McConnell and J. Strathdee: Mol. Phys. **2**, 129-138 (1959).
19. W. Derbyshire: Mol. Phys. **5**, 225-231 (1962).
20. M. Barfield: J. Chem. Phys. **53**, 3836-3843 (1970).
21. H.C. Box, E.E. Budzinski and G. Potienko: J. Chem. Phys. **69**, 1966-1970 (1978).
22. A. Hordvik, Acta Chem. Scand. **15**, 16-30 (1961).
23. S. Takagi and G.A. Jeffrey: Acta Crystallogr. Sect. **B33**, 3033-3040 (1977).
24. B.G. Ershov and A.K. Pikaev: Radiation Research Review **2**, 1-101 1969.
25. G.V. Buxton and G.A. Salmon: Chem. Phys. Letters **73**, 304-306 (1980).
26. P-O. Samskog, A. Lund, G. Nilsson and M.C.R. Symons: J. Chem. Phys. **73**, 4862-4868 (1980).
27. A.S.W. Li and L. Kevan: J. Chem. Phys. **76**, 5647-5648 (1982).
28. O. Claesson, M. Ogasawara, H. Yoshida and A. Lund, J. Phys. Chem. **88**, 5004-5008 (1984).
29. H. Berman and R.D. Rosenstein: Acta Cryst. **B24**, 435-441 (1968).
30. H.C. Box, H.G. Freund and E.E. Budzinski: Radiat. Res. in press.
31. H. Muto, K. Nunome, K. Toriyama and M. Iwasaki: J. Phys. Chem. **93**, 4898-4903 (1989).
32. R.M. Keyser and F. Williams: J. Phys. Chem. **73**, 1623-1624 (1969).
33. H.C. Box and E.E. Budzinski: J. Chem. Phys. **60**, 3337-3338 (1974).
34. H. Muto, K. Nunome and M. Iwasaki: J. Chem. Phys. **61**, 5311-5314 (1974).
35. S.M. Adams, E.E. Budzinski and H.C. Box: J. Chem. Phys. **65**, 998-1001 (1976).
36. E. Sagstuen, E.O. Hole, W.H. Nelson and D.M. Close: J. Phys. Chem. **93**, 5974-5977 (1989).
37. H.C. Box, E.E. Budzinski and H.G. Freund: J. Chem. Phys. **81**, 4898-4902 (1984).
38. E.E. Budzinski and H.C. Box: J. Chem. Phys. **88**, 3487-3490 (1985).
39. H. Muto, M. Iwasaki and Y. Takahashi: J. Chem. Phys. **66**, 1943-1952 (1977).
40. C.A. Naleway and M.E. Schwartz: J. Phys. Chem. **76**: 3905-3908 (1972).
41. M. Hilczer, W.M. Bartczak and M. Sopek: J. Chem. Phys. **85**, 6813-6814 (1986).
42. M. Tachiya and A. Mozumder: J. Chem. Phys. **61**, 3890-3894 (1974).
43. L. Raff and H.A. Pohl: Adv. Chem. Ser. No. **50**, 173-179 (1965).
44. D-F. Feng and L. Kevan: Chem. Res. **80**, 1-20 (1980).
45. L. Kevan, S. Schlick, P.A. Narayama and D-F. Feng: J. Chem. Phys. **75**, 1980-1983 (1981).

THE ELECTRON AS A REDUCING AGENT
A Topic of a Novel Anion

Tadamasa SHIDA
Department of Chemistry, Faculty of Science, Kyoto University, Kyoto 606, Japan

1. Introduction

Discoveries of seminal importance are sometimes made simultaneously. In 1962 when E.J. Hart and J.W. Boag first identified the absorption spectrum of the hydrated electron [1], W.H.Hamill and his coworkers stated "Gamma irradiation of pure tetrahydro-2-methylfuran produced a broad absorption band originating at @ 4000 Å and increasing monotonically to the limit of observation at 13,000 Å··· Electrons are trapped and solvated in tetrahydro-2-methylfuran···" [2]. The furan, more commonly called 2-methyltetrahydrofuran (MTHF), had been exploited for the study of radical anions by the group of G.J. Hoijtink, a pioneer of the spectroscopic study of the radical anions of aromatic hydrocarbons [3]. Going further back in 1954 H. Linschitz et al. [4], following G.N. Lewis [5-6], succeeded in observing the visible-near IR absorption spectrum of photolytically produced solvent-trapped electrons in a glassy mixture of ether-isopentane-triethylamine-methylamine. In retroscpect, the statement of Linschitz [4] that "Working in ether-isopentane or ether-isopentane-alcohol (EPA) solvents, Lewis and Lipkin could find no absorption which might be ascribed to the solvated or bound electron"[5] can now be understood by realizing that the trapped electron in ethers or paraffins only absorbs significantly in the near IR region, which is obviously out of the range of Lewis' measurements, and that the trapped electron in alcoholic media is subject to photodecomposition by the reaction,

$e^- + ROH = RO^- + H$ [7,8].

The finding by Hamill was fostered to a technique of observing the optical absorption spectrum of radical anions of various molecules [9-11]. Previous studies clearly show that high energy irradiation of frozen solutions in MTHF is indeed a versatile method for the formation of the desired radical anion of molecules dissolved as the solute in MTHF [10-11].

In this chapter I should like to present another study of a radical anion, which may appear eccentric but may have a potential significance to understand the effect of the electric charge on the electron spin alignment. The system is a monoanion of a high spin organic molecule which could be a basic model system to explore molecular ferromagnets with electric conductivity.

Starting from the first organic quintet molecule of *m*-phenylene(bisphenylmethylene) (*m*-PBPM) [12], K. Itoh and his coworkers have studied extensively aromatic polycarbene

A. Lund and M. Shiotani (eds.), Radical Ionic Systems, 427–432.
© 1991 *Kluwer Academic Publishers. Printed in the Netherlands.*

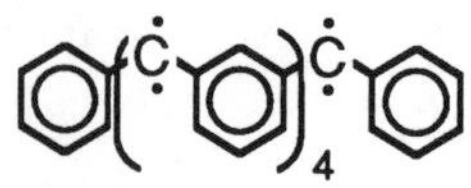

I

systems of high spin multiplicity, and have obtained up to an undecet molecule in Scheme **I** [13].

The spin alignment of the π electrons in these neutral polycarbenes is dictated by the topological degeneracy [14], a concept suggested by H.C. Longuet-Higgins in 1950 [15]. In the VB picture the electron spins of the out-of-plane π electrons are predicted to align alternately up and down so that the spins on the carbene sites are either all up or all down. In addition, the in-plane electrons around the divalent carbene sites possess the same polarization as that of the out-of-plane p electrons by Hund's rule due to the near degeneracy of the two orbitals on the carbene sites [14].

Our primary concern is whether these high spin molecules can be charged or not and what will be the effect of the charge upon the spin alignment. We present the result of our study on the simplest member of the family in Scheme **I** , *i.e.*, *m*-PBPM.

2. Production of the Anion

We choose the precursory diazocompound, 1,3-bis(a-diazobenzyl)benzene (1,3-BDB) as the solute in γ-radiolysis of frozen MTHF solutions. This is expected to give the target anion by analogy with the photolysis of 1,3-BDB to split off the two nitrogen molecules [12].

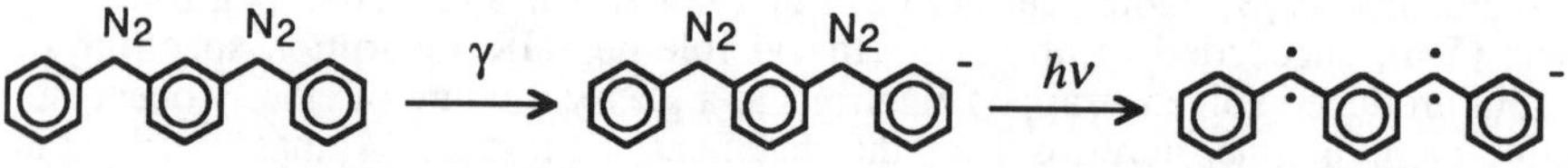

II

Thus, we dissolved the diazo precursor to obtain a pink-colored, glassy solid at 77K, and measured the ESR and optical spectra at the end of the irradiation and photolysis.

3. Results and Discussion

Curves 1 to 3 in Fig. 1 show the optical changes before and after irradiation, and after the successive photobleach with $\lambda > 620$ nm, all observed at 77K. The bump with $\lambda_{max} = 530$ nm in curve 1 is due to the $n\pi^*$ transition of the parent diazocompound whose molar extinction coefficient is determined to be 254 l mol^{-1} cm^{-1}[16]. This transition gives rise to the observed pink color. The extinction coefficient allows us to estimate the molar concentration of the diazocompound to be @ 10 mM which is just sufficient to scavenge most of the electrons ejected from the solvent MTHF molecules upon γ-irradiation (Note the absence of the well known spectrum due to the solvent-trapped electron in the near IR [9,11]).

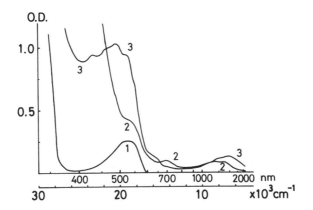

Fig. 1. Electronic absorption spectra of 1,3-BDB/MTHF solutions at 77K. Curve 1: Before γ-irradiation. Curve 2: After γ-irradiation. Curve 3: Same as curve 2 after successive photobleach with λ > 620 nm.

At this stage the ESR spectrum shows signals only at 0.30 - 0.35T along with the doublet spectrum due to the hydrogen atoms produced by irradiation. However, the succeeding photolysis in Scheme **II** caused a dramatic ESR spectral change, *i.e.*, the appearance of the peaks at < 0.30T and > 0.35T for the X-band ESR measurement shown at the top of Fig. 2. Concomitantly, the optical spectrum changed from curve 2 to 3 in Fig. 1.

Since the ESR peaks are spread over such a wide range of the field, it is obvious that the spectrum is due to the fine structure of a high spin molecule, and according to the predicted scheme it will be the target anion of high spin. Since the addition of an extra electron to the neutral dicarbene of spin multiplicity of five, *i.e., m*-PBPM, should yield species of the multiplicity of four or two, the appearance of the ESR peaks outside the range of 0.30 to 0.35T immediately indicates the formation of an anion of spin multiplicity of four.

Following the successful analysis of neutral high spin polycarbenes by Itoh et al. [12-14,17], we assumed the same simple spin Hamiltonian,

$$\mathcal{H} = \beta S \cdot g \cdot H + D \left[S_Z^2 - \tfrac{1}{3} S(S+1) \right] + E \left(S_X^2 - S_Y^2 \right)$$

and attempted to simulate the observed spectrum assuming a spin multiplicity of four and by fixing the microwave frequency to $\nu = 9.218$ GHz used in the experiment. Since the sample is an amorphous glassy solid, each radical should be randomly oriented so that we have to sum up the contribution of the ESR absorption broadened by the anisotropy of the fine structure tensor. The angles θ and ϕ at the bottom of Fig. 2 represent the polar angles of the applied magnetic field in reference to the principal axes of the fine structure tensor.

After several trials we found that the set of the parameters of $g = 2.003$, $|D| = 0.1200$ cm^{-1}, and $|E| = 0.0045$ cm^{-1} gives the simulated spectrum shown in the middle of Fig. 2 which compares quite favorably with the observed spectrum at 77K except for the region masked by the byproduct doublet radicals at 0.30-0.35T. Fig. 3 demonstrates the energy diagram and the transition intensities at the resonance fields calculated for a representative set of the angles of θ and ϕ.

430

Fig. 2

Observed (top) and simulated (middle)
X-band ESR spectra for the quartet state
of the anion of m-PBPM. The angular
dependence of the resonance fields for
random orientation is shown at the
bottom. Symbols F, X to Z, and A in
the middle panel denote the "forbidden",
the "allowed", and the extra lines [18]
which correspond, respectively, to the
points of $dH/d\theta = 0$ in the figure at the
bottom.

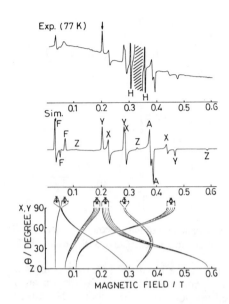

Fig. 3

Energy diagram of the quartet state as a
function of the magnetic field (top) and
the transition intensities at the
resonance fields (bottom) calculated for
a representative set of the polar angles θ
and ϕ.

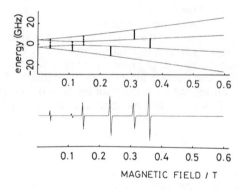

The vertical lines in the diagram indicate the three "allowed" $\Delta m_S = \pm 1$ (solid lines) and
the three "forbidden" $\Delta m_S = \pm 2$ and ± 3 (dotted lines) transitions. From Figs. 2 and 3 it is
seen that not only the allowed but also the forbidden and the off-axis extra lines [18], which
corresponds to a turning point at an angle other than the principal axes of the fine structure
tensor, are well reproduced.

However, with the ESR measurement at a single temperature of 77K we cannot be sure
whether the quartet state is the ground state or a thermally populated excited state.

Fig. 4

Temperature dependence of the ESR intensity measured at the derivative height of the low-field Y-axis canonical peak shown by the arrow in Fig. 2. The experimental data in circles are common to the two cases of $D > 0$ and $D < 0$. The curves are drawn by assuming the Boltzmann distribution in the first two states of the low-spin (LS) and the high-spin (HS).

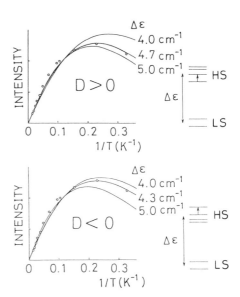

To get the answer we have carried out the temperature dependence of the intensity of the ESR signal. For convenience we monitored the derivative height of the peak marked by the arrow in Fig. 2. The measured intensities were plotted against the reciprocal of temperature to obtain the result in Fig. 4. The circles represent the peak heights measured down to 3K. The convex curve implies that at low temperatures the quartet state is depopulated which, in turn, indicates that the quartet state is not really the ground state but below that state there is a doublet state, which must be the real ground state. If we assume the Boltzmann distribution, we can estimate the energy gap $\Delta\varepsilon$ between the two states. Since we cannot know the sign of the fine structure parameters from the ESR spectrum alone, we have tested both cases of $D > 0$ and $D < 0$. As shown in Fig. 4, the best fit to the observed points is obtained for the gaps of 4.7 cm^{-1} and 4.3 cm^{-1} for the respective cases. In any case, we can conclude safely that the gap is about 4 to 5 cm^{-1} which is extraordinarily small compared to the high spin neutral molecules studied by Itoh and his coworkers [12-14,17]. Even more notably, we have reached a conclusion that the order of the high spin and the low spin states is reversed by adding an extra electron to m-PBPM whose ground state is the quintet.

4. Concluding Remark

We have presented the result of the study on the anion of a rather peculiar system which only can be formed by way of the technique established on the basis of accumulated wisdom in the past four decades. The system studied is special but there seems to be several problems worthy of pursuing further. First of all, what is the mechanism which controls the spin alignment when an extra electron is added to the spin aligned system ? Also, we would like to know the reason for such a small energy gap between the first two states and understand the ease of the denitrogenation upon photolysis. Furthermore we hope to analyze the optical absorption spectrum appearing the whole spectral region from IR to UV. As for the first question K. Yamaguchi is now endeavoring to get clues from the viewpoint of understanding the essentials of molecular magnetism and electric conductivity[19].

The essential part of the present work has been done in collaboration with K. Itoh, T. Takui, and Y. Teki at Osaka City University and M. Matsushita and T. Momose at Kyoto University[20].

References

1. E. J. Hart and J.W. Boag: *J. Amer. Chem. Soc.* **84**, 4090 (1962).
2. P.S. Rao, J.R. Nash, J,P, Guarino, M.R. Ronayne,and W.H. Hamill: *J. Amer. Chem. Soc.* **84**, 500 (1962).
3. G.J. Hoijtink and P.J. Zandstra: *Mol. Phys.* **3**, 371 (1960).
4. H. Linschitz, M.G. Berry, and D. Schweitzer: *J. Amer. Chem. Soc.* **76**, 5833 (1954).
5. G.N. Lewis and D. Lipkin: *J. Amer. Chem. Soc.* **64**, 2801 (1942).
6. G.N. Lewis and J. Bigeleisen: *J. Amer. Chem. Soc.* **65**, 2419 (1943).
7. T. Shida, S. Iwata, and T. Watanabe: *J. Phys. Chem.* **76**, 3683 (1972).
8. T. Shida and M. Imamura: *J. Phys. Chem.* **78**, 232 (1974).
9. T. Shida: *J. Phys. Chem.* **73**, 4311 (1969).
10. T. Shida and S. Iwata: *J. Chem. Phys.* **56**, 2858 (1972).
11. T. Shida: *"Electronic Absorption Spectra of Radical Ions"*, Physical Sciences Data 34, pp.446, Elsevier Science Publishers, Amsterdam (1988).
12. K. Itoh: *Chem. Phys. Lett.* **1**, 235 (1967).
13. I. Fujita, Y. Teki, T. Takui, T. Kinoshita, K. Itoh, F. Miko, Y. Sawaki, H. Iwamura, A. Izuoka, and T. Sugawara: *J. Amer. Chem. Soc.* **112**, 4074 (1990).
14. K. Itoh: *Pure Appl. Chem.* **50**, 1251 (1978).
15. H.C. Longuet-Higgins: *J. Chem. Phys.* **18**, 265 (1950).
16. K. Itoh, H, Konishi, and N. Mataga: *J. Chem. Phys.* **48**, 4789 (1968).
17. a)Y.Teki, T. Takui, H.Yagi, K. Itoh and H. Iwamura : *J. Chem. Phys.* **83**,539(1985).
    b)Y.Teki,T.Takui, K. Itoh, H. Iwamura, and K. Kobayashi: *J. Amer. Chem. Soc.* **108**, 2147 (1986).
18. Y. Teki, T.Takui, and K. Itoh: *J. Chem. Phys.* **88**, 6134 (1988).
19. K. Yamaguchi, Y. Toyoda, and T. Fueno: *Synthetic Metals,* **19**, 81 (1987).
20. M. Matsushita, T. Momose, T.Shida, Y. Teki, T. Takui, and K. Itoh: *J. Amer. Chem.Soc.* in press.

III COMPLEX SYSTEMS

RADICAL IONS AND THEIR REACTIONS IN DNA AND ITS CONSTITUENTS

Contribution of Electron Spin Resonance Spectroscopy

JÜRGEN HÜTTERMANN
*Fachrichtung Biophysik und Physikalische Grundlagen der Medizin,
Universität des Saarlandes, 6650 Homburg/Saar
Federal Republic of Germany*

1. Introduction

The rationale for the interest in radical ions in DNA derives not so much from the expectation of observing unique spectral and electronic parameters but rather from the vital role of this polymer in the manifestation of cellular damage induced by ionizing radiation which is mediated by radical ions and their ensuing neutral free radicals. This situation has given impetus to extensive studies of radiation chemical effects in DNA itself, its constituents and related molecules (e.g. carbohydrates, alkyl phosphates) for more than three decades. A wealth of information on free radical intermediates and radiolysis products from purine and pyrimidine bases, their nucleoside and -tide derivatives, from oligo- and polynucleotides as well as from DNAs of a variety of structures and origins has been accumulated under experimental conditions aimed at modelling the intracellular situation which itself is by and large inaccessible to mechanistic investigations.

The model system studies usually are divided into those performed in the solid state or those in the aqueous phase. From the latter, conclusions are drawn with respect to what is denoted the "indirect" action of radiation on cells in which the initial ionization event involves molecules of the matrix surrounding the DNA, i.e. mainly water. The reactive species produced there ($OH^{.}$, e^{-}_{aqu} and $H^{.}$) initiate cellular damage by subsequent reaction with DNA constituents. The mode of their action is inferred from analysis of products by chemical methods and/or by studying the free radical intermediates formed. The limited lifetime of free radicals in solution usually requires special techniques (e.g. in situ irradiation, fast flow) when applying the typically slow but sensitive and specific conventional steady-state Electron Spin Resonance (ESR) spectroscopy method. Quite frequently in such studies, fast optical measurements of radicals are employed instead.

The analysis of structural and quantitative aspects of free radical intermediates by ESR-spectroscopy and its refined extension, Electron Nuclear Double Resonance (ENDOR)-spectroscopy is conveniently performed on solid state samples. These are considered to be suitable models for the "direct" radiation action in which the primary ionizing event takes place at the biological target directly. Both modes of action, direct or indirect should be taken

Dedicated to the memory of Prof. Dr. K. G. Zimmer

A. Lund and M. Shiotani (eds.), Radical Ionic Systems, 435–462.

as operational descriptions rather than as required by physics laws; there is a floating borderline between them. In cells, their contribution is estimated roughly to be equal.

Over the past ten years, several review articles on the DNA radiation-chemistry have appeared which typically cover either solutions studies or solid state work. An effort to present both aspects of DNA radiation damage in more coherent way dates back to 1978 [1]. Since then, both fields have developed considerably leading to a considerable body of data. In this report we restrict ourselves to a discussion of radical ions and their reactions in solid phase systems containing DNA and its constitutents based on ESR/ENDOR results. Even though, a comprehensive account of the available literature as well as a listing of all spectral parameters are beyond the scope of this article.

We rather confine ourselves to a more general description of radical structures and reaction pathways aiming at coherency on one hand and on timeliness on the other. Therefore, well established reaction schemes will be included together with the most recent results which are influential in the author's opinion.

Another restriction to be mentioned is that only structural features of radical ions and neutral free radical intermediates will be considered. No account is given of quantitative aspects since little if any information on single radical yields after irradiation is available. Total radical yield measurements which give useful global damage estimates have been covered exhaustively in [1].

For DNA constituents, the structural aspects will not only be reported for powders or, preferentially, single crystals but also for aqueous low-temperature glasses and ices. The glassy system forms a bridge between solid-state and solutions studies since the radical lifetime is that in a solid if the temperature of the glass is sufficiently low but the mode of radical formation is more closely related to that in solution. For low temperature ices, phase separation upon ice formation renders the radiation action to be more of the direct type but in an aquated environment as far as the dissolved substrate molecule is concerned.

Of the previous reviews on free radicals intermediates as studied by ESR/ENDOR-spectroscopy we mention the work by Bernhard [2] as an eminent data source on spectral parameters in nucleic acid bases. Myers [3] emphasizes the influence of parameters like light and variation of radiation quality. A more global account of the solid-state radiation chemistry of DNA and constituents is given by Hüttermann [4] and coworkers [5] which has recently been updated specifically for DNA by Hüttermann and Voit [6]. Regular coverage of free radical intermediates in DNA and constituents is found in the Magnetic Resonance Review series under the headings ESR and ENDOR of irradiated solids of biological significance [7, 8, 9] and in the Specialist Periodical Reports on Electron Spin Resonance [10].

The chemical ascpects of free radical mediated product formation in irradiated aqueous solutions of purines and pyrimidines have been presented recently by Cadet and Berger [11], Steenken [12] and Von Sonntag and Schuchmann [13].

Several current articles deal with the consequences of free radical intermediates or products from DNA with respect to damage on a biochemical or biological level. Cullis and Symons [14] connect free radicals to strand breaks under direct radiation action conditions. Hutchinson [15] and Ward [16] report on the chemical changes induced in DNA by radiation. Von Sonntag et. al. [17] give a survey on the connection of chemical products and biochemically detected damage at the sugar-phosphate backbone.

2. Mechanistic aspects of radiation-induced free radical formation in DNA

It is one of the intriguing features of DNA-constituents primary radicals that their chemical structure scarcely provides for clear, measurable spectral parameters by which the mode of formation, e. g. via the charge state can be deduced without ambiguity. In addition, most of the secondary radicals observed at higher (> 77 K) temperatures differ from the parent compound simply by net gain or loss of a hydrogen, a result which can be obtained either by protonation and deprotonation of radical ions or by attack of e. g. diffusible H$\cdot$ atoms produced from primary homolytic C-H bond breakage. In this situation it is not surprising that there has been much discussion as to whether radical formation in DNA starts from charged species ("ionization pathway") or from excited molecules which undergo homolytic bond scission to produce H$\cdot$ ("excitation pathway") [1, 2].

Considering MH as the molecule under study which contains hydrogen in a C-H or N-H bond, one pathway can be written as in scheme *(A)*:

$$MH \xrightarrow{\quad\wedge\wedge\wedge\wedge\quad} MH^{+\cdot} + e^- \qquad (1)$$

$$MH^{+\cdot} \xrightarrow{\quad -H^+ \quad} M^{\cdot} \qquad (2)$$

(A)

$$MH \xrightarrow{\quad +e^- \quad} MH^{-\cdot} \qquad (3)$$

$$MH^{-\cdot} \xrightarrow{\quad +H^+ \quad} MH_2^{\cdot} \qquad (4)$$

Following primary electron ejection (1), the radical cation $MH^{+\cdot}$ may deprotonate to form the neutral radical $M^{\cdot}$ comprising net hydrogen atom loss (2). The electron which escapes recombination from the parent ion can form a radical anion (3) which can take up a proton from the matrix to form a radical showing net hydrogen gain (4).

The same net result (2) and (4) can be achieved by homolytic bond cleavage as in sequence *(B)* from an excited state MH^* (6) and subsequent attack of H$\cdot$ atoms at undamaged sites (7). The excited states in turn are

$$MH \xrightarrow{\quad\wedge\wedge\wedge\wedge\quad} MH^{**} \longrightarrow MH^* \qquad (5)$$

$$MH^* \longrightarrow M^{\cdot} + H^{\cdot} \qquad (6)$$

(B)

$$MH \longrightarrow MH_2^{\cdot} \qquad (7)$$

$$MH^{**} \longrightarrow MH^+ + e^- \qquad (8)$$

$$MH^+ \longrightarrow MH^* \qquad (9)$$

formed either form higher excited states MH^{**} by internal conversion (5) or by electron-hole recombination (9) after ionization from state MH^{**} (8).

One would be tempted to expect that the detailed structural information on free radical intermediates usually obtained from ESR/ENDOR-spectros-
copy would allow for an unequivocal distinction between e. g. a species M·
formed by (2) in the "ionization" path *(A)* or produced in reaction (6) of the
"excitation" path *(B)*. Unfortunately, this is frequently not the case. Nevertheless, the present status on radical formation mechanisms can probably be
summarized best by saying that the majority of available data indicates the
ionization path of scheme *(A)* being the relevant pathway under direct radiation action conditions. This view will be adopted throughout this report.

While the above discussion was concerned with specimen which are, at
300 K, crystalline solids and therefore are typically dry except e.g. for water
of crystallization or structural water, the system of frozen ices mainly introduced and utilized by Gregoli and co-workers [e.g. 18] and Symons and associates [e.g. 19] allows to partially study the influence of the water matrix
(bulk and hydration water) under conditions in which the primary radical formation apparently still obeys the "direct effect" rules. By freezing aqueous
solutions of DNA or its constituents to 77 K, phase separation will concentrate the majority of H_2O in the polycrystalline ice phase while DNA constituents are separated out into dry or hydrated puddles. When additives e.g.
electron or hole scavengers are included at 300 K (usually in mM concentration), the hydrated phase will tend to concentrate these even though they
may not bind to DNA or constituents. It is thus possible to study the influence of additives in a manner comparable to aqueous solution radiation
chemistry.

In such a system, irradiation at e.g. 77 K produces OH· radicals (at neutral pH the only species) from the ice phase together with what is reported
to be radical cations and anions from the substrate molecules [20]. Upon
warming, the OH· species recombine without interfering to any measureable
extent with the substrate. The presence of water is manifest from secondary
reactions usually induced upon warming, e.g. OH^--addition to radical cations

$$MH^{+·} \xrightarrow{\ +OH^-\ } MHOH· \tag{10}$$

The effect of additives typically is to interfere at an early stage of radical formation e.g. by suppression of production of a substrate ion radical
and formation of the corresponding additive ion.

Another solid system to be discussed is that of low temperature aqueous
or non-aqueous glasses. In contrast to aqueous ices, the radiation-chemistry
of aqueous glasses is comparable to that of aqueous solutions as long as the
glassforming agent (acid, base, salt) is in the concentration range of about
3 to 12 M at most. In this case, ionization of water still is the decisive initial
process which leads to electron gain and loss centers, their subsequent fate
being strongly influenced by the additives. Electrons as produced by initial

$$H_2O \xrightarrow{\quad \wedge\wedge\wedge\wedge\quad } H_2O^+ + e^- \tag{11}$$

ejection can become mobile species e^-_m which may be trapped as e^-_t in a polar environment or as anion of some sort. In neutral and alkaline aqueous
glasses as well as in several organic glasses electrons e^-_t are trapped at 77 K.

In strongly acidic glasses, however, the dominance of the reaction with pro-

$$e^- + H_3O^+ \longrightarrow H^. + H_2O \qquad (12)$$

ton donors (H_3O^+) gives rise to mainly $H^.$ atoms which are trapped at 77 K in glasses like H_2SO_4/H_2O and H_3PO_4/H_2O.

The primary hole formed in water, H_2O^+ is highly instable and has so far not been observed. Deprotonation yields $OH^.$ as free radical species which, however, is only stable in ice at 77 K. In acidic glasses, the reaction of $OH^.$ with the acid anion gives $SO_4^{-.}$. Together with $H^.$, these are the two

$$OH^. + SO_4^{--} \longrightarrow SO_4^{-.} + OH^- \qquad (13)$$

prominent species in H_2SO_4/H_2O glasses. If unwanted, $SO_4^{-.}$ formation can be suppressed by photolytic $H^.$ production from Fe^{++}-ions in acid.

$$Fe^{++} \xrightarrow{\quad h\nu \ (253\,nm) \quad} Fe^{+++} + e^- \qquad (14)$$

$$e^- + H_3O^+ \longrightarrow H^. + H_2O \qquad (15)$$

For stabilizing $OH^.$ in low temperature glasses, the glassforming agent must contain a non-oxidizable anion. F^- and ClO_4^- are suitable candidates [21].

By thermal annealing of the glass which is usually irradiated at 77 K, the trapped species become mobile and can react with substrate molecules to form substrate radicals. It is clear from the foregoing, that not only the reactions of the three water radiolysis radicals $OH^.$, $H^.$ and e^- can be studied but also, through the application of suitable additives in various concentrations, the reactions of a broad variety of oxidizing and reducing species like $SO_4^{-.}$, Cl_2^-, O^- etc. The system thus is suitable for comparing the reactions with those studied in aqueous solution at room temperature. Due to the homogenous distribution of H_2O and reactants, it models the indirect radiation-action under solid state conditions as long as substrate concentrations are below a level of interference with the H_2O radiation chemistry ($\leq$ 5 mM).

3. Free radical intermediates from DNA constituents in single crystals, frozen aqueous solutions and glasses

3.1 PYRIMIDINE BASES AND RELATED COMPOUNDS

The three nucleic acid pyrimidine bases thymine, **T** and cytosine, **C** for DNA together with uracil, **U** replacing **T** in RNA (cf. structural formulae) have been investigated in single crystal form either as such or as base components in nucleosides and -tides. With the reservation stated above for all three compounds as pure bases, we discuss free radical formation upon direct radiation action in terms of the "ionization pathway" as common route which is **oxidation** by ejection of an electron from the highest occupied molecular orbital (HOMO). The oxidized species is a π-radical cation or a neutral radical derived from it by deprotonation.The reaction applicable for **U** is depicted in Scheme I.

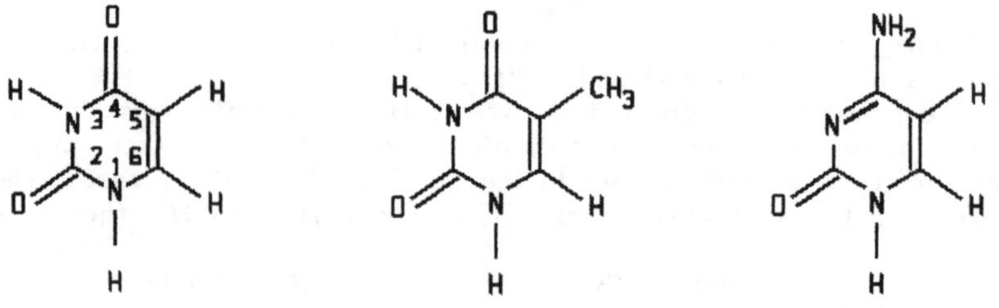

We mention that it is difficult or impossible to decide between the charged and the neutral oxidation-derived radical intermediate on the basis of the available spectral information alone. Theoretical calculations [e.g. 22]

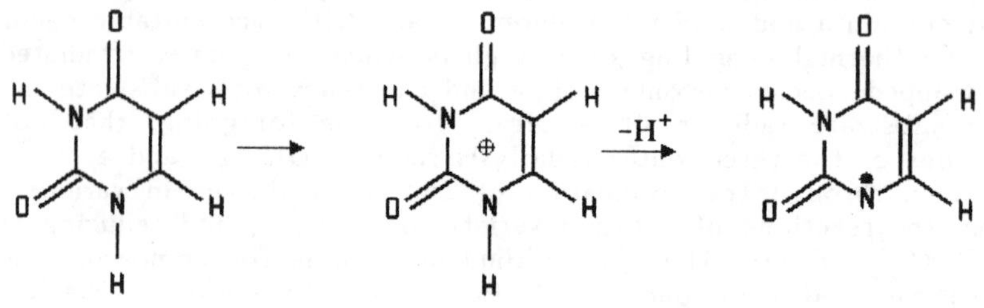

SCHEME I

show that N_1 and C_5 are the two major sites of unpaired electron spin-density in the **U** (and **T**) cation for which detectable hyperfine interaction is expected (together with O_2 and O_4 which would yield couplings only upon isotopical labelling). The calculations also imply that the main effect of cation deprotonation at N_1 is an enhancement of spin-density at that position by about 10 %. So far, there are not sufficient benchmark data available for **U** for both types of intermediates to allow for a more clearcut decision as to which intermediate is trapped from spectral parameters [23, 24].

The situation is changed when introducing substituted derivatives of **U**. Thymine itself is 5-CH_3-substituted **U** and the consequence of this substitution is a change in the deprotonation pattern which is clearly derivable from the spectra and is depicted in Scheme II.

The electron donating CH_3-group at C_5 efficiently competes with N_1 in terms of H^+-donation so that all deprotonation derives from this position to yield an allyl-type radical by delocalization between C_5-CH_2 and C_6-H [25].

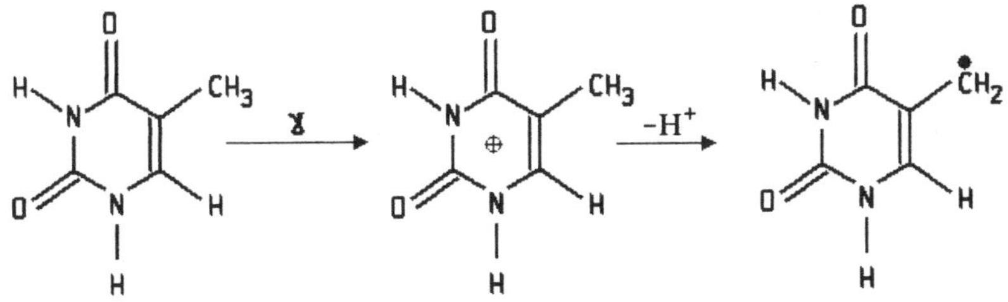

SCHEME II

Methylation at N$_1$ in **T** does not change the deprotonation pattern while in **U** deprotonation is from the N$_1$-CH$_3$ group [26] as depicted in scheme III.

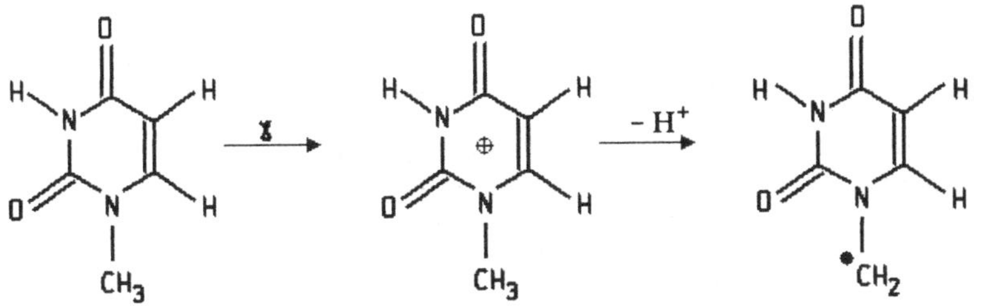

SCHEME III

The position N$_3$ never appears to be involved, perhaps due to a protection by the β-carbonyl groups at C$_2$ and C$_4$.

We have mentioned before that net loss of hydrogen at a methyl group could be expected not only to result from cation deprotonation but also from either attack of H·-atoms at the methyl group or from homolytic C–H bond cleavage, a process which liberates H·-atoms. It has been proposed that pairs of radicals resulting from net loss as well as from net gain of H·-atoms could be indicative of the "excitation" pathway of radiation damage. It is striking that methylated pyrimidine compounds like 1-CH$_3$-**U** (**T**) do exhibit radical pairs at low temperatures (77 K) [26, 27]. On the other hand, while it is true that ionic species are not expected to be trapped within the appropriate distance of about 10 A to qualify for a radical pair one cannot probably exclude the possibility of pairwise trapping from ionic precursors. In order to truly distinguish the excitation pathway from the ionization pathway more specific information from e.g. H/D exchange or trapping of thermalized H· (D·) at very low temperatures would be necessary. There appears to be no experimental proof so far for this mechanism.

Of other possible substitutions at the uracil core we mention the group of halogens for which the effect of 5-substitution has been extensively investigated due to its consequences for DNA damage under in vivo conditions [28]. Studies in single crystals have produced spectral parameters for the deprotonated radical cations in the pure bases, 5-chloro- and 5-bromouracil [29, 30]. (cf. scheme IV)

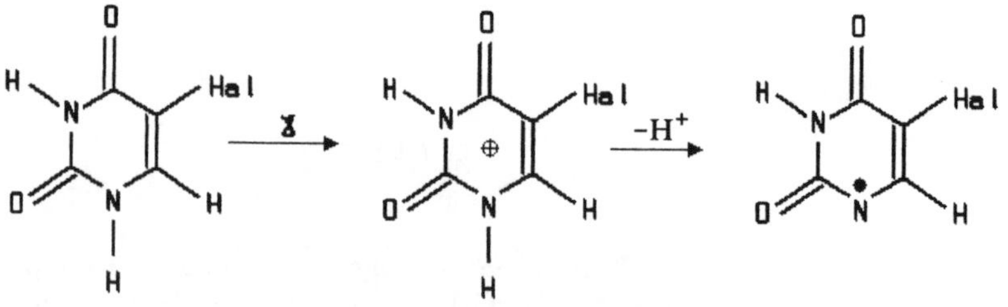

SCHEME IV

As we shall detail below, the electronic properties of the halogens in this case give a clear possibility to distinguish between the charged and the neutral oxidation product. The charged species in single crystals have been characterized for 5-chloro- [29] and 1-CH$_3$ -5-bromouracil [31].

For cytosine, **C**, oxidation and subsequent deprotonation as shown in scheme V also is a sequence of events with favourable support by experimental data. The cytosine π-cation has been studied in glasses and in single crystals [32,33]. In 5-CH$_3$-substituted **C**, the sites of deprotonation also is

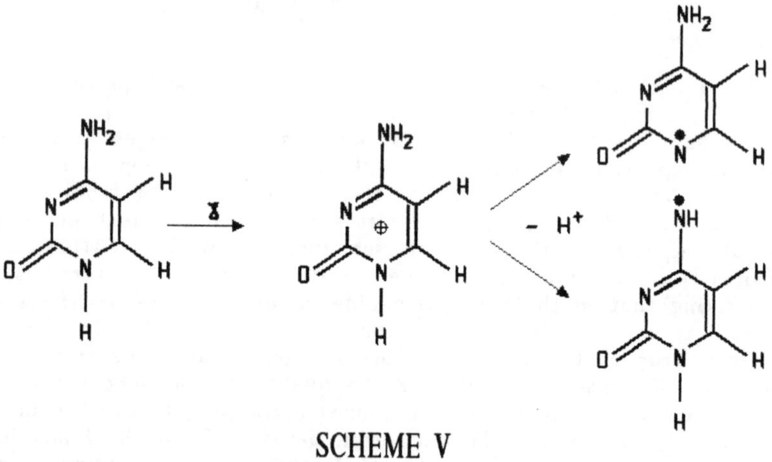

SCHEME V

established to be the 5-CH$_3$-group, as in **T** [34, 35]. For cytosine itself, deprotonation at N$_1$ has been proposed [36] but proton loss has also been inferred to occur at the exocyclic amino group [2]. Proof of this latter possibility has been obtained from work with glasses containing 5-halogenated cytosines [37].

The bulk of the experimental verification of the oxidation-derived schemes discussed has been obtained from single crystal studies. Little structural information came from irradiated powders due to usually unfavorable combinations of g- and hyperfine anisotropy of the radical involved which gave no clear spectral parameters. The system of low temperature aqueous glasses on the other hand, although subjected to the same problems of "powder"-type ESR-spectroscopy, has given more relevant information since the chemical nature of the free radical under study frequently was less ambiguous due to the knowledge of the way of is production. This system was pioneered by Sevilla and co-workers in numerous studies [e.g. 38]. Cations were produced either by biphotonic photoionization from suitable longlived triplet states of the substrate molecules or by reactions of substrates with oxidizing spezies like Cl_2^- of SO_4^- . Most of the relevant work involving photoionization has been discussed in detail in Bernhard's article [2] and is mentioned here when applicable. A more detailed presentation is given of recent work concerning reactions with Cl_2^- and SO_4^- to which, for the latter, contributions also have come from our own group.

The DNA base cations in aqueous glasses employing 12 M LiCl and containing 10^{-2} M $Fe(CN)_6^{3-}$ for electron- and N_2O for OH'-scavenging are formed by Cl_2^- attack upon annealing the glasses to about 150 K after irradiation at 77 K [39]. The spectral parameters of the cations studied, mostly pyrimidines, were found to agree favorably with those reported for single crystals although nothing could be said about the protonation state. Another oxidizing species, SO_4^- , was applied to T and U as well as to the range of 5-halogen substituted uracils and their nucleosides derivatives in acidic glasses [40]. The sequence of events observed is shown in scheme VI.

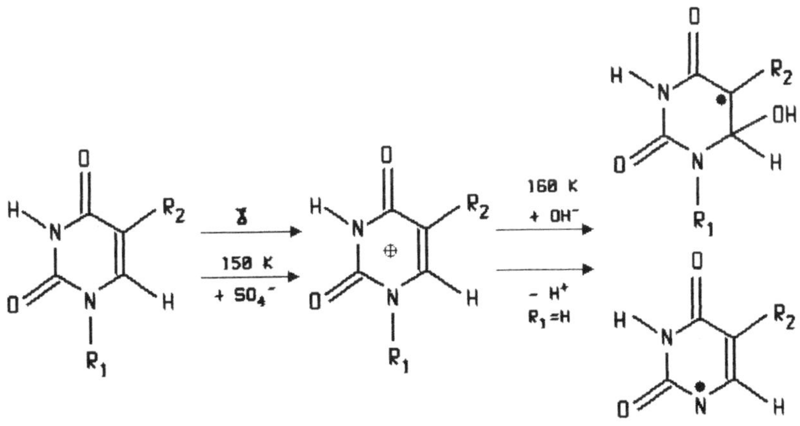

R_1 = H, CH_3, dR, R R_2= H, CH_3; F, Cl, Br, I

SCHEME VI

Base cation formation was found to occur both by excited hole transfer at 77 K and by $SO_4^{-\cdot}$ attack at elevated temperature. The latter species also led to addition at the unsaturated carbon-carbon pyrimidine bond. The fate of the cation in all cases was OH^--addition at carbon C_6 in nucleosides where the glycosidic bond protected the site N_1; the bases, on the other hand, were found either to deprotonate at N_1 or to add OH^- at C_6. The halogen hyperfine coupling turned out to be sensitive enough to allow for a distinction between the cation and its neutral, N_1-deprotonated successor. In addition, for the cation species, it even reflected the amount of water present in the matrix which changed using H_3PO_4 instead of H_2SO_4 as glassforming agent. Testing the reaction of $Cl_2^{-\cdot}$ with the range of 5-halouracils gave comparable results [37]. Interestingly, the halogenated cytosines, in that study, 5-fluoro- and 5-bromocytosine were found to deprotonate, from the base cation, at the exocyclic amino group as mentioned above.

The other aqueous low temperature matrix, that of frozen ices, has been employed only in the case of the nucleotide 2'-deoxy-thymidine-5'-mono-phosphate [18]. In principle, the sequence expected for oxidation, loss of electron from the base by direct radiation absorption was found to occur. Upon warming, OH^--addition was proposed to take place just in the same fashion as discussed for the glassy system above. It has been argued, however, that this latter assignment might have to be replaced by a radical resulting from intramolecular cyclisation or combination of an intact thymine with a thymine radical [41, 2]. Such reactions are conceivable on account of the relatively high mobility in the softening ices together with the close proximity of molecules in the separated substrate phase.

It is noteworthy that so far there is no report in the ice system on oxidation of the sugar alcohol group(s) which is, as will be shown below, competing efficiently with base cation formation in single crystals of nucleosides and -tides. If the radiation chemistry in the ice is indeed that of the direct model, one would expect large fractions of alkoxy-radicals to be stabilized at 77 K as they are e.g. in polycrystalline powders. Their ESR-spectrum is clearly assignable even in powder-type spectra.

The other class of radiation induced reaction to be considered involves **reduction** of the bases. In single crystals, again the question is not unequivocally settled whether the secondary radicals observed have base anions as precursors or are formed from H·-attack since often the kinetics and the stöchiometry of the reaction are not fully established; nor is it always possible, on the basis of spectral information, to assign the nature of the primary product correctly to anions as fas as the charge state is concerned. This is due to the fact that in all pyrimidines protonation can occur at exocyclic positions which need not but may give rise to spectral changes as compared to what is expected, from spin-density calculations, for the charged species.

The complex situation is exemplified for the case of thymine. It is not until very recently that the established knowledge in the literature would assign a low-temperature doublet feature due to spin-density at carbon C_6 to an anion, $T^{-\cdot}$ [42] and the well-known room temperature octet to its successor species formed from protonation at C_6 as is depicted in Scheme VII. We mention in passing that the octet species with its large spread (140 G) and dominantly isotropic interactions of β-methyl and β-methylene protons serves as fingerprint for thymine base radicals in DNA under many different experi-

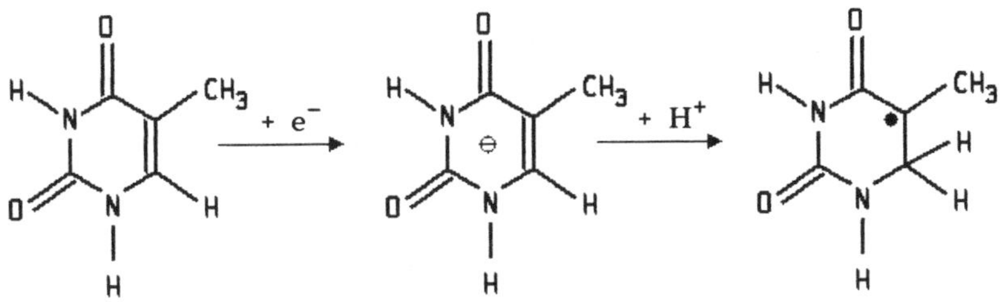

SCHEME VII

mental conditions. It was the species first to be structurally assigned in irradiated DNA [43, 44].

Recent work with thymidine single crystals showed, however, that the previously assigned the "doublet" species for the anion observed at 10 K was indeed
a triplet due to protonation at O_4 and that it existed together with the octet
species even at that low temperature [45]. Although the strong ESR-effect of
protonation in this case is assignable to crystalline environment conditions,
one could argue that the scheme of events should rather be written as for **U**
for which electron addition and protonation is known to yield two types of
radicals due to reaction at either carbon C_6, as above in **T**, or the exocyclic
C_4 carbonyl oxygen [24]. The situation is depicted in scheme VIII.

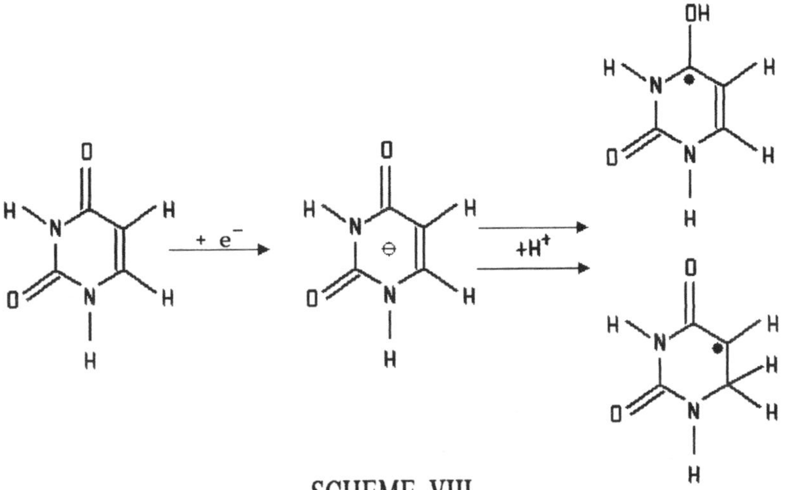

SCHEME VIII

The dominant reaction in **U** is that leading to the 5-yl radical. The other
species, which is in ESR nearly indistinguishable from the parent anion and
which was ascribed to the protonated species mainly due to its stability at
300 K has been detected (by ENDOR) in several 5-halogen-substituted uracils
[46].

The amino group at C_4 in cytosine appears to direct the site competing with C_6 for protons to the C_2 carbonyl group as is shown in scheme IX. Again, the majoritiy of results implies protonation at C_6 to be the dominant alternative pathway [47, 33]. A quite recent report on work in LiCl-glasses adds another possibility for protonation sites, the exocyclic amino group [48]. Interestingly, the resulting ESR-triplet pattern, which is also observed in BeF_2-glasses after electron addition [49], is not displayed in oligonucleotides in LiCl-glasses where a doublet feature prevails as it would be expected for the charged species [50]. It is not clear presently, whether this means that protonation does not take place under these conditions or that it leaves the spectral pattern unchanged. Interestingly, thymine oligonucleotides under the same conditions also showed only the familiar doublet feature for which the state of protonation is not clearly derivable.

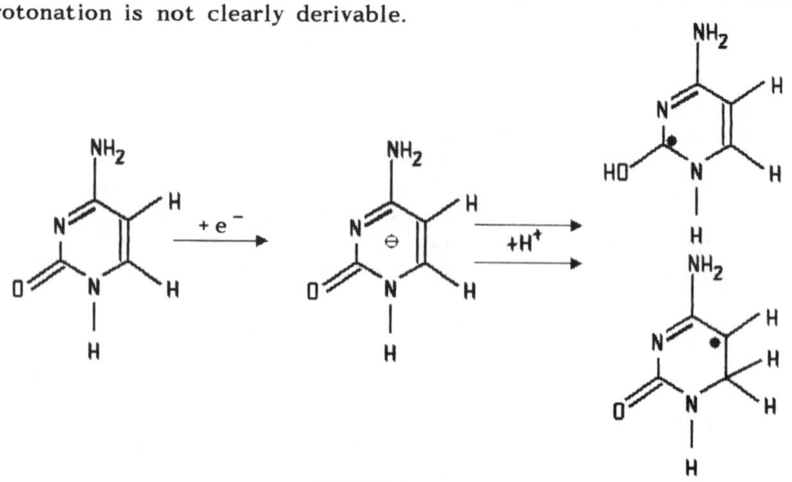

SCHEME IX

Surveying the state of affairs concerning the "anion" spectral features critically it appears conceivable that, as a rule, all pyrimidine anions so far reported in the literature are indeed neutral due to protonation at sites which only rarely reveal themselves in ESR-spectroscopy. Such a "silent" protonation at heteroatoms is expected to be observable in ENDOR but too few studies exist presently in which this aspect was investigated.

The symmetry of the 5,6-carbon double bond in **U** and **C** apparently allows for the study of interesting radical conversions. For example, as detailed in scheme X, the 5-yl radicals formed from protonation at C_6 can be converted to 6-yl radicals by heat. Alternatively, the population of 6-yl radicals can be reduced and 5-yl radicals be formed by illumination with visible light ($\lambda > 400$ nm) in single crystals [51]. It has been argued that 6-yl radicals might reflect the efficiency of the excitation scheme **(B)** since this radical is the major consequence of H˙ (or OH˙) reaction in aqueous solutions [52, 13]. This thesis would imply that 6-yl radicals should be formed by H˙-atom attack in solid systems, too. Alternatively, protonation of the anion which yields the 5-yl radical would, of course, reflect the ionization path **(A)**.

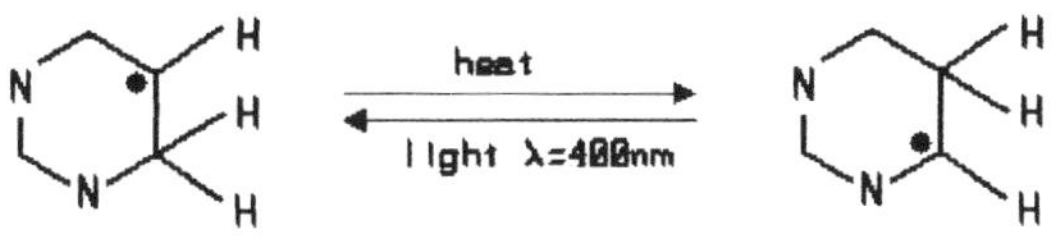

5 - yl   6 - yl

SCHEME X

While there have been many direct proofs, e.g. by H-D exchange for the latter pathway to exist, no experimental evidence could so far be gained for the H$\cdot$-attack model, however. Alternatively, one could argue, that the 6-yl radical just is thermodynamically the more stable radical configuration at the C_5 - C_6 site in **U** and **C** and is also formed from the 5-yl radical so that the anion would be the common precursor of both radical populations. This would explain, among others, why 6-yl radicals are absent e.g. in thymine single crystals although they may be formed in low temperature glasses containing thymine under conditions where only H$\cdot$-attack is prevalent [53].

A similar interconversion of "H-addition"-sites has been proposed recently to be involved in aqueous solution. In that model, the carbonyl oxygen O_4 would be the site for fast (kinetically controlled) protonation of the electron adduct to uracil (and, possibly, **T**) and conversion to C_6 to yield the more stable intermediate 5-yl species would take place on a much larger time scale (thermodynamically controlled) [54]. In continuation of these processes one might speculate that a still further conversion to produce the 6-yl radical is possible in **U** (not **T**) and in **C**, where the initial site of anion protonation, however, might be O_2 or a nitrogen.

Turning now to the group of 5-halogen substituted uracils, we note that the reaction of the bases with electrons has received considerable attention over the past 20 years. In aqueous solutions, electron addition was found to be dissociative leading to halide ions and a highly reactive, σ-type radical at the C_5-carbon position of the uracil core [55, 56]. It was argued, that this radical should, at least, in part be responsible for the in vivo radiosensitivity of 5-halouracil substituted DNA. On the other hand, under solid state conditions modelling the direct effect of radiation, no halide elimination was observed in fairly extensive single crystal studies performed in the authors laboratory (e.g. [5]). Instead, the radiation chemistry of 5-halouracils subsequent to electron addition followed the normal pathway comparable to **T** as depicted in scheme XI.

These findings prompted us to start a more systematic study of the influence of matrix parameters on the reactions involved in electron addition. For this, among others, the system of low temperature glasses was chosen. It was known before that at high pH (alkaline glasses) the π^*-anions are stable with respect to halide elimination only for fluorine as substituent. In 5-chlorouracil, dehalogenation can be followed upon annealing of the

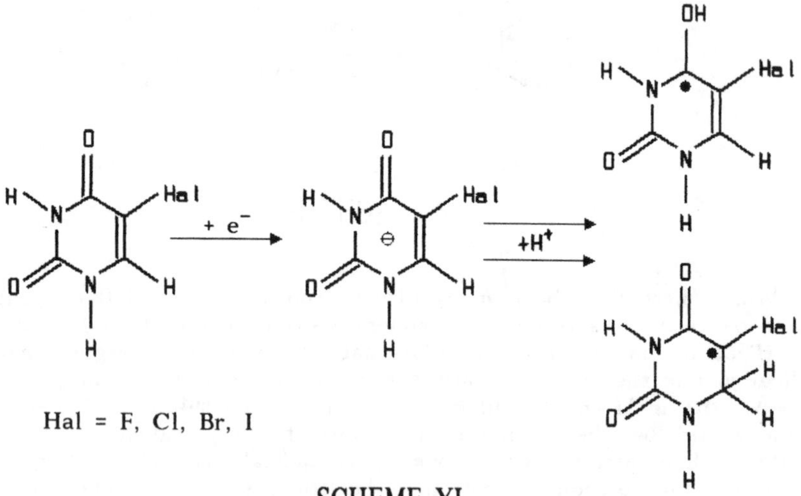

Hal = F, Cl, Br, I

SCHEME XI

π^*-anion and in 5-bromouracil the σ-type uracil-yl radical resulting from halide elimination as shown in scheme XII is already observed at 77 K [57, 58].

In acidic glasses, we found that for fairly high solute concentrations (200 mM), electron addition to form substrate anions competes with H· formation allowing therefore to study the anion spectra and their secondary products. For **U** and **T** as well as for all 5-halouracils, π^*-anions were found to be formed which were stable against halide elimination but rather protonated at carbon C_6 upon thermal annealing according to scheme XI. Experiments with LiCl glasses showed that this is not restricted to highly acidic glasses but also holds for neutral pH-values. In organic glasses (e.g. CH_3OH), π^*-anions are formed, too, but proton donation by the alcohol to add at C_6 is much less efficient than that from water. Indeed, the anions were found to be stable until matrix softening destroys the radicals [59]. It

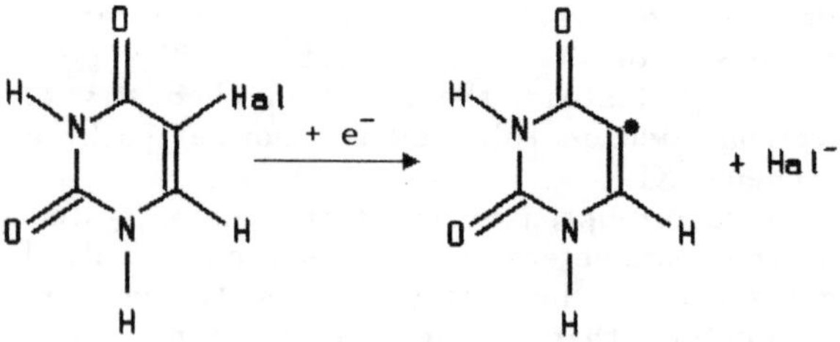

SCHEME XII

is interesting to note that especially very soft low temperature matrices like methyltetrahydrofuran (MTHF) stabilize not only π^*-halouracil anions for all halogens but also, only for 5-bromo- and 5-iodouracil, σ^*-anions. In these, the unpaired electron is located in the carbon-halogen anti-bonding σ^*-orbital (scheme XIII) instead of in the π-electron system of the uracil ring thus giving rise to unmistakable halogen couplings serving as fingerprints of these species [60]. One might argue that σ^*-anions should be precursors of halide elimination but this could not be verified so far experimentally.

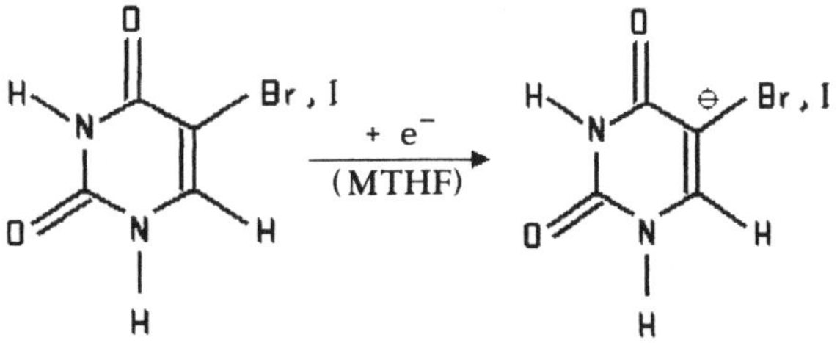

SCHEME XIII

In view of these findings it will be interesting to study the possibility of two different anions in DNA and to follow their fate if formed in order to get more information on possible implications of the 5-halouracil radical chemistry in vivo radiosensitization (see below).

The system of frozen ices has been applied only once to thymidine-5'phosphate. At low temperatures, $T^{-\cdot}$ formation and subsequent protonation according to scheme VII was observed in the absence of oxygen [18]. Oxygenation gave rise to large fractions of peroxy-radicals.

3.1 PURINE BASES AND RELATED COMPOUNDS

Definitive assignments of free radical structures and reactions in purines even from single crystal work have proven difficult to obtain. Most of the information concerning primary radicals (cations, anions) has been gained from low temperature glasses for which the chemistry, not the spectral parameters, has lended some security to the assignment. The situation has improved more recently for guanine , **G**, due to extensive efforts of a few laboratories in which, among others, ENDOR and specific deuteration were applied successfully to single crystals. Still, a coherent picture, is only emerging and there are more gaps than settled questions, not to speak about the situation concerning the other DNA purine base adenine, **A** .

For **G** under conditions of direct radiation action it appears to be clear now that **oxidation** yields a π-cation radical, the spectral parameters of

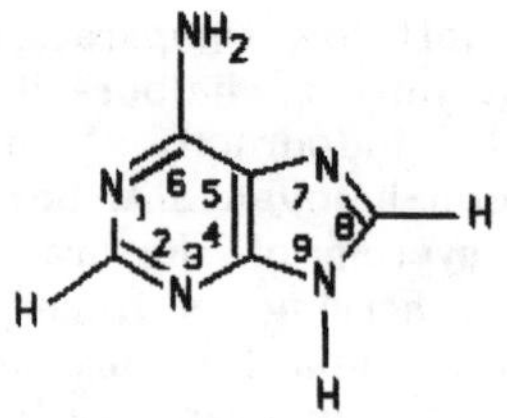

A 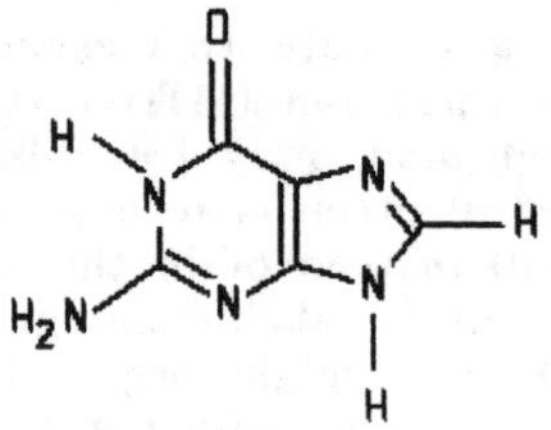   G

which have been isolated in part in single crystals of guanine HCl(H$_2$O) and of 2'-deoxyguanosine 5'-phosphate [61, 62]. The radical is dominated by two ^{14}N-hyperfine interactions which make it difficult to extract full tensor parameters due to extreme spectral broadening. Instead, numerical spectra simulation with estimated tensors is often an unavoidable amendment. The proton coupling tensor was derived by ENDOR quite recently in guanine HCl (H$_2$O) [63] and in guanosine cyclic monophosphate [64].

In low temperature glasses, photoionization of **G** gave rise to singlet type spectra for the π-cation radical in the LiCl-system [65]. More recently, the same spectral parameters were found upon photolysis in acidic glasses [66]. Similar results came from Gregoli and coworkers in γ-irradiated frozen ices of 2'-deoxy-guanosine-5'-phosphate [67].

The fate of the guanine cation radical is deprotonation but the site concerned is under debate . In single crystals of the 5'-nucleotide a conversion was found to occur starting around 40-60 K. The radical formed was obser-

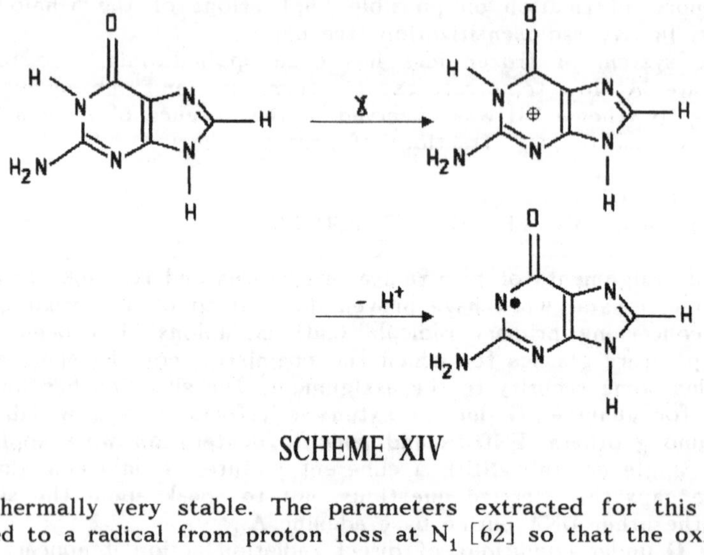

SCHEME XIV

ved to be thermally very stable. The parameters extracted for this species were assigned to a radical from proton loss at N$_1$ [62] so that the oxidative pathway in guanine should be written as in scheme XIV. This view was challenged in an ENDOR-study of the same crystal system in which the site of deprotonation was assigned to the excocyclic amino group as in XV [68].

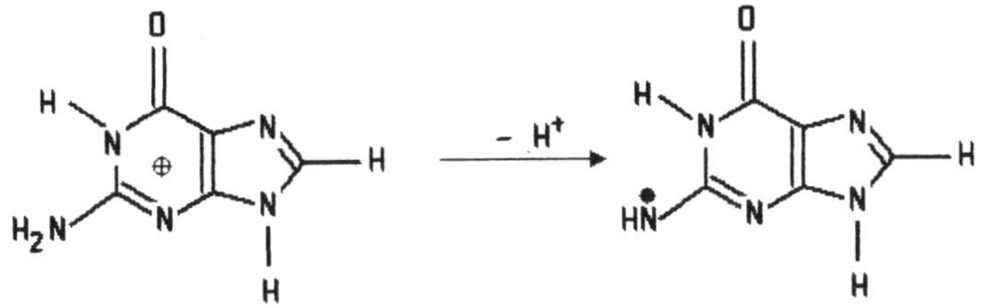

SCHEME XV

The work performed in frozen ices or aqueous glasses gives little information relating to this debate since for none of the alternatives one would expect to observe significant spectral differences in the powder-type spectra between the parent cation and its deprotonated successor. One-electron-oxidation of guanine nucleosides in aqueous solution supports the view that N_1 is the site of deprotonation [69].

An interesting alternative to oxidative reaction at the base has been proposed for a single crystal of guanine $HCl(H_2O)_2$ very recently in which is OH^--additon to carbon C_8 was observed on account of oxidation of the co-crystallized water [70]. In a similar fashion, oxidative reactions are considered to be responsible for a radical showing net hydrogen gain at carbon C_8 which is formed in guanosine-5'-monophosphate at 10 K and is stable until 300 K [71]. Such a radical is usually assigned to be formed from protonation of the reduced base. In that crystal, no base cation was observed as primary species.

Oxidation of **G** in hydrochloride monohydrate crystals is also thought to produce an imidazole ring-opened radical which decays at about 150 K by forming the hydrogen gain radical at C8. In this crystal, the deprotonated cation coexists with the ring-opened radical at low temperatures but decays without sucessor at about 60 K [72].

We note that there seems to be a considerable diversity of mechanisms depending on crystal structure. A reaction scheme valid for 2'-deoxyguanosine-5'-monophosphate differs more or less completely from that found in the ribonucleotide; a sequence for the base in the dihydrate hydrochloride crystal is not at all detectable in the monohydrate hydrochloride specimen. This holds for the schemes following base reduction in a similar way (see below).

In adenine (**A**), the main information on π-cations radicals came from work with low temperature glasses which gave an unresolved singlet as probable powder-type feature [73]. In single crystals, with only recent exceptions, species which at best could be related to oxidation had been accessable like H·-loss radicals at the alkyl-group in 9-CH_3-adenine and 9-Ch_3CH_2-adenine [74, 75]. Better access to the cation has been gained in a study of a co-crystal of the riboside adenosine together with 5-bromouracil [76]. In that work, as well as in a more recent study on crystals of the pu-

rine nucleoside alone [77], ENDOR indicated that the cation might be deprotonated at the exocyclic amino group. In frozen ices, OH⁻-addition to the
π-cation has been proposed to occur [78]. With the reservation that the
spectral parameters of the true cation have not yet been unravelled we can
summarize the findings as in scheme XVI.

There is a comparable paucity of data relating to the spectral features
of **electron gain intermediates** in purines. In guanine, photolytically generated
electrons in low temperature glasses give rise to a singlet upon reaction
[73]. Recent single crystal studies in guanine HCl(H$_2$O) [72,77] and guanine
HCl (H$_2$O)$_2$ [70, 79] indicate that the anion is protonated at O$_6$. In both
these crystals, the molecules are protonated at N$_7$ originally. Anion features
in a crystal containing a neutral guanine base were reported for 5'-monophos-

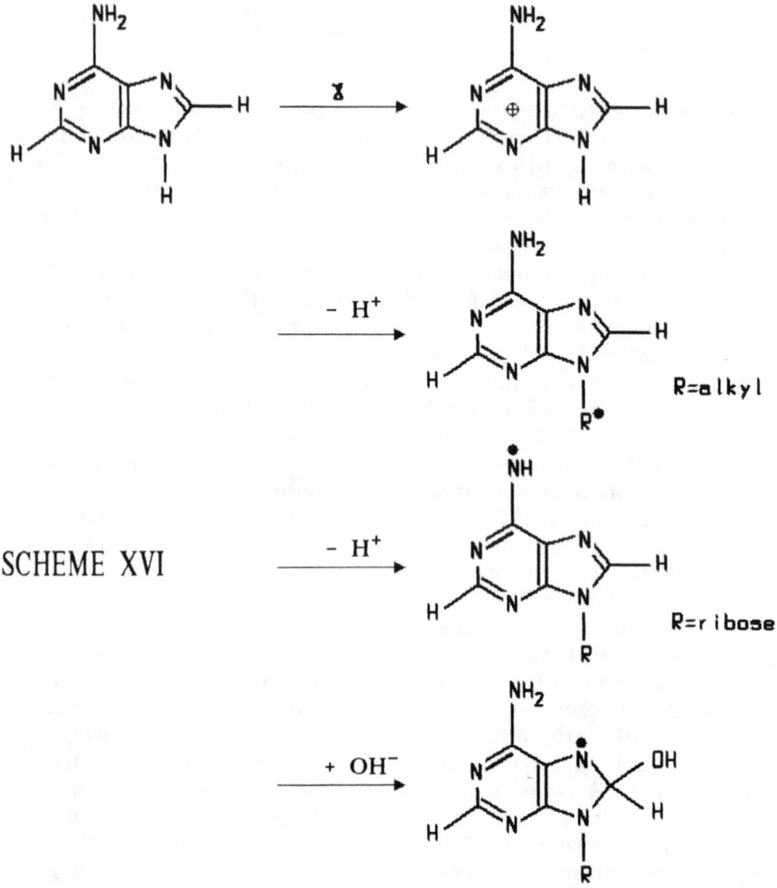

SCHEME XVI

phate but the information which could be extracted by ESR gave no information on the protonation state [63, 80].

The scheme of reactions following electron gain at the guanine base appears to be diverse. In the 5'-deoxynucleotide, there is a sequence leading to a fairly stable radical by H-addition at carbon C_8 [62]. This is a dominant species at 300 K which has been characterized in detail already some time ago in guanine derivatives [81]. It is interesting to note that the sequence contains an unusual intermediate, the structure of which is uncertain. No connection, however, was found between the H-addition radical and the anion in the 5'-nucleotide, which, when irradiated at 10 K and annealed to 25 K decayed without successor [72]. In guanine $HCl(H_2O)$ crystals [71, 72], the anion which was protonated at O_6 decayed upon annealing into a species for which so far a tentative structural model has been proposed which involves H·-abstraction from N_9 of a neighboring molecule. On the other hand, the H-addition radical has been found to be present already at low temperatures. Therefore, the authors claim that the radical might be formed by real H·-atom addition. I feel, however, the sequence should be rather written as in scheme XVII in which the anion either protonates at C_8 to form the well-known H-addition radical upon annealing and/or at the C_6 carbonyl oxygen. Both processes appear to be possible at low temperatures in crystals with a protonated base whereas protonation at C_6 is dominant for a neutral base and only occurs at elevated temperature. The situation could reveal some similarity to the discussion of protonation sites of pyrimidine anions.

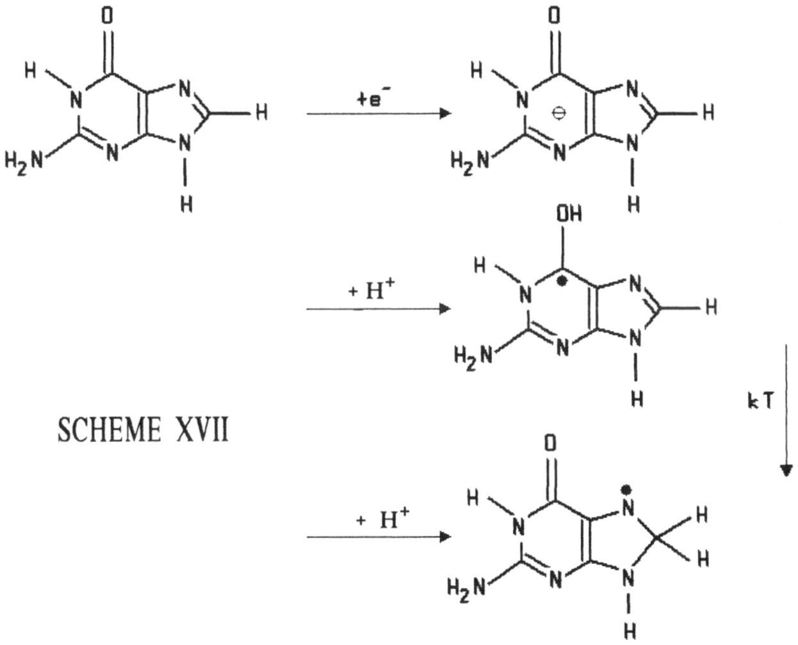

SCHEME XVII

For adenine, electron gain at the base has been studied in single crystals of adenine $(HCl)_2$ [82, 83], adenosine [76, 77] and deoxyadenosine [77].

In the two latter systems, the main proton hyperfine interaction came from the C_2-proton so that the anion was inferred to be protonated at that site. In the crystal containing the base alone, the main fraction of spin-density was found to reside on carbon C_8. It is not clear yet, if this difference reflects the differences in the protonation state of adenine in the various crystal systems.

In low-temperature glasses, adenine anion radicals have been produced by photoionization of ferrocyanide and gave a structureless singlet [73]. A similar pattern, although superimposed by a cation singlet, was gained in frozen ices of the 5'-nucleotide after irradiation at 77 K [78].

The fate of the adenine anions is only partially established. There appears to be unequivocal consent that protonation at carbon C_8 to form the corollary of the C_8-H-addition is a feasable reaction [84], although some crystal systems offer no observable connection between the low temperature species and those obtained after irradiation at room temperature or upon warming [77]. The latter radicals are characterized by net hydrogen gain at either carbon C_8 and/or C_2 [74, 84]. It has been shown that H'-atoms in acidic glasses can produce both types of addition radicals [66]. For the single crystals, an initial preference postulated for C_2 as site of H'-attack and C_8 as site of anion protonation [74] was not observed to hold throughout the range of adenine components studied [84].

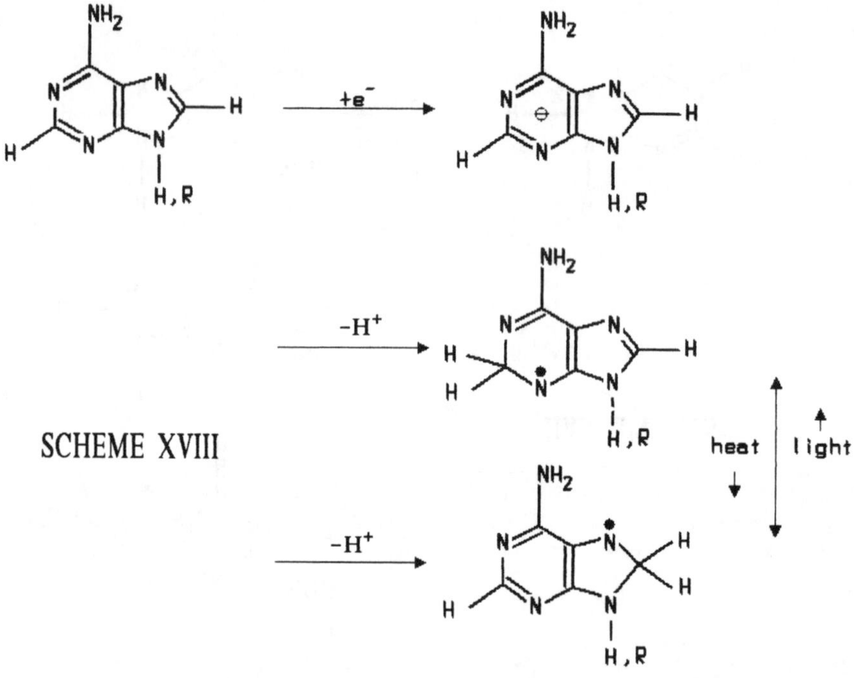

SCHEME XVIII

An interesting radical interconversion scheme induced light and heat between C_2- and C_8-addition radicals was observed, comparable to scheme X of the pyrimidines **U** and **C**. The radical site N_1 - C_2 is shifted to N_7 - C_8 by heat whereas the site C_8 moves to C_2 by treatment with light ($\lambda > 360$ nm) [52]. It is therefore tempting to speculate that C_8 is the thermodynamically more stable species and thus the ultimate consequence of electron gain in adenine derivatives as is shown in Scheme XVIII, in which, however, suggestions for protonation states have not been included.

4. Sugar-phosphate group

Free radicals on the sugar-phosphate group are important intermediates monitoring strandbreak-formation mechanisms. There are, however, problems with respect to model system studies. For one thing, in the presence of nucleic acid bases, there appears to be extensive shielding of the phosphoester bond and, of course, of the carbohydrate from capturing an electron. Despite extensive efforts, there has been no report of phosphorous containing radicals with only one recent exception [85].

Therefore, under conditions of direct radiation action, the only aspect to be considered appears to be sugar oxidation. Both in nucleosides (-tides) and base free carbohydrates, oxidation gives rise to alkoxy radicals as deprotonated cation radicals which have been characterized to occur at primary or secondary alcohol groups [86, 87, 88]. These radicals are initiating rather complex schemes of reactions upon warming most of which, however, depend on the proximity of other sugar moleculs as it is offered in single crystals [89, 90]. In base free sugars additional reactions perhaps resulting from initial homolytic bond scission take place. Such aspects have been studied in several carbohydrates by Bernhard and co-workers [91, 92].

We refrain from presenting these schemes in any detail since so far no indication of their importance in DNA has become available. Indeed, as we have hypothesized earlier, sugar oxidation appears to be suppressed completely in DNA due to the strong electron donor properties of the bases, specifically the purines [5]. Those radicals observed in DNA at the sugar-phosphate group thus very probably derive from initial base damage. Such processes have not been observed in model systems; they will be treated in the subsequent section on DNA.

We mention that free radical formation has also been studied in carbohydrates, nucleosides (-tides) upon reaction with $H^{\cdot}$-atoms in acidic glasses [93], with $OH^{\cdot}$ in BeF_2 glasses [21] and in aqueous solution [94, 95]. In accord with product analysis [96], abstraction of $H^{\cdot}$ from CH-bonds is the common pathway of both reactants which takes place at all possible sites. The resulting radical then is converted into a carbonyl-conjugated secondary species by acid or base catalyzed water elimination.

5. DNA

The contribution of ESR-spectroscopy to the elucidation of the structural aspects of free radical formation in DNA under direct radiation action conditions i. e. in the form of oriented fibers, frozen aqueous solution and lyo-philized powders has been summarized recently [6, 14]. In both articles, based for one thing on pioneering results obtained over more than a decade utilizing oriented DNA fibers [97, 98] and on re-investigations of this system expanding the chemical compositon of the DNAs by incorporation of 5-halouracils replacing thymine [6, 99] and, on the other hand, on detailed studies in the system of frozen aqueous solutions [19, 20] the view was held that the primary free radicals formed at low temperature (4 K - 77 K) were radical cations on the purine base guanine ($G^{+\cdot}$) and radical anions on the pyrimidine base thymine ($T^{-\cdot}$) (cf. structural formulae). Annealing to higher temperatures was

$$\mathbf{G^+} \qquad\qquad \mathbf{T^-}$$

thought to cause $T^{-\cdot}$ to protonate at carbon C_6 to form the well known octet species first identified in DNA in 1963 [44] and used since then as thymine damage fingerprint (cf. Scheme VII). In one of the articles, a proposal for the decay of the $G^{+\cdot}$ radical was presented for the first time which was considered to be deprotonation at fairly high temperatures ($\sim$ 220 K) at the site N_1 as detailed in Scheme VIII. While this was observed to be a common pathway in all DNA fibers independent of base composition [6], the anionic pathway was thought to be diversified when thymine residues were replaced by 5-halouracils [100]. Finally, in oxygenated frozen aqueous solutions, peroxy radicals were said to be formed at the cost of the $T^{-\cdot}$ reaction [20].

Since then, this picture has experienced both substantial support as well as criticism from model system studies. Support came from several results mentioned in **3.2** referring to the structure of $G^{+\cdot}$. At the time of its first proposal [97], no other spectroscopic means but theoretical spin-densities and simulation were available. Neglecting finer details concerning protonation states, all recent studies of the guanine oxidation in single crystals have presented spectral parameters supporting the DNA-assignment of $G^{+\cdot}$. Work with oriented DNA-fibers in which all protons, not only the easily exchangeable ones were replaced by deuterons gave additional evidence that the nitrogen couplings assumed for $G^{+\cdot}$ were correctly assigned [101, 102].

Considerable attention has been given on the question of $T^{-\cdot}$ as other primary species. It has long been critizized, that the basis for its identification was not unique but rather came from spin-density calculations and spectra simulation as well as other, indirect pieces of evidence like the presence of the 5-yl thymine octet at elevated temperatures [2]. Specifically, a methyl proton interaction in $^{\cdot}T^{-}$ was considered to be too large, a problem still increasing when its refined parameters were presented by us [99]. In order to cope with this argument we have studied oriented fibers of DNA containing deuterons instead of protons at the methyl group and at carbon C_6 of **T**. If correct, the previously assigned 10-line feature for T⁻˙ in the orientation parallel to the helix axis should change drastically in the modified DNA. However, the pattern remained unaltered. Very careful re-investigations involving kinetic measurements and a variety of new DNA-specimen gave the picture that T⁻˙ was indeed formed but as a minority species comprising a comparatively broad, underlying doublet in normal DNA and that it transformed into the 5-yl radical with an activation energy of 0.05 eV. A third primary component was distinguished which displayed the above mentioned, sharp 10-line pattern. It decayed at higher temperatures (> 180 K) with an activation energy of about 0.5 eV, comparable in kinetic behaviour with G⁺˙. This component can be well explained with a cytosine anion, especially on account of effects of deuteration on the spectral shape [101, 102]. However, as with T⁻˙, the assignment of the pattern to C⁻˙ is again based on very plausible but indirect

evidence and must remain tentative until we can present results from e. g. a DNA containing a modified cytosine or other direct spectroscopic indications (ENDOR, ESEEM). Work is in progress aiming at this goal. We mention that the powder-type spectrum for the species is a doublet, in contrast to what was expected on account of efficient protonation in frozen aqueous solutions of the base where a triplet is observed [48, 49]. The doublet resembles to what is observed in cytosine containing oligonucleotides [50]. This should not been taken to indicate that C⁻˙ is not protonated; the doublet could well be due to an ESR-silent protonation, probably at N_3 to yield CH.

The established secondary radicals in oriented DNA-fibers are the 5-yl octet species at the thymine moiety which makes up for about 10 % of the radical populations and, in our view, the N_1-deprotonated cation radical from **G**. The product formed from the assumed cytosine anion is not known. The radical is stable and decays only at elevated temperatures indicating that is should perhaps be neutral already at low temperatures. Spin-density calculations for the N_3-protonated C⁻˙ agree well with the main doublet feature. Such an ESR-silent neutralisation might then be attributed to G⁺˙ too. If so, the transitions seen at higher temperatures involve neutral species which for one, the guanine radical, is ESR-detectable.

Strong indications of peroxy-radicals which are found in oxygenated frozen ices [20] are not observed in the fiber DNA systems; in some, minority components were reported to occur [103].

458

There are other secondary (or tertiary) radicals in oriented DNA fibers for which the structure at present is only tentative. One of these giving an axial ESR-spectrum due to a $I = 1/2$ nucleus is restricted to DNA with Li as counterion and to D_2O-equilibration. Another one is more ubiquitous and could be assigned to a radical at the sugar group [101,102].

Non oriented DNA-systems have recently been extensively studied in frozen aqueous solution [19, 104, 105, 106]. As mentioned above, oxygen was found to enhance the yield of ROO˙ radicals at the cost of T^-. Likewise, nitroimidazole, iodoacetamide and H_2O_2 were reported to reduce the $T^{-\cdot}$ $\longrightarrow$ TH˙ channel by competing with T for electrons. However, only H_2O_2 did show an effect on the $G^{+\cdot}$ population.

We now turn to the system of DNAs in which the thymine residue is replaced a 5-halogen substituted uracil. Studies were performed with chlorine, bromine and iodine substituted uracils replacing 90 % of the thymine moieties [99, 103]. The cationic pathway was established to be identical to that of normal DNA. As far as electron gain centers are concerned, a π^*- together with a σ^*-anion were observed in 5-bromo- and 5-iodouracil-substituted DNA whereas only a π^*-species is formed in 5-chlorouracil DNA. It is interesting to note that radicals at the deoxyribose moiety are formed from the π^*-, not the σ^*-anions directly but after $\sigma^* \longrightarrow \pi^*$-transformation by transfer from base to sugar-phosphate. The degree of definition of sugar-centered radicals is necessarily poor ("powder"-spectra); thus, assignments cannot not be given presently. It is clear, however, that the amount of such species increases in the order of halogen-substitution: Cl < Br < I. There is no indication of dissociative electron attachment, in line with all solid state results [103].

Combining these findings with the recent results from normal DNA outlined above we can hypothesize that the effect of radiosensitization of DNA by halouracil incorporation might be due to a change in the anionic component from $C^{-\cdot}$ in normal to halouracil$^-$ in the substituted DNA. Whereas both $C^{-\cdot}$ and $G^{+\cdot}$ are fairly inert species halouracil anions are feeding into reactions producing potentially damaging radicals at the sugar-phosphate group, albeit by mechanisms differing from those postulated previously from aqueous solution radiation chemistry.

We close by stating that during the past five years substantial progress has been made in the understanding of free radical mechanisms in DNA following initial ionization. This is due, among others, to considerable success in the elucidation of purine solid and liquid phase radiation chemistry on one hand and, on the other, to an enhanced occupation with DNA itself under a variety of physico-chemical conditions. Nevertheless, although first, promising attempts of unifying approaches between aqueous-solution radiation chemistry and direct radiation action on DNA are given [12] and relevant biological endpoints of damage are connected to free radical detection [19], there is still some way to go until Electron Spin Resonance keeps its potential promise of monitoring cellular damage mechanisms in situ.

Acknowledgement

The fruitful cooperation in the DNA-fiber work with Drs. A. Gräslund, A. Rupprecht and W. Köhnlein is gratefully acknowledged. Work from the authors laboratory was supported by grants from the Deutsche Forschungsgemeinschaft.

References

1. J. Hüttermann, W. Köhnlein and R. Teoule (Eds.):
   Effects of Ionizing Radiation on DNA, Springer, Berlin (1978).
2. W. A. Bernhard: *Adv. Radiat. Biol.* **9**, 199-280 (1981).
3. L. S. Myers, Ir., in Pryor (Ed.): *Free Radical in Biology,* Vol. **IV**,
   Academic Press, New York, 95-114 (1980).
4. J. Hüttermann: *Ultramicroscopy* **10**, 25-40 (1982).
5. H. Oloff, H. Riederer and J. Hüttermann: *Ultramicroscopy* **14**,
   183-193 (1984).
6. J. Hüttermann and K. Voit, in Weil (Ed.): *Electronic Magnetic
   Resonance of the Solid State,* Canadian Society for Chemistry,
   Ottawa (Canada), 267-279 (1987).
7. J. H. Hadley, Ir.: *Magnetic Resonance Review* **6**, 59-84 (1980).
8. D. M. Close: *Magnetic Resonance Review* **11**, 41-80 (1986).
9. D. M. Close: *Magnetic Resonance Review* **14**, 1-47 (1988).
10. Electron Spin Resonance, A Specialist Periodical Report,
    The Royal Society of Chemistry, London, 1973 ff
11. J. Cadet and M. Berger: *Int. J. Radiat. Biol.* **47**, 127-143 (1985).
12. S. Steenken: *Chem. Rev.* **89**, 503-520 (1989).
13. C. Von Sonntag and H. P. Schuchmann: *Int. J. Radiat. Biol.* **49**, 1-34
    (1986).
14. P. M. Cullis and M. C. R. Symons: *Radiat. Phys. Chem.* **27**, 93-100
    (1986).
15. F. Hutchinson: *Progress in Nucleic Acid Research and Molecular
    Biology* **32**, 115-154 (1985).
16. J. F. Ward: *Adv. Radiat. Biol.* **5**, 181-239 (1975).
17. C. von Sonntag, U. Hagen, A. Schön-Bopp and D. Schulte-Frohlinde:
    Adv. Radiat. Biol. **9**, 109-142 (1981).
18. S. Gregoli, M. Olast and A. Bertinchamps: *Radiat. Res.* **65**, 202-219
    (1976).
19. P. J. Boon, P. M. Cullis, M. C. R. Symons and B. W. Wren:
    J. Chem. Soc., Perkin Trans. **2**, 1393-1399 (1984).
20. S. Gregoli, M. Olast and A. Bertinchamps:
    Radiat. Res. **89**, 238-254 (1982).
21. H. Riederer, J. Hüttermann, P. Boon and M. C. R. Symons:
    J. Magn. Reson. **54**, 54-66 (1983).
22. H. Oloff and J. Hüttermann: *J. Magn. Reson.* **40**, 415-437 (1980).
23. H. C. Box, G. Potienko and E. E. Budzinski: *J. Chem. Phys.* **66**,
    343-346 (1977).
24. H. Zehner, W. Flossmann, E. Westhof and A. Müller: *Mol. Phys.*
    32, 869-878 (1976).
25. J. Hüttermann, Int. J. Radiat. Biol. 17, 249-259 (1970).
26. W. Flossmann, J. Hüttermann, A. Müller and E. Westhof:
    Z. Naturforsch. C. **28**, 523-532 (1973).
27. A. Dulcic and J. N. Herak: *Mol. Phys.* **26**, 605-614 (1973).
28. W. Szybalski: *Cancer Chemotherapy Rep.* Part 1 **58**, 539-557 (1974).
29. H. Oloff and J. Hüttermann: *J. Magn. Reson.* **27**, 197-213 (1977).
30. H. Oloff, J. Hüttermann and M. C. R. Symons: *J. Phys. Chem.*
    82, 621-622 (1978).
31. H. Oloff and E. Sagstuen: *Int. J. Radiat. Biol.* **40**, 493-505 (1981).

460

32. M. D. Sevilla, C. Van Paemel and C. Nichols: *J. Phys. Chem.*
    76, 3571-3577 (1972).
33. W. Flossmann, E. Westhof and A. Müller: *Int. J. Radiat. Biol.*
    30, 301-315 (1976).
34. M. D. Sevilla, C. Van Paemel and G. Zorman: *J. Phys. Chem.*
    76, 3577-3582 (1972).
35. J. Hüttermann, J. F. Ward and L. S. Myers, Jr.: *Int. J. Radiat. Phys.
    Chem.* **3**, 117-129 (1971).
36. J. N. Herak and V. Galogaza: *J. Chem. Phys.* **50**, 3101-3105 (1969).
37. M. D. Sevilla, S. Swarts, H. Riederer and J. Hüttermann: *J. Phys. Chem.*
    88, 1601-1605 (1984).
38. M. D. Sevilla, in Pullman and Goldblum (Eds.): Excited States in
    Organic and Biochemistry D. Reidel (Dordrecht) 15-25 (1977).
39. M. D. Sevilla, D. Suryanarayana and K. M. Morehouse: *J. Phys. Chem.*
    85, 1027-1031 (1981).
40. H. Riederer and J. Hüttermann: *J. Phys. Chem.* **86**, 3454-3463 (1982).
41. M. D. Sevilla and M. L. Engelhardt: *Faraday Disc. Chem. Soc.* **63**,
    255-263 (1977).
42. H. C. Box and E. E. Budzinski: *J. Chem. Phys.* **62**, 197-199 (1975).
43. A. Ehrenberg, L. Ehrenberg and G. Löfroth: *Nature* (Lond.) **200**,
    376-377 (1963).
44. R. Salovey, R. G. Shulman and W. M. Walsh, Ir.: *J. Chem. Phys.* **39**,
    839-840 (1963).
45. E. Sagstuen, E. O. Hole, W. H. Nelson and D. M. Close:
    J. Phys. Chem. **93**, 5974-5977 (1989).
46. H. Oloff, E. Haindl and J. Hüttermann: *Radiat. Res.* **80**, 447 (1979).
47. J. N. Herak, D. R. Lenard and C. A. McDowell: *J. Magn. Reson.* **26**,
    189-200 (1977).
48. P. M. Cullis, I. Podmore, M. Lawson, M. C. R. Symons, B. Dalgarno
    and J. M. Clymont: *J. Chem. Soc. Chem. Comm.*, 1003-1005 (1989).
49. J. Ohlmann, Univ. d. Saarlandes, Diploma Thesis (1990).
50. W. A. Bernhard and A. Z. Patrzalek: *Radiat. Res.* **117**, 379-394 (1989).
51. W. Flossmann, E. Westhof and A. Müller: *J. Chem. Phys.* **64**,
    1688-1691 (1976).
52. E. Westhof, W. Flossmann, H. Zehner and A. Müller: *Faraday Discuss.
    Chem. Soc.* **63**, 248-254 (1978).
53. H. Riederer, J. Hüttermann and M. C. R. Symons: *J. Phys. Chem.* **85**,
    2789-2797 (1981).
54. S. Das, D. J. Deeble, M. N. Schuchmann and C. Von Sonntag:
    Int. J. Radiat. Biol. **46**, 7-9 (1984).
55. K. M. Bansal, L. K. Patterson and R. Schuler: *J. Phys. Chem.* **76**,
    2386-2392 (1972).
56. J. D. Zimbrick, J. F. Ward and L. S. Myers, Jr.: *Int. J. Radiat. Biol.* **16**,
    505-523 (1969).
57. M. D. Sevilla, R. Failor and G. Zorman: *J. Phys. Chem.* **78**, 696-699
    (1974).
58. L. D. Simpson and J. D. Zimbrick: *Int. J. Radiat. Biol.* **28**, 461-475
    (1975).
59.. H. Riederer, *PhD-thesis*, Univ. Regensburg (FRG) (1981).
60. H. Riederer, J. Hüttermann and M. C. R. Symons: *Chem. Comm.*
    313-314 (1978).

61. D. M. Close, E. Sagstuen and W. H. Nelson: *J. Chem. Phys.* **82**, 4386-4388 (1985).
62. B. Rakvin, J. N. Herak, K. Voit and J. Hüttermann: *Radiat. Env. Biophys.* **26**, 1-12 (1987).
63. D. M. Close, W. H. Nelson and E. Sagstuen: *Radiat. Res.* **112**, 283-301 (1987).
64. H. Kim, E. E. Budzinski and H. C. Box: *J. Chem. Phys.* **90**, 1448-1451 (1989).
65. M. D. Sevilla, J. B. D'Arcy, K. M. Morehouse and M. L. Engelhardt: *Photochem. Photobiol.* **29**, 37-42 (1978).
66. K. Sieber and J. Hüttermann: *Int. J. Radiat. Biol.* **55**, 331-345 (1989).
67. S. Gregoli, M. Olast and A. Bertinchamps: *Radiat. Res.* **72**, 255-274 (1977).
68. E. O. Hole, W. H. Nelson, D. M. Close and E. Sagstuen: *J. Chem. Phys.* **86**, 5218-5219 (1987).
69. L. P. Candeias and S. Steenken: *J. Am. Chem. Soc.* **111**, 1094-1099 (1989).
70. E. Sagstuen, E. O. Hole, W. H. Nelson and D. M. Close: *Radiat. Res.* **116**, 196-209 (1988).
71. W. H. Nelson, E. O. Hole, E. Sagstuen and D. M. Close: *Int. J. Radiat. Biol.* **54**, 963-986 (1988)
72. D. M. Close, E. Sagstuen and H. W. Nelson: *Radiat. Res.* **116**, 379-392 (1988).
73. M. D. Sevilla and P. A. Mohan: *Int. J. Radiat. Biol.* **25**, 635-638 (1974).
74. H. Zehner, W. Flossmann and E. Westhof: *Z. Naturforsch.* **31c**, 225-231 (1976).
75. W. Flossmann, E. Westhof and A. Müller: *Int. J. Radiat. Biol.* **25**, 437-443 (1974).
76. L. Kar and W. A. Bernhard: *Radiat. Res.* **93**, 232-253 (1983).
77. a) D. M. Close, W. H. Nelson and E. Sagstuen, in Weil (Ed): *Electronic Magnetic Resonance of the Solid State*, Canadian Society for Chemistry, Ottawa (Canada), 237-250 (1987).
    b) D. M. Close and W. H. Nelson: *Radiat. Res.* **117**, 367-378 (1989).
78. S. Gregoli, M. Olast and A. Bertinchamps: *Radiat. Res.* **60**, 388-404 (1974).
79. P. Strand, E. Sagstuen, T. Lehner and J. Hüttermann: *Int. J. Radiat. Biol.* **51**, 303-318 (1987).
80. E. O. Hole and E. Sagstuen: *Radiat. Res.* **109**, 190- (1987).
81. C. Alexander and W. Gordy: *Proc. Natl. Acad. Sci* (USA) **58**, 1279-1285 (1967).
82. J. Hüttermann: *J. Magn. Reson.* **17**, 66-88 (1975).
83. H. C. Box and E. E. Budzinski: *J. Chem. Phys.* **64**, 1593-1595 (1976).
84. H. Zehner, E. Esthof, W. Flossmann and A. Müller: *Z. Naturforsch.* **32c**, 1-10 (1977).
85. A. Celalyan-Berthier, T. Berchaz and M. Geoffray: *J. Chem. Soc.* (Farad. Trans. 1) **83**, 401-409 (1987).
86. H. C. Box, E. E. Budzinski and G. Potienko: *J. Chem. Phys.* **65**, 1966-1970 (1978).
87. W. A. Bernhard, D. M. Close, J. Hüttermann and H. Zehner: *J. Chem. Phys.* **67**, 1211-1219 (1977).
88. G. Radons, H. Oloff and J. Hüttermann: *Int. J. Radiat. Biol.* **40**, 345-263 (1981).

89. J. Hüttermann, W. A. Bernhard, E. Haindl and G. Schmidt:
    J. Phys. Chem. **81**, 228-232 (1977).
90. M. Höhn and J. Hüttermann: *Int. J. Radiat. Biol.* **45**, 99-116 (1984).
91. K. P. Madden and W. A. Bernhard: *J. Phys. Chem.* **83**, 2643-2649 (1979).
92. T. Horning and W. A. Bernhard: *Radiat. Res.* **99**, 262-271 (1984).
93. J. Krieger and J. Hüttermann: *Int. J. Radiat. Biol.* **48**, 893-915 (1985).
94. J. Krieger, PhD.-thesis, University of Regensburg, (FRG) (1985).
95. B. C. Gilbert, D. M. King and C. B. Thomas: *J. Chem. Soc. Perkin
    Trans. II,* 1186-1199 (1981).
96. M. N. Schuchmann and C. von Sonntag: *J. Chem. Soc. Perkin Trans. II,*
    1958-1983 (1977).
97. A. Gräslund, A. Ehrenberg, A. Rupprecht and G. Ström: *Biochem.
    Biophys. Acta* **254**, 172-186 (1971).
98. A. Gräslund, A. Ehrenberg, A. Rupprecht, G. Ström and H. Crespi:
    Int. J. Radiat. Biol. **28**, 313-323 (1975).
99. J. Hüttermann, K. Voit, H. Oloff, W. Köhnlein, A. Gräslund and
    A. Rupprecht: *Faraday Discuss. Chem. Soc.* **78**, 135-149 (1984).
100. K. Voit, H. Oloff, J. Hüttermann, W. Köhnlein, A. Gräslund and
    A. Rupprecht: *Radiat. Environ. Biophys.* **25**, 175-181 (1986).
101. I. Zell, J. Hüttermann, A. Gräslund, A. Rupprecht and W. Köhnlein:
    Free Rad. Res. Comms. **6**, 105-106 (1989).
102. I. Zell, PhD-thesis, University of Saarland (FRG) (1990).
103. K. Voit, PhD-thesis, University of Regensburg (FRG) (1988).
104. J. Boon, P. M. Cullis, M. C. R. Symons and B. Wren: *J. Chem. Soc.
    Perkin Trans II,* 1057-1061 (1985).
105. P. M. Cullis, M. C. R. Symons, B. Wren and S. Gregoli:
    J. Chem. Soc. Perkin Trans. II, 1819 (1985).
106. P. M. Cullis, M. C. R. Symons, M. C. Sweenly, G. D. D. Jonses and
    J. D. McClymont: *J. Chem. Soc. Perkin Trans. II,* 1671-1676 (1986).

CONDUCTING POLYMERS

LOWELL D. KISPERT
Department of Chemistry, The University of Alabama, P. O. Box 870336,
Tuscaloosa, AL 35487-0336, USA.

1. INTRODUCTION

1.1. Background Definitions

Polymers with conjugated π-electron backbones exhibit low ionization potentials, high electron affinities and low energy optical transitions. This results in a polymer that can be more easily oxidized or reduced than conventional polymers. Typically the oxidation or reduction is carried out using charge transfer agents such as AsF_5, I_2, SbF_5, $FeCl_3$, O_2 or alkali metal which changes a normally insulating polymer into a conducting polymer. The conductivity can vary from that of semiconducting (10^{-6} S/cm) to near metallic (10^3 S/cm). As a point of reference, mercury has a conductivity around 10^4 S/cm. In the early stages of conducting polymer research it was found that polyacetylene films could be "doped" by electron donors or acceptors to conductivity levels around 10^3 S/cm, a value considered remarkable since a high degree of disorder is present in polymers. Previously, conductivity of this magnitude was associated with the highly ordered organic charge-transfer crystals. Typically the "doping" concentration can reach 50% of the final weight of the polymer and not the parts per million found in inorganic semiconductors. Thus the word "doping" is hardly appropriate for the oxidation or reduction process but it is used nevertheless in most conducting polymer literature. The doping process increases the carrier concentration by producing a radical cation or anion on the chain. If the hole (+) or the negative charge can overcome the binding energy, it can then move through the polymer and contribute to conductivity. It has been the ultimate goal of conducting polymer research to combine the attractive properties of polymers such as processibility and lightweightness with the electronic properties of metals or semiconductors.

The number of conducting polymer papers is too numerous to be covered here. Therefore, only the EPR study of the radical cations and anions are examined. Thus the reader is referred to several other reviews [1-3] for a more complete coverage. In addition to polyacetylene (PA), poly(p-phenylene) (PPP), poly(p-phenylene sulfide) (PPS), polypyrrole (PP), polythiophene (PTP), and polyaniline (PAN) have been doped to high conductivity. The similarity between the experimental behavior of doped PPP and PPS to PA, made it necessary to generalize theories such as the soliton theory of transport and doping that are specific to PA. A large amount of the theoretical work has dealt with the charge defects formed on the polymer chain. EPR studies have been carried out in an attempt to examine the theoretical predictions or assignments.

A. Lund and M. Shiotani (eds.), Radical Ionic Systems, 463–478.

Trans-polyacetylene can be represented by two energetically equivalent representations (A) and (B).

(A)   **(B)**

A cation, anion or radical defect on a polyacetylene backbone divides the polymer into two parts (A) and (B).

$+, -, \cdot$

(A)   **(B)**

The defect acts as a divider between two equal-energy species. Therefore the defect can move in either direction without affecting the energy of the backbone assuming an infinite chain. The movement of such a defect is described mathematically as a soliton. A radical defect is called a netural soliton; cation and anion defects are charged solitons. The soliton is considered the mode of charge and spinless transport in PA. However, spinless transport has also been observed in other systems such as PPP and PPS which do not possess two energetically equivalent structures. This feature necessitates a change in nomenclature, the charged defects (cations, radical-cations and dications) in PPP and PA are known as, (a) polarons (radical cations, positively charged hole sites or negatively charged electron sites (radical anion)) such as

A^-

(b) dications or dianions (bipolarons) - caused by the exothermic reactions of two radical ions such as

A^-

or (c) solitons (cation) such as

A^-

A^- represents the dopant ion. To understand the conducting polymer structure, EPR studies have been undertaken to determine the number of radicals present in each system and to determine the structure of the radical species.

1.2. Difficulties Arising From Inhomogeneous Doping

Studies have shown that the reaction of electron donors or acceptors with monomers or polymers occurs inhomogeneously in solid materials [4]; the local radical concentration depends on the physical shape (film or bead) of the reacting polymer. Reactions on surfaces of films occurs along thin cracks, while for beads the reaction occurs in localized spherical areas. Decreasing the dopant concentration, decreases the number of defects but not the local radical concentration. The resulting EPR spectrum is an exchanged narrowed line where both the g and the hyperfine anisotropy can be averaged out. For instance, reaction of AsF_5 with p-terphenyl powder results in an EPR spectrum that exhibits a Dysonian lineshape (Fig. 1) with a 0.1 G - linewidth, among the narrowest observed. A Dysonian line occurs for conducting samples whose thickness is greater than the microwave skin depth. One way to eliminate the inhomogeneous doping problem is to carry out the reaction in solution such as using an SO_2 solvent in which to react SbF_5 with p-terphenyl. In this case I = 0 for both ^{32}S and ^{16}O nuclei so that possible EPR broadening by interaction with the solvent is eliminated [4]. As a result, resolved powder pattern exhibiting cylindrical symmetry is observed despite the small g anisotropy (Fig. 2). Reaction with an electron acceptor like AsF_5 gives rise to a higher conducting material than reaction with an electron donor like alkali metals.

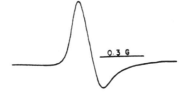

Fig. 1. The Dysonian EPR line shape observed for highly conducting AsF_5-doped p-terphenyl powder. The peak to peak linewidth increases to greater than 1 G for doped PPP polymer (taken from reference 4).

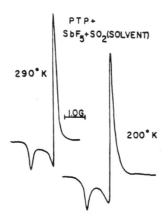

Fig. 2. Homogeneous solution doping of p-terphenyl with SbF_5 in a SO_2 solvent. The lack of possible magnetic nuclei, reduces line broadening from the solvent; the $g_{xx} = g_{yy} \neq g_{zz}$ powder pattern is observed as the conducting material precipitates out of solution (taken from reference 4).

1.3. Limitations of EPR Measurements

The EPR study of radical cations and anions in conducting polymer systems has been limited by the lack of spectral resolution. Typically the EPR spectrum consists of a single broad line exhibiting a peak-to-peak linewidth of 1 Gauss or larger when the polymer contains only C, H, O or N atoms. The proton hyperfine couplings are on the order of the dipolar anisotropy and similar to the linewidth - so the hyperfine couplings contributes only to the linewidth and not to a resolved spectrum. In addition, the g anisotropy (2.0031 - 2.00227), $\Delta g = 0.0008$ for carbon centered radicals such as in poly(p-phenylene), and even sulfur containing polymers like poly(p-phenylene sulfide) are too small for spectral resolution to be possible. Evidence also exists that exchange narrowing may occur, thereby causing a further reduction in resolution. However, it is possible to observe resolved hyperfine couplings if the conducting polymer can be formed in solution, such as was found possible for AsF_5-doped poly(p-phenylene sulfide) dissolved in liquid AsF_3 [5]. In some cases spectral resolution is possible in powder samples of conducting polymers, providing that the unpaired electron density for the radical cation or anion is largely centered on a sulfur or selenium atom. For example, AsF_5 doped poly(thiophene sulfide) and poly(thiophene selenide) exhibited an anisotropy of $\Delta g = 0.009$ ($g_{iso} = 2.0056$) and $\Delta g = 0.0565$ ($g_{iso} = 2.026$) respectively; a spectral spread sufficient to produce an anisotropic powder pattern [4]. Due to the lack of spectral resolution, *the EPR study of conducting polymers is limited to determining the radical concentration, linewidth as a function of temperature, effective g value for the spectral center of the pattern, spin susceptibilities, and establishing whether radicals are the carriers of conductivity.* This data is used as a basis for models of the spin system.

Polyacetylene (PA), polyparaphenylene (PPP) and its deuterated analogs, polypyrrole, and poly(p-fluorophenylene) represents the primary polymers containing C, H, O, N that have been studied by EPR techniques after the polymer has been reacted with electron donor or electron acceptor molecules. Similar EPR studies [4] have also been conducted for sulfur or selenium containing polymers, primarily poly(p-phenylene sulfide), PPS, poly(p-phenylene selenide) (PPSe), poly(thiophene sulfide) (PTS) and poly(thiophene selenide) (PTSe). The monomers 3-methylthiophene (3MT), p-terphenyl, (TP), p-quaterphenyl (QP), and p-sexiphenyl (SP) were shown to form polymeric materials upon reaction with electron acceptor or electron donors.

2. POLYMER SYSTEMS

2.1. Polyacetylene

Trans PA gives rise to an EPR spectrum which was first characterized as due to an extended π-electron defect approximately one per 3000 carbon atoms [6,7]. The Curie Law was found to hold ($\chi \sim 1/T$) indicating localized spins yet the EPR linewidth was narrow suggesting motionally narrowing of the hyperfine interaction. EPR and ENDOR studies [8-17] of PA and related polyenes [18-20] have examine the spin distribution which leaves little doubt that the unpaired electron resides in a highly delocalized p-orbital of the undoped polymer. The question is the extent of delocalization. Furthermore, "cis-rich" and "trans-rich" polyacetylene before doping have been studied by ENDOR techniques [21] in the temperature range 4.2 - 1.9 K. Both forms exhibit a spectrum similar to the high-temperature ENDOR spectrum suggesting the existence of a common delocalized π-electron defect or soliton. The hyperfine tensors for cis-rich polyacetylene are temperature independent while the hyperfine couplings for trans-rich polyacetylene are observed to be temperature dependent and characterized by a 1 K-activation energy. At higher temperatures (6 - 300 K) the trans-PA has a 400 K energy barrier.

The ENDOR measured ^{13}C and ^1H tensors for cis-rich polyacetylene (upper part of Table I) before doping [22] was found to be compatible with 2 spin densities, $\rho_{odd\ c}$ = + 0.06 (for 25 carbons) and $\rho_{even\ c}$ = -0.02 for 24 carbons. The unpaired density alternates along the 49-carbon polyene segment; this dependence arises from electron-coulomb interactions. Similar values were also obtained for the trans-rich polyacetylene. The observation of two unique hyperfine tensors for ^1H and ^{13}C establishes the wave function to an extended but well-defined region of the polyene backbone. A determination of the signs of the couplings was based on a Karplus-Fraenkel analysis and is given in the lower part of Table I.

On the other hand, Kuroda et al. [14] carried out an ENDOR study at 4°K of the undoped cis-rich stretched oriented PA, and found that the measured unpaired spin density ranged from 0.11 to 0.19; values that are consistent with the prediction of the soliton theory. In addition, a wide distribution of hyperfine fields was found (14, 15-17) consistent with a smooth falloff of the midgap-state wave function in contrast to the Thomann proposed delocalized over 50 sites. It was suggested [3] that Thomann's results are consistent with a spin-density profile predicted by the soliton theory if it is assumed that a residual center-of-mass motion with associated averaging is occurring. Current on-going studies by Dalton and coworkers have shown that the differences in the ENDOR results can be traced to the method in which PA if formed.

Table I. Electron Nuclear Hyperfine Interaction Tensors[a] Determined from ENDOR and TRIPLE Measurements and Assignments of Electron Nuclear Hyperfine Tensors[a] Based on Karplus-Fraenkel Analysis (taken from ref. 15).

	^1H	^{13}C
Electron Nuclear Hyperfine Interaction Tensors		
odd carbon positions	$A_x = + 0.64$	$A_x = (?)\ 1.10$
	$A_y = + 2.50$	$A_y = (?)\ 1.10$
	$A_z = + 1.39$	$A_z = \pm\ 2.50$
even carbon positions	$A_x = \pm 0.18$	$A_x = (?)\ 0.46$
	$A_y = \pm 0.75$	$A_y = (?)\ 0.46$
	$A_z = \pm 0.39$	$A_z = +\ 0.75$
Assignments of Signs Based on Karplus-Fraenkel Analysis		
odd	$A_x = -0.64$	$A_x = +1.10$
	$A_y = -2.50$	$A_y = +1.10$
	$A_z = -1.39$	$A_z = +2.50$
even	$A_x = +0.18$	$A_x = -0.46$
	$A_y = +0.75$	$A_y = -0.46$
	$A_z = +0.39$	$A_z = -0.75$

[a]In gauss. (?) indicates uncertainty in sign.

The temperature and frequency dependence of the electron spin lattice relaxation rate [23] for the defects in pristine trans-polyacetylene has been analyzed in terms of electron-electron-dipolar interaction between spins modulated by one-dimensional diffusion induced by phonon scattering of the spins. A model showed the symmetry between moving and stationary solitons scattered by phonons. Soliton-^{13}C interactions have been examined by ESE techniques in trans-polyacetylene [24].

Trans-polyacetylene (PA) can be used as the cathode and Li/Al alloy as the anode of a battery [19-27]. In situ EPR measurements have been carried out during the charge/discharge cycle and the resulting EPR spectrum exhibits an unresolved Dysonian line shape whose A/B ratio increases from 1.0 to 4.2 for 6% doping [27]. The reactions that take place upon discharge (undoping) are given by eqns 1 and 2 for the cathode and

$$[(CH)^{+y}(Z^-)_y]_x + yxe^- \longrightarrow (CH)_x + yx(Z^-) \tag{1}$$

$$yxLi \longrightarrow yxLi^+ + yxe^- \tag{2}$$

anode respectively [26] where the electrolyte was 0.5 M LiCl$_4$ in propylene carbonate. The detected spin concentration reversibly increases [27] by less than a factor of two and the maximum A/B ratio decreases with each repeated charge/discharge cycle. The decreasing A/B ratio is consistent with the observed decreasing conductivity, however the invariance of the EPR linewidth before and after doping suggests that massive disruption of chain conjugation does not occur during battery cycling.

2.2. Poly(p-phenylene)

2.2.1. *g value and linewidth, donor-doped*

An extensive study was undertaken [28] of the alkali metal (Li, Na, K, Cs and Rb) doped PPP and deuterated PPP (DPPP). Temperature dependent linewidth studies, g value measurements and anion radical concentrations were determined. The linewidth measurements show some interactions with Cs and Rb but almost no interactions with Na, Li and K. Upon deuteration, the peak-to-peak (ΔH_{pp}) linewidth at 300 K decreases from 5.0 G to 4.5 G for Cs, from 7.6 to 7.2 G for Rb from 2.0 G to 1.4 G for K-doped, from 1.2 to 0.6 G for Na and from 0.6 to 0.3 G for Li-doped PPP. This indicates that the linewidth is determined partly by unresolved proton couplings. The temperature dependence of the linewidth suggests that exchange narrowing occurs for the Rb, Cs and K-doped PPP and DPPP at low temperatures while exchange narrowing seems to occur at 300 K for Na and Li-doped PPP. Such is possible, if contraction of the lattice occurs for the Rb, Cs and K samples as the temperature is lowered. The g values for the Li, Na and K-doped samples (Table II) are equal to 2.0027. A small increase in g values occurs for the Rb, and Cs samples (Table II) due to weak spin-orbit coupling of Rb and Cs with the unpaired electron. The measured susceptibilities show that the unpaired spin concentration of the radical anions is much lower than the alkali dopant concentration (Table II). The temperature dependence of the susceptibility was found to fit an expression for an equilibrium between separated polaron (mobile radical anions) defects and singlet and triplet spin states which are formed intermolecularly via polaron pairing. The interaction is found to be antiferromagnetic and the binding energy between polarons is 2-3 meV.

Comparing the g value of 2.0027 for the radical anion in the alkali-doped conducting PPP, and potassium doped terphenyl (K-TP), quaterphenyl (K-QP), and sexiphenyl (K-SP) to the average g-value of 2.00249 observed for the cation radical [29,30] salts

Table II. Percent dopant, number of unpaired spins per phenyl, and g values for the alkali metal doped samples (taken from reference 28).

Sample	Dopant	% Dopant per phenyl	Unpaired[a] spins per phenyl 50 K	297 K	Unpaired spins per dopant 297 K	$g$value (Temp.)
PPP	Li	12.2	0.04_2	0.03_5	0.29	2.0027(11) 2.0027(297)
	Na	40.5	0.01_5	0.009	0.02	2.0027(9.4) 2.0027(297)
	K	38.8	0.01_7	0.02_0	0.05	2.0027(9.4) 2.0027_6(297)
	Rb	35.0	0.04_0[b]	0.02_1	0.06	2.0030(11) 2.0034(297)
	Cs	27.6	0.02_6	0.01_2	0.04	2.0029(12) 2.0029(297)
DPPP	Li	30.0	0.04_3	0.04_4	0.15	2.0028(8.5) 2.0027_6(297)
	Na	38.0	0.047	0.05_0	0.13	2.0027(8.3) 2.0027(297)
	K	19.9	0.02_0	0.01_2	0.06	2.0027(8.1) 2.0027_8(297)
	Rb	35.0	0.02_0	0.02	0.06	
	Cs	16.0	0.01_3	0.01_6	0.10	2.0028(7.8) 2.0029(297)
K-TP		27.3	0.02_0			2.0027_7(297)
K-TP[c]		23.7	0.006	0.007	0.03	2.0027_7(297)
K-QP		16.0	0.005	0.003	0.02	2.0027_0(297)
K-SP		13.5	0.02_0	0.02_3	0.17	2.0027_0(297)

[a]Experimentally measured by EPR.
[b]Spins measured on sample identically prepared to the one for which % dopant was measured.
[c]No naphthalene catalyst used in doping.

of p-terphenyl (TP) and p-quarterphenyl (QP), indicates a measureable difference in g values for the two conducting systems as noted in Tables II and III. The reason for this is represented by the difference between the g tensor of a monomer radical cation in a stacked array and the radical anion or cation formed in PPP (or in AsF_5-doped TP crystals) where polymerization has occurred. The g anisotropy is averaged along the stack in the cation radical salt crystal by rapid electron exchange so that $g_{av} = 2.0025$. As opposed to this, the conduction electron averages the g anisotropy along the chain axis of a polymer [28] to yield a g_{av} value of 2.0027.

The temperature dependence of the EPR linewidth for the alkali-doped PPP, TP, and QP was examined by electron spin echo measurements. The measured T_1 (spin lattice) and T_m (spin memory) values showed no significant dependence

Table III. EPR g values and linewidths for various arene radical salts (taken from reference 29).

Crystal	EPR	g linewidth	columnar value stacking
$(FA)_2PF_6$	<10 mG (stable at 25 °C	near 2.0023	FA twisted alternately by 180°.
$(C_{10}H_8)_2AsF_6$ $(C_{10}H_8)_2PF_6$	2 - 40 mG (unstable at 25 °C	2.0025 2.0025	$C_{10}H_8$ twisted alternately by 90°.
$(TP)_{\leq 2}PF_6$	70 - 125 mG (stable at − 40 °C) (unstable at 25 °C)	2.00312 2.00230 2.00206 <2.00249>	TP twisted alternately by 45°.
$(QP)_{\leq 2}PF_6$	150 - 220 mG (stable at 5 °C)	2.00310 2.00217 2.00217	1 magnetic site, QP parallel alignment

on temperature. This temperature independence is observed when spin exchange occurs and is attributed to exchange narrowing between anions located on neighboring chains with r < 10 Å and not to charge carriers of conductivity. This result suggests that EPR linewidth variation is due to differences in donor-donor distances. The spins observed by EPR reside largely in the organic moities and not in the metals. The lack of g values and linewidths varying with the spin-orbit coupling of the metal suggests that the EPR observed free radicals are not the major carriers of conductivity.

2.2.2. *Spin Exchange, Donor and Acceptor-doped.* The presence of exchange was studied further [32] by measuring the difference in EPR linewidth of the protonated and deuterated PPP samples donor-doped with Li, Na and K. The differences tend to zero for K-doped samples below 25 K and above 300 K for Li-PPP and Na-PPP. The removal of the linewidth broadening from protons suggests rapid spin exchange at these temperature limits. A fit [32] to the temperature dependent EPR linewidth ΔH_{pp} for Li- and Na-PPP samples using the empirical two term equation

$$\Delta H_{pp} = A + B^2/(B^2+v^2)^{1/2} \tag{3}$$

suggests a thermally activated exchange of $v = Dexp(-E'/kT)$ (fast exchange at 300 K) while the linewidth for K-PPP is consistent with a lattice contraction where $v = Dexp(E'/kT$ (fast exchange at low temperature) and E' is the energy gap between two energy levels corresponding to species giving rise to linewidths A and B. The calculated temperature independent linewidth A of 0.46 Gauss was detected by ESE measurements [31] for Li-doped at 9 K and for K-doped at 300 K. Furthermore, acceptor-doped (AsF_5) PPP samples give rise to materials which exhibit a higher conductivity and an ESE derived linewidth of 0.08 Gauss. The lower value of the linewidth is presumably due to an increase in spin exchange as the conductivity increases. This observation can be rationalized if the spin exchange increases as the mobility of the polaron, the carrier of conductivity, increases. In lightly doped (AsF_5) DPPP there is the absence of a temperature independent ESE linewidth [32] and thus no evidence for rapid spin exchange. In contrast, highly acceptor-doped (AsF_5) PPP gives rise to a Dysonian EPR lineshape and a temperature independent ESE linewidth of 0.08 Gauss. The appearance of an ESE line whose linewidth that does not vary with temperature but varies with the conductivity (decreasing with increasing conductivity) suggests that spin exchange plays a dominant role in donor and acceptor doped PPP samples.

The lack of any triplet state species being observed at low temperatures and the calculated coupling B for the second species between 0.52 and 5.0 Gauss being inconsistent with an exchange energy for a triplet species of 132 - 184 K, suggests that the second species is due to isolated radicals [32]. This suggestion is also consistent with the decrease in radical concentration upon long term reaction of AsF_5 with PPP - where the material continues to polymerize by a free radical mechanism.

The narrow line species detected by ESE measurements was deduced to be the carrier of conductivity. The EPR line is actually a composite of a narrow line component due to polarons superimposed by a broader line component that is the exchange interaction between isolated radicals in equilibrium with polarons and bipolarons with the equilibrium in favor of bipolarons at 4 K. The composite EPR line represents only approximately 5% of the dopant concentration - the rest (95%) exists in a singlet ground state as bipolarons.

2.2.3. Acceptor-Doped PPP. Reaction of excess AsF_5 dopant over a 24 hour period results in conductivities of 50 - 100 S/cm [33]. This represents approximately 2 orders of magnitude larger conductivity than donor-doped material. It was found that at high dopant concentration and high conductivity, most of the charge is in bipolarons (~95%). With an increase in doping time, the spin concentration decreased, presumably due to both polaron-polaron reaction and polaron ionization to form bipolarons [34-39]. The temperature dependence of the susceptibility [33] deviates from Curie-Weiss behavior at low temperature and can be explained by the low-temperature condensation of isolated polarons to form intermolecular polaron pairs in singlet ground states. The interaction between the spins is antiferromagnetic.

2.3. Oligomer Monomers, Acceptor Doped

Upon reacting a donor (alkali metal) or acceptor (AsF_5), with monomers such as p-terphenyl, p-quaterphenyl or p-sexiphenyl, a polymeric material is formed [1,2,4]. IR studies have shown that the chain length may be as much as $16°$. The mechanism of this polymerization process was examined by EPR measurements of the radical formation upon reacting AsF_5 or SO_3 at 300 K with single crystals of p-terphenyl [40]. It was observed that the resulting π-radical cation is displaced from its original molecular positions by a rotation of $64°$ about the rod axis and a tilt of $42°$ toward the c* crystal axis. This means that the polymerization occurs between monomers related by a 2_1 screw symmetry axis instead of a c-axis translation where little destruction of the crystal would result. Instead, the crystal does completely disintegrate upon complete reaction, in contrast to heat-treated or irradiated diacetylene crystals which retain their crystalline nature upon converting from the monomer to the polymer. The EPR study was aided considerably by strong spin

472

exchange that caused a narrowing of the EPR line so that the different non equivalent crystals sites could be followed.

2.4. Poly (p-perfluorophenylene)

Donor-doped (K-) poly(p-perfluorophenylene) gave rise to a conductivity that is slightly larger than that of the analogously potassium doped poly (p-phenylene) [41]. It was thought that fluorinated polymers would have higher conductivity attributed to increased electron affinities and smaller band gaps which resulted from substituting fluorine for hydrogen in the starting material. Additionally, the unpaired spin concentration is decreased by an order of magnitude. Even so, the measured EPR susceptibilities have a temperature dependence that suggests an equilibrium between separated polaron defects and the ground state bipolarons. This further supports the assignment of the bipolaron as the carrier of conductivity in doped PPP.

2.5. Poly(p-phenylene sulfide)

When poly(p-phenylene sulfide) dissolved in liquid AsF_3 is reacted in solution with a 0.5 molar ratio of AsF_5 to monomer equivalent (C_6H_4S) of the PPS polymer, $AsF_5/\phi S$, the solution EPR spectrum [5, 42, 43] exhibits a 1:4:6:4:1 resolved pattern centered at 2.0079 with a hyperfine coupling equal to 1.2 G (Fig. 3). An INDO calculation shows that such a spectrum can only be observed by a non-planar sulfur-centered radical cation with coupling of the electron to the four meta ^{1}H's on the adjacent phenyl rings (structure I). Cyclized structures such as II or III

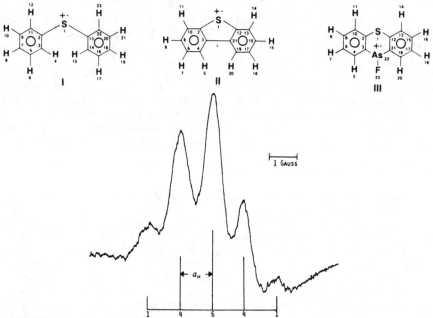

Fig. 3. EPR spectrum of AsF_5-doped PPS in AsF_3, solution No. 1. The calculated dopant/monomer molar ratio equals 0.5 $AsF_5/\phi S$, g equals 2.0079, and a_H equals 1.2 G (taken from reference 5).

yield g values or coupling constants which are inconsistent with observed g values, unpaired spin densities, and coupling constants; the INDO calculated g values and proton couplings are given in Table IV. The best described structure of I

Table IVa. Calculated g values and unpaired spin densities of sulfur-centered radical cations in PPS (taken from reference 4).

Radical cation [structural information	g_{iso}	g_{aniso}	ρ/atom orbital
I; Θ_{CSC} = 130°, ϕ planes at ± 45° relative to the C$_2$–S$_1$–C$_{13}$ plane.	2.0068	2.0023 2.0037 2.0143	0.008/S-p_x 0.36/S-p_y 0.37/S-p_z

Table IVb. Calculated unpaired spin densities (ρ) and hyperfine couplings (a_H) for ^{1}H's in sulfur-centered radical cations (taken from reference 4).

Radical cation; structure info.	^{1}H number	ρ_H	a_H(G)
I; Θ_{CSC} = 130°, ϕ planes at ± 45° relative to the C$_2$–S$_1$–C$_{13}$ plane.	4,8,15,19	-0.0007	-0.38
	6,17	0.0019	1.03
	10, 21	0.0015	0.81
	12, 23	-0.0006	-0.32

most closely resembles that of the neutral parent polymer PPS in which adjacent phenyl rings lie orthogonal to each other and at alternately +45° and -45° with respect to the C$_2$–S$_1$–C$_{13}$ plane. The radical cation charge carrier is highly localized on sulfur. The spin concentration indicates that approximately 0.01 spins occurs per monomer unit. At higher dopant concentration (2.8 AsF$_5$/ϕS), no resolved hyperfine structure is observed (Fig. 4). The resultant EPR line exhibits a Dysonian lineshape persumably due to increased carrier mobility and thus spin exchange.

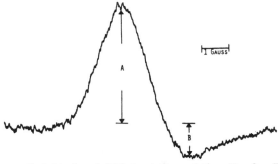

Fig. 4. EPR spectrum of AsF$_5$-doped PPS in AsF$_3$, solution No. 2. The calculated dopant/monomer molar ratio equals 2.8 AsF$_5$/ϕS, g equals 2.0076, and the A/B ratio equals 3.6 (taken from reference 5).

2.6. Sulfur and Selenium Analogs

It is always of considerable interest to relate the monomer structure of a polymer to its conductivity. An interesting series was the conducting polymers formed by the reaction of AsF_5 with the polymers PPS, PPSe, PTS and PTSe [4],

where a selenium atom has been substituted for a sulfur atom. These polymers had reported conductivities of 1, 8.2×10^{-8}, 2.6×10^{-5}, and 0.12 s/cm respectively. Unfortunately the conductivities did not follow any obvious order. A search for the reason involved measuring the g value for each cation radical formed upon doping and comparing it to that predicted by an INDO [44] calculation. The minimum energy structure for PTS was found to be planar with no cyclization occurring (structure IV). Upon substitution of a Se atom (PTSe), a planar structure was calculated with a C–Se–C angle equal to 115°, suggesting a possible cyclization could occur with a hydrogen-selenium interaction (structure V). On the other hand,

the minimum energy configuration of PPSe was calculated to be non-planar with no evidence for cyclization (structure VI). Thus the PTS cation radical is not predicted

to cyclize but may cyclize upon substitution of Se atom. On the other hand this tendency to cyclize does not tend to occur upon Se substitution for S in PPS. In fact experimental evidence exists that cyclized structures do occur as final products for PPS (structure I). Differences in structure are predicted and could influence the 3-dimensional packing structure of the material, thus the measured conductivity. The calculated g values for PTS, PTSe, PPSe, are given in Table V. The g values for AsF_5 doped PPS has been given previously.

Table V. Calculated (INDO/CI) and experimental g values for radicals formed from PTS, PTSe, and PPSe (taken from reference 4).

		g Values		
Cation radical		Calculated (INDO/CI)	experimental EPR	
PTS		2.0021	-	
		2.0023	-	
		2.0110	-	
	$<g>$	2.0052	2.0056[a]	
PTSe		(2.0022)[b]	2.0561	2.034
		(2.0023)	2.0266	2.0079
		(2.0029)	1.9996	2.0079
	$<g>$		<2.0261>	<2.0166>
PPSe		(2.0023)[b]	2.0577	
		(2.0023)	2.0266	
		(2.0023)	2.0015	
			<2.0286>	

[a]Center of spectrum - single broad EPR spectrum.
[b]Apparently, the current version of INDO fails to correctly calculate g values for Se-centered cation radicals.

2.7. 3-Methylthiophene

Most polymers before reaction with electron donors or acceptors possess a weak unresolved EPR signal. The EPR spectrum is usually due to radicals resulting from the polymerization process such as chain ends, left over catalyst, and unreacted short chains and is of little use in understanding the conducting polymer process. Exceptions exists, especially for systems like polyacetylene but in many systems, such spectra just generates confusion. An attempt to eliminate this problem was undertaken by reacting monomers with donors or acceptors, eliminating the initial polymer signal. Such a study involved using 3-methylthiophene as a monomer dissolved in AsF_3 and reacting with AsF_5 [45]. A conducting polymer solution is formed which upon solvent removal shows that a low molecular weight (4 repeat units) polymer has been formed. The advantage in studying a system which exhibits no radical signal initially permits a study by nmr of the non-radical processes and a study of the radical processes by EPR that may occur during the formation of the conducting polymer signal.

An EPR study showed that when the ratio of AsF_5 to 3 MT is greater than 1.5, an unresolved EPR solution spectrum with a g value of 2.0038 and a linewidth of 7.4 G is observed. The g values and linewidths are comparable to that reported by Schmidt for the tetrakis (4-hydroxyl phenyl) thiophene radical cation in concentrated H_2SO_4 [46]. Thus the solution spectrum results from the oxidation of the thiophene to yield a sulfur-based radical cation. However, when the solvent is removed and a

476

thin film is formed, a powder EPR spectrum is also formed due to a carbon based radical where g = 2.0023 and the linewidth decreases to 4.4 G. This surprising result is consistent with the XPS experimental data and theoretical calculations by Salaneck and Bredas [47] which indicate a carbon-based conductivity carrier. The decrease in EPR linewidth can be due to an increase in electron exchange effects. A measure of the radical concentration as a function of dopant level and as a function of increasing conductivity shows that the carrier of conductivity is not due to the radical species. Related work on polythiophenes [48, 49] has shown that diamagnetic bipolarons are the carrier of conductivity at high dopant concentration. At low dopant concentration, the polaron acts as the charge carrier.

2.8. Polypyrrole

Polypyrrole is a conjugated polymer with a non-degenerate ground-state. EPR measurements [50] at low doping levels show that the spin concentration detected is due to isolated polarons as charge carriers. At higher levels of oxidant, the spin concentration decreases rather than increases because the polarons bind in pairs to form bipolarons.

3. ACKNOWLEDGEMENTS

This work was supported by the Division of Chemical Sciences, Office of Basic Energy Sciences, Office of Energy Research of the U. S. Department of Energy under Grant DE-FG05-86ER13465. J. Thrasher is thanked for his assistance. I also thank E. Jackson for typing this manuscript.

REFERENCES

1. J. E. Frommer and R. R. Chance: In *Encyclopedia of Polymer Science and Engineering*, 2nd ed., **5**, pp. 462-507, Wiley: New York, (1986).
2. J. E. Frommer: *Acc. Chem. Res.* **19**, 2-9 (1986).
3. A. J. Heeger, S. Kivelson, J. R. Schrieffer and W.-P. Su: *Rev. Mod. Phys.* **60**, 781-850 (1988).
4. L. D. Kispert: *Reviews of Chemical Intermediates*, **7**, 45-70 (1986).
5. D. P. Murray and L. D. Kispert: *J. Chem. Phys.* **83**, 3681-3684 (1985).
6. I. B. Goldberg, H. R. Crowe, P. R. Newman, A. J. Heeger and A. G. MacDiarmid: *J. Chem. Phys.* **70**, 1132-1136 (1979).
7. B. R. Weinberger, E. Ehrenfreund, A. J. Heeger and A. G. MacDiarmid: *J. Chem. Phys.* **72**, 4749-4755 (1980).
8. S. Kuroda and H. Shirakawa: *Solid State Communications* **43**, 591-594 (1982).
9. S. Kuroda, M. Tokumoto, N. Kinoshita and H. Shinakawa: *J. Phys. Soc. Japan* **51**, 693-694 (1982).
10. S. Kuroda, M. Tokumoto, N. Kinoshita, T. Ishiguro and H. Shirakawa: *J. de Phys.* **44**, C3-303-C3-306 (1983).
11. H. Thomman, L. R. Dalton, Y. Tomkiewicz, N. S. Shiren and T. C. Clarke: *Phys. Rev. Lett.* **50**, 533-536 (1983).
12. M. T. Jones, H. Thomann, H. Kim, L. R. Dalton, B. H. Robinson and Y. Tomkiewicz: *J. Phys.* **C3-44**, 455-458 (1983).
13. J. F. Cline, H. Thomann, H. Kim, A. Morrobel-Sosa, L. R. Dalton, B. M. Hoffman: *Phys. Rev. B.* **31**, 1605-1607 (1985).

14. S. Kuroda, H. Bando and H. Shirakawa: *Solid State Commun.* **52**, 893-897 (1984).
15. A. Grupp, P. Hofer, H. Kass, M. Mehring, R. Weizenhofer, and G. Wegner in Electronic Properties of Conjugated Polymers edited by H. Kuzmany, M. Mehring and S. Roth, Springer, Berlin, P. 156, 1987.
16. H. Kass, P. Hofer, A. Grupp, P. K. Kahol, R. Weizenhofer, G. Wegner, and M. Mehring: *Europhys. Lett.*, **4**, 947 (1987).
17. S. Kuroda and H. Shirakawa, *Synth. Met.* **17**, 423 (1987).
18. L. R. Dalton, H. Thomann, A. Morrobel-Sosa, C. Chiu, M. E. Galvin, G. E. Wnek, Y. Tomkiewicz, N. S. Shiren, B. H. Robinson and A. L. Kwiram: *J. Appl. Phys.* **54**, 5583-5591 (1983).
19. D. Davidov, A. J. Heeger, F. Moraes, H. Kim, L. R. Dalton: *Solid State Commun.* **53**, 497-500 (1985).
20. D. Davidov, A. Heeger, H. Thomman, H. Kim, P. Bryson, L. R. Dalton, J. F. Cline and B. M. Hoffman: Unpublished results.
21. J. F. Cline, H. Thomann, H. Kim, A. Morrobel-Sosa, L. R. Dalton and B. M. Hoffman: *Phys. Rev. B.* **31**, 1605-1607 (1985).
22. H. Thomann, J. F. Cline, B. M. Hoffman, H. Kim, A. Morrobel-Sosa, B. H. Robinson and L. R. Dalton: *J. Phys. Chem.* **89**, 1994-1996 (1985).
23. B. H. Robinson, J. M. Schurr, A. L. Kwiram, H. Thomann, H. Kim, A. Morrobel-Sosa, P. Bryson and L. R. Dalton: *Molec. Cryst. Liq. Cryst.* **117**, 421-429 (1985).
24. H. Thomann, H. Kim, A. Morrobel-Sosa, C. Chiu, L. R. Dalton, and B. H. Robinson: *Molec. Cryst. Liq. Cryst.* **117**, 455-458 (1985).
25. P. Bernier, A. E. El Khodary and F. Rachdi: *Proc. IUPAC, 28th Macromolecular Sym.*, Amherst, MA, p. 407, July 12 (1982).
26. D. Macinnes, Jr., M. A. Drury, P. J. Nigrey, D. P. Nairs, A. G. Macdiarmid and A. J. Heeger: *J. Chem. Soc. Chem. Comm.*, 317 (1981).
27. L. D. Kispert, J. Joseph, T. V. Jayaraman, L. W. Shacklette and R. H. Baughman: *J. De Phys.* **44**, C3-317-C3-320 (1983).
28. L. D. Kispert, J. Joseph, G. G. Miller and R. H. Baughman: *J. Chem. Phys.* **81**, 2119-2125 (1984).
29. L. D. Kispert, J. Joseph, J. McGraw, L. Robinson and R. Drobner: *Synth. Met.* **19**, 67-72 (1987).
30. V. Enkelmann, K. Göckelmann, G. Wieners and M. Monkenbusch: *Mol. Cryst. Liq. Cryst.* **120**, 195-204 (1985).
31. L. D. Kispert, J. Joseph, M. K. Bowman, G. H. van Brakel, J. Tang and J. R. Norris: *Mol. Cryst. Liq. Cryst.* **107**, 81-90 (1984).
32. L. D. Kispert, J. Joseph, J. Tang, M. K. Bowman, G. H. van Brakel and J. R. Norris: *Synth. Met.* **17**, 617-622 (1987).
33. L. D. Kispert, J. Joseph, G. G. Miller and R. H. Baughman: *Mol. Cryst. Liq. Cryst.* **118**, 313-318 (1985).
34. J. L. Bredas, R. R. Chance and R. Silbey: *Mol. Cryst. Liq. Cryst.* **77**, 319-331 (1981).
35. J. L. Bredas, R. R. Chance and R. Silbey: *Phys. Rev. B.* **26**, 5843-5854 (1982).
36. R. R. Chance, J. L. Bredas and R. Silbey: *Phys. Rev. B.* **29**, 4491-4495 (1984).
37. G. Crecelius, M. Stanm, J. Fink and J. J. Ritsko: *Phys. Rev. Letters* **50**, 1498-1500 (1983).

478

38. J. Fink, G. Crecelius, J. J. Ritsko, M. Stanm, H. J. Freund and H. Gonska: *J. de Phys. Coll.* C3 **44**, 741 (1983).
39. M. Peo. S. Roth, K. Dransfeld, B. Trieke, J. Hocker, H. Gross, A. Grupp and H. Sixl: *Solid State Commun.* **35**, 119 (1980).
40. T. Robinson, L. D. Kispert and J. Joseph: *J. Chem. Phys.* **82**, 1539-1542 (1985).
41. L. D. Kispert, J. Joseph and R. Drobner: *Synth. Met.* **20**, 209-214 (1987).
42. D. P. Murray, L. D. Kispert, S. Petrovic and J. E. Frommer: *Synth. Met.* **28**, C269-C274 (1989).
43. J. E. Frommer, D. P. Murray and L. D. Kispert: *Synth. Met.* **15**, 259-263 (1986).
44. H. Oloff and J. Hutterman: *J. Magn. Res.* **40**, 415-437 (1980).
45. D. P. Murray, L. D. Kispert, S. Petrovic and J. E. Frommer: *Macromolecules* **22**, 2244-2252 (1989).
46. U. Schmidt: *Angew Chem. Int. Ed. Engl.* **3**, 602-608 (1964).
47. W. R. Salaneck, C. R. Wu, O. Inganas, J. E. Osterholm and J. L. Bredas: *Abstracts of Papers,* 192nd National Meeting of American Chemical Society, Anaheim, CA, INOR 375; American Chemical Society: Washington, DC (1986).
48. G. Harbeke, E. Meier, W. Kobel, M. Egli, H. Kiess and E. Tossatti: *Solid State Commun* **55**, 419 and references therein (1985).
49. J. Chen and A. J. Heeger: *Solid State Commun.* **58**, 251 and references therein (1986).
50. J. C. Scott, J. L. Bredas, J. H. Kaufman, P. Pfluger, G. B. Street and K. Yakushi: *Mol. Cryst. Liq. Cryst.* **118**, 163-170 (1985); J. H. Kaufman, N. Colaneri, J. C. Scott, K. K. Kanazawa and G. B. Street: *Mol. Cryst. Liq. Cryst.* **118**, 171-177 (1985).

Index

sulfides 30, 34
sulfuryl chloride 315
superhyperfine splitting 365
superhyperfine tensor 345
superhyperfine tensors of hydrogen bonded
    protons 344
superoxide anions 317
switching of the direction of distortions 380
symmetric configurations 367
symmetry-adapted 181
symmetry-adaptad form 179
symmetry-based selection 368
symmetry of intramolecular potentials 106
synsesquinobornene 148
synsesquinorbornene oxide 129

T^- 456

$t-C_4H_9$ radical 117

TCNE 320

TeF_{6-} 321
teraalkyl olefins 26
tertiary alkyl radicals 146
tertiary products 382
tetraalkyl 28
tetraalkyl hydrazines 26
tetracyanonickelate 318
tetrahedral molecule 108
tetrahydrofuran (THF) 134, 184
tetrahydrothiophene 135
tetrakis (4-hydroxyl phenyl) thiophene 475
2,2,3,3-tetramethylbutane 139
tetramethylcyclobutadiene radical cation 171
tetramethylethylene (TME) 99, 168
2-tetramethylhydrofuran glass 399
tetrametylsilane matrix 322
tetrazine 318
thermal electron-hole reaction 368
thermal excitation 117
thermal reactions 160, 172
thermalized 122
thermally activated migration of [⊕] 382
thermionic electron source 56
THF 186, 188
THF^+ 135, 144

thietan-2-yl 132
thietane 131, 132
thiirane 130
thiocyanates 319
thioether 131
thiopene anion 355
three-membered heterocycles 130
three-membered rings 126
through-bond interaction 38, 186
through-space 38
through space interaction 189
Tl^{2+} 318
Time-resolved Fluorescence Detected Magnetic
    Resonance 196
time-resolved magnetic resonance 195
2,2,3 TMB 112, 118
TMPD 325
TMS 322
toluene cations 66, 67
toluene radical cation 67
topological degeneracy 428
TP 469
track 362
trans C-H$_\beta$ 111
trans effect 162
trans-gauche isomerization 156
trans-polyacetylene 464, 467, 468
trans-rich polyacetylene 466, 467
transition-metal carbonyls 316, 321
transition moment 8, 13, 37, 39
trap depths 420
trap dimensions 420
trap geometries 421
trapped electron 314, 317, 385, 409, 427
trehalose 411
Triad molecules 40
trialkylphosphates 317
trifluoride 61
trifluoromethyl cation 59
triglyceride 318
trihalomethyl cations 57
1,3,5-trimethylcyclohexane 141
trimethylene oxide 132
1,3,5-trimethyl-substitution 141
TRIPLE 467
triplet state 316, 471